Teacher's Edition

Living Things

Harold E. Teter
Gabrielle I. Edwards
Frederick L. Fitzpatrick
Thomas D. Bain

HOLT, RINEHART AND WINSTON, PUBLISHERS
New York • Toronto • Mexico City • London • Sydney • Tokyo

AUTHORS

HAROLD E. TETER Formerly Head of the Science Department at Cooley High School, Detroit, Michigan

GABRIELLE I. EDWARDS Assistant Principal, Supervision, Biology and General Science, Franklin Delano Roosevelt High School, Brooklyn, New York

FREDERICK L. FITZPATRICK Late Professor Emeritus of Natural Sciences and Head of the Department of Science Education at Teachers' College, Columbia University, New York, New York

THOMAS D. BAIN Formerly a teacher of biology at Harding High School, Marion, Ohio

---

EDITORIAL DEVELOPMENT: William N. Moore, Karen K. Gotimer, Lorraine Smith-Phelan

PRODUCT MANAGER: Tom Rooney

FIELD ADVISORY BOARD: Joan Crist, James Donatelli, Linda Ludwic, David J. Miller, Rhenida Rennie, George Rybak

MARKETING RESEARCH: Erica Felman

EDITORIAL PROCESSING: Margaret M. Byrne, Richard D. Sime

ART AND DESIGN: Carol Steinberg

PRODUCTION: Beverly Silver, Ira Goldner, Maureen Noonan

PHOTO RESOURCES: Linda Sykes, Rita Longabucco

CONTENTS

COMMENTARY

Unit 1 Principles of Biology T34

Unit 2 The Continuation of Life T42

Unit 3 Ecology: Living Things in the Environment T49

Rationale

Living Things is an introductory life science course dynamically investigating and analyzing important biological concepts. Because it emphasizes fundamental concepts, it is well-suited to beginning science students on the junior high or high school level. The *Living Things* program attempts to instill in the beginning biology student a true interest and appreciation for the natural order of living organisms and their relation with their environment. *Living Things* stresses basic concepts rather than the technical jargon that often overwhelms beginning students. The language is kept simple so students concern themselves with the "why-and-how" questions of science and are not over-burdened with the task of learning complicated scientific terminology, which is not essential to the understanding of the scientific idea. This simplicity in language will provide confidence and allow for a better understanding of biology.

In the 1981 edition of *Living Things* an attempt has been made to relate the scientific material to students' everyday lives. Students living both in urban and rural areas will find a wide range of practical examples to which they can relate. Current issues, such as energy and pollution, have been added. All factual material has been updated.

The first three units of *Living Things* provide basic science information that will increase the students' understanding of the material that follows, as well as of life science in general. The following four units provide specific information about the structure and function of living organisms from the simple *ameba* to the most complex organisms, humans.

Components

The student text is supported by the following:
- Teacher's Edition to the Text (See page T8, "How to use the annotated Teacher's Edition.")
- *Exercises & Investigations for Living Things* is an activity and laboratory workbook that contains review exercises for each chapter as well as laboratory investigations. The review exercises help the student survey the concepts in the Text. The laboratory investigations permit more in-depth, concrete, "hands-on" experiences to enhance student understanding of these concepts. There are also exercises to help students improve science reading and writing skills, such as the writing of lab reports, as well as science safety and science careers.
- The Teacher's Edition of *Exercises & Investigations for Living Things* provides references to the Text and shows how this book can be correlated to the Text. It also provides answers to the exercises and points out laboratory safety procedures.
- Tests (duplicating masters) are for each chapter that test the students' command of the Objectives. These are made up of both short-answer questions and questions that require the students to write complete sentences and formulate ideas.
- The *Sample Booklet* contains reduced pages of the Tests.

Special Features

The following pages, reproduced from the text and reduced in size, illustrate the features of *Living Things*.

MOTIVATION:
Chapter opens with a four-color photograph that is tied into the opening motivational paragraphs; triggers students' interest.

PRESTATED OBJECTIVES:
Previews concepts to be covered; gives nature of expected performance.

KEY WORDS:
Newly introduced terms are keyed in the margin for easy reference.

CHAPTER 12 THE NONLIVING ENVIRONMENT

objectives

After you read this chapter, you should be able to:

___**List** some conditions necessary for life on earth

___**Diagram** the water cycle

___**Describe** how rainfall affects the production of different landscapes

___**Compare** the effects of altitude with the effects of latitude

___**Explain** the importance of light

ecology

The earth is not just made up of people, dogs, insects, trees, and other living organisms. It is also composed of nonliving things such as air, water, soil, and rocks. Each of the living organisms on earth is affected by the nonliving things around it and each particular living organism must be adapted to live in its own particular environment.

THE EARTH'S ENVIRONMENT

No living thing can exist, even for a few seconds, without being affected by its environment. You would realize this if you tried holding your breath while you read the rest of this chapter! Air is part of your environment. It affects you by supplying the oxygen you need. You affect it by removing some oxygen from it and adding some carbon dioxide to it. In what other ways do you and your environment affect one another?

The environment is everything that surrounds or affects the individual, including such nonliving things as light, heat, water, soil, and air. It also includes all the other living things in a region. In this chapter, we shall study the nonliving environment, and in the next, the living environment. This whole study of living things in relation to their environment is the science of *ecology*.

Let us think about the earth as a whole. The earth is the home of plants, animals, and people. What kind of environment must these living things have if they are to succeed on the earth?

First, the environment for living things must contain *water*. Protoplasm is made up largely of water. The other molecules in this living material can only meet and react with each other if they are dissolved in a liquid. There is probably no other liquid in the universe that can take the place of water in living things.

112

FUNCTIONAL ILLUSTRATIONS AND PHOTOS:
Aid concept development; reinforce ideas.

EASY-TO-READ TEXT:
Nontechnical language makes concepts easy to group. Practical examples are relative to students' interest.

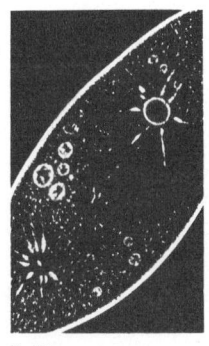
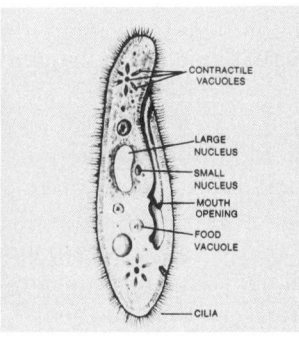

CONTRACTILE VACUOLES

LARGE NUCLEUS

SMALL NUCLEUS

MOUTH OPENING

FOOD VACUOLE

CILIA

Fig. 19-4
Left: An actual photograph of a magnified Paramecium; Right: A diagram showing the parts of Paramecium. (Walter Dawn)

Something else to look for in *Paramecium* is a star-shaped set of vacuoles near each end of the cell. These vacuoles act like pumps to get rid of extra water that comes in through the mouth opening and the cell membrane. Many protozoa have vacuoles of this general type. As these vacuoles fill with water they get larger and larger. Then one of them contracts suddenly and the water is forced out through an opening in the cell covering. For this reason these are called *contractile vacuoles*. In *Paramecium*, one vacuole contracts while the other vacuole fills with water. Then the second vacuole contracts while the first takes on water again.

Paramecium is larger than many other single cells. If you hold up a drop of water containing several *Paramecia*, you can just barely see them moving around. They look like tiny white specks in the water. Most other cells cannot be seen without a microscope. When you examine *Paramecium* under a microscope you soon see that it is not as simple as ameba. Besides the things we have mentioned, *Paramecium* has more than one nucleus. One is a small nucleus, which contains chromosomes and seems to control reproduction and heredity. The other is a very large nucleus containing many sets of the same chromosomes. These most likely control the growth and activities of the *Paramecium* cell.

196

SUMMARY

Matter is anything that has weight and occupies space. Air, water, protoplasm, and carbon are examples of matter. The basic substances from which matter is made are elements. Elements cannot be broken down into simpler forms of matter by ordinary chemical means. The smallest part of an element that has the properties of the element is an atom. An element is made of one kind of atom. Each element has its own atoms. When atoms of different elements combine, a chemical change takes place. Atoms are held together by the force of chemical bonds. When atoms of different elements are bonded together, a compound is formed. The smallest part of a compound that keeps the characteristics of that compound is known as a molecule. Sometimes substances are put together without chemical bonding to form a mixture. Chemical energy is stored in chemical bonds. The energy stored in food molecules may be released by oxidation and changed to the energy of heat or motion.

SUMMARIES:
Provide reinforcement; clarify concepts.

Activity

Formation of Compounds

A. Pour a little limewater into a test tube. Blow your breath through a drinking straw into the limewater. Your breath contains carbon dioxide.

    1. What change takes place in the limewater?

B. Place a short candle inside a large beaker. Your teacher will show you how to fix the candle to the bottom by using melted wax. Light the candle. Cover the top of the beaker with a glass plate. When the candle flame goes out, lift the lid of the beaker just enough to pour in a little limewater. Gently swirl the limewater in the beaker while carefully holding down the lid.

    2. What change do you notice in the appearance of the limewater?

    3. Where have you seen this change before?

    4. What gas turns limewater milky?

    5. What gas was produced by the burning candle?

C. Now dry out the beaker. Fix a candle to the bottom of the beaker as you did before. Hold a cold surface just above the flame for a moment. **CAUTION: Be careful not to burn your fingers.** A small jar containing an ice cube or very *cold* water will work very well. **Do not hold the jar low enough in the flame to deposit black soot on it.** Notice the moisture that collects on the glass.

    6. Where did the moisture come from? Candle wax is made of the two elements carbon and hydrogen. Air in the beaker contains oxygen.

    7. What gas is produced when 1 atom of carbon is joined chemically with 2 atoms of oxygen?

    8. What gas is produced when 2 atoms of hydrogen are bonded chemically to 1 atom of oxygen?

ACTIVITIES:
Can be easily done; are optional; enhance understanding of concepts and require relatively inexpensive materials.

SAFETY FEATURES:
Students are cautioned about potential hazardous materials and/or procedures that appear in Activities.

26

Word Quiz

Choose the letter from **Column B** that best matches each number in **Column A.** Do not write in this book.

Column A	Column B
1. ecology	**a.** maintaining an even body temperature
2. water cycle	**b.** moisture moving from the ocean to the land and back again
3. warm-blooded	**c.** body temperature varies
4. cold-blooded	**d.** cold region with stunted plant growth
5. hibernation	**e.** study of environment
6. tundra	**f.** slowing down of all body processes in winter

END-OF-CHAPTER QUESTIONS:
WORD QUIZ:
Reinforces new vocabulary.

Check Your Facts

1. What makes light very important to living things? Do all living things need the same amount of light?
2. Diagram the water cycle.
3. What general type of plant life covers the land in a wet region? In a fairly dry region? In a very dry one?
4. What does soil supply to living things? How can soil influence what grows in an area?
5. What is supplied to living things by the air?
6. How does temperature influence what grows in a region?
7. Why do animals hibernate?
8. What is meant by cold-blooded and warm-blooded? What advantage is there in being warm-blooded?
9. What causes winter kills in lakes?
10. What is the natural type of plant cover growing in the region where you live? What conditions are responsible for this?
11. "Changes in altitude have the same effect as changes in latitude." Look back at Fig. 12-7 and explain this statement.

CHECK YOUR FACTS:
Tests basic concepts and chapter objectives.

Thought Questions

1. In what ways are the living conditions in a cave similar to those deep in the sea?
2. During the Ice Age, glaciers reached as far south as the Ohio River. If this happens again, what living things will grow in Michigan, in Kentucky, and in Mississippi? What changes would it make where you live?

THOUGHT QUESTIONS:
Provide a further challenge for the more advanced students.

121

How To Use
The Annotated Teacher's Edition

On pages T34 to T96 there is a commentary for each chapter of the student text. In this commentary you will find the following information:

OVERVIEW   Brief summary of the concepts covered within the chapter.

HINTS   Information on particular approaches to covering the chapter material; demonstrations, films, filmstrips, or outside readings that would be particularly helpful in conveying certain concepts.

ACTIVITY SUGGESTIONS   List of needed materials, explanation of procedures, results, notes of caution, hints to aid in carrying out the activity, and additional or alternate activities where appropriate.

ANSWERS TO QUESTIONS   Word Quiz, Check Your Facts, Thought Questions.

SUPPLEMENTARY READING   List of references for teachers and students.

AV MATERIALS LIST   Films, filmstrips, etc.; enhances understanding of concepts.

In addition to the chapter commentary, marginal annotations can be found throughout the Teacher's Edition of the student Text. These annotations have been divided into categories and given headings as follows:

CAREERS   Points out places where discussions of various science careers could take place.

WORD STUDY   Introduces root words, suffixes, and prefixes to help the student develop and reinforce a new biological vocabulary.

REFERENCE   Cites books, articles, etc., that deal with the topic being discussed in the text.

QUESTION   Suggests questions for stimulating student thought.

SUGGESTION   Gives various suggestions to aid in carrying out the activities.

ACTIVITY   Mentions additional activities or demonstrations.

ENRICHMENT   Provides additional, more in-depth information about various science concepts; suggests additional scientific terminology that the teacher may wish to introduce to the students.

SCIENCE READING SKILLS   Points out areas where various reading skills can be reinforced; divided into the following categories: Reading Illustrations, Compare and Contrast, Cause and Effect, Classification, Sequencing, Generalization, and Problem Solving.

Approaches
to Teaching *Living Things*

Module and "Minicourse" Approach

The sequence of units in *Living Things* represents a systematic approach to covering biology. However, many of the units and chapters in *Living Things* may be taught out of sequence. The Plant Kingdom (Unit 5), for instance, might be left until last, in order to allow the students the opportunity to do field work. Unit 7 (Human Biology) could be taught at any point after the introductory units (1 and 2) have been covered. The first two units form a natural sequence. They present the basic concepts of biology. The remaining units build in an orderly way upon these concepts. The final unit applies all of this background to the understanding of human biology.

The format of *Living Things* is flexible enough to allow the teacher to use the modular approach. The number of modules and the organization of modules is very flexible. The following outline suggests a six-module course covered over a 35-week school year.

Module 1 (Unit 1)
Chapters 1–6
(4–5 weeks)

Major Concepts
scientific method, biochemistry, cell and its parts, photosynthesis, respiration, groups of cells

Module 2 (Unit 3)
Chapters 12–16
(4–5 weeks)

ecology

Module 3
Chapters 17–19, 7, 10, 11
(6 weeks)

classification of living things, bacteria, protozoans, simple forms of reproduction, heredity

Module 4
Chapters 20–24, 8
(5–6 weeks)

algae, fungi, mosses, true plants, leaves, stems, roots, and reproduction in plants

Module 5
Chapters 25–32, 9, 37
(6–7 weeks)

various phyla of the Animal Kingdom

Module 6
Chapters 33–44
(8–9 weeks)

human biology systems, diseases, drugs

Optional Module
Chapters 7–11

genetics and reproduction

Individualized Curriculum Approach

In order to develop a life science curriculum effectively, consideration should be given to the students' interests, ability, and background. For individualizing instruction to meet student needs, the following suggestions may be helpful in meeting objectives for (1) a minimal course, (2) an average course, and (3) an enriched course.

	MINIMAL	AVERAGE	ENRICHED
CHAPTERS	1, 2, 6, 7, 8, 9, 15, 17, 18, 19, 20, 21, 22, 25, 26, 27, 28, 29, 30, 31, 32, 35, 36, 37, 38, 39, 42	All chapters in text.	All chapters in text plus additional topics selected by teacher.
APPROACH	Class time for directed reading comprehension. Class discussions used for motivational purposes and to clarify concepts.	Class discussion to clarify concepts and expand ideas. Field trips and teacher demonstrations.	Chapter reading assignments given as homework. Class discussion to expand ideas. Field trips and student demonstrations.
QUESTIONS	Word Quiz and selected Check Your Facts	Word Quiz and Check Your Facts	Word Quiz, Check Your Facts, and Thought Questions
READING AND WRITING SKILLS (See TE pp. T17 to T26 and skills exercises in *Exercises & Investigations for Living Things*.)	Classroom time spent on how to use the text, reading for comprehension, and sentence writing.	Classroom time spent on improving sentence structure and how to write a lab report.	Written reports assigned as homework, answers to Thought Questions reviewed for written formulation of ideas, and outside readings assigned.
CHAPTER REVIEWS	Class review of text material, written assignments, and tests. (See Review Exercises—*Exercises & Investigations for Living Things*.)	Written assignment and discussion as needed. (See Review Exercises—*Exercises & Investigations for Living Things*.)	Assignments as homework. (See Review Exercises—*Exercises & Investigations for Living Things*.)
LABORATORIES (See *Exercises & Investigations for Living Things*.)	Selected part ones	All part ones and selected part twos	All part ones and part twos

Sample Flowchart for Using Living Things and Its Components

The following flowchart may be used to help the teacher plan and prepare daily and weekly class lessons, activities, homework assignments, and tests. A master of this chart can be made and copies run off for future chapters. The solid-line boxes represent the core materials found in the pupil's text. The broken-line boxes represent the Teacher's Edition, laboratory text (*Exercises & Investigations for Living Things*), and Test masters.

Chapter 1 What is Science?

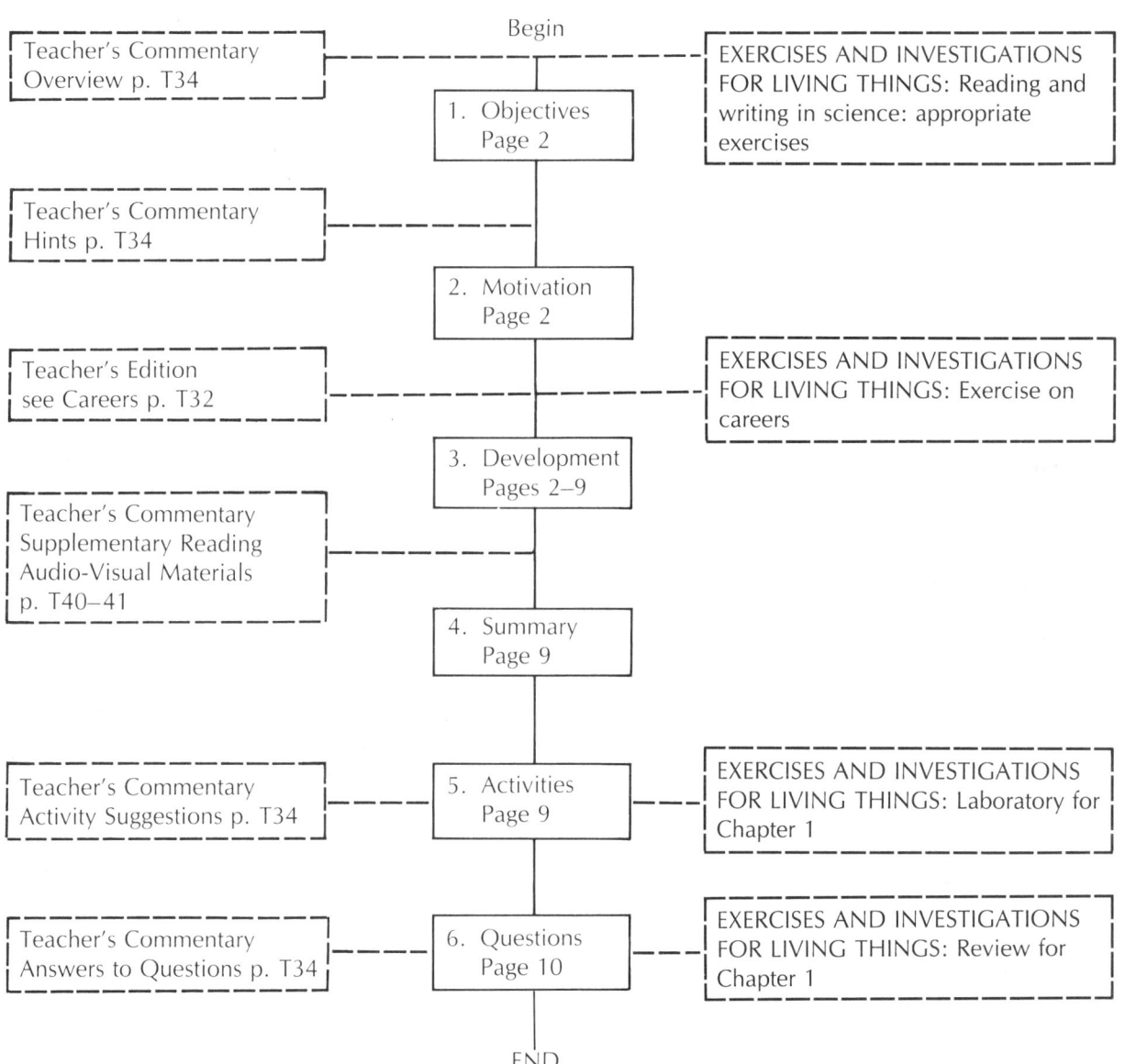

Begin

Teacher's Commentary
Overview p. T34

EXERCISES AND INVESTIGATIONS
FOR LIVING THINGS: Reading and
writing in science: appropriate
exercises

1. Objectives
Page 2

Teacher's Commentary
Hints p. T34

2. Motivation
Page 2

Teacher's Edition
see Careers p. T32

EXERCISES AND INVESTIGATIONS
FOR LIVING THINGS: Exercise on
careers

3. Development
Pages 2–9

Teacher's Commentary
Supplementary Reading
Audio-Visual Materials
p. T40–41

4. Summary
Page 9

Teacher's Commentary
Activity Suggestions p. T34

5. Activities
Page 9

EXERCISES AND INVESTIGATIONS
FOR LIVING THINGS: Laboratory for
Chapter 1

Teacher's Commentary
Answers to Questions p. T34

6. Questions
Page 10

EXERCISES AND INVESTIGATIONS
FOR LIVING THINGS: Review for
Chapter 1

END

Chapter Comparison of *Living Things* and *Modern Biology*

Material	**Amount needed for 30 students	Chapter in which used
agar* (or prepared plates)	30	18
alcohol, isopropyl*	250 mL	43
algae* (live or preserved)	variety of different species	20
animals* (live or preserved)	variety of different species	17
aprons	30	
aquarium (40 liters)	4	9,31,40
bag, garbage* (large)	1 box	16
band, rubber	30	38
beaker 100 mL	30	3
beaker 250 mL	30	3,4
beans, kidney*	1 large box	43
beans, lima*	1 large box	43
bone, chicken*	30	39
box (plastic shoe box)	30	14
bread, white*	15 slices	21
bromthymol blue solution*	2 liters	5
bucket (or glass aquarium)	4	37
burner, bunsen	15	39
can, tin (large)	30	39
candle, small*	30	3
chicken wing*	30	6
clips, paper	1 box	44
cotton	2 packages	36,38
corks (for 30-mL test tube)	100	5,43
cornstarch*	1 box	35
cover slip	2 boxes	2,8,19,20,21,22,23
crayfish*	15	28
culture, Daphnia*	1 portion	25
culture, Hydra*	class of 30	25
culture, Planaria*	class of 30	26
culture, pond*	2 liters	19,26
culture, vinegar eel*	class of 30	26
cylinders, graduated	15	37
dish, glass (6 cm deep)	30	21
dish, culture (glass)	30	24
dish, petri	30	18,36
dropper, eye	30	35
echinoderms* (living or preserved)	variety	29
eggs, frog*	several clumps	9
Elodea* (Anacharis)	3 bunches	5,13
fern*	several plants or leaves	22
fish* (goldfish)	30	36
flower* (any species)	30	8

Material	**Amount needed for 30 students	Chapter in which used
frog, live* (any species)	30	30
goggles	30	3,39
hand lens	30	22,25,26
ice cubes*	1 bag	3,30
incubator	1	12,18
ink, red or black*	1 bottle (200 mL)	4
iodine solution*	200 mL	2,35,43
jar, battery	2–4	45
jar, 4 liter	30	13,37
jar, mayonnaise (small)	30	13,27
juice, lemon*	1 bottle (200 mL)	42
leaf* (any deciduous species)	30	11
leaf, geranium*	15–30	23
lettuce*	3 leaves	27
limewater*	2 liters	3
moss*	class of 30	22
moss, club*	15 pieces	22
mushroom, fresh*	30	7
needle, dissecting	30	8,45
onion, medium*	2	2
pan, dissecting	30	6,45
pan, plastic	30	28
paper, lens*	8 packages	2
peroxide, hydrogen*	1 large bottle (250 mL)	43
pins, dissecting	1 box	45
plant, water* (i.e., Elodea)	variety	13
plants* (for classification)	variety	17,22
plants* (examples of flowering)	variety	22
plants* (examples of nonflowering)	variety	22
plaster of Paris	1½ kg	11
plate, aluminum (pie)	30	11
plate, glass (11.5 cm × 11.5 cm)	30	3,21
polish, silver	1 bottle	20
quinine* (tonic)	1 bottle	42
rod, glass	30	45
ruler, metric	30	33,37
salt, table*	1 box	42
sand* (or soil)	20 kg	12,13
scissors	30	6,7,45
seeds, bean*	1 large box	12
seeds, oat*	1 large box	12
seeds, radish*	1 large box	12,24
seeds, squash*	1 large box	12
shells, sea	30	11
slide, microscope	60	2,8,19,20,21,22,23,36

Material	**Amount needed for 30 students	Chapter in which used
slide, prepared— xs. of leaf	30	23
slide, prepared— xs. of stem (3 types)	30	24
slide, prepared— ls. of root tip	30	24
snail, land*	30	27
snail, water*	30	13,27
sod, grass*	several pieces	14
starfish* (living)	3–4	29
straws, drinking*	60	3,5
sugar*	2 kg	42
swabs, cotton*	100	42
tadpoles* (toad or frog)	20	40
tape, masking*	1 roll	30
test tubes (small—30 mL)	90	3,5,13,35,43
thermometer	30	38
toothpicks*	1 box	2
tray, plastic	30	12
turtle, land*	1 or 2	31
turtle, water* (painted turtle)	1 or 2	31
tweezers	30	6,21,23,45
twigs, tree*	30	11
tubing, rubber (45 cm)	30 pieces	37
tyroxin tablets*	30	40
vinegar*	1 bottle	39,45
wire (10 to 12 cm)	30 pieces	45

*Indicates the material is consumable and may have to be obtained each year.
**The amount of materials needed for each class can be reduced considerably if students work in small groups.

Sources of Equipment and Supplies

American Optical Co.
Instrument Division
Sugar and Eggert Roads
Buffalo, NY 14215

Baltimore Biological Laboratory
Hunt Valley Industrial Park
Cockeysville, MD

Calbiochem
3625 Medford Street
Los Angeles, CA 90054

Carolina Biological Supply Co.
Burlington, NC 27215

Central Scientific Co.
2600 South Kostner Ave.
Chicago, IL 60623

Connecticut Valley Biological
Supply Co., Inc.
Southampton, MA 01073

Eduquip – Macalaster Corp.
1085 Commonwealth Ave.
Boston, MA 02215

Fisher Scientific Co.
633 Greenwich Street
New York, NY 10014

LaPine Scientific Co.
6001 S. Knox Avenue
Chicago, IL 60629

Macmillan Science Co.
8200 S. Hoyne Ave.
Chicago, IL 60620

Nasco
901 Janesville Avenue
Fort Atkinson, WI 53538

or

P.O. 3837
1524 Princeton Avenue
Modesto, CA 95352

Sargent-Welch Scientific Co.
7300 North Linder Avenue
Skokie, IL 60076

Ward's Natural Science
Establishment, Inc.
P.O. Box 1712
Rochester, NY 14603

Suppliers (Audio-Visual)

Aims Instructional Media (AIMS)
P.O. Box 1010
Hollywood, CA 90028

American Educational Films
331 North Maple Drive
Beverly Hills, CA 90210

*BFA Educational Media
2211 Michigan Avenue
P.O. Box 1795
Santa Monica, CA 90406

Carousel Films
1501 Broadway
New York, NY 10036

McGraw-Hill, Inc.
1221 Avenue of the Americas
New York, NY 10020

Coronet Films
65 E. South Water Street
Chicago, IL 60601

Current Affairs Films
24 Danbury Road
Wilton, CT 06897

*Denoyer-Geppert
5235 Ravenswood Avenue
Chicago, IL 60640

Doubleday Multimedia
P.O. Box 11607
Santa Ana, CA 92705

*Ealing Films Corporation
2225 Massachusetts Avenue
Cambridge, MA 02140

*Educational Dimensions Corp.
Box 126
Stamford, CT 06904

*Encyclopaedia Britannica
425 North Michigan Avenue
Chicago, IL 60611

(FDI) Film Distributors International
2223 S. Olive Street
Los Angeles, CA 90007

Filmstrip House
6633 West Howard Street
Niles, IL 60648

Filmstrip of the Month Club
355 Lexington Avenue
New York, NY

Guidance Associates of
Pleasantville, NY
23 Washington Avenue
Pleasantville, NY 10570

Hawkhill Associates, Inc.
Black Earth, WI 53515

Houghton Mifflin Co.
Pennington-Hopewell Road
Hopewell, NJ 08525

*Hubbard Science
2855 Shermer Road
Northbrook, IL 60002

McIntyre Educational Media, LTD
716 Center Street
Lewiston, NY 14092

Modern Learning Aids, Inc.
1212 Avenue of the Americas
New York, NY 10036

Mobil Oil Corporation
150 East 42nd Street
New York, NY 10017

Moody Institute of Science
12000 E. Washington Avenue
Whittier, CA 90606

*National Geographic Society
Educational Services Dept. 77
P.O. Box 1640
Washington, DC 20013

New York Times Education
Division
229 W. 43rd Street
New York, NY 10036

Pathescope Educational Films Inc.
71 Weyman Avenue
New Rochelle, NY 10802

Popular Science Pub. Co., Inc.
355 Lexington Avenue
New York, NY 10017

Prentice-Hall Media (PH)
150 White Plains Road
Tarrytown, NY 10591

The School Times Teaching
Resources Films
Bedford Hills, NY 10507

*Scott Educational Division
104 Lower Westfield Road
Holyoke, MA 01040

Shell Film Library
1433 Sadlier Circle West Drive
Indianapolis, IN 46239

(SVE) Society for Visual Education
1345 Diversey Parkway
Chicago, IL 60614

Thorne Films
1229 University Avenue
Boulder, CO 80302

Time-Life Education
Box 834
Radio City Post Office
New York, NY 10019

Universal Education and
Visual Arts
100 Universal City Plaza
Universal City, CA 91608

Valiant I.M.C.
237 Washington Avenue
Hackensack, NJ 07602

*Walt Disney Educ. Media Co.
500 South Buena Vista Street
Burbank, CA 91521

Ward's Natural Science
Establishment, Inc.
P.O. Box 1712
Rochester, NY 14603

NOTE: Sources are free unless
marked with *

T16

During the past few years, science teachers, as well as teachers in other content areas, have been faced with a decline in student interest and ability in reading. This has led to an increasing concern about readability in student materials.

Two methods of determining the reading level of a text are generally accepted. One method is based on graded word lists and the other on a readability scale, such as Fry, which is based on average sentence length and total number of syllables. (To learn how to use the Fry Readability Scale, please refer to "The Science Teacher," March 1974, pages 26–27.)

These methods focus on the mechanical aspects of reading rather than on the meaning of the words. Materials rated at a higher "reading level" according to some mechanical tests might actually be easier for junior high school students to understand than materials rated at a lower "reading level." However, these types of tests are useful in eliminating texts that are well out of the range of the students' abilities.

Diagnosis

One method of determining whether a student understands the text material is the Cloze test. The results of this test can help you decide if the student can work independently, requires instruction, or, even with instruction, will have reading difficulties.

To make your own Cloze test, select a 250-word passage from the material you would like to assign. This material must be new to the student. Leave the first sentence complete. Starting with the second sentence leave a uniform blank for every fifth word. Continue to delete every fifth word until there are fifty spaces. Give this material to your students with the instruction to fill in each blank with the word that makes the most sense. The following small sample from page 12 will give you an idea of what the test looks like.

If living things are so different from things that are not alive, they must be made of something very special. This special material is _____ protoplasm. The activity of _____ protoplasm gives organisms the _____ to do the things _____ have just read about. _____ are alive, so you _____ be made of protoplasm. _____ plants and animals contain _____ . But it is a _____ different in each living _____ . Sometimes it is thin _____ water. At other times _____ may be thick, like _____ . It may have color _____ usually it is clear _____ colorless. It feels slippery _____ raw egg white. Even _____ from different parts of _____ same body may be _____ .

To Score:
1. Award 2 points for each exact word. Award no points for synonyms.
2. Independent level: A score *above 60* usually indicates that the material is easy enough for the student to read without assistance.
3. Instructional level: A score *between 40 and 60* usually indicates that the

material is at the student's instructional reading level, that is, with teacher guidance.

4. A score *below 40* indicates that an examination for synonyms is necessary. At this point, the student should not be penalized for synonyms or appropriate answers. The teacher may find that the student has chosen better or more appropriate answers than the author. If the student has not written appropriate synonyms, the teacher can expect that the student will have problems reading this material.

Reading Science

Science teachers sometimes feel that they do not have time to teach reading if they are to cover the science for which they are responsible. Actually, in the normal handling of science materials and in emphasizing scientific methods, they spend a great deal of time helping their students develop reading and thinking skills. Consider, for instance, the ability to distinguish between fact and generalization or between cause and effect, or the ability to recognize and make use of sequence, pattern, and classification. Such analytical and critical thinking skills are as much a part of science as of reading. Once acquired, these skills will help students solve problems and zero in on the main idea in a paragraph or article long after they have forgotten such things as the symbols for the elements or the composition of protoplasm.

Reading skills are important if science materials are to be read successfully. This text has been designed to aid both teacher and pupil in the development of student reading skills in science. The information that follows is intended to aid the teacher in evaluating student reading skills and in developing techniques to be used in the science classroom to improve and enhance these skills.

Parts of the Textbook

Even though your students have been in school for a number of years, they may not be using their textbooks efficiently. Very early in the course it is a good idea to review the parts of the text and the function of each. This would help reduce the frustration that students may occasionally feel in trying to find information.

Point out the special aids in the front and the back of the text: Introduction, Table of Contents, Glossary, and Index. Show the students how to use these sections for locating information and definitions of terms. Drill them on locating lesson topics and definitions of terms quickly.

Spend a little time familiarizing the students with the basic format of the chapters. For this, you will find the appendix of the student text and the corresponding exercises found in *Exercises & Investigations for Living Things*, especially useful.

In order to evaluate the level of the skills of your students in use of the textbooks an *Informal Reading Inventory* (IRI) can be used. The purposes of an IRI are to:

1. determine whether the students know how to use a specific textbook
2. determine whether the students can read that textbook with sufficient understanding to help them learn the skills and content of the particular course.

Ability to Read the Textbook:

The students are directed to read a section of the textbook. After the students have read the section, questions are presented on the material just read. Depending upon the section read, questions similar to the following could be asked.

1. What would be a good title for this passage?
2. Put these four events in chronological order.
3. What are four reasons for _____ ?
4. What conclusion can be drawn from this section?
5. What resulted from _____ ?
6. What is the meaning of _____ ?
7. In what two ways does this compare and contrast with that?
8. What does the author mean by the last sentence in the third paragraph?
9. Write a two-sentence summary of the selection.

A reading inventory sample like the one shown above is not a test. It is used to determine whether students have the necessary skills needed for comprehension of the subject area. After the answer sheets are corrected, a table similar to the one below can be constructed for the class. Using this record enables the teacher to determine at a glance areas of general weakness.

SAMPLE INFORMAL GROUP READING INVENTORY,
where "x" signifies problem areas

NAME	USING THE TEXT			VOCABULARY SKILLS			STUDY SKILLS			COMPREHENSION SKILLS					
	INDEX	GLOSSARY	TABLE OF CONTENTS	CONTEXTUAL CLUES^A	TECHNICAL VOCABULARY	STRUCTURAL CLUES^X	USE OF REFERENCES	USE OF GRAPHS AND DIAGRAMS	FOLLOWING DIRECTIONS	MAIN IDEA	SEQUENCING	COMPARE AND CONTRAST	CLASSIFICATION	PROBLEM SOLVING	GENERALIZATION
	1	2	3	4	5	6	7	8	9	10	11	12	13	14	15
John	x			x	x			x							
Susan								x						x	x
Phyllis	x	x	x		x		x		x			x			
Robert	x	x				x		x							
etc.															

See following page for explanatory notes.

— when students recognize a pattern in a sentence or phrase that enables them to deduce a probable meaning
ˣstructural clue — when students subdivide an unknown word into known prefixes, suffixes, and roots that enable them to guess the meaning.

Following Directions

In science class more frequently than in other classes, the students have the opportunity of following directions and knowing immediately whether they have followed the directions correctly. The importance of following directions cannot be stressed too much in a science class. At times, even the safety of the experimenter is at stake if proper procedures or techniques are not followed.

The teacher can develop some warmup exercises for the students to do in order to help them to see the importance of following directions. The example given here of making an envelope can be an interesting way to see immediate results and to correct errors. Duplicate the directions and ask students to read them and carry them out a step at a time. The envelopes can be used later in the course to keep various activity materials.

The answer to each question is "yes." If you do not answer "yes," go back and redo the previous step or steps.
A. Obtain a clean square piece of paper.
B. Fold the paper in half from top to bottom; unfold.
C. Fold the paper in half from left to right; unfold.
    1. Is the paper divided into four squares?
D. Label the corners in clockwise order A, B, C, D.
E. Fold the paper diagonally, so A covers C; unfold.
    2. Is the diagonal BD?
F. Fold the paper diagonally, so D covers B; unfold.
    3. Is the diagonal AC?
G. Position the paper on the desk with corner A at the top.
    4. Is corner C at the bottom?
H. Draw a dot on the center of the paper.
    5. Is the dot where all the folds cross?
I. Draw four dots around the center dot as follows: 1st — one cm above, 2nd — one cm to the right, 3rd — one cm below, 4th — one cm to the left.
J. Label the dots: above (E), right (F), below (G), left (H).
K. Fold D to the dot next to F; unfold.
L. Fold B to the dot next to H; unfold.
M. Fold C to the dot next to E; unfold.
N. Fold A to the dot next to G; unfold.
O. Repeat steps K, L, and M without unfolding.
P. Tape these three triangles together.
Q. Repeat step N.

Reading Illustrations

The many photographs and graphic illustrations in the text are designed to help the students understand the concepts better or to help them follow directions in the activities more easily. Some students may not have mastered the skill of reading illustrative materials.

One way to make the students more aware of the role of illustrations is having them work with unit openers in the text. First, have the students locate these pages, but do not give them the page numbers. Let them use the Table of Contents by themselves. Then direct the students to study the illustrations one at a time. Ask them to tell you or to list what they see. After all of the facts are discussed or listed, ask students to interpret the illustration. What is the main point? How does the illustration relate to the title of the unit? What do students think the unit will be about, based on a study of the illustration?

Reading Graphs and Diagrams

Have the students practice reading a diagram. This can be done in a lesson not yet covered, or as one is being covered.

Have the students follow these steps:

1. Read the title of the lesson that contains the diagram and, if it has one, the title of the diagram.
2. Review the symbols or terms used in the diagram.
3. Think about what the diagram shows and why a diagram is a good way of showing it.
4. Read the text surrounding the diagram and see how they relate to each other. Write the steps on the board. Have students copy the steps in their notebooks for reference.

Finding Patterns in Data

Scientists spend a lot of time collecting data and then looking for patterns in the data. In certain activities, the students collect the facts or data. To look for patterns, they will arrange the data in tables. For an example of this, see the Activity at the end of Chapter 34. From the data collected a pattern should develop. When students read other tables they should be aware that they should go from the facts to a relationship, from which they can form a pattern.

Studying Vocabulary

Many new words are introduced to the student in the science class. When important new terms are introduced in *Living Things* they are printed in boldface type and also appear in the margin nearby. At the end of each chapter is a Word Quiz. The Glossary is the alphabetical listing of the words in the margin.

The problem of vocabulary is one of the greatest difficulties in dealing with science materials. The following is a list of many of the difficulties of the scientific vocabulary:

1. Scientific words for the most part are new and have little relationship to the past experiences of the student.

2. The vocabulary is precise and exact. There is no room for generalizations or hazy ideas about meanings.
3. Many words are familiar to the student in general use but do not have the same meaning in science.
4. The vocabulary is technical. Some of the vocabulary is solely applicable to the science field, for example, the term *cytoplasm*.
5. Large polysyllabic words cause problems with meaning and pronunciation.

The following is a list of suggestions that will help the student find science vocabulary a little bit easier:

1. Select a scientific term, such as *echinoderm*. Ask the students to study its root and discuss its origin.
2. Call attention to the frequently recurring prefixes and suffixes that are so abundant in science; for example, <u>ex</u>crete, <u>ex</u>tinct.
3. Call the students' attention to the new vocabulary listed for each chapter.
4. Include the already introduced words in classroom charts and bulletin boards.
5. Encourage a search of newspapers and magazines for articles on topics discussed in class.
6. Teacher-made and student-made crossword puzzles using science vocabulary are helpful.

Reading for Cause and Effect

One of the thinking and reading skills scientists rely on is that of relating an effect to its cause. Scientists are constantly relating cause and effect in their studies and experiments. Students should do the same in their reading and activities. They will learn facts faster if they look for cause/effect relationships among them, which will thus clarify their study of life science. Throughout *Living Things*, the student will be called upon to practice this skill.

Many exercises can be given to the students to test their understanding of cause and effect. Given two statements, have the students identify which is the cause and which is the effect. Also have students find other examples of cause and effect in the text.

Reading to Compare and Contrast

Reading comprehension is improved if the reader has a definite purpose in mind. Scientists, for instance, might be thinking of comparing and contrasting information as they read. They may plan to analyze results of experiments by looking for similarities and differences. Such an analysis may be the key to a discovery. If the student knows how to look for similarities or differences in the reading or activity work, it will be easier to remember the facts, relate them to one another, and apply them to new situations.

Classification

In our everyday lives, we use classification constantly. Because of classification systems, we know our way around new supermarkets and can find books in libraries. We can predict to some degree what a record will sound like before playing it, if, for instance, we know it is jazz as opposed to rock or classical music. We can make the prediction because we keep in mind the various characteristics of the types of music.

In science, knowledge is systematized so that it can be studied. When you consider the vastness of this subject, you can realize how necessary it is to have classification systems. An organism is classified according to similarities and differences with other organisms. The elements that an organism is made up of are classified and arranged in a Periodic Table. For a classification Activity, see Chapter 17. Other exercises devised to help students group substances or processes are found in some of the Word Quizzes, such as the one in Chapter 36.

Generalization

A generalization is a statement or concept that encompasses common characteristics of a group of observations or statements. The observations could be made in the Activities. The statement could be found in the reading material. Through analysis of their experiences, observations, and the facts given, the students should be able to form a generalization. One example of where they are asked to do this is in the Check Your Facts section at the end of Chapter 2.

Sequencing

Sequencing is an important part of science. The continuity of many scientific facts, concepts, and experimentations relies on sequencing. Being able to follow one thing after another in a logical order is a valuable skill that will help students perform activities and recognize various concepts. An example of sequencing can be found in the Word Quiz at the end of Chapter 17.

Once the students become familiar with the importance of sequencing in science, it will help them to predict outcomes of activities and to apply concepts logically to new situations.

Problem Solving

Briefly, problem solving is defined as thinking of ways to combine observations with facts to find a solution. Scientists use the scientific method to solve problems. Students, too, should be encouraged to use the scientific method to solve problems in everyday matters, as well as in science. It is important to realize, however, that every problem doesn't warrant a long, systematic approach for its solution. Some problems are so simple that the solution is easily discovered. The scientific method of problem solving should be employed only where there is a genuine problem to be solved. Students are encouraged to develop this practice in Chapter 1.

The following is a list of steps that can be used by students to solve a problem:

1. Define the problem. The problem must be understood before it can be solved.
2. Gather information about the subject of the problem. Examine possible ways of solving the problem.
3. Make a hypothesis. Make a prediction based on the available information.
4. Test the hypothesis. A hypothesis can be tested by performing experiments, making observations, and collecting data.
5. Interpret the data. In light of the data collected, choose the best solution. Was the suggested hypothesis correct? What are the conclusions?
6. State the conclusion (solution).

The student is faced with a variety of problems in *Living Things*. Problems in the form of questions may appear in the opening motivations of a lesson. Captions offer questions that can be easily solved by reading the text and observing the illustrations. Questions at the end of lessons and chapters present the student with the opportunity to use the scientific method of problem solving. Activities can be approached as problems.

Writing in the Science Classroom

Just as science teachers are faced with the problem of the decline in student reading ability, they are also faced with a decline in students' ability to communicate ideas in writing. Studies have indicated that the problems students encounter with writing fall into one or more of the following categories:

- A difficulty to convert mental ideas into written expressions resulting in a lack of written content.
- Inability to properly understand a specified task.
- Lack of organization.
- Lack of familiarity with writing conventions, sentence sense, standard usage, and spelling.

The only cure for writing difficulties is to provide the students with frequent opportunities to write. This can be done in the science classroom, as well as in any other classroom.

In *Living Things* the students are given this opportunity to write by answering the end-of-chapter questions. In some of the Word Quiz sections, such as the one for Chapter 20, the students are asked to formulate a definition in their own words. In the Check Your Facts sections, the students are asked to give specific information, which can be presented in two or three well-worded sentences. In the section entitled Thought Questions, the students must express and develop in writing an original idea in such a way that it can be completely and readily understood by others.

In the science classroom there are many other times when the students are called upon to write. Some of these are:

- Note Taking
- Collecting Data
- Writing a Lab Report
- Reporting on Outside Reading or Projects
- Test Taking

In all of these instances the teacher should respond to the students' written expressions with reactions, exchanges of ideas, and helpful advice.

This type of response can be time consuming; therefore, many teachers may wish to develop a system that will minimize the amount of time needed to review and comment on the students' written work. One such system would be to familiarize students with certain editorial-type symbols and abbreviations and then use these consistently to correct student assignments. An example of this is as follows:

COMPOSITION CORRECTION Symbols

Symbol	Item Being Corrected	Example
S.F.	Sentence Fragment	If you go to the zoo and see a lion. S.F.
R.O.S.	Run-on Sentence	She went to the doctor the doctor gave her medicine she felt better. R.O.S.
SP.	Spelling Error	The lungs take in (oxigen) S.P.
P.	Punctuation Error	What causes a heart attack○ P.
CAP.	Capitalization Error	This organism belongs to the genus CAP ⁄paramecium
PW.	Poor choice of Words or wrong homonym	P.W. The whether is very mild today.
AWK.	Awkward Construction	Mendel discovered everything we know about heredity almost. AWK.
Sense	This does not make sense. Read aloud.	The United States has a lot of overweight people and exercise also. SENSE
∧	Something has been left out.	The circulatory system has arteries and. ∧
₵	Incorrectly Paragraphed	
REP	Undesirable Repetition	REP. He knew that if he went he would be in trouble, but he went anyway because he wanted to.
FACTS	Facts are questionable.	The man held his breath for 20 minutes. FACTS
⬭	Spell this out.	She was in the hospital for ① week.

V.T.	Incorrect Verb Tense	The doctor seen [V.T.] that they were ill.
AGR.	Lack of Agreement between verb and subject or pronoun and antecedent	The stomach and intestine is [AGR.] parts of the digestive system. The student carried their book to class. [AGR.]
FORM	Incorrect format for footnote or bibliography	
DIR.	You did not follow directions.	

Writing an Organized Report

Since the organization of information is so critical to science and since many students have difficulty in writing information in an organized manner, the teacher may wish to use the following exercise to help teach students how to organize information.

Preparing a Report

AIM: To organize a report in logical form.

Brainstorm a topic such as "Reasons for a Visit to the Doctor."

Students may offer the following:

Flu	Skin rash
Chicken Pox	Severe cut
Yearly check-up	Vaccination
Broken arm	Blow to the head
Check-up during Pregnancy	

To illustrate how to organize these apparently random ideas, demonstrate organization by classification. This can be done by organizing the contents of a wallet or purse, a desk drawer, a wastebasket, or a classroom closet. Group the items according to type.

Now return to the list of reasons for visiting the doctor and group as follows:

Disease
Injury
Preventive Care

Now list the items under the appropriate heading.

Prepare a cooperative report on the board following the outline just prepared on "Reasons for a Visit to the Doctor."

Another exercise to aid students in writing and organizing information is to refer them to a photograph such as the ones that open Chapters 1 and 2 of *Living Things*. Ask the students to write an investigative report on what is happening in the photograph. Their reports should answer, in complete sentences, the questions, Who?, What?, When?, Where?, and Why?

For additional exercises to improve writing skills see *Exercises & Investigations for Living Things*.

PL 94-142

Mainstreaming has come to mean the integration of handicapped children into the mainstream of our society and into the educational mainstream wherever feasible. Some students with handicaps have always been mainstreamed. Their exceptionalities are not severe, or they are given the opportunities to achieve normal educational goals under conditions in which their exceptionalities are not thought to be handicapping. Others attend special education classes or receive related services that meet their educational needs.

Federal Law PL 94-142 states that handicapped children are to be educated in the "least restrictive environment." For some this means partial or total integration into regular school programs with nonhandicapped peers. In science classes where handicapped students are mainstreamed, the following information will be useful.

PL 94-142 requires all students with handicapping conditions to be identified, tested, and evaluated by the appropriate agencies or school services. (If you suspect that a student has a handicap never before identified, contact the school health department immediately.) The school service, in agreement with the student and his parents, selects the most appropriate and least environmentally restrictive educational plan. The plan can include total integration with related special services, such as speech therapy; or partial integration, with enrollment in selected courses for part of the day and enrollment in special education courses for the remainder.

Handicapping Conditions

Physical impairments include those of an orthopedic nature, visual impairments, blindness, and other health impairments.

Communicative handicaps are any of the following conditions that adversely affect a student's ability of effective use of language: hearing impairments, deafness, speech impairments, and specific learning disabilities. The term "specific learning disability" refers to a dysfunction in any one or more of the basic psychological processes involved in understanding or using language. Specific learning disabilities include dyslexia, minimal brain injury, perceptual impairments, and developmental aphasia.

Mild mental retardation refers to slowed intellectual functioning.

Emotional disturbance refers to an emotional condition that hinders a student's overall functioning.

Instructional Management Systems

Be aware of the educational plan selected for each handicapped student in the science class. If a student is partially integrated and returns to the special

EXCEPTIONALITIES

STRATEGIES	Orthopedic and Other Health Impairment	Visual Impairment	Hearing Impairment Deafness
Classroom management	Restricted movement Special furniture	Front row seating "Buddy" system for making carbon copies of notes, blackboard assignments	Front or 2nd-row seating for lip reading Sign language translator
Basic reading	Comprehension Word and sentence analysis Reference and study skills	No adjustment needed	Expanded interpretation Rephrasing of complex sentences Well-outlined lesson presentations
Textbook management	No adjustment needed	*Text read orally to students Text read with use of magnifier	No adjustment needed
Additional instructional materials	No adjustment needed	Text reproduced in large print with permission of the publisher	Captioned films** for the deaf
Achievement monitoring	Answers to exams Exam administered orally	Exam administered orally or on tape Exam Sheet enlarged on closed-circuit TV Exam translated into Braille	Committee assignments for oral reports

education teacher or resource teacher for supportive and additional instruction, the resource teacher writes an Individualized Educational Plan (IEP) with input from the course instructors. Select those concepts and skills listed at the beginning of each chapter that you consider the most important long-term goals for each student. If a handicapped student is totally integrated with no related services, no IEP is necessary.

STRATEGIES

Management strategies are necessary for optimal student achievement and are individualized depending on the degree of the exceptionality.

Regardless of the system selected, the student is maintained as a full-fledged member of the class, expected to participate in all activities and assignments.

EXCEPTIONALITIES

Speech Impairment	Specific Learning Disability	Mild Mental Retardation	Emotional Disturbance
No adjustment needed	Structured daily procedures	Structured daily procedures	Calm atmosphere, low in tension, high in motivation
No adjustment needed	Expanded interpretation Rephrasing of complex sentences Well-outlined lesson presentations	Expanded interpretation Rephrasing of complex sentences Well-outlined lesson presentations	No adjustment needed
No adjustment needed	No adjustment needed	No adjustment needed	No adjustment needed
No adjustment needed	Easy-reading health reference material	Easy-reading health reference material	No adjustment needed
Committee assignments for oral reports	Committee assignments for oral reports Written exam questions rephrased Oral exam administered	Committee assignments for oral reports Written exam questions rephrased Oral exam administered	No adjustment needed

*American Printing House for the Blind, 1839 Frankfort Ave., Louisville, Ky. 40206
**Films can be borrowed from the Distribution Center nearest your school and shown in any classroom where there is one hearing impaired student. Write to the Dept. of Health, Education & Welfare.

CAREERS FOR PERSONS WITH HANDICAPPING CONDITIONS

As educational opportunities for handicapped persons have increased, so have opportunities for employment. The Rehabilitation Act of 1973, Public Law 93-112, Section 504, provides for the nondiscrimination of qualified handicapped individuals for employment opportunities in any program receiving federal financial assistance. The passage of this law has made careers available to people with handicapping conditions who were once discriminated against or thought to be unemployable.

*Safety In The Science Classroom

Introduction

Any modern science program has a responsibility for teaching students about the nature of the scientific enterprise. Students have a right to learn how to think scientifically and solve problems through the active use of science processes. Students should have ample opportunities to engage in scientific work through laboratory or other "hands-on" activities. Safety should be an integral part of the planning, preparation, and implementation of any science program. Unless safety precautions are taken to protect the student and others during the performance of certain science activities, science programs could eventually be considered too dangerous to be included in our school curricula.

The responsibility for safety and the enforcement of safety regulations and laws in the science classroom and laboratory is that of the principal, teacher, and student—each assuming his or her share. Carelessness and negative or apathetic attitudes toward safety are the major causes of accidents. According to the National Safety Council, about 32,000 school-related accidents occur each school year; about 5,000 of these are science-related.

The prevention of accidents can be accomplished through a science safety education program that places emphasis on teacher and student awareness of the potential dangers in science-related activities. "SAFETY FIRST" should be the basic motto for the school science program. However, safety considerations should seldom rule out a science lesson. Developing and maintaining positive attitudes toward safety requires continual involvement in safety education. It is hoped that safety training in the science program will instill in students an awareness of the importance of safety in all areas of work and play.

The activities suggested in *Living Things* have been kept as simple as possible. The amount of materials required has been kept to a minimum. Wherever possible, activities have been suggested that require simple materials readily available from local drugstores, supermarkets, stationery stores, etc. This means that most of the activities in *Living Things* can be done in the regular classroom rather than in a formal laboratory setting.

If these activities are to be done in the regular classroom, certain precautions should be taken and various adaptations made in order to make the classroom a safe place in which to work. For example, fire extinguishers or first-aid kits may have to be brought into the classroom, a source of running water may be needed, and so on. Therefore, before the students engage in any of the activities, the teacher should make sure that all necessary safety devices are available and all necessary precautions have been taken.

The following list of laboratory safety hints should be used as a guideline in determining what steps should be taken in order to prepare the regular classroom for use as a temporary laboratory facility. Of course, if an actual laboratory facility is available for use, the guidelines should also be followed for that facility at all times.

*Taken, in part from "Safety First in Science Teaching." Division of Science Education, State Department of Public Instruction, Raleigh, NC 27611.

Laboratory Safety Hints

- All science laboratories should have and use the following safety items: fire extinguishers, first-aid kits, fire blankets, sand buckets, eyewash facilities, emergency shower facilities, safety goggles, laboratory aprons, gloves (asbestos, etc.), tongs, respirators, wire or asbestos gauze.
- Good housekeeping is essential for maintaining safe laboratory conditions.
- Confine long hair and loose clothing. Laboratory aprons should be worn.
- Never have students conduct experiments alone in the laboratory.
- Proper eye protection devices should be worn by all persons engaged in, supervising, or observing science activities involving potential hazards to the eye.
- Make certain that all hot plates and open burners are turned off when leaving the laboratory.
- Every laboratory should have two unobstructed exits.
- There should be an annual, verified safety check of each laboratory.
- One teacher should not have to supervise more than twenty-four students engaged in laboratory activities at one time.
- Master cutoff valves or switches are advisable for laboratories with gas, water, and electricity.
- Emergency instructions of procedures for fires, explosions, chemical reactions, spillage, and first aid should be conspicuously posted near all storage areas.
- Always perform an experiment or demonstration prior to allowing students to do the activity. Look for possible hazards. Alert students to potential dangers. Safety instructions should be given each time an experiment is begun.
- Horseplay or practical jokes of any kind are not to be tolerated!
- Never eat or drink in the laboratory or from laboratory equipment.
- Exercise great care in noting odors or fumes. Use a wafting motion of the hand.
- Never use mouth suction in filling pipettes with chemical reagents. (Use a suction bulb.)
- Never "force" glass tubing into rubber stoppers. Constant surveillance and supervision of student activities are essential.
- Student attitude toward safety is imperative. Students should not be afraid to do experiments or use reagents and equipment, but should respect them as potential hazards.
- Teachers should set good safety examples when conducting demonstrations and experiments.
- Never allow the open end of a heated test tube to be pointed toward anyone.
- Broken or chipped glassware should not be used.
- Breathing gases, especially in high concentrations, can be very dangerous. Carbon dioxide is no exception, since unconsciousness can result within seconds if high concentrations are breathed. Breathing pure nitrogen, argon, helium, or hydrogen is dangerous.
- Chemicals should not be tasted for identification purposes.
- When heating materials in glassware by using a gas flame, the glassware

should be protected from direct contact with the flame though use of a wire gauze.
- Don't pour water into acid.

First-Aid and Emergency Tips

- Have first-aid procedures established in the event of an accident.
- All students and teachers should know the location of fire extinguishers, eyewash fountains, safety showers, fire blankets, and first-aid kits.
- Safety signs should identify the location of safety equipment.
- Student aides should be fully aware of potential hazards and know how to deal with accidents.
- It is an excellent idea to display safety posters in science laboratories.

For further information on safety in the science classroom, consult the following:

Articles published by the National Science Teachers Association (NSTA) 1742 Connecticut Ave., N.W., Washington, DC 20009.
Brown, Walter R. "Hidden Hazards in the Science Laboratory." *Science and Children*, January–February 1974, pp. 11–13.
Fawcett, H. H. "Health Aspects of Common Laboratory Chemicals." *The Science Teacher*, December 1966, pp. 44–45.
Livingston, H. K. "Safety in Schools." *The Science Teacher*, September 1966, pp. 25–26.
Mitchell, John, Jr. "Handling of Glassware." *The Science Teacher*, January 1967, pp. 66–68.

Enrichment Features _____

At the end of each unit, additional information is presented that connects the material in the text to everyday issues and decisions.

The Working World describes career orientations in fields broadly related to the subject matter covered in each unit. The information in this section is designed to help the students think innovatively about the job market. Not every career included here represents a conventional alternative. We have tried to widen the students' view of their future jobs or careers.

Your Leisure Time relates career orientations to recreational and volunteer activities that can provide learning experiences helpful for developing job skills. The value of sharing one's skills and abilities in a nonprofit way is stressed as a necessity for the enrichment of the social community. Part-time jobs appropriate for this age group are also described.

The *Consumer Science* section uses actual excerpts from news articles that have appeared in national newspapers to illustrate how the need to understand scientific information is a fact of daily life. Articles are relevant to the subject matter of each unit, and a brief commentary, which is designed to generate discussion in any of several directions, is provided. A typical page from the *Consumer Science* section appears on the following page.

The perspective offered in this section is that of the newspaper reader as consumer, voter, and decision maker. Articles have been chosen to highlight the multiplicity of factors that influence everyday events and the need to understand basic scientific data in order to interpret these events. Similar events, or even the same events, are likely to have been reported in local newspapers through news services. The teacher is, therefore, provided with a ready extension of his or her source materials.

CONSUMER SCIENCE IN...

THE KNOXVILLE
JOURNAL

Waste expert warns U.S. 'still creating Love Canal'

By TERRY McWILLIAMS

As much as 90 percent of America's toxic garbage is dumped into landfills without treatment.

Years from now, a hazardous waste expert says, that practice may haunt us the same way decades-old toxic chemicals did when they oozed from burial pits at Love Canal, N.Y., into a subdivision.

"We have far and away more hazardous waste than any (country) — 60 million tons," says Gary Davis, a former California hazardous waste specialist and now a Knoxville environment lawyer.

"We're still creating Love Canals, legally, because current regulations say you can dump anything into landfills less secure than Love Canal. And Love Canal was pretty secure as it goes."

Davis traveled through Europe last summer and talked to government officials, environmental groups and industry consultants to learn what system those countries use to dispose of hazardous waste. He found a discernible difference from the way America disposes of its waste.

"They (Europeans) don't even consider putting hazardous waste in the ground," said Davis. "They don't consider it a viable technology...They know that liquid waste can migrate into the groundwater and they depend on groundwater for drinking water."

Two of the leading countries, West Germany and Denmark, burn or detoxify more than 60 percent of their special wastes. He said the reasons are simple.

"Europeans recognized years ago that landfills leak. They recognize they do not have as much land as we do to waste. They decided the technology was there and they were going to use it. They pioneered (new) technology and put the weight of government behind it," he said.

Waste disposal is an important global issue. As the technologies of industry become similar from country to country, so do the waste products of these processes. However, since social and political conditions differ, different alternatives for disposal are being chosen. Comparison through observation is a good method for learning how these different options work.

What factors in our own society influence our methods of disposal for waste products? What waste products do we produce? Is there any method of production which does not produce waste material? How do you think methods might differ if we were a country the size of Switzerland?

Only the residues left behind from incineration or other treatment are buried in landfills, Davis said.

Davis, one of the authors of a California assessment on alternatives to burying hazardous waste, believes Denmark is 10 years ahead of the United States in the way it disposes of waste. He blames lax U.S. environmental regulation for the lag.

"There's no God-given right to pollute," he said. But business will take more regulation "as an affront."

Most European governments establish quasi-public entities to finance and build hazardous waste treatment facilities for industry. Then all industry generating toxic waste must pay to use the plants, he said.

In Denmark, municipalities are responsible for collecting and processing hazardous wastes — just like U.S. cities maintain sewage treatment plants, he added.

The government in Denmark subsidizes new treatment systems by 25 percent. Other countries, like Germany, discourage the production of new hazardous wastes by requiring more efficient processes. There, a panel reviews all new processes to determine if the industry can reduce pollution even more, Davis said.

Environmental regulation of industry is less in Europe, he said, but the attitudes of industry officials are generally such that factories "are better citizens." The relationship is "less adversarial" and more informal, he said.

One of the problems in the United States, he contends, is the financial advantage some companies gain by rejecting incineration for the less expensive dumping into the ground.

172D

The *Consumer Science* section can be found on pages 54D, 110D, 172D (shown above), 216D, 248D, 330F, and 472F.

The Working World is found on pages 54 A and B, 110 A and B, 172 A and B, 216 A and B, 248 A and B, 330 A through D, and 472 A through D.

Your Leisure Time appears on pages 54C, 110C, 172C, 216C, 248C, 330E, and 472E.

Unit 1　Principles of Biology

Chapter 1　What Is Science?

OVERVIEW　Science is an integral part of the world in which we live. This chapter provides an explanation of the kind of work that a scientist does and discusses the types of problems that science can solve. Scientists approach problem solving through use of the scientific method. The tools of measurement extend the range of vision of scientists in the various fields of science.

HINTS　Initiate the work in this chapter with the theme "Science is all around us." Have a group of pupils prepare a bulletin board on current topics in science, health, and technology. Develop a chalkboard model of Pasteur's anthrax experiment. Help pupils to identify the steps in the scientific method. Show how this method is useful in all branches of science.

Career information can be significantly helpful to students. This chapter lends itself nicely to discussions and activities dealing with careers in science. For further information and suggestions on career activities see the section entitled "Careers in Science."

ACTIVITY SUGGESTIONS　No special materials are needed for this activity.

After pupils learn how to use the metric ruler, you may wish to have them measure some common objects, such as the length and width of *Living Things*, a notebook, and the table on which they are working. In this case metric rulers, or meter sticks will be needed.

ANSWERS TO ACTIVITY QUESTIONS　1)　10 mm 2)　10 mm 3)　10 mm 4)　Both are equal to 10 mm. 5)　4.5 cm 6)　55 mm 7)　100 mm 8)　10 cm

ANSWERS TO QUESTIONS

WORD QUIZ　1a)　life b)　elements and compounds c)　forms of energy d)　solar system e)　rocks and minerals. 2a)　Botany is the study of plants. Ecology is the study of living things in their nonliving environment. b)　Zoology is the study of animals. Physiology is the study of cell processes. c)　Anatomy is the study of the structure and form of the body. Physiology tells us how the body works.

CHECK YOUR FACTS　1.　Science is a body of knowledge that is discovered by scientists. 2.　A scientist's job is to find out how things work. 3.　Scientists work in laboratories of all kinds. Scientists might be found in hospitals, universities, and in manufacturing plants. 4.　Experiments help scientists to find answers to their problems. 5.a)　An hypothesis is a guess. b)　A result is data obtained from an experiment. c)　A conclusion is a summary statement that answers the problem question.

THOUGHT QUESTIONS　1. Some modern problems are the energy crisis, a cure for cancer, increasing mine safety, etc. 2. See problems clearly before trying to solve them, accept new ideas that are presented, check all known facts before acting on a problem. 3. So that scientists all over the world can communicate with each other using measurements that they all can understand. 4. Answers will vary.

CHAPTER 2   THE BASIC UNIT OF LIFE

OVERVIEW   This chapter introduces the concept of the cell as the unit of life. Life is defined in terms of life functions that are related to the cell and the living protoplasm of cells. Similarities and differences in plant and animal cells are presented.

HINTS   This chapter lends itself to a variety of demonstration lessons. Set up a demonstration of specimens; some living, some nonliving, and others dead. Encourage pupils to define life by identifying those processes that are necessary for life. Demonstrations of living protozoans will help to convey the idea that a single cell carries out life functions.

ACTIVITY I SUGGESTIONS   Materials needed are microscopes, clean microscope slides, cover slips, forceps, a dropper, lens paper, an onion, enough toothpicks for the class and some iodine stain.

Learning to use the microscope is an important skill for the beginning biology student. It takes several days for pupils to learn the names of the parts of the microscope. Very often, it takes time to train pupils to see through the microscope.

ANSWERS TO ACTIVITY II QUESTIONS   1.   Onion cells resemble rectangular boxes. 2.   The nucleus, cytoplasm, and cell wall are visible. 3.   The nucleoli and cell wall absorb stain. 4.   Cheek cells do not have cell walls. 5.   The nucleus of a cheek cell is quite visible. 6.   The shapes of cheek cells differ from those of onion cells, but both have nuclei, cytoplasm, and cell membranes. The latter are not visible under the light microscope.

ANSWERS TO QUESTIONS

WORD QUIZ   1.   Definitions will vary. They should approximate those in the text. 2.   Have pupils label a drawing of a cell.

CHECK YOUR FACTS   1.   Living organisms carry on certain activities that are necessary for life. Nonliving or dead things cannot move, reproduce, use food, grow, or get rid of wastes. 2.   Examples of nonliving things are a rock, mineral matter, a metal key, window glass, the wind. Five examples of living things are a pupil in a class, a barking dog, a flower in a field, a frightened mouse, a galloping horse. Five examples of dead things are a stuffed bird, a preserved frog, a mounted fish, a dried earthworm, a pressed flower. 3.   A knowledge of cells is basic to the understanding of biology. The proper functioning of each cell in the body is important to the well-being of the entire organism. 4. and 5.   Pupils' drawings of a plant and animal cell should reflect their knowledge of cell structure. 6.   the endoplasmic reticulum, the ribosomes, the mitochondria 7.   Plant cells have cell walls, chloroplasts, and large storage vacuoles. Animal cells have centrosomes. The vacuoles in most animal cells (with the exception of fat cells) are quite small.

THOUGHT QUESTIONS 1. The movement of the ball, log, and automobile are made possible by an energy source that is outside of and not a part of the object. 2. The mitochondria are too small to be resolved clearly by the light microscope. 3. Protoplasm is a colorless, semisolid material.

Chapter 3   Some Simple Chemistry

OVERVIEW This chapter directs the attention of the reader toward the conceptual meanings of element and atom; compound and molecule. As pupils understand these terms, they then can visualize what takes place during chemical change.

HINTS It is a good idea to teach the concept of mixture by moving from the familiar to the unfamiliar. Mix some salt and black pepper together. Each substance can be identified. Now mix together some powdered sulfur and iron filings. Separate by using a magnet. Next you can demonstrate compound formation by heating the iron and sulfur together. Show that iron sulfide resembles neither sulfur nor iron and does not respond to a magnet.

Bring to class samples of elements. Pupils cannot collect these unless given specific directions. You have shown them oxygen, hydrogen, sulfur, and iron. Now show them carbon, zinc, lead, tin, and other elements.

ACTIVITY SUGGESTIONS Materials needed are several test tubes, limewater, drinking straws, several large beakers, glass plates, small candles, ice water.

This activity demonstrates that some compounds are formed by burning. A birthday candle works well in this exercise. Whenever pupils use a flame, there is potential danger. **CAUTION: There must be sufficient space between pupils to prevent accidents. Long hair should be tied back. Matches are to be dipped in water before discarding. Insist that long, loose sleeves be rolled up and secured with rubber bands. Students should wear goggles.** The second part of this activity has special considerations. **The candle should be affixed to the bottom (inside) of a beaker, so that if the candle topples, it will be inside a safety unit.** Condensation of water vapor will take place if the jar that is being treated with the flame is very cold. This can be accomplished by partially filling it with very cold water or by using an ice cube.

ANSWERS TO ACTIVITY QUESTIONS 1. The limewater turns milky. 2. Limewater turns milky. 3. This happened when the breath was blown into the limewater. 4. carbon dioxide 5. carbon dioxide 6. Hydrogen from the candle wax and oxygen from the air formed water. 7. carbon dioxide 8. water vapor.

ANSWERS TO QUESTIONS

WORD QUIZ 1. matter 2. element 3. mixture 4. vapor 5. compound 6. atoms 7. molecule 8. bond energy 9. energy 10. oxidation

CHECK YOUR FACTS 1. Chemistry is the study of matter. 2. gold, silver, lead, oxygen, sulfur 3. Water is formed from 4 atoms of hydrogen and 2 atoms of oxygen. 4.a) water b) It is composed of the elements hydrogen and oxygen in chemical combination. c)   carbon dioxide d)   Light and heat are given off. e)   Oxygen is joined chemically to another substance.

THOUGHT QUESTIONS   1.   The gases listed retain their own identities. They are not joined in chemical combination. 2.   Hydrogen would burn and cause explosions as it does in the sun. 3.   Chemical energy in food is released in the cell. The energy enables cells to do work.

Chapter 4   The Chemistry of Protoplasm

OVERVIEW   Protoplasm is the living material of cells. The purpose of this chapter is to explain the chemical nature of protoplasm. It is pointed out that the compounds in protoplasm are used by the cell to aid in a particular life function, that water is necessary to the life of the cell, and that carbohydrates, proteins, fats, and mineral salts are useful to a cell only when they are in solution.

HINTS   To teach the meanings of the words "dissolve," "solution," and "solvent" demonstrate how solutions are made. Dissolve some sugar in water. Squeeze some lemon into tea. Try dissolving starch in water. Have the pupils explain the differences they see in the starch and water as compared with sugar and water. Use this information to explain diffusion, especially through a cell membrane. Diffusion is an important concept that should be explained and demonstrated.

ACTIVITY SUGGESTIONS   The materials needed are several 100-cc beakers or baby food jars, medicine droppers, red or black ink and water.

This activity is simple and self-explanatory. As an *alternate activity* you may wish to do (or have the students do) the following demonstration. Fill the bulb of a thistle tube with a solution of molasses and water. Tie a semipermeable membrane over the mouth of the thistle tube. Invert the bulb in a large beaker of water, affixing the stem to a ringstand. Show pupils how water rises in the stem.

ANSWERS TO ACTIVITY QUESTIONS   1.   The ink falls to the bottom of the beaker. It is heavier than water. 2.   Ink particles move through the water, traveling upward. 3.   The water becomes the color of the ink. 4.   Particles of ink move from the ink drop in which ink is concentrated to areas of the water where there is no ink.

ANSWERS TO QUESTIONS

WORD QUIZ   1.   dissolved 2.   solution 3.   diffusion 4.   glucose 5.   ATP 6.   fatty acids 7.   amino acids 8.   enzymes 9.   DNA

CHECK YOUR FACTS   1.   carbohydrates, proteins, fats, nucleic acids, mineral salts 2.   Water dissolves other compounds. 3.   Table 4-1 lists elements. Table 4-2 lists compounds. 4.   Carbohydrates are broken down for energy. 5.   Protoplasm is able to combine elements into compounds that are used by cells. 6.   The cell makes proteins from amino acids. The cell gets energy from fatty acids. 7.   Energy is used by the cell to move molecules and to do other work. 8.   ATP releases energy to the cells. 9.   Enzymes speed up chemical reactions that take place in cells.

THOUGHT QUESTIONS   1.   A knowledge of elements and compounds is necessary so that we can understand how the cells take apart molecules and build up others. 2.   Elements and compounds must be dissolved in water before they can cross cell membranes. Chemical activities in cells occur when molecules are in solution. 3.   Proteins contain the element nitrogen.

Chapter 5   Energy and Photosynthesis

OVERVIEW   Several concepts are presented in this chapter. Photosynthesis and respiration are explained and compared. Methods by which plant and animal cells obtain food are compared also. To grasp these ideas pupils have to understand how materials enter and leave cells. The importance of diffusion in the processes of photosynthesis, respiration, and food-getting by animal cells is emphasized. *Chlorella* and the ameba are used as examples of independent, living, single-celled organisms.

HINTS   To show how water gets into and out of cells, demonstrate plasmolysis in *Elodea*. Put a few blades of the water plant, *Elodea*, into some salt water. Show the shrunken cytoplasm. Explain. Reverse this process by placing these *Elodea* cells in distilled water. Use the definition of diffusion to explain. To show that chlorophyll is needed for photosynthesis, remove the chlorophyll from a variegated leaf, such as coleus, using warm alcohol. **CAUTION: Warm alcohol over a water bath. Do not heat alcohol over a naked flame. Alcohol is very flammable.** Test the bleached leaf with iodine for starch. Show that starch is present where the leaf was green.

ACTIVITY SUGGESTIONS   Materials needed are several test tubes fitted with stoppers, some sprigs of *Elodea*, bromthymol blue solution, soda straw, and beakers.

Pupils should work in groups of four. Prepare a 0.1 percent stock solution of bromthymol blue in 500 mL of water. Add a drop of ammonium hydroxide to turn the solution deep blue. Pupils use a scant teaspoon of bromthymol blue in ¾ test tube full of water.

As an *additional activity* you can have the students observe a single-celled organism containing chloroplasts. *Chlorella* is such an organism, which is mentioned in the text. However, *Chlorella* cells may be too small for beginning pupils to examine. You may wish to substitute the larger cells of *Pleurococcus*. You can scrape this alga from the sides of trees and suspend it in some water. Make several slides and demonstrate these cells to the class, pointing out the large chloroplast. You can also set up several demonstration microscopes showing an ameba. Be sure to cut down the light so that the nearly transparent body becomes visible.

ANSWERS TO ACTIVITY QUESTIONS   1.   Carbon dioxide changes bromthymol blue to bromthymol yellow. 2.   The tube with the plant in sunlight should lose its yellow color and the bromthymol blue should reappear. 3.   The tube without the plant should remain yellow, just as the tube in the dark. 4.   As carbon dioxide is used up, the blue color returns. 5.   The test tube without *Elodea* serves as a control.

ANSWERS TO QUESTIONS

WORD QUIZ   Definitions will vary. They should approximate those in the text.

CHECK YOUR FACTS   1.   Photosynthesis means putting together in the presence of light. 2.a)   Light is necessary for the manufacture of chlorophyll by cells. Chlorophyll traps light energy. b)   ATP stores chemical energy in cells. c)   Glucose is used by cells for energy. Glucose is a product of photosynthesis.

3.		photosynthesis	respiration
a.	glucose	synthesized	used up
b.	ATP	used up	built up
c.	carbon dioxide	used	released
d.	oxygen	released	used
e.	light	used	not used

4.   Producers make their own food. 5.   The cell membrane controls the passage of materials into and out of the cell. 6.a)   Chlorophyll is a pigment. A chloroplast is a structure. b)   *Chlorella* is an alga cell. A chloroplast is a structure in *Chlorella*. 7.   An ameba is a consumer because it cannot make its own food. 8.   Pupils' diagrams should indicate that they have learned the parts of the ameba. 9.   Pupils' diagrams should indicate their knowledge of the *Chlorella* cell.

THOUGHT QUESTIONS   1.   The carbon dioxide cycle should show recycling of the compound from plant to animal. 2.   Fertilizers add nitrogen to the soil. Proteins require nitrogen. 3.   The raw materials of photosynthesis are carbon dioxide and water. The source of energy is sunlight. The end products are carbohydrates, proteins, and fats.

CHAPTER 6   Cells in Groups

OVERVIEW   In the last chapter the life activities of cells that live independently were examined. This chapter shows how cells that live together function. Mitosis as a process of cell division is presented. Why "daughter" cells resemble the parent cell is explained through a discussion of genes. Cells in groups serve special functions, as shown through the introduction of tissues and organs.

HINTS   Use the illustrations in the text to full advantage. Have pupils study the chapter-opening paragraph that shows mitosis in the onion root. Use this photograph, microscope demonstrations, or prepared slides and chalkboard diagrams to show the outcome of mitosis. Explain that mitosis takes place in animal and plant cells. Use pipe-cleaner "chromosomes" to demonstrate replication and gene distribution.

ACTIVITY SUGGESTIONS   Materials needed are microscopes, lettuce leaves, forceps, clean microscope slides, cover slips, and lens paper. If you have some methylene blue stain, use it to color the nuclei. Show pupils how to strip off a small piece of transparent membrane from a cracked vein.

As an additional activity or demonstration you may wish to use any or all of the following: Feulgen stain can be used to demonstrate DNA in tissue cells. Use frog skin to show epithelial tissue. A thin smear of hamburger meat stained with methylene blue shows muscle cells to advantage. The ends of beef bones can be used to demonstrate cartilage and gristle.

ANSWERS TO ACTIVITY QUESTIONS   1.   It is yellowish, thick, and bumpy. 2.   The inner surface is smoother and lighter in color. 3.   It is slippery and spongy in texture. 4.   The muscles expand and contract. 5.   It feels softer than the bone and it is more elastic (flexible).

ANSWERS TO QUESTIONS

WORD QUIZ   1.   a   2.   c   3.   f   4.   g   5.   e   6.   h   7.   d   8.   a   9.   b   10.   nerve   11.   epithelial   12.   vascular   13.   muscle   14.   blood   15.   connective

CHECK YOUR FACTS   1.   There are no chromosomes. 2.   To produce more cells. A mature cell either dies or reproduces. 3.   Mitosis is equal division of the chromosomes. Student answers should describe major changes. 4.   Genes are made of DNA molecules. Chromosomes hold the genes. 5.   Enzymes speed up chemical reactions. 6.   A cell is the basic unit of structure and function of living things. Tissues are made of cells. Organs are made of tissues. Systems are made of organs. 7.   Energy is obtained from food molecules by cells. 8.   Water diffuses into the cell through the cell membrane. 9.   Carbon dioxide leaves the ameba by way of the cell membrane. 10.   A higher animal is one in which the body is made of many cells.

THOUGHT QUESTIONS   1.   Scientists know that genes exist because of the work that they do in transmitting traits from one generation to the next. 2.   Skin is made of different types of tissues. 3.   An ameba is an organism, a complete living thing. A skin cell is not an organism.

AV Materials   Unit 1

(Numbers in parentheses refer to chapter numbers.)

"Atoms & Molecules," from *Spaceship Earth*. Hawkhill, sound filmstrip. (3)
"The Cell," from *Spaceship Earth*. Hawkhill, sound filmloop. (2)
Cell Division—Mitosis. Encyclopaedia Britannica, filmloop. (6)
The Cell: Structural Unit of Life. Coronet, B/W–color, film. (2)
Cell Structure. Filmstrip House, filmstrip. (2)
Cells: A First Film. BFA, B/W–color, film. (2)

Cells of Onion Root Tip—Mitotic Division. Doubleday Multimedia, filmloop. (6)
Diffusion Explained. Doubleday, filmloop. (5)
Diffusion & Osmosis. Coronet, color, film. (5)
The Dividing Cell. Ealing, filmloop. (6)
End Products. Thorne, filmloop. (5)
Food: Energy from the Sun. BFA, color, film. (5)
Green Plants & Sunlight. Encyclopaedia Britannica, color, film. (5)
How Green Plants Make and Use Food. Coronet, B/W–color, film. (5)
"Living Things," from *Spaceship Earth*. Hawkhill, sound filmstrip. (2)
Molecules: A First Film. BFA, color. (3)
Motion of a Molecule. Ealing, filmloop. (5)
Osmosis: Movement of Water. Thorne, filmloop. (5)
Osmosis: Effect on Cell Membranes. Thorne, filmloop. (5)
Photosynthesis. BFA, set of 6 filmstrips. (5)
Testing for Starch in Green Plants. Encyclopaedia Britannica, filmloop. (5)
The Metric System—Universal Language of Measurement. Pathscope, sound filmstrip. (1)
The New Concept of the Cell. Filmstrip of the Month, filmstrip. (2)

Supplementary Reading   Unit 1

Asimov, Isaac. *Breakthroughs in Science*. New York: Scholastic Book Services, 1959.
——— . *The World of Carbon*. New York: Collier, 1962.
——— . *The World of Nitrogen*. New York: Collier, 1964.
Baker, Jeffrey, and Garland Allen. *The Process of Biology: Primary Sources*. Palo Alto, CA: Addison-Wesley, 1970.
Berger, Melvin. *Tools of Modern Biology*. New York: Thomas Y. Crowell Co., 1970.
Davson, Hugh. *A Textbook of General Physiology*. Baltimore: Williams and Wilkins, 1970.
DeRobertis, E.D.P., W. W. Nowinski, and F. A. Saez. *General Cytology*. Philadelphia: Saunders, 1965.
Fawcett, Don W. *The Cell: Its Organelles and Inclusions*. Philadelphia: Saunders, 1966.
Galston, Arthur W. *The Life of the Green Plant*. Englewood Cliffs, NJ: Prentice-Hall, 1964.
Giese, Arthur C. *Cell Physiology*. Philadelphia: Saunders, 1968.
Greene, Jay E. *100 Great Scientists*. New York: Washington Square Press, 1964.
Hecht, Selig. *Explaining the Atom*. New York: Viking Press, 1960.
Mader, Sylvia S. *Inquiry Into Life*. Dubuque, IA: Wm. C. Brown Co., 1976.
Morrison, Thomas F., et al. *Human Physiology*. New York: Holt, Rinehart and Winston, Publishers, 1977.
White, Abraham, Philip Handler, and E. L. Smith. *Principles of Biochemistry*. New York: McGraw-Hill, 1973.
Wilson, Carl, and Walter Loomis. *Botany*. New York: Holt, Rinehart and Winston, Publishers, 1971.

Unit 2   The Continuation of Life _____

Chapter 7   Reproduction in Simple Organisms

OVERVIEW   The basic differences between sexual and asexual reproduction are discussed. Examples of asexual reproduction discussed are fission in the ameba, spore formation in mushrooms, vegetative propagation in various plants (such as strawberry, onion, potato, etc.), grafting of fruit trees, and budding in *Hydra*. Sexual reproduction is discussed using conjugation in *Spirogyra* and the union of sperm and egg in *Hydra* as examples. The advantage of genetic recombination in sexual reproduction is pointed out in simple terms.

HINTS   The different kinds of living things are not studied until Units 4–7. Mushrooms, ferns, algae, coelenterates, etc., will be taught as whole organisms in those later units. It is not necessary, therefore, to teach any of the organisms in this chapter in detail. Each has been chosen simply to illustrate a particular type of reproduction. Stress the reproduction, not the organism.

When discussing vegetative propagation it would be helpful to use some indoor plants, such as *Sanseviera* or Boston ferns, as examples, since some students may have limited opportunity for outdoor observation.

ACTIVITY SUGGESTIONS   The materials needed are fresh mushrooms, sheets of paper (construction paper is good), clean microscope slides, and microscopes.

If microscopes are not available the spore prints can be done and the second part of the activity omitted. **CAUTION: Since wild mushrooms can be poisonous, only fresh mushrooms obtained from a grocery store should be used. Do not encourage students to collect mushrooms on their own.**

Spore prints can be used around the classroom for decorative purposes.

As an *alternate activity* you may wish to order the water mold *Allomyces arbusculus* from a supply house for the purpose of a microscopic investigation. If the mold is placed in water with boiled radish seeds (cracked open) it will form a fuzzy growth on the seeds in about one week. This can then be observed under the microscope. Spores and sex cells can be seen. It is possible that the process of fertilization may be viewed under the microscope.

ANSWERS TO ACTIVITY QUESTIONS   1.   because the mushroom, which is relatively small, must produce many spores 2.   Many of the spores will not land in an area with favorable conditions and therefore will not develop. 3.   asexual.

ANSWERS TO QUESTIONS

WORD QUIZ   Definitions will vary. All definitions should approximate those given in the text.

CHECK YOUR FACTS   1.   Asexual reproduction does not involve fertilization. The offspring are identical to the parent. Sexual reproduction involves the union of two cells. Offspring are not identical to either parent. 2.   Cell division results in growth when the new cells remain attached to the other cells in a multi-cellular organism. When two cells are formed from the division of a single-celled organism they separate. This is reproduction. 3.   A spore is a cell capable of growing into a new organism. Algae, fungi, mosses, ferns, and some other protists produce spores. 4.   Vegetative reproduction is the production of new plants from stems, roots, or leaves. Strawberry runners, bulbs (tulip, onion), and potato "eyes" are some examples. 5.   Grafting insures the type of fruit that will be obtained from the tree. 6.   Pairs of *Spirogyra* cells form cytoplasmic "bridges" between themselves through which one cell moves to unite with the other. 7.   Sperm are small and motile. Eggs are larger and nonmotile. 8.   *Hydra* sperm swim to the egg. 9.   In asexual reproduction there is no chance for sperm and egg to fail to meet, as in sexual reproduction. Offspring results can be definitely predicted. Sexual reproduction is of advantage to the long-term survival of the species due to the selective advantages of genetic recombination.

THOUGHT QUESTIONS   1.   No, this would take too long to produce mature trees. Grafting is a quicker method. 2.   No, grafting does not produce anything new.

Chapter 8   Reproduction in Flowering Plants

OVERVIEW   The structure and function of the various parts of a typical flower are explained. The processes of cross-pollination and self-pollination are described, and the events leading up to fertilization are discussed. Also explained are the function of the fruit and the structure and function of the seed. Mention is made of the life cycles of annual, biennial, and perennial plants.

HINTS   Point out that we take advantage of the life cycle of such biennials as carrot and beet. They store food in their large roots during their first growing season. We eat these roots instead of allowing the plant to use that food to produce a seed crop in its second year.

ACTIVITY SUGGESTIONS   Materials needed are several live flowers, clean microscope slides, cover slips, and microscopes. You may wish to time this chapter for a season when flowers are in bloom in yards in the neighborhood. Suburban students can often be stimulated to bring flowers to school. In winter, or in inner city neighborhoods, make contact with florists or funeral directors for surplus flowers. These greenhouse types are often fancy varieties with extra petals and reduced pistils or stamens. Simple large flowers like the tulip make the best specimens. In tulips the sepals and petals are alike.

Students may be asked to draw, label, and write a function for each flower part they see. Alternatively they may dissect a flower and paste the dried parts in order on a sheet of paper along with appropriate labels.

If microscopes are not available, Procedure D of this activity can be omitted. As an *alternate activity* you may wish to demonstrate seed germination. Beans, radishes, corn, sunflower seeds, and many other vegetable and flower seeds may be planted in moist soil. The cut-off bases of milk cartons can serve as flower pots.

ANSWERS TO ACTIVITY QUESTIONS 1.   Descriptions will vary. They should be reasonably accurate. Sepals cover flower buds before they open. Petals make the flower attractive. Stamens are the male reproductive structures, which produce pollen. Pistils are the female reproductive structures. 2.   pollen 3.   The pollen grain will not be blown off. 4.   Descriptions will vary depending on the species of plant used.

ANSWERS TO QUESTIONS

WORD QUIZ 1.   f 2.   j 3.   a 4.   h 5.   b 6.   m 7.   l 8.   k 9.   c 10.   d 11.   e 12.   g 13.   i

CHECK YOUR FACTS 1.   A flower produces seeds for the reproduction of the plant. 2.   Refer to Figure 8-1. Make allowance for the use of different flowers as examples. 3.   Sperm are in the pollen grain. 4.   Eggs are in the future seeds (ovules). 5.   The sperm reaches the egg by way of a pollen tube, which grows out from the pollen grain and down through the pistil to the opening in the future seed. 6.   Pollination is the arrival of pollen on the pistil. Fertilization occurs only after the pollen tube reaches the future seed and the sperm nucleus unites with the egg nucleus. 7.   Important parts of a seed are seed coat (protection), stored food (nourishment), and embryo (immature plant). 8.   The ripened ovary is the fruit. 9.   Pollen is most often carried to the flower either by insects or by wind. 10.   One type lives for only one growing season. Another type develops roots, stems, and leaves one year, flowers and seeds the second year, and then dies. The third type lives through several growing seasons and produces seeds each year.

THOUGHT QUESTIONS 1.   They are insect pollinated. The showy flowers that attract us are designed to be easily found by bees. Wind-pollinated flowers are inconspicuous, and they do not make good bouquets. 2.   The grassy field; grasses are wind-pollinated. Their pollens blow long distances and are inhaled.

Chapter 9   Reproduction in Higher Animals

OVERVIEW External fertilization is explained using fish and frogs as examples. Internal fertilization is explained using birds, reptiles, and mammals as examples. The basic points of development and birth (function of placenta, umbilical cord, etc.) are discussed using mammals (not specifically humans) as examples. There is a section on care of the young, which explains methods of caring for offspring used by various animals, such as birds, elk, oxen, and humans. The importance of child rearing and care for the unborn human fetus are stressed.

HINTS Some teachers will wish to introduce extra information. More detailed descriptions of the reproductive cycle in a variety of fish, reptiles, birds, and mammals can be used. More specific coverage of the reproductive organs and their functions can be included. Teachers doing this should consider the position of the school administration and the community as regards anything that might be construed as sex education.

Various phyla of animals are mentioned in this chapter, but classification should not be stressed at this point. This is dealt with later in the text.

ACTIVITY SUGGESTIONS   Materials needed are frog eggs, an aquarium, and several hand lenses.

In spring, frog eggs may be found in local ponds. If not available, they can be ordered from a supply house. Toad eggs are excellent for this activity. You can also obtain already hatched tadpoles and observe their development to adults.

Reproductive habits of local species can be used as an *alternate activity*. Northern city students might report on the nesting of robins or sparrows. Southern students could study the behavior of female alligators. The possibilities are endless.

ANSWERS TO ACTIVITY QUESTIONS   Answers will vary depending upon species of frog used and on geographic location. Toad eggs will hatch most quickly.

ANSWERS TO QUESTIONS

WORD QUIZ   1.   d   2.   c   3.   e   4.   g   5.   h   6.   b   7.   a   8.   k   9.   f   10.   j   11.   i

CHECK YOUR FACTS   1.   Eggs are laid in the water by the female and sperm are discharged over them by the male. 2.   In these land animals the male deposits semen in the lower oviduct of the female during mating. The sperm swim in the film of moisture lining the oviduct to reach the eggs in the upper oviduct and to fertilize them. 3.   Frog and fish eggs are small and moist. They would dry out if exposed to air. Bird and reptile eggs are well covered, so they do not dry out. They are large, producing large young that are capable of surviving on land. 4.   The birds care for their young. Reptiles do not. A few reptiles guard the eggs and newly hatched young, but no reptile feeds and rears the young. 5.   Mammals include egg-laying, pouched, and placental types. 6.   Care of the young gives protection from enemies and from weather. It provides food, so the young do not have to find their own before they are able to do so effectively. To varying degrees, in different species, the young can learn survival techniques from their parents. 7.   Children require a very long time to grow up. They must learn nearly everything they will need to know from parents and other adults.

THOUGHT QUESTIONS   1.   Neither birds nor mammals can be said to have a superior form of reproduction. They simply use two different ways of accomplishing the same thing. 2.   The rabbits are preyed upon by many natural enemies. Their losses are far too great to be offset by only two young every other year. They would become extinct.

Chapter 10   Heredity

OVERVIEW   The terms "dominant," "recessive," and "hybrid" are defined in light of the results of Mendel's experiments with pea plants. Incomplete dominance is also explained and the process of reduction division is outlined. Basic Punnet squares are introduced so that the students may learn to predict genetic possibilities. Other concepts discussed are sex determination, sex-linked and sex-influenced traits, the development of fraternal and identical twins, variation within a species, and the effects of environment on the development of an organism.

HINTS The concept of reduction division can be difficult for students to comprehend. It is of great value to use as many visual aids as possible to teach this. In the absence of more sophisticated visual aids you can hold your two hands together in a double fist. Let the students imagine that your fingers are chromosomes. Your double fist is a body cell with ten chromosomes. There are, therefore, two of each kind of chromosome—two thumb chromosomes, two forefinger chromosomes, etc. Now extend the fingers with hands together, prayer fashion. This represents the pairing of like chromosomes. Next separate the hands into two separate fists. Reduction division has taken place, and each new cell has only one of each kind of chromosome. These would be sperm or egg cells. Union of two such cells (fertilization) will restore the original chromosome number (the double fist).

ACTIVITY SUGGESTIONS No special materials are needed. The students should be encouraged to obtain information from as many people as possible for this experiment. The larger the sample, the more informative the results. The gene for tongue rolling is dominant and this should be indicated by the results.

As an *additional activity* you may wish to show the students how to do a pedigree chart and ask them to chart a particular trait in their families, such as eye color, attached ear lobes, etc.

ANSWERS TO ACTIVITY QUESTIONS These will all vary depending on the information collected by the students.

ANSWERS TO QUESTIONS

WORD QUIZ 1. reduction division 2. X-chromosome 3. fraternal twins 4. hybrid 5. dominant 6. incomplete dominance 7. recessive trait 8. Y-chromosome 9. sex-influenced 10. sex-linked 11. identical twins

CHECK YOUR FACTS 1. Mendel crossed peas with contrasting traits. He planted the resulting seeds and recorded the results. Then he interpreted the results to disclose laws of heredity. 2. A trait that develops in an individual even when carried in only one chromosome is said to be dominant. Examples are tallness in peas, blackness in guinea pigs, whiteness in hogs, hornlessness in cattle, etc. 3. A recessive trait is one that fails to develop when carried in only one chromosome. Examples are shortness in peas, whiteness in guinea pigs, dark color in hogs, horns on cattle, etc. 4. Incomplete dominance is a situation in which an inherited trait develops as a compromise between the effects of a pair of contrasting genes. 5. Reduction division resembles mitosis but results in a pair of cells, each of which contains only one of each kind of chromosome instead of two. It takes place during the formation of sperm and eggs. 6. A baby with two X-chromosomes will be a girl. One with an X and a Y will be a boy. It depends upon which kind of sperm fertilizes the egg. 7. Traits on the X-chromosome are sex-linked. In females they can be either dominant or recessive. In males they all act as dominants. 8. A trait controlled by a gene on a nonsex (not X or Y) chromosome that affects males and females differently is said to be sex-influenced. 9.   Fraternal twins develop from two fertilized eggs. Identical twins develop from a single fertilized egg that splits apart. 10.   Answers will vary. Watch for an understanding on the student's part that heredity sets the basic pattern of physical development but that environment supplies the materials and condition needed by the individual in order for the hereditary pattern to be expressed.

THOUGHT QUESTIONS 1.   No. He will learn speech from his adoptive parents and will speak with their accent. 2.   0%, 0%, 0%, 100%, 50%, 25%

CHAPTER 11   MUTATION AND ADAPTATION

OVERVIEW Mutation is defined as the alteration of the DNA molecule. Causes of mutation are given as various chemical substances (formaldehyde, mustard gas, creosote, etc.), and radiation. The concept of natural selection is introduced as a possible explanation of why the best adapted individuals survive in a given population. Artificial selection is explained as a means of obtaining desirable characteristics in milk-producing cows, steers bred for beef, etc.

HINTS The idea that species can change will disturb some students. Try not to upset them. Point out that they might study the beliefs held by a religious group other than their own. They would not accept those beliefs. They would simply try to understand the other person's point of view. In the same way, students whose personal beliefs will not let them accept the idea of species changing can study the idea as a theory. They will be finding out what it is that those other people believe.

ACTIVITY SUGGESTIONS Materials needed are several aluminum pie plates, an assortment of leaves, plaster of paris.

The making of these "fossil" leaves can be fun, but be sure the students understand the idea behind the exercise. (Mud hardens into rock over long periods of time.) Students can think of their casts as being real fossils. After all, they are a record of the fact that a particular species of tree lived in that area at a particular time. The same can be said for prints in a sidewalk.

You may find that some leaves need light oiling to prevent their sticking to the plaster, but don't overdo it. In most cases it will not be necessary. If a wire loop (made from a paper clip) is cast into the edge of the plaster the finished product can be used as a wall decoration.

ANSWERS TO ACTIVITY QUESTIONS 1.   An imprint of the leaf is left. 2.   yes 3.   Answers will vary. Since it is a preserved record of a living species, it could be considered a fossil, though not in the classical sense. 4.   Yes, they can be considered as such.

ANSWERS TO QUESTIONS

WORD QUIZ 1.   d 2.   e 3.   b 4.   a 5.   c

CHECK YOUR FACTS 1.   Mutation is a change in a gene. 2.   vibration of molecules, chemicals, radiation, and other unknown causes. 3.   They are usually harmful. They happen by chance, and chance is not likely to cause improvement. 4. Harmful genes handicap the individuals who carry them. On the average, they do not live as long or reproduce as often. So the genes become more scarce. 5. Some genes will adapt a species to its environment better than other genes will. Individuals carrying the more adaptive genes survive in larger numbers and produce more young. In effect, the environment has selected these genes, hence the name *natural selection*. By the same process, nonadaptive genes tend to die out. 6. Land has risen above the sea, and land has sunk under the sea. Mountains have been pushed up, and mountains have been worn away. Climate has changed back and forth between cold and warm and between wet and dry. 7. Changing environment will cause extinction of many species. In others, natural selection will favor new genes and eliminate old ones so that the species change. They become adapted to the new environment. 8. People have kept the animals that were most useful to them. Under these conditions genes that made the cattle more useful to people survived. Others disappeared.

THOUGHT QUESTIONS 1. Dark-haired parents often have light-haired children. This is because the parents are hybrids. They carry recessive genes for light hair. A child who gets these genes from both parents will have light hair. 2. A dwarf has two genes for the trait. If the farmer breeds the cows to a bull that is pure for the dominant (normal) gene, there will be no dwarf calves. Each will receive at least one dominant gene (from the bull).

AV Materials   Unit 2

(Numbers in parentheses refer to chapter numbers.)

Amoeba-Fission. Encyclopaedia Britannica, filmloop. (7)
Budding of Yeast Cells. Ealing, filmloop. (7)
Fertilization. Thorne, filmloop. (9)
Flowers and Fruits. Encyclopaedia Britannica, filmstrip. (8)
Genetics: Mendel's Law. Coronet, B/W–color, film. (10)
Gregor Mendel. Encyclopaedia Britannica, filmstrip. (10)
How Flowers Make Seeds. Coronet, color, film. (8)
How Plants Reproduce. McGraw-Hill, color, film. (8)
Human Reproduction and Birth. Parts I and II, Time-Life, filmstrips. (9)
Hydra. BFA, filmloop. (7)
Life Before Birth. Parts I and II, Time-Life, filmstrips. (9)
Mechanics of Inheritance. McGraw-Hill, B/W–color, film. (10)
Molds and How They Grow. Coronet, color, film. (7)
Plants and Their Environment. SVE, filmstrip. (8)
Plants that Grow from Flower, Stems, and Roots. Coronet, color, film. (7)
Reproduction by Simple Fission. Doubleday Multimedia, filmloop. (7)
Reproduction in Animals. Coronet, color, film. (9)
The Reproductive System. McGraw-Hill, filmstrip. (9)
Thread of Life. Bell System, color, film. (10)

Supplementary Reading   Unit 2

Asimov, Isaac. *The Genetic Code*. New York: The New American Library, 1962.

Bold, Harold C. *The Plant Kingdom*. 3rd ed. Englewood Cliffs, NJ: Prentice-Hall, 1970.

Fried, John. *The Mystery of Heredity*. New York: John Ray, 1971.

Heintz, Carl. *Genetic Engineering*. Nashville: Thomas Nelson, Inc., 1974.

Hutchins, Ross E. *This is a Flower*. New York: Dodd, Mead, & Co., 1963.

Jenkins, Marie M. *Embryos and How They Develop*. New York: Holiday House, 1975.

Keeton, William T. *Elements of Biological Science*. New York: W.W. Norton and Co., 1973.

Kraft, Kent Pat. *Luther Burbank: The Wizard and the Man*. New York: Meredith Corp., 1967.

Lehrman, Robert L. *The Reproduction of Life*. New York: Basic Books, Inc., 1964.

Lipke, *Heredity*. Minneapolis: Lerner Publishing Co., 1971.

Morholt, E., et al. *A Sourcebook for the Biological Sciences*. New York: Harcourt, Brace, Jovanovich, Inc., 1966.

Went, Frits, and the editors of *Life*. *The Plants*. New York: Time, Inc.. 1963.

Unit 3   Ecology: Living Things In The Environment

Chapter 12   The Nonliving Environment

OVERVIEW   The importance of the nonliving components of the environment is pointed out. Water is discussed in terms of the effects of the availability of water on plant and animal life on land, as well as the importance of water as an ecosystem for aquatic and marine life. The water cycle is explained, as is the carbon-dioxide–oxygen cycle. The cycling of minerals between water plants and animals and the water they live in, between land plants and the soil, and between animals and the food they eat is described. Other physical factors of the environment discussed in relationship to living organisms are temperature, altitude, and light.

HINTS   You may wish to stress the physical properties and characteristics of the area in which your school is located. Show how these factors influence the flora and fauna indigenous to your area, as well as the physical appearance of your region, and even possibly the life styles of the local residents. (For example, you may live in a ski town, mining town, etc.)

When discussing the water cycle, all that is needed is that the students understand the basic idea of how water reaches the land. More detail on this topic will be given in Chapter 14. The idea of cycling, as such, should be stressed. Other examples will continue to turn up throughout the ecology section of the book.

ACTIVITY SUGGESTIONS   Materials needed are radish seeds, three trays of moist sand or soil, access to a refrigerator freezer and an incubator (or other device for keeping a temperature of 38° C.).

The "trays" used here can be small. The base of a milk carton will hold many radish seeds. The different rates of growth give a visual example of the effect of variations in an environmental factor (temperature) upon the growth of a living plant.

As an *alternate activity* you may wish to keep several slices of potato in a hot and dry environment in order to evaporate the water from them. Weigh the slices before and after drying. The loss in weight corresponds to water loss. This will illustrate that water is an important part of a living organism.

ANSWERS TO ACTIVITY QUESTIONS   All information will vary based on students observations of the experimental results.

ANSWERS TO QUESTIONS

WORD QUIZ   1. e   2. b   3. a   4. c   5. f   6. d

CHECK YOUR FACTS   1.   Light supplies energy for nearly all food making (photosynthesis). Green plants must have light. Animals can live without it, but they must have food that was originally made by plants that used light. 2.   Refer to Figure 12-1. 3.   forest, grasses, desert plants, respectively 4.   Water and minerals. Poor soil grows plants adapted to grow with a poor mineral and water supply (example: a jack pine plain). Good soil grows a much richer mass of vegetation (example: a hardwood forest). These two examples could be in sight of each other. 5.   Air supplies carbon dioxide, oxygen, and moisture. 6.   Living things adapted to a hot climate would lack the ability to live in a cold one. The difference between a temperate prairie and a tundra is a matter of temperature. 7.   Cold-blooded animals hibernate because their protoplasm will not function at freezing temperatures. Some mammals hibernate because their food supply disappears during the winter. 8.   Cold-blooded animals cannot maintain body heat in cold weather. Warm-blooded ones can. Therefore, the warm-blooded ones can be active in winter. 9.   oxygen depletion 10.   This will vary with the region. 11.   The changes in climate that take place as you travel from south to north are similar to those that take place as your altitude increases. Therefore, vegetation that is found at lower altitudes in the north can also be found at higher altitudes in the south.

THOUGHT QUESTIONS   1.   A cave is dark, like the deep sea. In both places animals would have to receive food from outside the local environment, since light is needed for food making. 2.   Michigan and other northern states would be tundra until the ice covered them. Kentucky would be similar to present-day Michigan or southern Canada. Alabama would be like present-day Kentucky. The answer to the last question depends upon the region.

Chapter 13   The Living Environment

OVERVIEW   The concept of producers and consumers and their relationships to each other in food chains is discussed. Emphasis is put on the cycling of food and energy-producing materials in nature. This idea is expanded to cover the concept of natural enemies, predator-prey relationships, protective adaptations, and competition and co-operation between species. Other topics discussed are succession within a community, balance within a species, and the possible consequences of outside forces upsetting the balance of nature.

Try to find local examples of areas thrown into succession by some sort of disturbance. Around cities, human disruption will be the usual cause.

Open the question of whether our human community, made up of farms, mines, forest enterprises, fisheries, cities, etc., is a permanent new climax condition or a stage in a succession. Do not try to answer the question yet. Just set the students to thinking about it so they will read the conservation chapters that follow with more interest.

ACTIVITY SUGGESTIONS   Materials needed are several snails, some *Elodea* leaves, four test tubes per group, four rubber stoppers, aquarium water, a one-liter mason jar, a four-liter mason jar, a small amount of soil, and sand.

The smaller "communities" will not last as long as the larger ones. The one placed in the dark will, of course, be the shortest lived. Without light there can be no food made. The snail could, of course, eat the decaying plant, but without photosynthesis, there will be no source of oxygen. Carbon dioxide from snail and bacterial respiration will accumulate. The snail will die of suffocation. The larger aquarium has a better chance of achieving a reasonable balance.

Oxygen is evolved during photosynthesis, and plant growth shows that food is being produced. Animals feed on the food. Animal wastes are used by the plants in making more food (as a source of minerals). There is a balance, also, between respiration and photosynthesis in the production and use of carbon dioxide and oxygen.

As an *alternate activity* you may wish to demonstrate some type of competition. For example, you can plant a large number of seeds (any fast-growing variety will do) very close together in a small area and observe the results. A control group, having reasonably spaced seeds, should be planted also, for comparison.

ANSWERS TO ACTIVITY QUESTIONS   Answers will vary depending on experimental results. However, they should approximate those given here. 1.   The plants in community "C" died due to lack of CO_2; the plant in community "D" died due to lack of sunlight. (Plants in community E may die also, if balance is not achieved.) 2.   the ones in communities "A" and "F" (and possibly "E") 3.   food and oxygen 4.   from plants 5.   The dead plants and snails show evidence of decay.

ANSWERS TO QUESTIONS

WORD QUIZ   1.   d   2.   m   3.   a   4.   l   5.   b   6.   f   7.   e   8.   k   9.   c   10.   h   11.   g   12.   i   13.   j   14.   o   15.   q   16.   p   17.   n

CHECK YOUR FACTS   1.   Food groups are producers, plant eaters, flesh eaters, scavengers, variety eaters, parasites, and decomposers. Examples will vary. 2.   Each one will be different. Watch for correctness of concept. Do not fault those who cannot draw well. 3.   Animals are usually colored to match their background (protective coloration). This makes it hard for enemies to see them. Some animals match the shapes of things around them also. 4.   Answers will vary. Some examples of cooperation (which is beneficial to all organisms involved) and competition (which is disadvantageous to some of the organisms involved) are: Plants compete for space (taller trees shade out shorter ones);

hawks and owls live largely on mice, which would otherwise destroy our crops; wolves help each other to catch prey; bees and flowers help each other. 5. all of the living things in an area 6. Temperature, water supply, and soil are some of the most important. Of course, the actual species available in the area are also important. European hedgehogs could probably live in New Jersey, but they have no way of getting there. 7. Succession is a series of changes in the kinds of living things in an area. One community replaces another, only to be replaced in its turn. Many examples can be given. Most students will use one from the text. 8. The final, permanent community at the end of a succession is the climax community. It is in balance when the size of any given species of organism in the community can remain about the same over a long period of time.

THOUGHT QUESTIONS 1. Human hunters take the place of wolves. Without hunting, deer would overbreed their range, destroy the food plants, and starve. There are more deer where there is a properly regulated amount of hunting. 2. There would be a succession that would establish the same communities now found in the southern states.

Chapter 14   Conservation of Soil and Water

OVERVIEW The formation and composition of soil is discussed in order to give the students an appreciation of how long it would take to replace eroded soil. The erosion of soil by water and methods of preventing this, such as contour plowing, strip cropping, and terracing, are explained. Also discussed is the problem of wind erosion due to overgrazing of land. The point is made that soil erosion is also directly related to loss of water. The importance of soil and water conservation programs is mentioned.

HINTS Students in rural areas can report on local problems and practices, both good and bad. The teacher will need to furnish extra information to city students to help them understand the chapter. Field trips are ideal, if they can be arranged, but usually the teacher will have to depend upon pictures, student reports, films, slides, and the like. State and national agricultural offices can furnish useful pamphlets.

If other AV materials are not available, you can draw on the chalkboard a profile of farm land, starting with flat land, which gives way to steeper and steeper slopes as it goes up a hillside. Discuss which plants are appropriate. Sketch in corn on the flat areas, small grains on the gentler slopes, then hay, permanent pasture, and forest as the slope gets steeper. Next, ask how the hillside can be protected further. Alter the profile to show terraces, the furrows of contour plowing, and the alternation of different crops, representing strip cropping.

ACTIVITY SUGGESTIONS Materials needed are two greenhouse flats per group, pitchers of water, and a sink. To make this dramatic be sure that the sod used forms a good, tough mat. Let the bare soil have a rather loose texture. It will help if there is little or no lower side on the flats. The water should be poured, or run from a hose, to create a fairly vigorous stream. There should be no visible damage to the sod. A very visible gully should be apparent in the unprotected soil. In discussion, relate the experiment to the effect of rain and runoff on vegetated and on bare soils of hillsides.

ANSWERS TO ACTIVITY QUESTIONS 1. the bare soil surface 2. Answers will vary. 3. Planting certain crops on steep hillsides can protect the soil from erosion.

ANSWERS TO QUESTIONS

WORD QUIZ 1. f 2. m 3. j 4. a 5. c 6. l 7. h 8. i 9. k 10. b 11. d 12. e 13. g

CHECK YOUR FACTS 1. wise management of resources; Soil is the basis of our food supply. 2. through the weathering of rock 3. Topsoil has more humus, more living organisms (roots, insects, bacteria, worms, etc.), more dissolved minerals, and a darker color. 4. the upper surface of the water in the water-filled (saturated) zone; Roots are generally found above the water table. 5. loss of minerals, usually carried away in crops going to market; Depletion can be avoided by using crop rotation and fertilizer. 6. the wearing away of soil by wind, water, etc. 7. Plants break the force of raindrops and slow down runoff. They also ''bind'' the soil to prevent wind erosion. 8. Contour plowing: Each furrow acts as a little dam to hold back the water. Terracing acts as a bigger dam to hold the water and lead it to a controlled runoff channel. Strip cropping lets water run only short distances before being slowed down by the more protective crop covers, such as hay. 9. Floods deposit good topsoil on the land. They also wash away soils, depending on the location. 10. Forests hold back the water, allowing it to sink into the ground instead of running off rapidly into streams. 11. controlled grazing, maintaining of plant cover, strip cropping, north-south plowing (across the prevailing wind), planting shelter belts (strips of trees)

THOUGHT QUESTIONS 1. Soil erosion will cause a decrease in useable farm land for planting crops and cattle grazing. This will cause a decrease in food supply and an increase in food prices at the supermarket. 2. The trash protects the surface from splashing by raindrops and it slows down runoff. The soil is kept looser, so water soaks in more readily.

Chapter 15   Conservation of Forests and Wildlife

OVERVIEW In the section on forests, the seven different forest regions of the United States and Canada are described and their usefulness is pointed out. The dangers of forest fires as well as various methods of forest management, such as block cutting, selective cutting, reforestation, etc., are described.

The section on wildlife describes the usefulness of various animals to the environment as well as some of the problems of survival faced by these organisms. Conservation practices regarding big game, birds, and fish are outlined.

HINTS The visiting of actual forests is an ideal activity, but it is possible only in certain locations. You may treat the trees of your neighborhood as a forest. Learn the species. Determine what characteristics make them desirable for city planting. Have students make tree leaf collections. Show them how to dry the leaves flat by placing them between magazine pages for a few days. They can then be mounted on notebook pages (using strips of transparent tape) together with name of tree, place collected, etc. Warn students not to pick leaves on private property without permission. Emphasize that animals can only survive in an appropriate environment. They cannot be stockpiled. Only such numbers can survive as can be sustained by that environment. Hunting and predation can do harm only if they reduce breeding stock to a point where losses cannot be replaced. The function of protective laws is to maintain the breeding stock. Beyond that, all meaningful conservation measures are aimed at protecting and improving environments.

ACTIVITY SUGGESTIONS   This activity can be done without special materials; however, it will save time if you can obtain copies of the game laws and supply them to the class. In many states hunting licenses are sold in hardware stores or other neighborhood locations. They can generally supply printed summaries of the game laws. If this is not the case in your area, write to the appropriate agency in your state.

The questions can be used simply as a basis for discussion or they may be used as a written assignment. Numbers 1–6 are factual, based on the information at hand. Number 7 requires some thought. Note that fully protected species are often endangered ones or are species with small food value and high aesthetic appeal (songbirds, for instance). The unprotected ones are usually very abundant or are considered vermin. Note that length of season and size of bag limit reflects abundance and/or susceptibility to depletion. Fall hunting seasons reflect a desire to allow species to raise their young and to harvest surplus populations before they are lost through winter kill.

ANSWERS TO ACTIVITY QUESTIONS   All answers will vary depending on local information obtained.

ANSWERS TO QUESTIONS

WORD QUIZ   1.   thinning 2.   improvement cutting 3.   reforestation 4.   selective cutting 5.   block cutting 6.   clear cutting followed by replanting 7.   wildlife 8.   extinct

CHECK YOUR FACTS   1.   See the listing on text pages 146–149. 2.   protect water sheds, home for wildlife, places for recreations; produce lumber, paper pulp, etc. 3.   fire protection 4.   Improvement cutting is useful where undesirable trees are competing with useful ones. Reforestation is used where natural seeding fails. Selective cutting is used where trees of many ages are mixed in the stand. Block cutting is appropriate in equal-aged stands and where the species involved can grow from seed on open land. 5.   Fire is used in some areas (especially in the South) to maintain the forest at a stage in succession where the most valuable species grow. Also, fire is used to burn the slash (limbs, tops, etc.) after logging. This removes fuel from the forest and reduces the danger of hot fires in the future. 6.   Wildlife can survive only if the environment supplies all requirements of the species. 7.   Big animals cannot survive in settled country because they compete with people and domestic animals for the food supply. 8.   They would overbreed the range, destroy food plants, and starve. 9.   Answers will vary.

THOUGHT QUESTIONS   1.   Selective cutting, in all three cases: It maintains a continuous cover, which protects soil and wildlife. The permanent forest favors recreational use. Other methods produce a churned-up waste that takes many years to become beautiful again. 2.   These species would become too abundant. Competition for food would become too great and the big game animals would die of starvation.

Chapter 16   Pollution and Energy

OVERVIEW   In general, this chapter describes the causes and possible remedies for various types of pollution, such as sewage, solid wastes, chemical pollution, pesticides and herbicides, oil spills, forms of air pollution, smog, and acid rain. The depletion of the ozone layer and the phenomenon known as the greenhouse effect are also discussed. Emphasis is placed on both urban and rural pollution problems.

The need for conservation of energy is stressed and possible alternate sources of energy, such as energy from plant life, geothermal energy, wind energy, nuclear energy, and solar energy are explained.

HINTS   It would be a good idea to keep a file of news articles relating to the topics covered here. They can be used in class to illustrate the subject. Items of local interest are especially useful. Chemicals at Love Canal and smog in Los Angeles are mentioned in the book. Problems near home will be much more impressive to students living in other parts of the country. Students can be asked to make pollution or energy scrapbooks.

It follows that some topics will seem more relevant than others, depending on locality. Rural students may see problems with pesticides as more important than air pollution. Smog will seem important to those who see and smell it while doing their homework. Sewage disposal may seem a serious problem to those living downstream from a city. Adequate coverage should be given to these local issues in class. At the same time, do not lose sight of the nationwide problems. The students should learn to see the situation as a whole.

ACTIVITY SUGGESTIONS   Materials needed are several large plastic trash bags and a spring scale.

Before collecting the trash, be sure you have a place where it can be stored until the end of the week. At the outset, have students take notice of the condition of the classroom in order to make them more aware of the trash problem. You may wish to have a student take a survey of the amount of trash usually found in the hallways, cafeteria (especially immediately after a lunch period), and so on. During the course of this activity try to instill a positive attitude in the students toward keeping the school building, as well as other places, clean and as free of pollution as possible.

ANSWERS TO ACTIVITY QUESTIONS   These will vary depending on the results of the activity.

ANSWERS TO QUESTIONS

WORD QUIZ   1.   b   2.   e   3.   c   4.   d   5.   f   6.   a   7.   g   8.   j   9.   i   10.   l   11.   k   12.   h

CHECK YOUR FACTS 1. garbage burned (with some generation of power) or buried; metals and glass: salvaged and re-used; other trash: buried in sanitary land fills 2. Chemicals must be destroyed or buried where they can never enter ground water or air. 3. They may kill many species besides the pest. Fish killed by agricultural runoff are an example. 4. an increase in the air temperature of the earth due to reradiation of heat caused by an increase in the amount of CO_2 in the atmosphere 5. Smog forms when sunlight reacts with unburned hydrocarbons. It is harmful because it damages respiratory tissue when inhaled. 6. Answers will vary. A basic understanding of the fact that present resources cannot meet current levels of consumption should be evident. 7. Answers will vary. 8. because it shields out harmful ultraviolet rays that come from the sun

THOUGHT QUESTIONS 1. If we could use sunlight to separate water, the hydrogen could then be burned in car engines, power plants, and home heating furnaces. 2. In nature the pollutants, such as nitrogen oxides (from lightning and bacteria) or sulphur dioxide (from decay) are diffused throughout the environment. Natural processes remove them. (Carbon monoxide, for instance, is absorbed by soil bacteria.) Our problem is that we produce pollution in concentrated amounts in the areas in which we live. Nature cannot handle these large concentrations.

AV Materials   Unit 3

(Numbers in parentheses refer to chapter numbers.)

"The Biosphere," from *Spaceship Earth*. Hawkhill, sound filmstrip. (12)
The Case of the Bighorn Sheep. PH, sound filmstrip. (15)
The Deciduous Forest. PH, sound filmstrip. (15)
"The Earth," from *The Universe*. NGS, sound filmstrip. (12)
An Ecosystem: Struggle for Survival. NGS, film. (13)
Environmental Pollution. Wards', filmstrip. (16)
Land: Uses and Values. PH, sound, filmstrip. (16)
The Living Desert. Walt Disney, sound filmstrip. (13, 14)
Living Things & Their Habitats. Wards', filmstrip. (13)
"Man in Space," from *The Universe*. NGS, sound filmstrip. (13)
Nature's Half-Acre. Walt Disney, film. (15)
Our Mountains of Trash. Time-Life, sound filmstrip. (16)
Pollution: Problems & Prospects. NGS, sound filmstrip. (16)
Pond Life—Food Webs. NGS, film. (13)
The Story of the Everglades. PH, sound filmstrip. (13)
"The Swamp," from *Exploring Ecology*. NGS, sound filmstrip. (13)
Waterways or Sewers. Time-Life, sound filmstrip. (16)
The Wetlands. PH, sound filmstrip. (13)
White Wilderness. Walt Disney, film. (12)

Supplementary Reading   Unit 3

"Arctic Walk." *Natural History*, vol. LXXVII, no. 5 (May 1969).

Audubon, vol. 78, no. 4 (July 1976, entire issue).

Brenchley, D., and J. R. Towler. *Resource & Handbook: The Environment—Activities & Explorations*. Boston: Houghton-Mifflin, 1975.

"Canada's Heartland: The Prairie Provinces." *National Geographic*, vol. 138, no. 4 (Oct. 1970).

Couchman, J. K., et al. *Miniclimates*. Toronto: Holt, Rinehart and Winston, Publishers, 1971.

Examining Your Environment Series. Toronto: Holt, Rinehart and Winston, Publishers, 1971.

"Food, Will There Be Enough?" *National Geographic*, vol. 148, no. 1 (July 1975).

"First Colony in Space." *National Geographic*, vol. 150, no. 1 (July 1976).

Leen, Nina. *And Then There Were None: America's Wildlife*. New York: Holt, Rinehart and Winston, Publishers, 1973.

"The Metric Forest." *Natural History*, vol. LXXXII, no. 9 (Nov. 1973).

Schlichting, Harold E., Jr., and May Southworth. *An Introduction to Pollution*. Austin: Steck-Vaughn College, 1972.

Szulc, Tod. *The Energy Crisis*. New York: Franklin Watts, Inc., 1974.

"The Tundra." *National Geographic*, vol. 141, no. 3 (March 1972).

Watson, Jane Werner. *Our World Tomorrow*. New York: Western Publishing Co., 1973.

"Wolves *vs*. Moose on Isle Royale." *National Geographic*, vol. 123, no. 2 (Feb. 1963).

Unit 4 Classification and Simple Forms of Life

Chapter 17 Scientific Names

OVERVIEW This chapter introduces the students to the basic principals of scientific classification. The need for a standardized system of naming living organisms is stated and the system of binomial nomenclature is explained, using common examples. The terms kingdom, phylum, class, order, family, genus, and species are introduced. It is pointed out that the basis for classification is structure, but that it is sometimes difficult to determine into which kingdom an organism should be placed. A four-kingdom system of classification is used here—The Bacteria and Blue-green Algae Kingdom, the Protist Kingdom, the Plant Kingdom, and the Animal Kingdom. A representative chart of these kingdoms is included within this chapter.

HINTS The idea of the system should be made clear. A natural feeling for its use will come through practice during study of the following chapters. Emphasize that the system is for our own convenience. It was not a natural law waiting to be discovered. Therefore, there is no one "correct" classification system. This becomes apparent when we try to agree (only to end by disagreeing) upon which kingdoms to use, or whether to include algae in the plant kingdom. Our use of four kingdoms is a reasonable one, though not the only reasonable one.

ACTIVITY SUGGESTIONS   Materials needed are an assortment of organisms, either preserved, living (plants), or artificial. **CAUTION: Do not encourage students to bring living or dead animals into class.** The animals for this activity can be supplied in the form of preserved specimens from a supply house, models of animals, or even photographs or realistic stuffed toys! The plants used can be any common house plants or plants obtained from a florist, etc. Artificial plants can also be used.

Lead the class in a decision-making process. How many groups should be used? Which items belong in each group? What basis should be used for classification? Color? Size? Texture? Do not try to lead them into "correct" decisions. Let them learn by experience that there is no one true way to do it.

ANSWERS TO ACTIVITY QUESTIONS   These will all vary based on the results of the activity.

ANSWERS TO QUESTIONS

WORD QUIZ   kingdom, phylum, class, order, family, genus, species

CHECK YOUR FACTS   1.   so that scientists from all regions can understand which species are under discussion; It is a uniform cataloguing system. 2.   Linnaeus 3.   genus and species names, together 4.   Some one language needed to be agreed upon. At the time the system was established everyone who went to school, even for just a few years, learned Latin. If it were being done today English would probably be used. 5.   Examine its structure and try to determine to what already known organism it was most closely related in structure. It might then be given the same genus name and a new species name.

THOUGHT QUESTIONS   1.   Plant and animal kingdoms would be all that was needed. The puzzling species are too small to see. 2.   Plants and animals do not care what we call them. The system is for our benefit: to help us to remember and to show relationships we see as being significant.

Chapter 18   Bacteria

OVERVIEW   The bacteria and blue-green algae are placed in the same kingdom, so their structural similarities are noted here. The three basic shapes of bacteria (round, wide, and spiral) are shown. The food-getting methods of various bacteria are discussed, as is the significance of bacteria as part of the nitrogen cycle. The conditions for bacterial growth are discussed and this is related to the causes of food spoilage and certain diseases. Various practical uses for bacteria, such as in food production and sewage disposal, are mentioned. The general characteristics of blue-green algae are stated. Viruses are also discussed in this chapter.

HINTS   You might wish to describe the localized deep-sea communities along the Galapagos ridge. Vulcanism gives rise to hot water vents. Water issuing from these vents is loaded with sulphur compounds. Sulphur bacteria derive energy from the sulphur to manufacture food (chemotrophy). These bacteria are the producers in numerous food chains. Great numbers of clams, starfish, fish, and other marine animals cluster around these vents. It is a complete ecosystem living entirely independent of sunlight or photosynthesis. You might also mention here the current attempts to introduce selected genes into bacteria. It is hoped, for instance, that we may place certain human genes into bacteria so that they will manufacture insulin, or some other useful hormone. Then this product can be extracted from the culture medium and used in treating people.

ACTIVITY SUGGESTIONS Materials needed are several agar plates, an incubator (or other suitable warm container), and an autoclave (or bath of boiling water).

CAUTION: If you are preparing your own agar plates, do not allow students to handle the hot culture medium or hot Petri dishes.

A plate exposed to the air for a few minutes in the hall during class can be compared with one similarly exposed during the passing of classes. The indicated increase in dust is impressive.

After the plates have been inoculated, tape them shut and make sure no student opens them. **CAUTION: There is a remote possibility that pathogenic bacteria will grow in quantity. Give these no chance to infect anyone. When the experiment is over, boil or autoclave the cultures without opening them. Then they can be cleaned.**

As an alternate activity you may wish to soak several lima beans in a beaker of water. After a few days, bacteria will grow in this water. (A characteristic odor will develop.) Have students stain a drop of this water with methylene blue and observe it under the microscope at high power.

ANSWERS TO ACTIVITY QUESTIONS 1.   Answers will depend on individual results. 2.   No growth of bacteria is visible. 3.   so that we can pinpoint the cause of the observed results 4.   to kill the bacteria; Yes, these bacteria were present in the environment already, but not in such great concentrations.

ANSWERS TO QUESTIONS

WORD QUIZ   1. f   2. d   3. g   4. h   5. c   6. b   7. a   8. e

CHECK YOUR FACTS   1.   They have no nuclei or chloroplasts. 2.   shapes: round, rod-shaped, spiral; size: very small, even for cells 3.   as decomposers; by extracting energy from chemicals (such as sulphur) and using it to make food; also by a form of photosynthesis 4.   dissolve (digest) them 5.   Refer to Fig. 18-4. 6.   refrigeration: slows down bacterial activity; freezing: stops bacterial activity; salting: removes water from bacterial cells; drying: likewise; also deprives any new bacteria of necessary moisture needed for growth. Canning kills all bacteria in the food and prevents others from entering. In pickling, acids inhibit bacteria. 7.   See list under heading, "Practical Uses for Bacteria," in text. 8.   any species that has a simple body structure and chlorophyll 9.   most often blue-green, but other colors possible; one-celled, or cells in clumps, or cells in chains 10.   a bit of DNA or RNA enclosed in a protein cover; They make us ill. 11.   The DNA or RNA enters a particular kind of host cell. Its genes take over the cell's equipment and direct it to take more viruses.

THOUGHT QUESTIONS 1.   8,   64,   4,096,   16,777,216   2.   Answers will vary. There would be a steady reduction in the number of living species as essential materials failed to become available for reuse in building the bodies of new plants and animals. 3. The botulism toxin is destroyed by heat, so cooking will prevent poisoning. Only steam pressure is safe in the canning of vegetables. Fruits may be canned by any method. *Botulinus* bacteria do not grow in acid environments.

CHAPTER 19   PROTOZOA

OVERVIEW   The organisms discussed in this chapter fall into four groups: the Ameba phylum, the Flagellates, the Ciliates, and nonmotile spore-forming protozoa. Besides describing the main characteristics of these protozoa, such as structure and function, methods of food getting, etc., the importance of protozoa as members of food chains, producers of oil deposits, and agents of disease is also discussed.

HINTS   Remind students that bacteria comprise more species than do all other living things and that they are the principal decomposers. The protozoa eat bacteria. Small crustaceans eat protozoa and larger animals eat these crustaceans. Students can draw some circular food chains based on this idea. Animals are decomposed by bacteria, which are eaten by protozoa, which are eaten by a succession of animals, etc. Point out that food is steadily diminished along the line through respiration and that new food must constantly be fed into the chain by the plants.

ACTIVITY SUGGESTIONS   Materials needed are a hay infusion pond culture, clean microscope slides, cover slips, and microscopes.

Bacterial action soon uses up the oxygen in the water, so the protozoa will be concentrated in the surface film. A rod or a match stick can be dipped into this film (which, unfortunately is rather stinky) and touched to the microscope slide. This should deliver a good supply of protozoa. Note a background of tiny specks seen in the microscope field of view. Some are in motion. These are the bacteria upon which the protozoa are feeding.

ANSWERS TO ACTIVITY QUESTIONS   1.   the hay 2.   bacteria 3.   Protozoa 4.   Answers will depend on the particular case. In general, there will be a succession. First, bacteria will increase. Then a few kinds of protozoa will appear. Then there will be a peak, with numerous species, mainly ciliates. *Paramecia* or similar large types will be common. Gradually numbers will decrease, as living conditions deteriorate. 5.   Probably not, but it may. Algae or photosynthetic bacteria (usually purple colored) may become established. These manufacture food, so depletion of the food supply in the original hay no longer matters. Bacteria will be the decomposers; and small worms, protozoa, etc., will be the other consumers.

ANSWERS TO QUESTIONS

WORD QUIZ   1. d   2. g   3. a   4. f   5. b   6. h   7. c   8. e

CHECK YOUR FACTS   1.   Refer to listing in the Summary. 2.   The characteristic shapes and the presence or absence of flagella and cilia will be the main features visible. 3.   *Euglena* has chloroplasts, like a plant. It does not eat like an animal (but other flagellates do). 4.   malaria, amebic dysentery, African sleeping sickness, and others 5. As bacteria eaters, they are part of many food chains. They cause some disease in animals and humans. They help in the formation of petroleum and limestone deposits. 6. toward; to get light for photosynthesis

THOUGHT QUESTIONS 1. It is assumed that cells would have to develop first, and then groupings of cells could come next. The fossil records shows simple bacteria and algae (mainly blue-green) for three billion years before higher animal life appeared. Higher plants come still later. 2. Food-eating flagellates can be called protozoa. Those that decompose food and absorb the digested nutrients can be called fungi. Those with chloroplasts can be called algae. Some have more than one of these adaptations. The flagellates have a good deal to do with the decision to put all simple types into the Protist Kingdom.

CHAPTER 20 ALGAE—THE PLANT-LIKE PROTISTS

OVERVIEW This chapter begins by giving some general characteristics of four phyla of algae; the green algae, the yellow-green and golden-brown algae, the brown algae, and the red algae. After this, there is a discussion on various forms of reproduction in algae. The importance of algae both as members of food chains and in commercial uses (such as in the manufacture of toothpaste, cleansers, ice cream, etc.) is emphasized.

HINTS Two main things should be accomplished here. One is to give students an idea of what algae are and what they look like. Use the figures and any reference books that you have available. Especially, give students an opportunity to examine algae in the laboratory if possible.

The second main thing to accomplish is to impress students with the worldwide importance of algae as the main food makers of the sea. Point out the importance of seafood to humans. Point out, also, the importance of carbon dioxide absorption and oxygen release by algae. There is somewhat less food made in the sea than on land, but the amount is very large, nevertheless.

ACTIVITY SUGGESTIONS Materials needed are samples of various algae, a jar of silver polish, several clean microscope slides, cover slips, and microscopes.
Procedure A: This is written as if the students are expected to bring in the specimens. That is fine if it can be done. Most of the time, teachers will have to supply the algae, as discussed above.

Watch especially for diatoms. Many of these will be found attached to underwater surfaces, including the surfaces of filamentous algae. These collected diatoms can be compared with those in section B of the activity.

Protococcus can be scraped from tree bark in almost any neighborhood. Its cells come alone or in clusters. Cell structure is nearly identical to that of *Chlorella*.
Procedure B: Wright's Silver Cream is usually used for this. Some jars of it will prove better than others. Find a good one and it will last for a very long time. **CAUTION:** Warn students not to get the polish on the microscope lens. It is abrasive and will injure the lens in time.

ANSWERS TO ACTIVITY QUESTIONS 1. Check to see that student's labels are accurate. 2. This will vary, depending upon which diatomaceous earth deposit was used in making that particular jar. 3. They lived in the sea in the geological past. Many millions of years ago, in most cases. 4. It *does* scratch. That is its function. It scratches off the tarnish on the silver surface. The diatoms are very small, so the scratches are too fine to be seen. They blend to give a polished surface.

ANSWERS TO QUESTIONS

WORD QUIZ Definitions will vary. They should approximate those given in the text.

CHECK YOUR FACTS 1. They are simpler in structure. They lack the outer coverings, roots, leaves, and two-way vascular tissue found in the higher land plants. 2. Refer to the descriptions in the text on pages 201–204. 3. Their food making is the basis for nearly all sea life. 4. Some are eaten in the orient. Irish moss is used for food. Agar is used in bacteriology as a base for nutrient plates. Algae are used commercially in ice cream, iodine, cleanser, etc. In the future algae may be grown for cattle feed. 5. They foul reservoirs.

THOUGHT QUESTIONS 1. This is actually done on a small scale in Japan and Korea, where they eat sea-weed. To do it on a large scale would require the solving of many problems. Student answers will vary. Possible ideas would include: Finding palatable algae. Learning to grow the algae. Protecting them from underwater algae-eaters. Developing harvesting methods, etc. 2. They need no such tissue, since their entire surface is in contact with water. Land plants need it because their leaves are not in contact with their water supply.

Chapter 21   Fungi—The Decomposer Protists

OVERVIEW Various characteristics of fungi are discussed. The importance of yeast in cooking and because of its fermentation properties is explained. The structure of molds (such as bread mold) is depicted and their importance in the production of certain foods and drugs is pointed out. The structure and function of higher fungi are discussed, using mushrooms as the main example. Diseases caused by fungi (ringworm, athlete's foot, etc.) are also mentioned. An explanation is given of the relationship between a fungus and an alga in a lichen.

HINTS You may want to point out to the students that lichens are very sensitive to air pollution. They will not grow in polluted areas. The amount of lichen growth in an area can be used as a type of informal air pollution index.

Slime molds have not been introduced in this chapter, but you may wish to do so in class. Collections of slime mold fruiting bodies can be kept glued into the bottoms of large match boxes. These will keep for years.

ACTIVITY SUGGESTIONS Materials needed are several slices of bread, glass dishes, and glass plates to cover these with, (plastic margarine containers, etc., which have lids, can be substituted here), tweezers, clean microscope slides, cover slips, and microscope.

If possible, get bread that contains no mold inhibiter. Other bread will mold, but it takes much longer.

CAUTION: Some people are allergic to mold spores. Excuse from this activity any students who are aware of such a problem. Keep the inhaling of spores at a minimum for all students.

If microscopes are not available, Procedure B of this activity can be omitted. For Procedure C use the *Rhizopus* growth.

ANSWERS TO ACTIVITY QUESTIONS

1. More or less variety may be present. The common *Rhizopus nigricans* should appear. 2. The starting point of each patch of mold is the point where the original spore landed. 3. It contains an abundant amount of spores. 4. Spores should be easily visible. *Rhizopus*, with its globular sporangia will be most common. You may get *Penicillium* with spores in chains, or *Aspergillus*, with chains of spores radiating from a center. 5. Check drawings for accuracy. 6. You should get a pure stand of the one kind of mold (*Rhizopus nigricans*, if that is what you used). 7. Only one kind of mold was "seeded" on this bread.

ANSWERS TO QUESTIONS

WORD QUIZ 1. fungi 2. molds 3. antibiotics 4. fruiting body 5. shelf fungus 6. lichen 7. fermentation

CHECK YOUR FACTS 1. baking; Yeasts growing in the dough release carbon dioxide, causing the bread to rise. *Industrial* and *beverage* alcohol are both produced by allowing yeast to ferment glucose in some food material. 2. Spores carried in air or water land by chance on food material. In some water molds the spores swim with flagella. 3. Molds get on food, leather, or other useful material and spoil it. Some cause disease. The use of mold in blue cheese or the production of penicillin are examples of our use of molds. 4. the fruiting body 5. Fungi first secrete digestive enzymes to dissolve the food. Then they absorb the dissolved material. 6. Lichens have bodies made up of alga and fungus cells in close association. They are air "plants," growing on rocks, trees, or any other solid support.

THOUGHT QUESTIONS 1. at the grocery store; Do *not* depend on anyone claiming to know the wild mushrooms. No "test" is reliable. Sure identification of species takes long study. 2. The mold contains many compounds. Penicillin must be extracted and purified.

AV Materials   Unit 4

(Numbers in parentheses refer to chapter numbers.)

Classifying Plants & Animals. Coronet, B/W–color, film. (17)
"The Coral Reef," from *Small Worlds of Life*. NGS, sound filmstrip. (20)
Growth of Molds. McGraw-Hill, filmloop. (21)
Growth of Mushrooms. McGraw-Hill, filmloop. (21)
"Life at the Seashore," from *Interactions in a Population*. Houghton-Mifflin, sound filmstrip. (20)
Life in a Drop of Water. Coronet, B/W–color, film. (19)

Life Science: Responses in a Simple Animal. BFA, color, film. (19)
Microorganisms That Cause Disease. Coronet, B/W–color, film. (19)
Microscopic Fungi. McGraw-Hill, B/W–color, film. (21)
Protists: Threshold of Life. NGS, film. (19)
Simple Plants: Algae & Fungi. Coronet, color, film. (20, 21)
Simple Plants: Bacteria. Coronet, B/W–color, film. (18)
Spirogyra: Structure & Life Functions. Coronet, color, film. (20)
Viruses, Threshold of Life. Coronet, B/W–color, film. (19)
Why Foods Spoil. EBF, B/W–color, film. (18–21)
World of Little Things. Moody, film. (19)

Supplementary Reading   Unit 4

Alexopoulos, C., and H. Bold. *Algae and Fungi*. New York: Macmillan, 1967.

Brooks, Stewart M. *The World of the Virus*. New York: A. A. Barnes & Co.

Carpenter, P. *Microbiology*. 2nd ed. Philadelphia: W.B. Saunders & Co., 1967.

Christensen, Clyde M. *The Molds and Man: An Introduction to the Fungi*. Minneapolis: University of Minnesota Press, 1962.

Cosgrove, M. *Wonders Under a Microscope*. New York: Dodd, Mead & Co., 1959.

Coulter, M. C., and H. J. Dittmer. *The Story of the Plant Kingdom*. Chicago: University of Chicago Press, 1964.

Dubos, Rene. *The Unseen World*. New York: The Oxford University Press, 1962.

Dolan, Edward F. *Adventures with a Microscope*. New York: Dodd, Mead & Co., 1964.

House, Homer D. *Wild Flowers*. New York: Macmillan Co.

Kavaler, Lucy. *The Wonders of Fungi*. New York: The John Day Co., Inc., 1964.

Lange, Morten, and Hora Bazard. *A Guide to Mushrooms and Toadstools*. New York: E. P. Dutton, 1963.

Simpson, George, and W. Beck. *Life: An Introduction to Biology*. New York: Harcourt, Brace, Jovanovich, Inc., 1969.

Wilson, C.; W. Loomis; and T. Sleeves. *Botany*. 5th ed. New York: Holt, Rinehart and Winston, Publishers, 1971.

Zahl, Paul. "Where Would We Be Without Algae." *National Geographic*, March 1974.

Unit 5   The Plant Kingdom

Chapter 22   Types of Plants

OVERVIEW   Mosses and liverworts are discussed as examples of nonvascular plants. The life cycle of a moss is given and the ecological and economic importance of mosses explained.

Club mosses, horsetails, ferns, and seed plants are discussed as examples of vascular plants. The structure and function of each of these is explained as is the importance of both the flowering and nonflowering seed plants. A brief explanation of the postulated history of vascular plants is given.

HINTS Most students will not have ever really looked at mosses. Help them distinguish between the various green coverings seen in nature—mosses, lichens, liverwort, and algae. It is very important that specimens be available for observation in the classroom. They can be collected, even in winter; but it is better to get them in the fall, summer, or spring and have plenty growing in terrariums. Stacked laboratory finger bowls or dishes can hold lots of moss in small window-shelf areas.

The vascular plants are far more highly organized than any organisms we have studied so far. Details of this organization will be developed in the next two chapters. At this point, simply show students how stems, roots, leaves, and vascular tissue are all adaptations to land living.

ACTIVITY SUGGESTIONS Materials needed are a collection of mosses, club mosses, horsetails, and ferns. Seed plants may be collected as prepared specimens also, but they are easily brought in fresh from the neighborhood by students and teacher. Fresh mosses are best in A and B.

As an *alternate activity* you may wish to have the students set up some terrariums for the classroom. Place some woodland soil in the bottom of empty aquariums. Cover the soil with your moss "carpets." Sink a dish level with the surface of the moss and fill it with water. This will look like a tiny pond. Cover the aquarium with a piece of glass. Keep the moss layer damp, but do not flood it with water. These terrariums make nice decorations for the room. They are also good places for keeping small frogs and salamanders if they are kept from becoming too hot.

ANSWERS TO ACTIVITY QUESTIONS 1. No. Answers to the second part of the question will depend upon which mosses were available. 2. The spores will be easily observed. Most students will describe them as tiny balls or spheres. 3. Some moss leaves will prove to have only one layer of cells. Others will have varying numbers, but generally only a few. Flat-leaved species work best for this study.

ANSWERS TO QUESTIONS

WORD QUIZ A. Definitions will vary. They should approximate those given in the text. B. Vascular plants are ferns, seed plants, and horsetails. Non-vascular plants are mosses and liverworts. Descriptions of each will vary.

CHECK YOUR FACTS 1. central stalk; root-like cells at lower end; leaf-like structures along the stem 2. The liverworts are flat, ribbon-like plants that grow flat on the ground and have root-like cells growing into the soil. They grow in the same shady, moist places as mosses, but are less common. 3. Roots can bring water from the ground. Stems provide transport for this water and hold leaves up to the sunlight. Leaves are able to catch this sunlight for use during photosynthesis. 4. They look like small evergreens (most nearly like cedar) or like overgrown mosses; scale-like green leaves; clubs (spore-producing cones). Horsetail has a ribbed, rough-feeling stem in sections; no leaves; finely branched, green stem system or reed-like. 5. They use spores. This means they cannot get started in a new location unless there is a long wet period during which the spore and tiny young plant can get established.

6.   The bracken is a dominant plant in many northern cut-over areas. It is a major producer and soil builder in these places. 7.   seed plants, especially the flowering group; It supplies food for us and our animals. It also supplies building materials, fibers, drugs, dyes, chemicals, and many other products. 8.   First, the spore-producing vascular plants formed the forests of the Coal Age. Cold weather (and, no doubt, competition) ended this period. Second, nonflowering seed plants predominated. Third, flowering plants displaced their predecessors by direct competition.

THOUGHT QUESTIONS   1.   Most of the mosses would die. Exposure to sunlight would dry out the soil. 2.   They are algae, growing on the rocks. 3.   Spores could not survive long enough to establish ferns in the desert. Wet periods never last long enough. This being the case, ferns have never developed the other special adaptations for survival in a desert climate.

Chapter 23   Leaves of Flowering Plants

OVERVIEW   The structure and function of the leaf is explained in detail. The various topics covered are the movement of molecules in the leaf during photosynthesis, the action of the guard cells in controlling water loss, the reasons for the change of color and falling of the leaves from the tree in autumn. There is a brief discussion of the various shapes and arrangements found in leaves.

HINTS   Vocabulary has been kept very simple. Some teachers may wish to teach the terms "palisade tissue," "spongy tissue," "stomata," etc.

Point out the dilemma of the plant during dry weather. It must "choose" between water shortage and photosynthesis. To put it another way, food can only be made at the cost of losing water by evaporation through the open pores (stomata). As far as food production goes, the leaf must "close shop" during periods of high evaporation. It does this by closing the pores by guard cell action. Many of the sunniest hours of a summer day may be lost for photosynthesis.

ACTIVITY SUGGESTIONS   Materials needed are geranium (or other type of) leaves, tweezers, clean microscope slides, cover slips, and microscopes.

Peeling the surface (epidermis) from a geranium leaf is made easier if you first tear the leaf as you would tear a sheet of paper, moving one hand downward and the other one upward. Places can then be observed where the clear lower surface cells hang out over the torn edge. Grasp one of these places with tweezer points and peel. Only a small bit is needed. If other plants are used, the cell pattern will be different. *Sansevieria* and wandering Jew are favorites.

Leaf hairs will be very evident on the geranium leaf. Explain that their function is to reduce wind velocity across the leaf surface and over the stomata (pores). This reduces the rate of evaporation from the leaf.

ANSWERS TO ACTIVITY QUESTIONS   1.   Drawings will vary. Check for accuracy. 2.   Veins will show because they are brightly stained. Some will be cut at any angle, giving elongated sections. 3.   Most slides show one or two layers. This indicates growth in strong sunlight. Shaded leaves produce no such cells. 4.   Stomates should be visible. The location will vary depending on the type of plant used. In land plants look for the stomates on the lower epidermis. 5.   Guard cells may or may not be visible depending on the commercial preparation of the slide.

ANSWERS TO QUESTIONS

WORD QUIZ   1.   modified 2.   pigment 3.   food-making cells 4.   pores 5.   guard cells 6.   epidermis

CHECK YOUR FACTS   1.   Drawings will vary. They should be similar to (but not exact copies of) Fig. 23-1. 2.   Chloroplasts make food. Surface layers have strength to bind the leaf together and they resist evaporation of water from the inner cells. Food-making cells carry on photosynthesis. Veins carry water into the leaf and food from the leaf to the rest of the plant. Pores allow carbon dioxide to enter and oxygen to leave. Guard cells close the pore during dry periods. 3.   Leaves in winter would lose water by evaporation when no new water could be obtained from frozen ground. The tree would dry out and die. 4.   When water pressure in the plant is high, the guard cells "open" and expose the stomate so that transportation can take place. When water pressure is low, the guard cells "close" over the stomate so that water is held in the plant. 5.   Yellow color becomes visible when chlorophyll breaks down. The yellow pigments were there all the time. They assist in photosynthesis. 6.   Red forms under the influence of sunlight from breakdown products of chlorophyll.

THOUGHT QUESTIONS   1.   These are water-storing leaves. They are adapted to a dry climate. This type of plant should not be watered too frequently. 2.   Yes. Leaves make the food that is stored in the potato. Reduced leaf area will mean less food to store.

Chapter 24   Stems and Roots

OVERVIEW   The structures and functions of various stems and roots are explained. Cross sections of woody and herbaceous dicots and herbaceous monocots are diagramed (although this terminology is not introduced), showing the arrangements of the vascular tissue. The parts of the stem are described, as is the structure of a bud. In the section on roots a cross section of a typical root is diagramed and the functions of the root cap and root hairs are explained. A brief comparison is made between various types of root systems.

HINTS   The rise of water in a stem can be demonstrated by placing a cutting from any convenient potted plant (such as a geranium) in a glass of colored water. Red ink in water will do nicely. Later, a cut across the stem will reveal red dots where the strands of water-carrying vascular tissue (xylem) pass up through the stem. The same can be done with celery, but explain that celery is actually a leaf stalk (petiole), not a stem. It is, nevertheless, a demonstration of vascular bundles.

The winter twig, with its buds, leaf scars, etc., can be examined in the classroom by students. Hickory, lilac, maple, and other species with prominent buds are the most useful. Fresh ones are best, but old twigs can be used if fresh ones are out of season.

ACTIVITY SUGGESTIONS   Materials needed are microscopes, prepared slides of cross sections of the three basic stem types, prepared slides of onion root tips, several radish seeds soaked in water, and hand lenses.

If microscopes are not available procedures A and B can be omitted, leaving procedure C to be done by the students. The root hairs on sprouting seeds show up beautifully. Emphasize that these are projections from single cells. Also point out how enormously they increase the water-absorbing surface of the root. As a variation of this experiment, some seeds can be sprouted in damp sand. When the shoots are an inch or so high the plants can be pulled up and shaken. Sand grains will cling to the roots. They are held there because root hairs wrap around them so closely. This helps to picture how roots are able to absorb the water film from the surface of a soil particle.

ANSWERS TO ACTIVITY QUESTIONS   1. and 2.   Drawings will vary. Check for accuracy of labels. 3.   root hairs 4.   They increase the plant's ability to absorb water.

ANSWERS TO QUESTIONS

WORD QUIZ   1.   D   2.   A   3.   E   4.   B   5.   C

CHECK YOUR FACTS   1.   to hold up the leaves and to carry liquids between leaves and roots 2.   food carrying and water carrying 3.   They are the vascular tissue-in-a-ring type, vascular bundles scattered throughout the stem, and woody stems. Diagrams or style of descriptions will vary with the student. 4.   The wood carries water and the inner bark carries food. 5.   by counting the growth rings in a cross section of the trunk 6.   miniature leaves (or flowers) waiting to grow 7.   Growth in length is from the tips. Growth in thickness is at the cambium (between wood and bark). 8.   Roots anchor the plant and they absorb water and minerals. 9.   at the tips 10.   It is a covering on the root tips. It takes the wear as the root tip pushes through the soil, thereby preventing damage to the growing cells. 11.   They are slender projections from the surface cells of the young root. They absorb water.

THOUGHT QUESTIONS   1.   one meter; only the stem tips are growing in length 2.   because the root system of a tree growing in a swamp does not grow as deep as the root system of a tree growing on higher ground; this means that the tree growing in the swamp is not anchored as well.

A-V Materials   Unit 5

(Numbers in parentheses refer to chapter numbers.)

The Development of Plants. PH, sound filmstrip. (22–24)
Kingdom of Plants. NGS, sound filmstrip. (22–24)
Pathways of Water in Herbaceous Plants. Popular Science, filmloop. (24)
Pathways of Water in Woody Plants. Popular Science, Filmloop. (24)
Plant Tropisms and Other Movements. Coronet, color, film. (22)
Roots of Plants. Encyclopaedia Britannica, filmstrip. (24)
Stems of Plants. Encyclopaedia Britannica, filmstrip. (24)
Trees Grow Through The Years. Coronet, color, film. (24)

Supplementary Reading   Unit 5

Crafts, A. S. *Translocation in Plants*. New York: Holt, Rinehart and Winston, Publishers, 1961.

Flowers, Non-flowering Plants, Trees. Golden Nature Guides. New York: Golden Press.

Hutchins, Ross E. *This Is a Leaf*. New York: Dodd Mead & Co., 1962.

Morholt, E., et al. *A Sourcebook for the Biological Sciences*. 2nd ed. New York: Harcourt, Brace, Jovanovich, Inc., 1966.

Raven, Peter H. *Biology of Plants*. New York: Worth Publishers, 1970.

Went, Frits, and the editors of *Life*. *The Plants*. New York: Time, Inc., 1963.

Unit 6   The Animal Kingdom

Chapter 25   Sponges and Coelenterates

OVERVIEW   The student is familiarized with various types of cells found in sponges. These include epithelial cells, collar cells, and wandering cells. Simple vocabulary is used in referring to these. Reproduction in sponges is explained. Also discussed are the existence of various types of sponges, the division of labor among cells of a multicellular organism, and the commercial uses of sponges.

Hydra, jellyfish, sea anemones, and corals are given as examples of coelenterates. These are discussed in reference to structure, function, methods of food getting, and habitat. Hints: The more actual examples that can be shown in the classroom the better. A school should build a collection. A variety of dried sponges can be stored. Coelenterates will need to be preserved (4% formaldehyde is best). Of course, the stone coral bases ("skeletons") will need no preservatives, and they make an attractive display.

Live *Hydra* can be ordered from supply houses. Teachers living near rocky coasts can collect sea anemones in tidal pools. Some can be kept alive for a while in salt-water aquariums. Inland schools can buy live sea-water aquarium species from supply houses, but they are expensive and difficult to maintain for very long. Films and slides can give some of the experience of a field trip. Teachers in favorable locations may be able to arrange actual field trips.

ACTIVITY SUGGESTIONS   Materials needed are cultures of live *Hydra* and *Daphnia* (water flea), petri dishes (or similar containers), microscopes (or hand lenses).

This activity can be easily carried out in one class period. Students usually enjoy observing *Hydra*. If a stereo microscope is available it can provide an excellent view of the feeding *Hydra*.

This activity can be extended to observe budding in *Hydra*. To do this keep the *Hydra* culture thriving by adding some crustacean culture to it each day. Examine the *Hydra* culture each day or so for the presence of budding *Hydra*. It is possible that sperm or egg production may be observed also.

ANSWERS TO ACTIVITY QUESTIONS   1. Yes. The tentacles move. 2. Sizes vary; usually one or two millimeters 3. It contracts into a ball. 4. It is stopped cold, stuck to the tentacle, and slowly moved into the mouth.

ANSWERS TO QUESTIONS

WORD QUIZ   1.   division of labor 2.   larva 3.   filter feeder 4.   reef 5.   coelenterates

CHECK YOUR FACTS   1.   Different kinds of cells do different work in the body. We listed about six kinds of cells in the sponge. Some take food, some cover the body, some build skeleton, etc. 2.   Food particles are taken into the feeding cells from water entering the sponge through pores in its side. The food is digested in the individual cells and passed from them to other cells of the body. 3.   Hooks on long poles are used for shallow water. Most are gathered from deeper water by divers. Then they are allowed to die. The cells decay, and the sponge "skeleton" is cleaned, trimmed, and sold. 4.   The *Hydra* stings live animals to death and then pushes them into its mouth. 5.   two; In sea anemones they are folded to give a thicker, stronger effect. 6.   Jellyfish: body like an umbrella, with jelly between the cell layers; tentacles hang down around the mouth. Sea anemones: bodies like giant *Hydra*, sometimes with hundreds of tentacles. Corals: same structure as sea anemones, only smaller; they build limestone bases around their bodies, with tentacles reaching out into the water. 7.   Corals build reefs, which give hiding places to sea animals. These reefs can be a hazard to ships (or a protection, when they create a protected lagoon that can be used as a harbor).

THOUGHT QUESTIONS   1.   They cut up sponges and scatter the pieces over favorable sea bottom. These grow into large sponges. 2.   The few layers of cells do not make a big enough mass of possible food material to be worthwhile. The sponge skeleton would be a further complication. 3.   The thin layers of cells would dry up. The weak movements of coelenterates could not operate against the pull of gravity. The sponges filter feeding would not work in air.

CHAPTER 26   WORMS

OVERVIEW   In this chapter the structure and function of flatworms, round-worms and segmented worms are discussed. *Planaria*, the classic example of a free-living flatworm, is used to introduce the presence of systems in organisms. The digestive and nervous systems of *Planaria* are described. The tapeworm and the fluke are given as examples of parasitic flatworms. The life cycles of these organisms are given and emphasis is placed on the potential health hazard of the blood fluke *Schistosoma*.

The roundworm *Ascaris* is introduced as an organism having the "tube-within-a-tube" type of digestive system. Other roundworms discussed are the hookworm and the trichina worm.

The earthworm is examined in detail as an example of a segmented worm. The well-developed digestive system is described as are the reproductive, circulatory, and excretory systems. The economic importance of the earthworm is mentioned also. Other examples of segmented worms (leech, sandworm, etc.) are cited.

HINTS   In studying this chapter students should begin to see clearly that the life style of an organism is directly related to its structure. Parasites are parasites because they do not have the structures required to be free living. The earthworm can live on land and eat soil because it has the necessary structures for locomotion and digestion. None of the worms mentioned have respiratory systems, therefore their outer skin must somehow remain moist for diffusion or they will die.

It is important that these relationships between structure, function, and life style are clearly established in the students' minds at this time. It is these relationships that will be developed throughout the rest of the text.

ACTIVITY SUGGESTIONS   Materials needed are cultures of *Planaria* and the vinegar eel, pond culture, and several hand lenses (or microscopes).

The live *Planaria* and vinegar eel can be ordered from supply houses. The trash in old pond cultures often contains roundworms. Sometimes there are also flatworms. Roundworms often appear in protozoan cultures. A stereo microscope provides an excellent view of *Planaria*. If *Planaria* is placed in one or two drops of water on a microscope slide and insect larvae or a drop of protozoan culture is added, *Planaria* can be observed evaginating its pharynx while feeding.

As an *alternate activity* you may wish to have students examine living or preserved earthworms. Behavior of living specimens can be observed while preserved specimens can be dissected.

ANSWERS TO ACTIVITY QUESTIONS   1.   Paragraphs will vary. Check for credibility and degree of observations. 2.   Answers may vary. Shape and type of intestine will be mentioned most often. 3.   Yes. The peristaltic movements are easily seen. 4.   Descriptions will vary.

ANSWERS TO QUESTIONS

WORD QUIZ   1.   The worms are flatworm, fluke, roundworm, and segmented worm. Check descriptions of structure for accuracy. 2.   Feces is a waste material. 3.   The body parts are the intestine (for absorption of nutrients), the anus (for elimination of solid wastes), the crop (for storage), and the gizzard (for grinding).

CHECK YOUR FACTS   1.   Flatworm: flat; one opening to the intestine. Roundworm: round; two openings to the intestine (mouth and anus). 2.   muscular, nervous, excretory 3.   parasitic flatworms not forming chains, as the tapeworms do 4.   soil, water, host plants and animals 5.   Tapeworm: a. reaches alternate host as eggs from feces of main host; reaches main host in meat of alternate host. b. Cook meat. c. Proper sewage disposal will prevent infection of alternate host. *Ascaris:* a. swallowed as eggs b. Keep clean. Do not place dirt in mouth. c. proper sewage disposal; Hookworm: a. Eggs in feces hatch on ground. Larvae enter new host through skin. b. Wear shoes. Be careful of where you swim. c. proper sewage disposal; Trichina: a. Host eats host. Humans ingest it with meat of a meat eater (usually pig). b. Cook meat, especially pork (and bear), thoroughly. c. Require cooking of garbage fed to pigs. Eliminate rats around pigpens. Blood flukes: a. Larvae enter snails and leave as adults in water. These adults bore through skin of host and enter blood. b. Wear shoes. Be careful of where you swim. c. Eliminate pollution of water by fecal material. Remove snails from infested waters. 6. The flaps along its side aid in swimming and in rapid oxygen absorption. The sense organs in the head enable it to hone in on food. 7. blood of vertebrates or juices of insects; They clamp onto the victim with their suction cups, cut the skin with their jaws, and drink the blood flowing from the wound.

THOUGHT QUESTIONS 1. There is no circulatory system to carry oxygen. A flat shape puts all cells near the surface, so oxygen has only a short distance to travel to reach any cell in the body. 2. Egyptian farmers often wade in irrigation canals where snail hosts are present. Sanitation is poor, so snails are commonly exposed to worms coming from eggs in human feces. This blood fluke matures in humans and causes severe, chronic weakness. 3. This is generally not helpful. If a garden has enough humus to support a worm population they will already be there. Putting worms into sterile soil will simply kill them.

Chapter 27   The Mollusks

OVERVIEW In order to present the general structure of mollusks, the chapter begins with a detailed description of the structure and function of a typical clam. After this the student is presented with a comparison of the structural and functional similarities and differences among various mollusks in what are referred to as the clam class (clams, mussels, oysters, scallops), the snail class, and the octopus class (octopus and squid). Reasons for the success of the mollusks phylum and the economic importance of these organisms are given.

HINTS Stress the importance of the mollusks as a numerous and varied group in the sea. Display of shell collections will help to stimulate interest. Live snails are easily kept in aquariums. They can be bought in pet stores and from biological supply houses. Wild ones can be gathered from ponds. Small clams for the aquarium may also be bought. Do not change the water too often or filter it too thoroughly if you keep clams. Remember that they are filter feeders.

The fresh-water clam is used to illustrate mollusk structure, because it is found throughout the country. Preserved ones are available if you wish to use them for dissection.

The economic importance of mollusks can be developed further. Students in coastal communities will be interested in which ones are good to eat and how they may be gathered. Oyster farming is an important industry in some areas. Its vulnerability to pollution can be a topic for reports.

ACTIVITY SUGGESTIONS Materials needed are one or more aquariums (can be various sizes), several water snails, lettuce leaves, several jars, a few land snails, and a metric ruler. The snails for this activity can be collected by the teacher. They can be gathered ahead of time. Aquatic ones will keep indefinitely in the same aquariums where algae are kept. Land snails are found by looking under the dead leaves in almost any deciduous forest. They eat lettuce and other vegetables. They may be kept in any sort of covered glass jars. These will need cleaning from time to time. If you cannot collect your own you can buy a variety of aquatic snails from pet stores or biological supply houses. Procedures B and C can be performed with some of these, though land snails work better.

ANSWERS TO ACTIVITY QUESTIONS 1.   The lung indicates that they may be descended from land snails. 2.   The tentacle (or "feeler") is pulled back into the head. 3.   The snail withdraws its head and foot into the shell. Some pull a round, hard "trap door" (the operculum) in after them to close the shell entrance. You can explain to the class that land snails seal themselves into their shells and wait out dry spells. They can live that way for months, waiting for a return of favorable weather. 4.   This figure will vary with the snail. The distance will not be very great.

ANSWERS TO QUESTIONS

WORD QUIZ

	FOOT		SIPHON		MANTLE		MANTLE CAVITY	
	LOCA-TION	FUNC-TION	LOCA-TION	FUNC-TION	LOCA-TION	FUNC-TION	LOCA-TION	FUNC-TION
CLAM	protrudes from shell	locomo-tion	between outer edges of opened shell	feeding	lines inside of shell	produces shell (protection)	space within shell	houses organs
SNAIL	Beneath shell on posterior end	locomo-tion	absent		lines inside of shell	produces shell (protection)	space within shell	houses organs
SQUID	attached to head	locomo-tion	absent		covers most of body	protection	within body wall	houses organs

CHECK YOUR FACTS 1.   Fresh-water clams hitchhike for a while by riding on fish as parasites. Then they drop off and become adult clams. Marine clams have free-swimming larvae. 2.   Clams close their shells. Snails draw into their shells. Squid fight or flee. 3.   no shell for the squid; free swimming; foot divided into tentacles armed with suction cups; jaws in mouth; head with eyes present; The clam shows opposite development in all these traits. The snail crawls on a combined head-foot. Mantle cavity and vital organs are in a spiral shell on the back. 4.   Snails eat the slime layer from underwater surfaces. Land snails eat vegetation. Drills eat clams. Squid are predators that eat such other animals as fish and crabs. Clams are filter feeders. 5.   See the section entitled "A successful phylum." 6.   As food: clam, oyster, scallop, octopus, squid, etc. Shell is used for buttons and for mother-of-pearl. 7. Pearls form when mantle cells get dragged into the deeper parts of a mollusk, where they deposit shell material around the parasite that brought them there.

THOUGHT QUESTIONS 1. To be better land animals the mollusks would need a tougher skin to protect them from drying out, and they would need legs with a jointed skeleton to enable them to walk. 2. This is caused by buried clams or other bivalves. They feel the ground quake as you approach. They shut their shells for protection. This causes water to jet out through the siphon from the mantle cavity.

Chapter 28   Arthropods

OVERVIEW Jointed appendages and an external skeleton are pointed out as two major characteristics of arthropods. It is also noted that small size and rapid rate of reproduction make Arthropoda a very successful phylum. The structure and function of various classes of arthropods are discussed. These classes are referred to simply as the centipede and millipede classes, the crustacean class (crabs, lobsters, shrimp, prawns, and crayfish), the spider class (spiders, mites, ticks, and scorpions), and the insect class. The outstanding characteristics of each class are given along with information about the habitat, economic importance, and potential health hazard associated with each, if any. Greatest emphasis is placed on the insect class, which is further broken down into seven orders. Control of insects and insect study as a hobby are also covered.

HINTS Some of the less common arthropods will not be familiar to many students. Therefore, the more direct experience that students can receive, the better they will understand these animals. Small plankton-eating crustaceans like *Daphnia* and *Cyclops* make good subjects for laboratory study. They can be ordered in pure culture from supply houses or they can be found in aquariums and pond water collections. Magnification is needed to study them. Collections of larger arthropods can be maintained in the classroom.

Insects come within everyone's experience, so there is nothing remote about this topic. You can draw upon the students' experiences for anecdotal illustration of each idea as you go along.

ACTIVITY SUGGESTIONS Materials needed are living crayfish, one or more small fish tanks (or similar containers) containing 5 cm of water, rubber gloves, and a few pieces of meat (or worms). **CAUTION:** Supervise the handling of crayfish by the students, so that they are not bitten. Students handling crayfish can wear rubber gloves for protection.

Living crayfish can be bought in some fish stores or ordered from supply houses. If you wish to have the students do a close anatomical study, preserved specimens should be used.

As an *alternate activity* you may wish to have the students examine a grasshopper as an example of an insect. The preserved specimens of the lubber grasshopper sold by supply houses are excellent for examining external structure and for dissection. **CAUTION:** Wash preserved specimens before distributing to students. The preservative can be irritating to skin and eyes. When doing dissections caution students about the use of scalpels, etc.

ANSWERS TO ACTIVITY QUESTIONS 1. They may suffocate in deeper tanks. 2. Yes. They will try to pinch. 3. Yes. It rears up and faces the "enemy" with claws poised. 4. Unless cornered they will retreat by swimming away backwards. They do this by flipping their tail fins under their bodies. In the wild, they retreat into a protected hide-away. 5. Descriptions will vary. Check for accuracy. 6. compound 7. There are 19 pairs. Students may not find them all, especially on live animals.

ANSWERS TO QUESTIONS

WORD QUIZ 1. b 2. c 3. e 4. f 5. k 6. i 7. j 8. a 9. d 10. g 11. h

CHECK YOUR FACTS 1. They have segmented bodies, which have all of the body systems, an exoskeleton, and jointed appendages (legs, wings, feelers, mouth parts, etc.). 2. They can hide easily, use small food sources, and recover losses by high reproductive rates. 3. The external skeleton (exoskeleton) gives great strength, but only in small sizes. It also serves as a protective armor. 4. Answers will vary. The following points may be mentioned: millipedes: segmented, with two pairs of legs on each segment; centipede: similar, but with one pair of legs per segment. Crustaceans: many forms; tendency toward specialized body parts and appendages. Spiders: two body parts and eight legs. Scorpions: two body parts; eight legs; enlarged mouth parts with pincers; stinger on end of abdomen. Insects: three body parts; six legs; four wings (usually); feelers; compound eyes; breathing tubes. 5. Millipedes: humus eaters; of minor importance. Centipedes: of some small importance as insect eaters. Crustaceans: of great importance in the sea; primary algae eaters, at start of marine food chains; Some are eaten by people. Spider class (arachnids): Spiders are major predators of insects; scorpions less so. Ticks and mites include bothersome parasites. Some carry serious diseases. Insects: very numerous and important on land; compete with humans for food and many other materials; Some carry diseases. Some pollinate plants. 6. They catch insects with their pincers and sting them to death. Their stings are generally just an annoyance to people, but small children can be killed by some species. 7. Spiders are of primary importance in keeping insects under control. They catch the insect, bite it to death with their poison fangs, and suck out its body juices. 8. Grasshoppers and dragonflies illustrate the arrangement in which a nymph, which is obviously insect-like, grows up into an adult. Beetles, butterflies, and flies illustrate the arrangement in which a worm-like larva eats and grows. It enters an inactive, resting stage (pupa) and finally emerges as an adult. 9. See listings in the chapter (kinds of insects). There are about eighteen other orders also. 10. Natural enemies keep them in check. We control insects by using poisons, by the release of sterile males, by attractive odors (pheromones), by the release of natural enemies, and by other manipulations of the environment.

THOUGHT QUESTIONS 1. No. Small size means that individuals are numerous. Individual life is cheap and easily replaced. The species survives, but many individuals are lost. Species whose members are large in size "invest" more in each individual (of which there are fewer). Large individuals are adapted for longer survival, since (from the species' point of view) "more eggs have been put in one basket." 2. It might, if it wiped out all harmful insects; but it might not. It might kill the ladybugs without controlling the aphids as well as they would.

Chapter 29   Echinoderms

OVERVIEW The major echinoderm discussed in this chapter is the starfish. The uniqueness of this organism is pointed out through explanations of the way it moves by means of tube feet, the way it can evaginate its stomach to catch and digest food, and its spring exoskeleton. There is also a brief discussion of the sea urchin, sea lily, and sea cucumber.

HINTS Use the common starfish as the example of the phylum. Use any specimens that you have available. Use collected specimens of other species also, such as sand dollars, sea urchins, etc.

The value of films and slides is especially important for a group like this. Students may have little contact with live echinoderms in their natural habitats.

ACTIVITY SUGGESTIONS Materials needed are several preserved starfish, living starfish, sea urchins, and any other echinoderms that may be available.

Procedure A: Ordinary preserved starfish can be used for this, but the ones mounted in museum jars are better. They can be bought with tube feet extended by the injection of colored liquid.

Procedure B: This activity will be more or less possible for your particular class depending upon what kind of collection your school has.

Procedure C: Sea-water aquarium collections may be bought from supply houses. They are expensive and difficult to maintain, but you may wish to try using them. The showing of good films may be substituted for this activity.

ANSWERS TO ACTIVITY QUESTIONS 1. Answers may vary. Check for accuracy. 2. It is hard and rough. 3. on the underside, in the center 4. They look like small suction cups. They are located on the undersides of the arms. 5. and 6. Answers will vary depending on the echinoderms studied. 7. The tube feet will stick to the sides of the aquarium like suction cups and unattach and reattach themselves alternately as the organism moves along.

ANSWERS TO QUESTIONS

WORD QUIZ Definitions will vary. They should approximate those given in the text.

CHECK YOUR FACTS 1. It has a round central disk with several arms radiating outward. The exoskeleton is hard and spiny. On the underside is a mouth and many tube feet. 2. on tube feet; These are hollow tubes that reach out when water is pumped into them and contract when it is withdrawn. It pulls a clam or oyster open with its tube feet. Then it turns its stomach inside out through its mouth, presses it against the flesh of the mollusk, and digests it. 3. sea urchin, sea lily, sand dollar, basket star, sea cucumber 4. numerous, important members of the sea communities; Sea cucumbers are eaten in some countries.

THOUGHT QUESTIONS 1. No. The half-starfish would grow replacements for their missing parts. The result would be twice as many starfish (in time). 2. Sea otters eat sea urchins, among other things. The urchins multiplied when their natural enemies were exterminated. Now the sea otters have returned, and there is a program for replanting the kelp beds.

Chapter 30   Fish and Amphibians

OVERVIEW   This chapter introduces the vertebrates. Fish are discussed in three groups, the jawless fishes (e.g., the lamprey), the sharks class (sharks, skates, and rays), and the bony fishes (perch, bass, mackerel, etc.). The special structures and functions of each of these groups are discussed. The importance of fish in marine and aquatic communities, as well as their commercial importance, is examined. Amphibians are also discussed in this chapter. Their adaptations to living partially on land and partially in water are explained. Fossil evidence of a transition from life in the water to life on land is mentioned.

HINTS   The jawless fishes will be new to most students. Your school should acquire preserved specimens of at least one species. Students could make reports on the problem of sea lampreys in the upper Great Lakes.

Sharks offer obvious material for reports. Almost any library will contain source material for several of them.

The bony fishes are, of course, the ones of greatest economic importance. Extra reports could deal with commercial fisheries, sport fishing, fish conservation, local species, etc.

ACTIVITY SUGGESTIONS   Materials needed are one living frog per group, masking tape, chalk, a meter stick (or other measuring device), a container of ice water, and a clock (or watch) with a second hand.

Be careful not to leave the frog in the ice water too long. Do this experiment in an area where the frog cannot jump into an inaccessible area. For easy management this activity can be done as a demonstration.

As an *alternate* (and ongoing) *activity* you may wish to set up a terrarium containing frogs and salamanders and/or a fish aquarium. Students can be assigned to maintain these after they are set up. Reports on various fish and amphibian activities can be assigned.

ANSWERS TO ACTIVITY QUESTIONS   1. Answers will vary. 2. Answers will vary. 3. It should travel farther when warm. 4. Answers will vary. 5. It would be a disadvantage because it could not move about as quickly to obtain food, escape danger, etc.

ANSWERS TO QUESTIONS

WORD QUIZ   1. vertebrates 2. lamprey 3. salamanders 4. amphibians 5. chordate 6. swim bladder

CHECK YOUR FACTS 1. They are the animals with backbones. They have internal, jointed skeletons and all of the body systems. Most have a head, neck, body, tail, and two pairs of appendages. Chordates are all members of the phylum to which vertebrates belong. Some chordates live in the sea and do not have backbones. 2. jawless fishes, sharks, bony fishes; descriptions of differences will vary. Information should approximate that given in the text. 3. through the thin outer membranes that cover the gills; Blood in the gills carries it to all parts of the body. 4. to regulate the fish's body weight so that it can drift in the water without rising or sinking 5. vertebrates with smooth, moist skins and a life cycle in which the larvae have gills and the adults have legs and lungs (with some exceptions) 6. Salamanders: smooth, moist skins; no claws. Lizards: dry, scaly skin; claws present. 7. They help control insect numbers. 8. They may have developed from lobe-finned lungfish hundreds of millions of years ago.

THOUGHT QUESTIONS 1. They cannot stand dry air for very long. Most of them must live in water during their larval stage. 2. Fish are well adapted to life in the water, so they succeed well in water environments. Amphibia are not fully adapted to either land or water, so they must live mainly in the borderline environments.

Chapter 31   Reptiles and Birds

OVERVIEW This chapter opens by describing the general characteristics of reptiles. Attention is then given to reptiles that are common today, such as various lizards, alligators, crocodiles, and snakes. Some common misconceptions about these animals are dispelled. Information is given about the eating habits of various snakes and whether or not they are poisonous. Reproduction and treatment of the young is discussed also. The economic importance of reptiles is mentioned.

The birds are discussed in much the same way as the reptiles, with information on their general characteristics, the life styles and adaptations of various birds for food getting, and reproduction and care of the young. Migration is also discussed as is the importance of birds to various communities.

HINTS   CAUTION: Do not encourage students to bring their pet reptiles to class or to collect reptiles in the wild. Films and slides are a desirable substitute for field work.

Student reports on extinct reptiles are popular. Some students study this topic until they can reel off the names and descriptions of dozens of dinosaurs. You can capitalize upon their enthusiasm.

Bird study is worth encouraging. You can keep a collection of bird guides in the classroom and show students how to use them. Point out which ones are easiest to use. Also point out which ones are low in price. Students who learn to identify songbirds may develop a lifetime hobby. They may tend to become citizens who support conservation programs.

ACTIVITY SUGGESTIONS Materials needed are a few living box turtles and a suitable place to keep them.

It is best for the teacher to obtain these turtles from a pet shop. Always supervise the handling of these turtles by students. Turtles should be kept in an area with ventilation as they may tend to give off an odor.

ANSWERS TO ACTIVITY QUESTIONS 1. They usually pull their heads and limbs into their shells in response to fright, when touched, or when sleeping. 2. They protect themselves in this way. 3. Students will probably describe the feet as looking like swim fins. 4. This type of feet would indicate that they live in the water. 5. It does this by pushing on its head and neck to right itself. 6. a long strong neck

ANSWERS TO QUESTIONS

WORD QUIZ Definitions will vary. They should approximate those given in the text.

CHECK YOUR FACTS 1. land reproduction (type of egg); dry, scaly skin, stronger legs, better lungs and circulatory system 2. Birds live in a different type of environment. They reproduce differently, care for their young, eat different types of food, and migrate. 3. lizards, snakes, turtles, and the crocodile group 4. grabbing and swallowing harmless species; suffocating prey by squeezing (constriction); killing with poison 5. rear-fanged serpents, cobra type, and vipers (Some will describe the fangs in each group.) 6. Snakes are part of the general balance of nature. Since most feed on insects or rodents we would count them as useful. 7. insects and seeds; There are many exceptions, which some students may list. 8. Refer to the discussion and to Fig. 31-8 in the textbook. Perching feet, webbed feet, scratching feet, clawed feet for killing (hawks), and feet for wading (heron) are a few examples. Beak types may include slender beaks for catching insects, heavy ones for cracking seeds, hooked ones for tearing flesh, and spear-like ones for catching fish. 9. to find food that is seasonally available in different places, especially to find abundant food for feeding the young 10. They are very important in controlling insects. Other things may be mentioned, such as eating weed seeds and furnishing us with eggs, meat, and feathers.

THOUGHT QUESTIONS 1. The danger here is that people will feed birds long enough to induce them not to migrate. Then they get careless and stop supplying food later in the winter. The birds do not fly south then and they starve. Once started, feeding should be kept up until spring. 2. The snake senses heat coming from bodies warmer than its own. A warm-blooded animal would have difficulty detecting another one in this way. There would be no contrast between the two warm bodies.

Chapter 32   The Mammals

OVERVIEW First, the general characteristics of the mammals are listed, and then various orders of mammals are discussed in more detail. The orders discussed are the egg-laying mammals (spring anteaters and platypuses), pouched mammals (kangaroos, koalas, wombats, etc.), the shrews and moles, the bats, the rodents (mice, rats, squirrels, woodchucks, beavers, etc.), the flesh eaters (dogs, cats, bears, weasels, raccoons, and seals), the whales, the elephants, the hoofed mammals (cattle, tapir, horses, etc.), and the primates (lemurs, monkeys, apes, and humans). The structure, life style, and habitat of each of the above is explained. Mention is also made of fossil evidence that has been found regarding early primates.

HINTS The catalogue of mammals in this chapter can have considerable appeal to students. Student reports, films, and slides can all be used. It is especially desirable that students know about the mammals in their local region.

A wild animal is an organism in its own right, with its own ecological requirements. Stress that each species must have all of its needs satisfied or it cannot survive. The interests of wild animals can be damaged when the need to balance populations to the carrying capacity of the range is not recognized.

ACTIVITY SUGGESTIONS No special materials are needed for this activity.

If possible, it is a good idea to take students on a field trip to a local zoo or game preserve. If this cannot be done, students can visit such places on their own or simply study mammals found in the neighborhood, as suggested in the activity. **CAUTION:** Discourage students from handling or taunting unfamiliar animals. These observations can be made from a distance.

You may wish to have a collection of resource books on mammals available in your classroom. Some zoos and game preserves also publish informative literature that could be of interest to the students. You could order several copies of these ahead of time to be distributed to the students or kept in a classroom reference file.

ANSWERS TO ACTIVITY QUESTIONS 1. and 2. Student reports will vary. Check for accuracy.

ANSWERS TO QUESTIONS

WORD QUIZ Definitions will vary. They should approximate those given in the text.

CHECK YOUR FACTS 1. warm blooded, hairy, intelligent, live bearing (with one exception); produce milk, care for young, etc. 2. Warm bloodedness enables them to be active all year and in all weather. Intelligence helps in adjusting to new situations. 3. See listings in text. 4. They are small, numerous, and hide easily. Rapid reproduction makes up losses. Rodent-type food is available in most environments. 5. See listings in text.

THOUGHT QUESTIONS 1. Elk and buffalo (bison) would compete with domestic animals for grass. Bear and wolf would directly attack domestic animals. In other words, these animals would compete with us for our food supply (domestic animals). We are the dominant species in the farm lands. We do not tolerate this competition. 2. The marsupials (pouched mammals) would suffer badly from competition with better adapted placental mammals. Most would probably become extinct. A few might survive and successfully invade Asia.

A-V Materials   Unit 6

(Numbers in parentheses refer to chapter numbers.)

Alligator. McGraw-Hill, film. (35)
Animals With Backbones. Coronet, film. (30–33)
Animals Without Backbones. Coronet, film. (25–30)
Arthropods: Insects and Their Relatives. Coronet, film. (28)
Birds and Their Characteristics. Coronet, film. (31)
Birds: How We Identify Them. Coronet, film. (31)
The Chordate: Diversity in Structure. Coronet, film. (30)
Coelenterates and Sponges. Doubleday Multimedia, filmloops. (25)
Echinoderms and Mollusks. Coronet, film. (27–29)
The Frog, Encyclopaedia Britannica, film. (30)
Mammals and Their Characteristics. Coronet, film. (32)
Mammals Are Interesting. Encyclopaedia Britannica, film. (32)
The Invertebrates. Coronet, film. (25–30)
Kingdom of the Animals. NGS, sound filmstrip. (25–32)
Life Story of the Earthworm. Encyclopaedia Britannica, film. (26)
Life Story of A Snake. Encyclopaedia Britannica, film. (31)
The Oceans. PH, sound filmstrip. (25)
Parasitism. Encyclopaedia Britannica, film. (26)
Sea Life. NGS, sound filmstrip. (25)
Sponges and Cup-like Animals. Coronet, film loop. (25)
Stinging-celled Animals—The Coelenterates. Encyclopaedia Britannica, filmstrip. (25)
Worms: Flat, Round, and Segmented. Coronet, film. (26)

Supplementary Reading   Unit 6

Audubon, vol. 73, no. 2 (March 1971, entire issue).
Boorer, M. *Mammals of the World*. New York: Grosset & Dunlap, Inc., 1967.
Buchsbaum, R. *Animals Without Backbones*. Chicago: University of Chicago Press, 1948.
——— and L. J. Milne. *The Lower Animals—Living Invertebrates of the World*. New York: Doubleday, 1960.
Callahan, Phillip, *Insect Behavior*. New York: Four Winds Press, 1970.
Carington, and the editors of *Life*. *The Mammals*. Time-Life Books. New York: Time, Inc., 1963.
Farb, Peter, and the editors of *Life*. *The Insects*. Time-Life Books, New York: Time, Inc., 1962.
Goodall, Jane. "Jane Goodalls' Investigations." *National Geographic* (August 1963).
Newman, L. H. *Man and Insects*. New York: Doubleday, 1967.
Snow, Keith R. *The Arachnids*. New York: Columbia Univ. Press, 1970.
"Starfish Threaten Pacific Reefs." *National Geographic* (March 1960).

HUMAN BIOLOGY

CHAPTER 33   HUMAN HISTORY

OVERVIEW In this chapter it is explained that modern humans belong to one species, *Homo sapiens*. Scientists believe that people, like other living species, have gone through changes. Attention is called to the fact that humans have traits that are not found in other species. It is because of these special characteristics that humankind dominates the earth. It is postulated that gene mutations and natural selection are the forces that caused appearance of the various racial groups. The differences between racial groups are not great. All humans belong to *Homo sapiens* because of genetic and reproductive compatibility.

HINTS Use the illustrations in the chapter to increase students' understanding of ideas that you are presenting. To explain how evidence is obtained from fossils show the film *Fossils: Clues to Prehistoric Times*. B/W, Coronet. Carefully develop the concept of primate characteristics using commercial charts, skeletons of vertebrates, illustrations from the chapter, and other material that is available to you. You may wish to have pupils do reports on the races and cultures of humans.

ACTIVITY SUGGESTIONS The only materials needed are Fig. 33-5 in the text and a metric ruler.

This type of comparison can be done between bones of various other organisms, using illustrations from the text or other sources. If class time does not permit, this activity can also be done as a homework assignment.

ANSWERS TO ACTIVITY QUESTIONS 1. The head, shoulder bones, forelimbs, and pelvic girdle are much longer in the gorilla than in the human. The rib cage of the gorilla is rounded and barrels out. Encourage the use of descriptive words so that pupils can present logical comparisons of the two skeletons. 2. The measurements should substantiate the visual comparisons. 3. The shortening of the shoulder bone and the hip bone makes upright posture possible in humans. The flattening of the rib cage is a contributing factor also.

ANSWERS TO QUESTIONS

WORD QUIZ 1. savanna 2. ground-living primates 3. *Homo erectus* 4. culture 5. *Homo sapiens* 6. Neanderthal 7. dating

CHECK YOUR FACTS 1. The climate became drier and the trees could not survive in an environment of reduced water. 2. The trees were replaced by a grassland. 3. The early ground-living primates had feet that were adapted for walking instead of climbing. Their hands were used for carrying things instead of swinging from trees. They had to find food that lived on the ground. 4. *Homo erectus* had larger brains than the early ground-living primates. They were probably able to talk and pass on information. They probably developed a culture. 5. Humans walk erect and can live anywhere in the world. The feet are used for walking and the hands for work. The kinds of motor skills and hand maneuvers that humans can perform outnumber those of any other primate. The human brain is well developed, permitting speech and storage of information. 6. All people have the same number of chromosomes and they are capable of crossbreeding and producing viable offspring. 7. Humans live all over the world. They are not confined to one type of climatic environment.

THOUGHT QUESTIONS 1. Apes have a way of life, but not a culture. Culture is a set of ethical values that is passed from parent to offspring by word of mouth. An ape's behavior is governed by instinct, an inborn set of reflexes. 2. Tool making indicates that thought processes dictate the need for a particular tool. The use of the hands permits the use of tools.

Chapter 34   Food and Nutrition

OVERVIEW This chapter explains that a Calorie is a measure of the energy in food. Carbohydrates and fats contain the greatest numbers of Calories. These nutrients are known as fuel foods. It is pointed out that proteins are tissue builders, while vitamins and minerals are chemical regulators. Deficiency diseases are symptoms of poor nutrition. Suggestions are made for a balanced diet. The various food groups are explained and examples given.

HINTS This chapter lends itself to a good deal of pupil activity. It is a good idea to start out by having pupils keep a three-day record of what they eat. You can then use this information as you teach the nutrition facts in class. Let a group of students prepare a class bulletin board on the balanced diet, on interesting lunch menus, or on nutritious snacks. You can also demonstrate food tests. (See Morholt, E., et al. *A Sourcebook for the Biological Sciences*. New York: Harcourt, Brace and World, 1966 for specific directions.) Impress upon pupils the need for good nutrition. The meaning of starvation can be demonstrated by showing pictures of the starving children of the third world countries.

ACTIVITY SUGGESTIONS No special materials are needed for this activity. Pupils follow the directions given to compute the number of Calories that they use each day. You may wish to have students report on how they can improve their eating habits. They can then continue to keep a daily Calorie chart to keep track of their progress.

ANSWERS TO QUESTIONS

WORD QUIZ Definitions will vary. They should approximate those given in the text.

CHECK YOUR FACTS 1. carbohydrates, fats, proteins, minerals, and vitamins 2. Calories measure the energy in food. 3. Carbohydrates provide the cells with energy. Fats provide energy and build fat tissue. Proteins are tissue builders. 4. Vitamin A prevents night blindness. Vitamin B complex prevents degenerative nerve and muscle diseases. Vitamin C prevents scurvy. Vitamin D prevents rickets. 5. Iron builds red blood cells. Iodine permits effective working of the thyroid gland. Calcium and phosphorus build strong bones and teeth. 6.   Malnutrition is poor nutrition caused by a lack of essential nutrients in the diet. Malnutrition causes a breakdown of body tissues. Malnutrition can be prevented by eating foods that provide a variety of minerals, vitamins, and the other major classes of nutrients. 7.   A balanced diet contains the right amounts of the nutrients needed to permit healthy functioning of body cells, tissues, and organs.

THOUGHT QUESTIONS   1.   Our daily nutrition is mainly responsible for the health of the body. 2.   Snack foods that are rich in proteins and vitamins are better for the health of the body than those that are full of sugar and starch. Proteins and vitamins are tissue builders; the carbohydrates produce energy. Excess energy is changed to fat and stored under the skin. 3.   Infants need vitamin D for the growth of bones and teeth. They need vitamin C for healthy blood vessels. Infants grow rapidly and their bodies go through major changes. They need daily amounts of certain vitamins to regulate the rapid growth of cells and tissues.

Chapter 35   Digestion

OVERVIEW   This chapter begins by explaining that the function of the teeth is to grind the food to make it ready for digestion. It is then explained that the digestive system consists of the alimentary canal and those glands that pour their digestive juices into the small intestine. The chapter points out that each part of the digestive system contributes to the major task of digestion. Only those nutrient molecules that can pass through the cell membranes are useful to cells. The structure and function of the organs of digestion are discussed in detail.

HINTS   It is a good idea to introduce this chapter by showing the filmstrip *The Digestive System*. This is one in a series of seven called *Understanding Your Body,* produced by Encyclopaedia Britannica. Have pupils look at their teeth by using hand mirrors. Emphasize that care of the teeth is important. You can use commercial charts and models to teach the parts of the digestive system. Demonstrate the function of enzymes, such as salivary amylase, pepsin, and rennin. These can be purchased from biological supply houses. To demonstrate selective permeability you can show that starch cannot pass through a membrane whereas dextrose can.

ACTIVITY SUGGESTIONS   Materials needed are cornstarch, test tubes, saliva or synthetic salivary amylase, iodine solution, warm water, and a warm-water bath.

   CAUTION: If using a flame to heat the water bath, students should wear goggles, tie back long hair, and secure loose clothing:
   In order to induce salivating you can have students chew on clean rubber bands.
   CAUTION: If natural saliva is to be used, have students handle only tubes containing their own saliva. Sterilize test tubes after use. If a student has a cold or other respiratory infection, do not use his or her saliva.

ANSWERS TO ACTIVITY QUESTIONS 1.   Iodine turns starch a blue-black.
2.   At each testing the color change should become lighter until no change
occurs at all. 3.   As the enzyme causes the digestion of starch, starch molecules
disappear. 4.   The end product of starch digestion is sugar, usually maltose.
5.   Benedict's solution might render a color change to light green.

ANSWERS TO QUESTIONS

WORD QUIZ   The sentences written by pupils will vary. Saliva is an enzyme.
A duct is a passageway. The esophagus is a tube that pushes food into the
stomach. The stomach is an organ where digestion begins. The small intestine
is the organ in which the major tasks of digestion and absorption occur. The
pancreas secretes pancreatic juice. Bile emulsifies fat. The liver synthesizes
bile and stores sugar. The gall bladder stores bile. The villi function in absorption. The large intestine absorbs water. The appendix has no function in
man.

CHECK YOUR FACTS   1.   Digestion changes large nutrient molecules into
smaller molecules. 2.   Teeth grind up the food, increasing the surface area on
which enzymes can work. 3.   Saliva moistens the food and contains a digestive
enzyme. 4.   mouth, esophagus, stomach, small intestine, large intestine, liver,
and pancreas 5.   Large molecules cannot go through the cell membranes.
6.   Digested foods are transported by the blood to all cells in the body.
7.   Mechanical digestion is the grinding of food. Chemical digestion is the
changing of molecules into different ones.

THOUGHT QUESTIONS   1.   The roots of teeth and the blood vessels and
nerves that service the teeth are living. The outer enamel is nonliving.
2.   Babies must have those enzymes that will digest the proteins and sugars
in milk. 3.   As a result of digestion, millions of glucose, amino acid, and fatty
acid molecules are produced. These would pile up in the intestines if provision
were not made for their removal.

Chapter 36   Blood and Circulation

OVERVIEW   This chapter describes the parts of the blood and gives the function of each. An overall view of the entire circulatory system is presented. The
organs of circulation, the heart, and the blood vessels are discussed with clarity
so that the reader can understand their importance. The significance of circulating lymph is also discussed.

HINTS   You may wish to introduce this chapter by showing the film *Work of
the Blood*, Encyclopaedia Britannica Films, color. You can obtain some blood
from a slaughterhouse or dated blood from a hospital laboratory. Have the
blood treated to prevent clotting. Spin in a centrifuge to show separation of
blood into its parts. Make some smears to demonstrate blood cell types. Use
commercial charts, films, and filmstrips to help make the material taught in
class more interesting.

ACTIVITY SUGGESTIONS   Materials needed are goldfish with full tails, absorbent cotton, Petri dishes, microscopes, and glass slides.

Show pupils how to wrap the fish so that the gills are covered with the wet cotton but the tails are exposed. **Do not leave fish out of water too long.** As an *alternate activity* students can study prepared slides of human and amphibian blood for comparison.

ANSWERS TO ACTIVITY QUESTIONS   1.   No. When viewed through the microscope, blood in the artery *appears* to be traveling upward. Blood in the vein appears to be moving downward. 2.   Blood in an artery moves in spurts. Blood in veins moves smoothly. 3.   Red blood cells move through a capillary in single file.

ANSWERS TO QUESTIONS

WORD QUIZ   Words to be circled: arteries, veins, capillaries, atrium, ventricle, lymph vessels, lymph nodes, spleen. Words to be underlined: plasma, lymph. Word meanings: Hemoglobin is the iron-containing compound in red blood cells. White blood cells resemble ameba. Platelets function in clotting of blood.

CHECK YOUR FACTS   1.   Plasma transports blood cells and nutrients. Red blood cells carry oxygen. White blood cells destroy bacteria. Platelets function in clotting. 2.   Red blood cells lack nuclei. White blood cells are nucleated. 3.   Arteries carry blood away from the heart. Veins carry blood to the heart. Capillaries connect arteries with veins. 4.   The heart is a muscular organ that is divided into four chambers. Pupils should use such words as atrium and ventricle. 5.   Blood flows from the right atrium into the right ventricle. It then flows to the lungs and travels back into the heart through the left atrium into the left ventricle and then to the body organs. 6.   Lungs: Blood exchanges carbon dioxide and water vapor for oxygen. Small intestine: Blood gives up oxygen nutrient molecules and receives food molecules and carbon dioxide. Kidneys: Blood exchanges urea for reabsorbed nutrients and oxygen. 7.   Lymph is a dissolving medium and keeps the tissues moist. 8.   Blood typing is a procedure used to determine the kinds of special proteins in the blood. It is dangerous to transfuse incompatible blood types. 9.   Proper amount of exercise, eating a balanced diet, and the avoidance of cigarette smoking help guard the health of the circulatory system.

THOUGHT QUESTIONS   1.   The circulating blood transports oxygen and nutrient molecules to the body cells. It carries away from the cells the wastes of respiration and protein metabolism. 2.   Lymph nodes are not vital organs. For example, tonsils are removed and the individual lives. 3.   The loss of 2 liters of blood is a 40% loss. This would cause a dangerous drop in blood pressure and the transport ability of the blood would be diminished.

Chapter 37   Breathing and Respiration

OVERVIEW   In this chapter the student is taught how the organs that compose the respiratory system are adapted for the movement of oxygen from the external environment to the blood stream. It is also pointed out that these same organs transport the wastes of cellular respiration, carbon dioxide, and water vapor from the cells to the external environment. It is explained that the need for a circulatory system is related to the fact that all cells in the body must have a means of obtaining oxygen.

HINTS   Use a dissectible mannikin to demonstrate the organs of respiration and to show the path of air through the nasal passages into the lungs. You can obtain a pluck from a butcher or a supply house and blow up the lungs with a bicycle pump to show their expanding potential. Use chalkboard diagrams to demonstrate the structure of the alveoli. Review the concept of diffusion.

ACTIVITY SUGGESTIONS   Materials needed are a 4-liter jar, a graduated cylinder, a metric ruler, and a length of flexible tubing (approximately 1.5 meters long).

Tubing having a diameter of approximately 5 cm works best. Students should be encouraged to blow into the tubing as forcefully as possible. **CAUTION: Any student who has a respiratory disease or infection should be excused from this activity.**

As an *alternate activity* you may wish to have students examine slides of the ciliated cells from the lining of the roof of a frog's mouth.

ANSWERS TO ACTIVITY QUESTIONS   1.   Four thousand mLs are equivalent to 4 liters. 2.   The breath replaces the water in the bottle. 3.   Pupils can mark the bottle indicating the level of the water remaining. They empty out the bottle and refill with water to the mark. Measuring the amount of the water will provide a measurement of the volume of breath that was expelled.

ANSWERS TO QUESTIONS

WORD QUIZ   Definitions will vary. They should approximate those given in the text.

CHECK YOUR FACTS   1.   nose, throat, windpipe, bronchioles, lungs, and air sacs 2.   The air is warmed, filtered by nasal hairs, and passes through the throat into the windpipe. Cells with cilia filter out dust particles from the inhaled air. 3.   Air in the windpipe passes into the lungs. Oxygen from this air diffuses into the blood stream. The other components in the air are exhaled. 4.   The lungs resemble two large, porous (spongy) sacs. They are covered with epithelial tissue, are pink in color, and are quite moist. 5.   The air sacs increase the surface area of the lungs for rapid diffusion of oxygen into the lungs. 6.   Breathing refers to inhaling and exhaling. Respiration is the entire process in which oxygen is used to extract energy from nutrient molecules. 7.   When the rubber diaphragm is pulled down, air rushes into the bell jar filling the lungs. This model resembles the way in which the chest cavity is enlarged when the diaphragm is pushed down and the ribs push outward. Exhaling is the opposite. The diaphragm pushes upward, decreasing the size of the chest cavity. 8.   Emphysema, tuberculosis, lung cancer, pneumonia, whooping cough, croup, and pleurisy are a few of the lung diseases.

THOUGHT QUESTIONS   1.   Young children might stuff a small bead in the nose or ear. 2   Exercise causes breathing to become deeper and more rapid.

Chapter 38   Excretion

OVERVIEW   It is pointed out that excretion is a vital activity of the body. The skin, lungs. and liver are presented as organs of excretion. The kidneys, bladder, urethra, and ureter are listed as the parts of the urinary system. This chapter explains that carbon dioxide, water, and urea are the waste products of cellular respiration, which must be excreted from the body. The methods by which this is done are discussed simply, but thoroughly.

HINTS   You will find the illustrations in the chapter useful in your daily lessons. For example, the tissues that compose the skin are shown quite clearly in the illustration. Dissect a beef kidney to show pupils the internal structures. Use prepared slides if you wish to demonstrate the fine structure of the kidney. National Teaching Aids, Garden City, NY, has developed a micro-slide viewer program that consists of a plastic viewer and strip film. Set 52, *Excretion*, provides a clear overview of the importance of excretion.

ACTIVITY SUGGESTIONS   Materials needed are a Celsius thermometer, cotton, rubber bands, and a warm-water bath. **CAUTION: It is best to heat the water in a pan over an electric hot plate. The ring stand with beaker and Bunsen burner setup requires special precautions and vigilance to see that the beaker of hot water does not fall or spill. If using a flame, students should wear goggles.**

ANSWERS TO ACTIVITY QUESTIONS   1. If the room temperature is 20°C, then the reading on the thermometer will be 30°C. 2.   The temperature will drop by 10 degrees or more. 3.   Evaporation is cooling. 4.   Evaporation of perspiration cools the skin. 5.   The reading on the thermometer will increase. 6.   The temperature does not decrease; it may stay the same or increase slightly. 7.   Humidity increases temperature.

ANSWERS TO QUESTIONS

WORD QUIZ   Student sentences will vary. Check to see that they have used the new words correctly.

CHECK YOUR FACTS   1.   carbon dioxide, water vapor, urea, mineral salts 2.   Excretion is the removal from the body of metabolic wastes. 3.   skin, lungs, liver, kidneys and kidney tubules, urethra, bladder, ureter 4.   The kidneys filter out the wastes and reabsorb useful materials. 5.   The ureters collect urea and water from the kidneys and transport these products to the bladder. Urine leaves the body through the urethra. 6.   Excess water, oil, and mineral salts leave the body through the skin.

THOUGHT QUESTIONS   1.   Without the kidneys, the wastes of cellular metabolism would pile up in the blood and poison the body cells. 2.   A person can live without sweat glands. People with reduced numbers of these glands are very uncomfortable during hot, humid weather.

Chapter 39   Bones and Muscles

OVERVIEW   This chapter teaches the students that the body framework consists of bones and connective tissues. We call this internal support system the skeleton. The skeleton gives shape to the body, permits locomotion and other movements, and protects the internal vital organs. The specific structure and function of the skeletal system is discussed. Hints: The most dramatic beginning is to show a human skeleton, if possible. Compare the bones in the human skeleton with those in the skeleton of other vertebrates: cat, bird, fish, or whatever you have available.

ACTIVITY SUGGESTIONS   The materials needed for the activity are chicken bones, vinegar, a tin can (or other metal container), and a Bunsen burner.

As an *alternate activity* you may demonstrate bone cross section using prepared microscope slides. Use hamburger meat to make slides of muscle tissue. Stain with methylene blue.

ANSWERS TO ACTIVITY QUESTIONS   1.   The bone feels rubbery. 2.   The bone is flexible. 3.   It has lost its hardness. 4.   Bone lightens in color. It becomes more brittle. 5.   Heating destroyed the proteins and the mineral matter.

ANSWERS TO QUESTIONS

WORD QUIZ   The sentences of pupils will vary. Have them write their sentences on the chalkboard for class discussion.

CHECK YOUR FACTS   1.   Bone cells do everything that other cells do. In addition, they secrete compounds of calcium and phosphorus that build up a hard matrix. Blood cells and plasma transport oxygen and nutrients to bone cells. 2.   gives shape to body, permits movement and locomotion, protects inner vital organs 3.   Bone is hard; cartilage is flexible. Ligaments connect muscles; tendons connect bones to muscles. 4.   Not all muscles are attached to bones. Heart muscle and the muscles of vital organs are not attached to bones. Muscles used for locomotion and voluntary activity are attached to bones. 5.   Muscles pull bones in many directions. Pairs of muscles attach to bones. When one set contracts, the other relaxes. 6.   The long bones are in the arms and legs. The flat bones are in the pelvis and ribs. 7.   The long, slender proteins in muscle cells respond to stimuli and cause the muscle cells to shorten and thicken. A great deal of energy in the form of ATP is needed for muscle contraction.

THOUGHT QUESTIONS   1.   Proteins help to build muscle tissue. 2.   Exercise helps to prevent tendon and ligament pulls and tears.

Chapter 40   The Ductless Glands

OVERVIEW   This chapter introduces the students to endocrine glands, which are the ductless glands that empty their secretions directly into the blood stream. These secretions are known collectively as hormones. It is pointed out that each hormone is specific in the work that it does monitoring some phase of cellular activity. The absence of a particular hormone is dramatized by the effect that it has on the body.

HINTS   One method of introducing the topic of ductless glands is to show the class pictures of people who have been affected by too little or too much of ductless gland secretion. A pituitary giant, a pituitary dwarf, a victim of acromegaly, or a person suffering from toxic goiter demonstrates the effect of hormones on the body. Use a commercial chart and the illustrations in this chapter to show the location of the ductless glands. It would be helpful to draw chalkboard diagrams to illustrate the difference in structure between a duct gland and a ductless gland. The films suggested in the A-V list help to make the information in the chapter more meaningful.

ACTIVITY SUGGESTIONS Materials needed are thyroid hormone tablets, two fish tanks, and ten tadpoles per group.

You can obtain the thyroid hormone needed from Carolina Biological Supply Company. You will receive directions on its dilution and use. **CAUTION: Carefully supervise the use of this hormone. Do not allow students to consume any or to remove it from the classroom.**

If tadpoles are not available from nearby ponds, you can purchase them from a biological supply house. The tadpole should be maintained in spring or pond water.

ANSWERS TO ACTIVITY QUESTIONS 1.   Tadpoles measure about 3 to 4 centimeters in length. They are fish-like, having fins and no limbs. 2.   The control tadpoles grow larger; the tail very gradually shortens. The experimentals do not increase in body length, but prematurely metamorphose. They lose the tail and grow limbs. They turn into very small frogs. 3.   The thyroxin stimulated the premature body changes.

ANSWERS TO QUESTIONS

WORD QUIZ 1. c   2. f   3. a   4. g   5. j   6. d   7. i   8. f   9. e   10. h

CHECK YOUR FACTS 1.   A ductless gland secretes hormones, emptying these directly into the blood stream. A duct gland pours its secretions into an organ. 2.   Ductless glands make hormones. 3.   Hormones are called chemical messengers because they stimulate other organs to work. 4.   Too much hormone causes overactivity of an organ or chemical process. Hyperthyroidism is caused by too much thyroxin; acromegaly by too much growth hormone. 5.   Too little hormone causes body disorders. Too little thyroxin causes hypothyroidism; too little growth hormone, dwarfism. 6.   In their drawings pupils should show the approximate locations of the endocrine glands.

THOUGHT QUESTIONS 1.   The monkey would experience disorders in the following systems: thyroid, reproductive, adrenal glands, pancreas; also, possible interference of growth of long bones. 2.   The bearded lady has too much male hormone.

Chapter 41   The Nervous System

OVERVIEW In general this chapter explains that the nervous system has a double function. It keeps the organism in contact with the external environment by integrating stimulus-response reactions. It also maintains the internal environment in steady-state control. The brain and central nervous system are the mediators between the external and internal environments. More specifically the chapter deals with the structure and function of the various parts of the nervous system. These include the parts of the brain, the spinal cord, and various types of neurons. Types of behavior are also discussed.

HINTS You may wish to introduce this chapter by showing the film *Behavior in Plants and Animals*. This is a short film that contrasts the difference between learned and unlearned behavior. You can demonstrate the structure of the vertebrate brain by using fresh and preserved specimens. You can also use a brain model that is dissectible. The illustrations in the chapter show types of nerve cells.

ACTIVITY SUGGESTIONS No special materials are needed for this activity.

You may wish to serve as timekeeper and have the class do this together. Students who are rapid learners and who are able to concentrate will have dotted fewer i's and crossed fewer t's than those who cannot concentrate as well. This exercise shows the strength of habit. However, pupils with poor writing skills who have not developed the habit of automatically crossing t's and dotting i's may do surprisingly well in this exercise.

As an *alternate activity* you can set up a nerve-muscle preparation from a freshly killed frog to demonstrate muscle response to nerve stimulation.

ANSWERS TO QUESTIONS

WORD QUIZ 1. cerebellum 2. stimulus 3. sensory 4. response 5. dendrites 6. nerve 7. cerebrum 8. medulla 9. axon 10. motor 11. reflex 12. end brushes 13. connecting

CHECK YOUR FACTS 1. A stimulus is a change in the environment. A response is a reaction to this change. 2. Behavior is the sum total of what an organism does. 3. A sensory nerve cell picks up the stimulus from the environment. A motor nerve cell transmits a nerve impulse to a muscle or a gland. Both nerve cells transmit impulses. 4. Pupil drawings should resemble those in chapter illustrations. 5. cerebrum, cerebellum, medulla 6. The spinal cord extends from the brain stem down the back. The spinal cord has nerves that radiate to all parts of the body. 7. Reflex and instinctive behavior are unlearned, inborn types of responses. The organism does not learn how to make these responses. These responses are inborn. 8. Memory, manual skills, and special motor skills, such as bicycle riding, are learned. 9. Intelligence is the ability of an organism to solve problems on the basis of reasoning.

THOUGHT QUESTIONS 1. Most habits are good: tying one's shoelaces, the skill in combing one's hair, penmanship, skating, etc., are examples of good habits. 2. The robin would not be able to maintain the balance necessary for flight.

CHAPTER 42 THE SENSE ORGANS

OVERVIEW This chapter points out that the sense organs keep us in contact with the external environment. The smell receptors in the nose and the taste buds in the tongue help us to smell and taste. The ears enable us to hear and permit the sense of balance. The eyes are light receptors that form images, making it possible for us to see. We feel pressure and pain because of the special receptors in the skin. These structures are all discussed in greater detail in the chapter.

HINTS   Without calling attention to it, open a bottle of ammonia in a far corner of the room. Note pupils' reactions as the odor reaches their nostrils. Elicit from the class the advantages of the sense of smell. Have pupils sit at their desks quietly. For two minutes they are to record every sound that they hear. Compare pupil listings.

ACTIVITY SUGGESTIONS   Materials needed are several cotton swabs, table sugar, salt, lemon juice, and quinine water (tonic).

Best results are obtained when the test substances sugar, salt, and lemon juice are in solution. Quinine should not be diluted. Between each test pupils should rinse their mouths with water.

As an *alternate activity* you can dissect a cow's eye. Show how the lens magnifies newsprint. You can also use glass rods (fire-polished ends) cooled in ice water to find the cold receptors in the back of the hand. Warmed rods (not too hot) can be used to find the heat receptors in the skin. **CAUTION: Do not allow students to handle glass rods that are too hot. Burns can result from this.**

ANSWERS TO ACTIVITY QUESTIONS   1.   The sweet receptors are at the front and sides of the tongue, extending inward about 5 millimeters on three sides. Salt receptors are in a narrow band around the front and sides of the tongue. These do not extend as far in as the sweet receptors. The receptors for sour taste occupy a very narrow band of space at the sides of the tongue. These are not found at the front or back of the tongue. The receptors for bitter taste are found only at the back of the tongue.

ANSWERS TO QUESTIONS

WORD QUIZ   1.   optic. 2.   auditory 3.   iris 4.   taste buds 5.   cornea 6.   cochlea 7.   eardrum 8.   retina 9.   semicircular canals 10.   pupil

CHECK YOUR FACTS   1.   Pressure, pain, and temperature receptors are embedded in the skin. They signal unusual conditions in the environment. 2.   sour, sweet, bitter, salty 3.   Outer ear collects sounds; middle ear passes on vibrations to inner ear, which transmits vibrations to auditory nerve. 4.   Lens forms an image; retina receives the image. The pupil regulates the amount of light that enters the eye. The cornea admits light into the eye. The optic nerve transmits the image from the retina to the brain. 5.   Nearsightedness is caused when the eyeball is too long and the image falls in front of the retina. Farsightedness is caused when the image falls behind the retina. These are corrected with concave lenses and convex lenses, respectively. 6.   Semicircular canals control balance.

THOUGHT QUESTIONS   1.   The cerebrum has to interpret whatever the sense organs perceive. 2.   They live their entire lives in darkness. The eye structures remain undeveloped.

CHAPTER 43   DISEASE

OVERVIEW   This chapter deals with infectious diseases. It explains that humans can become infected by pathogenic microorganisms and certain insect species. It is further explained that diseases are spread by contaminated water, food, and air. Emphasis is placed on the fact that although the body has natural defenses against certain pathogens, good health practices are our best protection against disease. Various diseases are discussed specifically with reference to cause, effect, cure, and prevention.

HINTS   Ask students about the illnesses they have had. Classify these as to their being infectious or noninfectious. It is helpful to use commercial slides to show pupils some infectious microorganisms. You may also wish to exhibit preserved specimens of parasitic worms, or to show a film about the ways in which the body is able to defend itself against disease. Have pupils report on ways in which their communities can improve sanitation.

ACTIVITY SUGGESTIONS   Materials needed are four test tubes (for each group), plugs of cotton, iodine, a tablespoon measure, dried lima or kidney beans, water, alcohol, and peroxide.

If the class is divided into groups of four, you can reduce the number of test tubes and dried beans that you need. Emphasize the value of Tube A as a control. Try to obtain denatured ethyl alcohol for this exercise. **CAUTION: Keep the handling of bacteria-laden test tubes to a minimum. Sterilize tubes in boiling water when finished.**

ANSWERS TO ACTIVITY QUESTIONS   1.   Tube A has the strongest odor of all. The strength of odors from the other tubes from strong to weak are as follows: Tube D, Tube C, Tube B. 2.   This odor means that bacteria are growing in the bean-water medium. 3.   Alcohol is the most active. 4.   Cotton plugs allow air to enter the tubes.

ANSWERS TO QUESTIONS

WORD QUIZ   Definitions will vary. They should approximate those given in the text.

CHECK YOUR FACTS   1.   infectious, deficiency, functional 2.   A parasite is an animal or plant that lives on or inside of a host organism and does harm to the host. 3.   Infectious organisms live on the body or inside of a host and they make the host sick. 4.   bacteria: tuberculosis; virus: measles; protozoan: giardiasis; fungus: ringworm; rickettsia: Rocky Mountain spotted fever 5.   Diseases may be spread through food, water, air. 6.   chemical treatment of the water supply and pasteurization of milk 7.   build antibodies 8.   developed a technique of using cowpox to immunize against smallpox 9.   Proper disposal of sewage prevents the outbreak of typhoid fever. 10.   Rodents that carry fleas have to be kept under control; mosquito-control methods used to kill off disease-carrying mosquitoes; meats have to be government inspected for tapeworm cysts; other methods as suggested by pupils. 11.   The list drawn up by various classes will vary.

THOUGHT QUESTIONS 1. Germs are microorganisms. Worm parasites can be seen without the aid of a microscope. 2. Drink only bottled or boiled water. Do not drink milk. Eat meat that is very well cooked. Do not eat rare meat. Do not eat raw vegetables unless they are plunged into boiling water and peeled.

CHAPTER 44   DRUGS, ALCOHOL, AND TOBACCO

OVERVIEW This chapter clearly points out that the use of drugs for other than medical reasons often results in damage to body organs. It is explained that brain cells are most vulnerable to the destructive effects of drugs. The students are told that opium and its derivatives are addicting. This means that the body develops a tissue tolerance and requires an increasing amount of the drug. Barbiturates are also addicting. It is further explained that most of the other drugs are mood changers, stimulants, or depressants that temporarily alter the user's reactions. Alcohol abuse is discussed as a major problem among adults and a growing one among teenagers. The symptoms, consequences, and possible solutions for drug and alcohol abuse are discussed. The negative effects of cigarette smoking are enumerated also.

HINTS It can be useful to begin the discussion of drug abuse by showing a film of realistic situations. *The People Next Door*, CBS Playhouse Production, distributed by BFA, depicts the effects of teenage and adult drug abuse on two families. You may wish to invite speakers from Alcoholics Anonymous to talk about alcoholism. The American Cancer Society will provide speakers on the tobacco question. You can obtain posters for the classroom and pamphlets for distribution concerning drug abuse from government agencies such as the National Institute of Mental Health in Bethesda, Maryland.

ACTIVITY SUGGESTIONS Materials needed are twelve paper clips, a clock with a second hand, and a swivel chair.

The purpose of this exercise is to have pupils see firsthand how dizziness affects reaction time. People who have consumed too much alcohol or used other drugs may become dizzy and therefore uncoordinated in motor skills. Have one pupil become dizzy by turning around in a swivel chair. **CAUTION Have other pupils positioned around the subject to prevent the pupil's falling from the chair. Do not select a student who may have a physical problem that could be aggravated by dizziness.** After several spins in the chair, have the pupil carry out the exercise. Two or three pupils can serve as timekeepers.

ANSWERS TO ACTIVITY QUESTIONS 1. These tests show that motor skills are impaired by dizziness. 2. Since alcohol makes one dizzy, it decreases one's ability to do things. 3. Alcohol remains in the bloodstream until processed by the liver. While in the blood, it affects the brain.

ANSWERS TO QUESTIONS

WORD QUIZ 1. l 2. d 3. f 4. i 5. j 6. h 7. k 8. g 9. e 10. b 11. a 12. c

CHECK YOUR FACTS 1. A medicine is prescribed by a medical doctor to treat some specific illness. An illegal drug may not be formulated by reliable manufacturers and is used for reasons other than the treating of illness. 2. Drug abuse means the using of drugs and other substances for the purpose of inducing a "high." 3. A hallucinogen is a drug that induces fantasies. Hallucinations are those fantasies induced by certain drugs. 4. Some mind-altering drugs are marijuana, LSD, mescaline, methamphetamine, cocaine. Mind-altering means that the drug controls the person's senses. 5. Marijuana is harmful to the lungs, induces hallucinations in some, makes a person lethargic, and removes the will to accomplish; it is an intoxicant. 6. Volatile compounds destroy the lining of the nasal passages and the lungs. They depress the brain cells and damage the liver. 7. Stimulants arouse the sensory neurons excessively. They make a person hyperactive. The family of amphetamines has a large number of stimulants. 8. The nervous system is depressed by depressant drugs. Reactions in individuals are slowed down. The family of barbiturates and alcohol are depressants. 9. Alcohol is a depressant. It distorts the way in which messages are carried to the brain. Alcohol can destroy the liver. 10. Tobacco causes lung cancer, induces high blood pressure, causes heart disease, cancer of the bladder, etc. 11. There are programs run by government agencies and private sponsoring agencies that help in the recovery of drug addicts and drug abusers.

THOUGHT QUESTIONS 1. Cocaine destroys the linings of the nasal passages. It induces paranoia in users. It makes some people drug dependent. 2. Drugs may cause psychological dependence or tissue tolerance in users. 3. a. for the friend not to follow the example of drinking b. to encourage the parent to go to Alcoholics Anonymous or other similar facilities for help c. to join the chapter of AA for teenagers who have problem parents

CHAPTER 45   INVESTIGATING THE FROG

OVERVIEW This chapter is set up as a lab activity. However, if laboratory facilities are not available, the chapter can be treated just as the preceding chapters. The illustrations can be observed in place of the actual frog. Both the internal and external structures of the frog are examined.

HINTS Before beginning the frog dissection, you should prepare the class for it. This chapter is optional. If the community in which you teach frowns on the dissection of vertebrate animals, this chapter should be omitted. If the use of the frog for dissection does not pose a problem, your students will benefit from the investigation. They will be able to see firsthand how a vertebrate animal, which has some structures that are similar to humans, is put together and functions.

ACTIVITY SUGGESTIONS The materials needed are one battery jar (or any widemouth jar) for each frog, two glass rods for each group (Make sure that the ends are fire polished.), flexible wire, tiny meat cubes, vinegar, wax bottom pans or dissecting boards, scissors, scalpels, forceps, straight pins, string (if a board is used), and perhaps a microscope and clean slides, if you wish pupils to look at some of the tissues under the microscope.

ANSWERS TO QUESTIONS   A. 1.   The eyes push back into the sockets and a transparent membrane covers the eye from the bottom up. B. 2.   The tympanic membrane vibrates in response to sound. E. 3.   The submerged frog is able to obtain oxygen from the water through its skin. F. 4.   The hind limbs are longer than the forelimbs. 5.   There are five toes on each hind limb. 6.   The toes are connected by a webbing. 7.   The males are usually smaller and thinner with enlarged thumbs on the forelimbs. The females are "fatter" because the body cavity is filled with eggs. 8.   Frogs do not have claws or nails. 9. and 10.   The frog turns in the direction opposite to that which the battery jar is turned. 11.   The frog makes a scratching response when vinegar is applied to the thigh. 12.   The frog may flip its tongue outward. 13.   The small intestine is about 15 centimeters long. 14.   yellowish bodies 15.   The throat opens into the esophagus. 16.   You can feel the teeth.

A-V Materials   Unit 7

Alcohol: Choices for Handling It. Coronet, color, film. (44)
Biology Dissections: The Frog. Prentice-Hall, filmstrip. (45)
The Blood. Encyclopaedia Britannica, B/W–color, film. (36)
Drugs and Medicines Series. Coronet, each a color film. (44)
The Endocrine System. McGraw-Hill, filmstrip. (40)
Fossils: From Site to Museum. Coronet, color, film. (33)
Frogs: An Investigation. BFA, color, film. (45)
The Human Body: A series of films on body systems. Coronet, color, films. (34–39)
The Human Brain. Encyclopaedia Britannica, B/W, film. (41)
Infectious Diseases and Man-Made Defenses. Coronet, B/W–color, film. (43)
Infectious Diseases and Natural Body Defenses. Coronet, B/W–color, film. (43)
Smoking and Health. BFA color, film. (44)
Your Ears; Your Eyes. McGraw-Hill, B/W, films. (42)

Supplementary Reading   Unit 7

Adams, John. *Viruses and Colds: The Modern Plague*. New York: American Elsevier Publishing Co., 1967.
Arlin, Marian. *The Science of Nutrition*. New York: Macmillan, 1977.
Avoy, Donald R. *Blood: The River of Life*. The American Red Cross, 1976.
Ayars, Albert L., and Gail C. Milgram. *The Teenager and Alcohol*. New York: Richards Rosen Press, 1970.
Boettcher, Helmuth. *Wonder Drugs: A History of Antibiotics*. Philadelphia: J.B. Lippincott Co., 1963.
Brecher, Edward M. *Licit and Illicit Drugs*. New York: Consumers' Union, 1972.
Edwards, Gabrielle. *The Student Biologist Explores Drug Abuse*. New York: Richards Rosen Press, 1980.
Fleck, Henrietta. *Introduction to Nutrition*. New York: Macmillan, 1977.
Klein, Aaron E. *Trial by Fury: The Polio Vaccine Controversy*. New York: Charles Scribner's Sons, 1972.
Navarra, John G. *Drugs and Man*. Garden City: Doubleday, 1973.
Street, Leroy. *I Was A Drug Addict*. New York: Arlington House, 1973.

Living Things

Harold E. Teter
Gabrielle I. Edwards
Frederick L. Fitzpatrick
Thomas D. Bain

HOLT, RINEHART AND WINSTON, PUBLISHERS
New York • Toronto • Mexico City • London • Sydney • Tokyo

AUTHORS

HAROLD E. TETER Formerly Head of the Science Department at Cooley High School, Detroit, Michigan

GABRIELLE I. EDWARDS Assistant Principal, Supervision, Biology and General Science, Franklin Delano Roosevelt High School, Brooklyn, New York

FREDERICK L. FITZPATRICK Late Professor Emeritus of Natural Sciences and Head of the Department of Science Education at Teachers' College, Columbia University, New York, New York

THOMAS D. BAIN Formerly a teacher of biology at Harding High School, Marion, Ohio

Photo credits for unit and chapter openers appear on page viii; all other photo credits appear in the captions.

Cover photo by John S. Flannery/Bruce Coleman

The animal on the cover is a cougar (*Felis concolor*), also called mountain lion, panther, or puma. Cougars were once common throughout North and South America. Today, they are found mainly in low-population areas of the Southwest, the Rocky Mountains, and western Canada. They are widely hunted for sport and by ranchers who believe that cougars are responsible for killing livestock. As a result, the cougar is now classified as an endangered species.

Literary acknowledgments appear with the materials used.

Copyright © 1985 by Holt, Rinehart and Winston, publishers
Copyright © 1981, 1977, 1970, 1966, by Holt, Rinehart and Winston, Publishers
All Rights Reserved
Printed in the United States of America
ISBN 0-03-071909-7

456789 039 987654321

ACKNOWLEDGMENTS

The authors wish to express their appreciation to the following critics who read parts of the manuscript and offered assistance by their helpful criticism.

Dr. K. Fred Curtis, Ed.D., Associate Professor, Chairman, Department of Professional Field Experiences, School of Education, Baylor University, Waco, Texas

Robert Friedel, Science Department Chairman, Riverview School, Silver Lake, Wisconsin

Michael Gerald Savage, Biology Teacher, North Rose-Wolcott Central School, Wolcott, New York

Virginia A. Way, Director, Office of Resource Education, Colorado School of Mines, Golden, Colorado

Dr. Sandra Z. Winicur, Associate Professor of Biology, Indiana University at South Bend, South Bend, Indiana

Preface for the Student

You could say that you are about to begin the study of life science, the study of *Living Things*. But think for a moment. Are you really only *beginning* the study of life science? Haven't you always been studying life science? Aren't you, in fact, a part of life science? You are, after all, a living thing, one of the living things you are going to learn more about in this text. And you have, whether you are aware of it or not, been studying the other living things around you from the time you were born. That is how you learned to walk and to talk and that birds fly, fish swim, and leaves change color and fall off the trees in autumn. You know that dogs are living things and that automobiles aren't, and you know that you are able to do things that other animals cannot do.

All of these things you have learned by observing. You have observed with your eyes, your ears, and your other senses. What you could not learn through observation you probably questioned, and you learned from the answers you received. Or perhaps you learned answers by doing some investigation on your own. By doing all of these things you were acting as a scientist does, and most of the information you were collecting was about life science.

So you see, you are not about to *begin* to study life science, but you are about to begin to organize and to build on the information you already have.

This text will provide you with new information and it will teach you new ways to investigate on your own. As you study this text, you will learn answers to questions you could not answer before and you will learn to ask questions you could not ask before. For example, you will learn how a water environment can become land, how starfish eat by pushing their stomachs out of their bodies, how a shark can drown, and how humans can sometimes grow to be 2½ meters tall.

Before you begin reading about these and the many other interesting things in this text, look at pages 473 to 478 of the appendix. The information there will explain how this book is organized and will help you to learn to use it more effectively. Then you can begin to enjoy further your study of *Living Things*.

CONTENTS

Unit 4 Classification and Simple Forms of Life   173

Unit 5 The Plant Kingdom   217

Unit 6 The Animal Kingdom   249

Unit 7 Human Biology 331

Appendix 473

Glossary 481

Index 491

Photo Credits

Art Credits

UNIT I The Principles of Biology

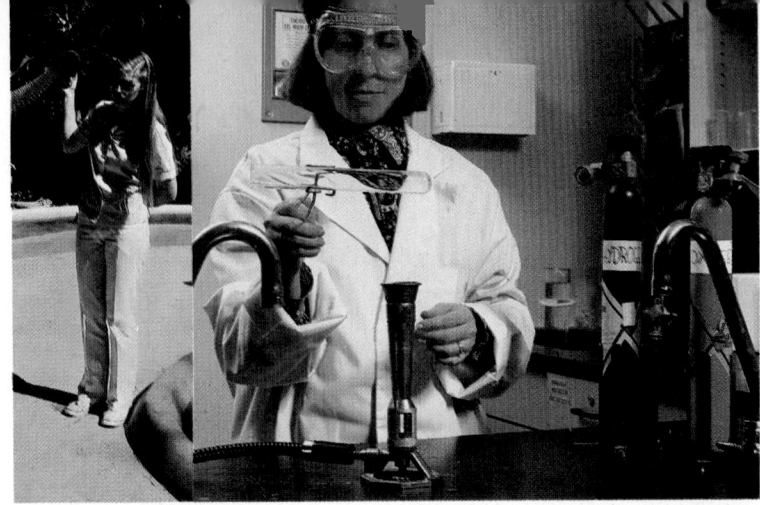

CHAPTER 1 WHAT IS SCIENCE?

Do you know what science is? All of the people in the photographs above are scientists. Look closely at the photographs. Can you tell by looking at them what a scientist does? It is difficult to tell just from a photograph. As you read this chapter you will begin to find out much more about what science is and what a scientist does.

WHAT IS A SCIENTIST?

The scientist's job. Suppose you want to know the answer to a question. Where do you go to get the answer? You probably would try asking someone who you think might know the answer. This can get answers for many of your questions, but not for all of them. If no one you know can give you the answer you may decide to go to a library. You could go to your neighborhood library, your school library, or a special science library. In all of these places you will find librarians who are glad to help you. In this way you can find the answers to most of your questions.

Once in a while your attempts to find the answer may fail. The answer you seek is not in any book nor is it known by any of your friends. No one seems to know the answer to your question.

At this point you may feel you have to give up. However, the scientist's work begins when they cannot find the answer to a question in any book. The job of scientists is to search for new information. Scientists are in the business of increasing human knowledge. They are trying to understand and to describe the universe in which we live. The work of scientists may be useful to us in some particular way, or it may merely increase our understanding of the world around us.

We have begun to answer our question of what a scientist's job is and what a scientist does. Before we find out more about the work of scientists, let's see what science is. It is several things. *First*, it is the activity of all scientists as they seek new knowledge. *Second*, it is the whole mass of knowledge already discovered by scientists. *Third*, it attempts to explain what newly discovered facts mean. Often new facts lead to new understandings.

Where does the scientist work? Scientists do their work in many places. Since the time of the early scientists, much of the best scientific work has been done in the science departments of universities. These schools not only carry on scientific research, but train new young scientists. A country that does not have good schools cannot keep up its scientific progress.

Another place where scientists work is in industry. All of the large manufacturing companies hire scientists. The scientist in a university often does basic research to discover new facts. Many scientists in industry apply these facts to the development of new products. The government also hires scientists of many kinds, both for basic research and for applied work. The Department of Defense, the Department of Agriculture, and the Department of Health, Education and Welfare all hire scientists.

The tools of a scientist. Scientists are always trying to find the answers to new questions. Scientists cannot always depend on their senses to give them the right answers. Look at Fig. 1-1. Which is longer, line AB or line CD? How can you make sure? You would have to use a means other than sight to be sure of the length of the lines in Fig. 1-1. Scientists also need tools to help with the sorting out of facts. They use rulers for measuring length, and scales for weighing. Microscopes are needed to see very small things and telescopes to study distant objects. A thermometer is used to measure temperature and a clock to measure time. A compass tells the direction, while a speedometer measures the rate at which an automobile moves. These instruments and others provide reliable information. All of these tools are useful to scientists and to students. The reason

Careers: Emphasize the career opportunities in the different fields of science.

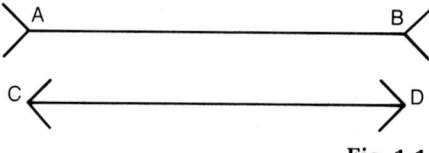

Fig. 1-1
Which line is longer?

Fig. 1-2
Measurements are important to all scientists. This one works in agriculture. (HRW photo by Paul Gignac)

3

References: *Tools of Modern Biology* by Melvin Berger and *The Changing Tools of Science: From Yardstick to Cyclotron* by Irving Adler.

hypothesis

experiments

Fig. 1-3
An artist's idea of Galileo dropping the two iron balls from the Leaning Tower of Pisa. (The Bettman Archive)

they are useful is because they measure certain things. Measurement requires numbers. Numbers help us to state facts accurately. Scientists all over the world make measurements in *metric* units. This makes it easier for these scientists to exchange information with each other.

Some early experiments. Scientists start with some sort of guess as to what the answer to a problem may be. Then they set out to test this guess and prove whether it is true or false. A scientific guess is called an **hypothesis** (hi-POTH-uh-sis).

Scientists often use **experiments.** An experiment is an activity used by a scientist to test an hypothesis. A famous experiment took place in Italy nearly 400 years ago. An Italian named Galileo (gal-i-LAY-oh) made a statement about how objects fall. He said that all objects dropped from the same height fall at the same speed. Many very important people argued with him. They said heavy objects fall faster than lighter ones. They showed him books to prove they were right. Galileo took two iron balls. One weighed several times as much as the other. He dropped them from the Leaning Tower of Pisa at the same time. They fell side by side and hit the ground together. Both balls had fallen at the same speed. This was just what Galileo had said would happen. Others tried the same kind of experiment and got the same result. They were forced to agree with Galileo. See Fig. 1-3.

Galileo's experiment was simple. It was also very important. It made educated people all over the world take notice. Here was a firsthand way of finding answers! If they wanted to know about something they could do an experiment to find the answer. They did not need to depend only on books and opinions. Galileo was not the first scientist, but he did a great deal to advance science. He showed people how to use experiments and observations.

Another very famous experiment took place in 1881. A Frenchman named Louis Pasteur had developed a new vaccine. He thought the vaccine would keep sheep from getting a disease called anthrax. How could he be sure? Here is how he set up his experiment.

Pasteur put 24 sheep from a flock in one pen and 24 sheep from the same flock in another pen. The sheep in one pen were given shots of the new vaccine. The sheep in the other pen were not vaccinated. Otherwise both pens of sheep were treated exactly alike. Later all the animals in both pens were given anthrax germs to see if they would get sick. All the vaccinated sheep remained healthy. All the unvaccinated sheep died.

The important thing to notice in this experiment is the way Pasteur used the two groups of sheep. They were as alike as two groups of sheep can be. They were treated exactly alike

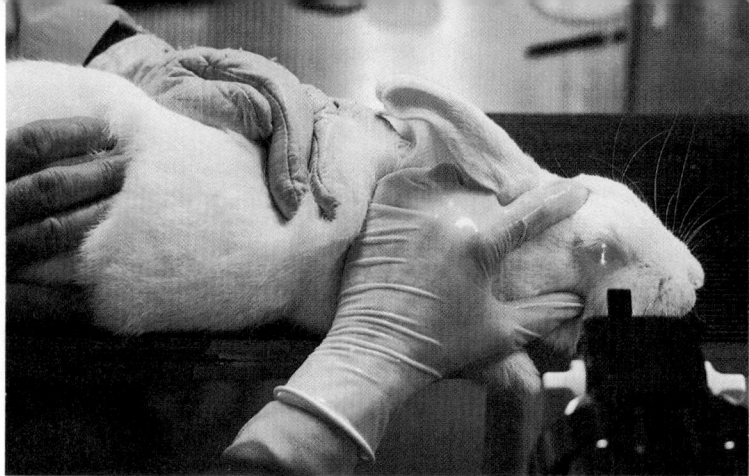

Fig. 1-4
Modern day scientists also experiment with animals, as Pasteur did. (Dan McCoy/Rainbow)

except for one thing—only one group was vaccinated. This was the *experimental group*. Since vaccination was the only thing that was different, it must have been the reason why the experimental group lived. The other sheep were the *control group*. They showed what would have happened to the experimental group if they had not been vaccinated. Using controls is important in many experiments.

WHAT IS SCIENCE?

The scientific method. Scientific method is the basic set of steps that scientists use in getting information. These steps are as follows:

Reference: For brief and interesting accounts of Galileo and Pasteur see *Giants of Science* by Phillip Cane.

Assignment: Have students look for science news articles from magazines and newspapers.

1. First, they get clearly in mind what it is that they want to find out.
2. Next, they use books and magazines to find out what is known.
3. Then they make the best hypothesis they can as to just what they believe the answer may be. Perhaps they can only guess that certain activities may lead to the answer.
4. Then they perform experiments or make observations to see if their guess is correct. If not, they make a new guess and try again. The outcome of these experiments or observations are their *results.*
5. Now, they try to decide what the results prove. In other words, they draw *conclusions.* They try to be honest about this. They will not claim that their conclusions are facts. They will simply say that their conclusions are what they think the facts mean.
6. After seeing the results of their experiments and coming to a conclusion, the scientists will retest their hypothesis. This is done by repeating their experiment many times. The scientists may even try new experiments that will test the same hypothesis. If the scientists come up with the same results every time they do their experiment, they may have good reason to believe their conclusions may become facts.

results

conclusions

Enrichment: Stress the difference between observation and conclusion.

Enrichment: Show examples of magazines such as *Science, Scientific American,* etc.

7. Finally, scientists write a description of their work and have it published. There are many special magazines which contain only the results of scientific research. These magazines are read by other scientists all over the world. They use new discoveries to help them make more discoveries.

It is not always necessary for a scientist to go through all of the steps above to find the answer to a question. Scientists can find the answers to some of their questions in other ways. One other way to find some answers is by *observation.* If scientists want to know what pheasants eat they may examine contents of the stomachs of dead pheasants to see what the birds have eaten. They may watch the pheasants hunt for food. They will do both of these things at different times of the year in case the birds' eating habits change. They will keep careful records of all this and, when finished, they will know a great deal about the eating habits of pheasants. Many problems in biology can be solved by observation, but many cannot be. It is up to the scientists to decide the best way to try to solve the problems they are working on.

Limitations of science. Scientific methods can only solve questions that have definite answers. "What is the center of the earth made of?" "Why is steel stronger than lead?" "How does a muscle get energy from food to do work?" Questions like these have definite answers and can be solved. Scientists are often successful at solving such problems. They may develop experiments and repeat them to see if their discoveries are really correct. It is possible to get nearly everyone to agree on the true answer.

But what about questions without such definite answers? "What is the best form of government?" "What type of music is the best?" "Who is the greatest actor or actress in the movies?" These questions do not have such definite answers, do they? Answering them must include a great deal of personal judgment. You may have some very strong feelings about them, but other people may not agree with your opinions. There are no observations or experiments that will make everyone agree on the answers.

So you can see that science does not claim to answer all questions. There are some areas that affect our lives that are outside the field of science. Science does help us understand the living and nonliving world around us. Science even includes the workings of our own bodies. It has been very successful at this, and it continues to make important discoveries.

What can science mean to you? Most scientists are professionals. They have been trained in college. They make their living as scientists. There are also some amateur scientists.

Remember, a scientist is anyone who uses scientific methods to learn new facts. This could be you. Young people your age have built their own telescopes to look at the stars. Others have studied the plants and animals that live in their neighborhoods. Insect study could give an amateur scientist a good opportunity to do scientific work. There are so many kinds of insects that no one has studied how all of them live, grow, and function.

Even if you never become a scientist, you can learn to think like one. You can learn to see your problems clearly before you try to solve them. You can check all the known facts before you act on a problem. You can try different ways of solving a problem to see what works best, as a scientist does when experimenting. You can learn to accept new ideas and give up old ones when new facts make this necessary. This approach is part of what is called the *scientific attitude*. If you develop a scientific attitude toward your everyday problems, your study of science will be worthwhile.

The branches of science. It is possible to divide science into several different fields of study.

Biology is the study of living things. *Chemistry* is the study of what materials are made of and how they are put together. *Physics* includes the study of energy in its various forms—such as electricity, light, heat, and mechanics. *Astronomy* deals with the sun, the planets, and the stars. *Geology* is the study of the earth itself—things like rocks, hills, valleys, and volcanoes.

biology

chemistry

physics

astronomy

geology

Fig. 1-5
What field of biology do you think this scientist works in? (HRW photo by Russell Dian)

anatomy

botany

ecology

physiology

zoology

Assignment: Have students report on biologists such as Jonas Salk, Walter Reed, Alexander Fleming.

What is biology? As we have said, biology is the branch of science that deals with living things.

Men and women who work in the science of biology are called biologists. Because there is so much to know about living things, the field of biology is divided into many branches. *Anatomy* is the study of the structure of living things. *Botany* is the science of plants. *Ecology* is the study of living things in relation to their surroundings. Heredity is the branch of biology that explains why offspring look like their parents. *Physiology* deals with the life activities of plants and animals. *Zoology* is the science of animals.

The work of the biologist is important to life in a modern society. Biologists have learned a great deal about plants and animals, how they live, how their bodies work, and how they affect one another. The whole field of medicine is really just a special branch of biology. A better understanding of biology will help you to take good care of your body.

Biology has helped to improve farming. Better breeds of farm animals and food plants have been produced. Better ways of fertilizing crops have been developed. A good farmer must understand biology. Even the home gardener can make good use of biological knowledge.

Conservation is a field in which biology plays an important role. Soil, water, forests, and wildlife can be destroyed if we are not careful. Biologists learn how to use these resources wisely.

Nature study is a pleasant hobby for many people. Perhaps you have gone out on hikes or camping trips. Perhaps you enjoy hunting or fishing. You will enjoy them more if you know something about the living things around you. Many biologists spend their time finding out how plants and animals live and grow. You can learn about nature study from biology books; not just textbooks like this one, but also from many kinds of field guides and nature study handbooks.

Science has also produced knowledge and inventions to make our work easier. We like the practical, useful things in

Fig. 1-6
Investigating isn't always done inside a laboratory. (HRW photo by Ken Karp)

our modern, scientific world. This book will help you understand some of the useful things that scientists have produced.
It will do something else, too. Human beings are curious creatures. We are always asking why things are the way they are.
This book will help to satisfy this human need. It will help you
to understand yourself and also to appreciate this wonderful
world of which we are all a vital part.

SUMMARY

A scientist is a man or woman who works to solve problems.
A scientist can solve those problems that have definite answers.
Because the field of science is so broad, there are many different
areas of study. Scientists often do their work in steps that are
known as the scientific method. Measurement and numbers are
necessary to the scientist in the search for accurate answers.
A very important part of the scientist's work is the publishing
of facts that are found. Biology is that branch of science that
deals with living things.

ACTIVITY

Using the Metric Ruler
Look at the ruler that is drawn at the bottom of this page. This ruler is divided into metric units. Locate line A and line B. The distance between these lines is one centimeter. The abbreviation for centimeter is cm. Count the number of smaller divisions between A and B.

1. How many are there?
Each of these divisions is one millimeter. The abbreviation for millimeter is mm. Now count the number of millimeters between line C and line D.

2. How many are there?

3. How many millimeters are in one centimeter?

4. Why is the distance between A and B equal to the distance between C and D?

5. How many centimeters are between A and C?

6. How many millimeters are between A and D?

7. What is the distance in millimeters between A and E?

8. How many centimeters are between A and E?

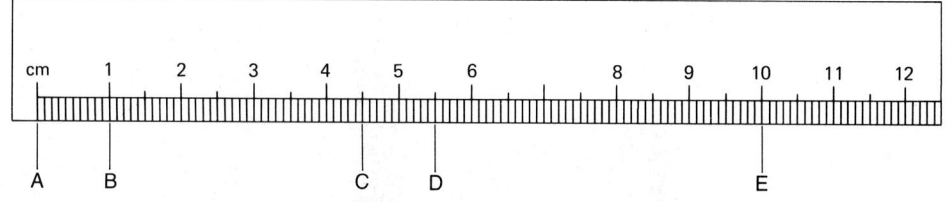

Fig. 1-7

Word Quiz

1. Complete each of the following statements: Do not write in this book.
   a. Biology is the study of _____.
   b. Chemistry is the study of _____.
   c. Physics is the study of _____.
   d. Astronomy is the study of _____.
   e. Geology is the study of _____.

2. Explain the difference between
   a. botany and ecology;
   b. zoology and physiology;
   c. anatomy and physiology.

Check Your Facts

1. What is science?

2. What is a scientist's job?

3. Where do scientists work? Where might you find scientists at work nearest to your school?

4. Of what use are experiments to scientists?

5. How are each of the words that follow related to an experiment?
   a. hypothesis   b. result   c. conclusion

Thought Questions

1. Look through your local newspaper.
   a. Make a list of present-day problems that might be solved by scientists.
   b. List those problems that cannot be answered by science.

2. What are some scientific attitudes that we can use in everyday life?

3. Why is it important for all scientists to use metric measurements?

4. Make a list of hobbies that require a knowledge of biology. Use magazines, books, and newspapers to help you. Which of these hobbies interests you the most? Why?

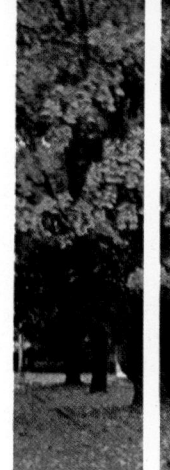

CHAPTER 2 The Basic Unit of Life

Look at the photograph above. Can you decide which things are living and which are not? As you read these pages, you will find out about the requirements of living things.

How can you tell when things are alive? A living thing usually has a definite size and shape, but this is not enough to prove it is alive. A statue has definite size and shape. Is it alive? Movement often helps us to tell if things are alive. But some nonliving things move also. A kite flies through the air, but it is not alive. Animals and plants are living things. They carry out activities that set them apart from things that have no life. These life activities allow living things to function in ways that nonliving things cannot.

LIFE ACTIVITIES

Movement is just one of the many activities that living things can carry on. Most animals move from place to place. Many walk, some fly, while others swim. Moving from one place to another is called *locomotion*. Plants do not have locomotion, but there is movement inside their living substance.

Response is another life activity. If a cat sees a mouse, the cat jumps at the mouse. If a potted plant is in a window, the leaves grow toward the light. If you see a car coming toward you, you get out of the way. These are examples of responses. This ability of living things to respond to their surroundings helps them to avoid danger and to get food.

Reproduction is another life activity. Each kind of living thing has some way of reproducing. No one animal or plant can live forever. The ability of living things to make more of their kind has made it possible for life to continue on earth for billions of years.

Four other important life activities are carried on by living

After you read this chapter, you should be able to:

—**List** and **Define** life activities

—**Describe** the appearance of protoplasm

—**Explain** why protoplasm is important

—**Name** the parts of a plant cell

—**Name** the parts of an animal cell

—**Explain** the function of each part of the cell

movement

response

reproduction

nutrition

respiration

growth

excretion

protoplasm

things. These are nutrition, respiration, growth, and excretion.

Nutrition is the taking in and using of food. One use for food is to supply energy. When living things release energy from food, the process is called *respiration.* There are several forms of respiration, but in the most common one oxygen is needed. You breathe in order to get oxygen from the air. This oxygen helps release energy from food. Breathing in oxygen makes it possible for you to carry on respiration.

Another use for food is in *growth.* When living things grow, they first break food down into simple chemicals. Then they build these chemicals into more complicated materials until these materials become a part of their own living substance.

During the processes of respiration and growth, there are usually wastes left over that the body does not need. Getting rid of these wastes is called *excretion.*

The process of being alive is simply a matter of carrying on these life activities. Each kind of living thing must be able to respond to its surroundings by movement of some kind, reproduce others like itself, get food, grow, and get rid of waste materials. It takes energy to do all of these things. This makes respiration an especially important life activity.

There is no single "right" way to list these life activities. We have listed seven of them. Some biologists add others. Some lump them together to get a shorter list. The important idea is that living things carry on a number of activities that help them to survive. This is what makes them different from nonliving things.

WHAT ARE LIVING THINGS MADE OF?

If living things are so different from things that are not alive, they must be made of something very special. This special material is called *protoplasm.* The activity of this protoplasm gives organisms the ability to do the things you have just read about. You are alive, so you must be made of protoplasm.

All plants and animals contain protoplasm. But it is a little different in each living thing. Sometimes it is thin like water. At other times it may be thick, like syrup. It may have color but usually it is clear and colorless. It feels slippery like raw egg white. Even protoplasm from different parts of the same body may be different.

We have said that you are made of protoplasm. If you are just a big lump of this slippery, jelly-like stuff, what keeps you from sliding off your chair? How do you hold your shape? To answer this, think first about air. Air has no particular shape. But if air is used to blow up a basketball, the air takes the shape of the basketball. The basketball is like a round package of air with a solid cover around it.

In much the same way, protoplasm comes wrapped in small packages. Each package has enough pressure inside its cover to give some firmness. So you are not just one large lump of protoplasm. You are a large mass of little lumps, all bound together.

Cells—units of protoplasm. The little packages of protoplasm we have been talking about are called *cells*. Protoplasm is usually organized in the form of cells. Living things are made up of cells. A bird's egg is a cell that you can see without a microscope. But most cells are much too small to be seen with the naked eye. If you lined up some of the smallest cells next to each other, you could fit one thousand of them in the space of one millimeter!

Some living things are made up of just one cell. Others are made up of many cells growing together. Your body contains

Enrichment: The name "cell" was first used by Robert Hooke to describe the microscopic "little rooms" that he observed in cork cells.

cells

Fig. 2-1
Left: Diagram of an animal cell; Right: Diagram of a plant cell. Both are drawn as they would appear through a light microscope.

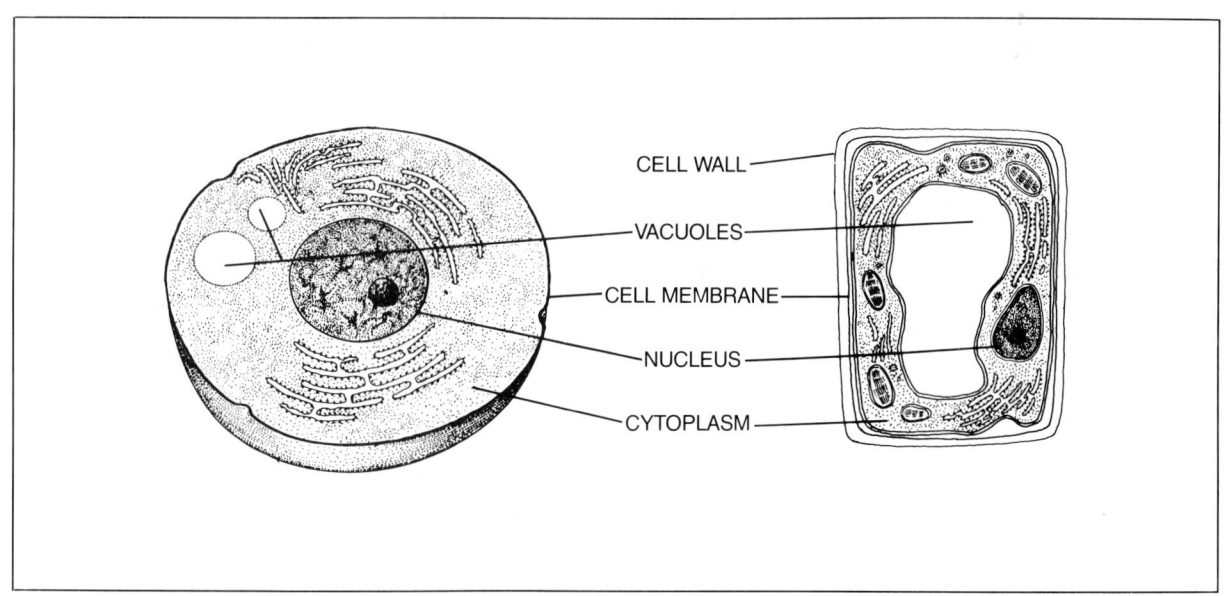

CELL WALL

VACUOLES

CELL MEMBRANE

NUCLEUS

CYTOPLASM

trillions of cells. One droplet of your blood the size of a pinhead contains millions of living cells, with room in between them for liquid. Look at Fig. 2-1. It shows two cells as you would see them through a light microscope. As the drawings show, these cells have thickness. They are not flat as they might seem to be in some drawings. One cell is a type you might find in an animal. The other is a type of plant cell. As you see, the differences are slight. The important parts of a cell are the same in all living things.

There is a round object inside the cell which is called the *nucleus* (NEW-klee-us); plural, nuclei (NEW-klee-eye). This is

Activity: Set up a microscope demonstration of cork.

Science Reading Skill: Have students compare and contrast the two diagrams.

nucleus

a very important part of the cell because it controls most of the cell's activities. We shall learn more about the nucleus later on.

You can see in Fig. 2-1 that the main mass of the cell is the **cytoplasm** (SY-toh-plaz-um). In most cells cytoplasm is clear and colorless, but some cells have colored cytoplasm. Often it looks

cytoplasm

Word Study: "cyto" means cell

Fig. 2-2
The electron microscope allows scientists to see very tiny structures inside the cell. (Ken Lax)

Enrichment: Living material must be killed and prepared before observing it with an electron microscope. In 1980, electron microscopes cost between $50,000 to $225,000. The median cost of a microscope for observing biological material is $150,000.

mitochondria

Reference: For more information see *The Electron Microscope: A Tool of Discovery* by Aaron E. Klein.

Question: More mitochondria are found in some cells, such as muscle cells. Ask students why this might be true.

ribosomes

Enrichment: You may want to introduce the term endoplasmic reticulum.

Enrichment: Other structures such as the lysosome and the Golgi apparatus can be discussed.

quite smooth, but in other cells it has a grainy appearance. Much of the cell's work takes place in the cytoplasm.

There are many special structures in the cytoplasm to do special jobs. Some of these special structures are so small that they can be seen clearly only through an electron microscope. Fig. 2-2 shows an electron microscope. It looks much different from an ordinary light microscope. One example of special structures in the cytoplasm are the **mitochondria** (mite-oh-KON-dree-ah). They are small, oval objects scattered throughout the cytoplasm. Their job is to carry on the life activity of respiration. The mitochondria are called the "powerhouses of the cell" because food and oxygen combine to produce energy inside them.

The way a cell appears through an electron microscope is shown in Fig. 2-3. You can see that the mitochondria have structures inside them. You can see also the *canals* that wind through the cytoplasm. Materials move from place to place in the cell through these canals. As you study Fig. 2-3, you notice that some of these canals are dotted with many small, rounded bodies. These are called **ribosomes** (RIBE-oh-sohms). Ribosomes manufacture chemical substances called *protein* (PROH-teen). The cell is made partly of protein. In Chapter 4 you will learn more about the production of proteins and several other kinds of chemical materials in the cell.

There are other special structures found in the cytoplasm. Research is going on to learn what each of these structures do. Remember that cytoplasm looks much like jelly. It has surprised everyone to learn that it has so many detailed structures.

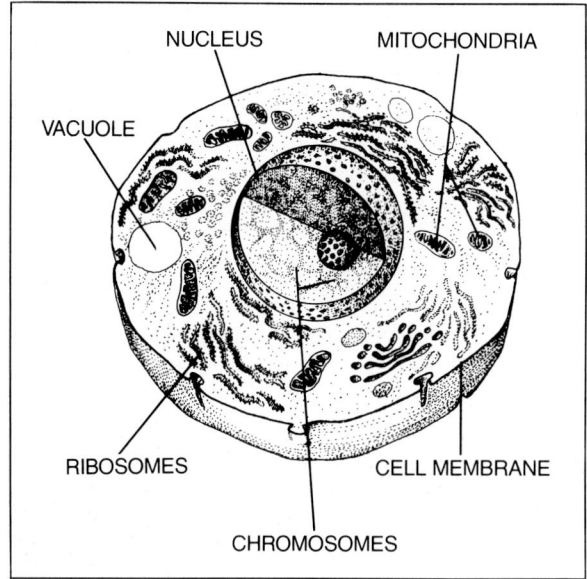

NUCLEUS   MITOCHONDRIA
VACUOLE
RIBOSOMES   CELL MEMBRANE
CHROMOSOMES

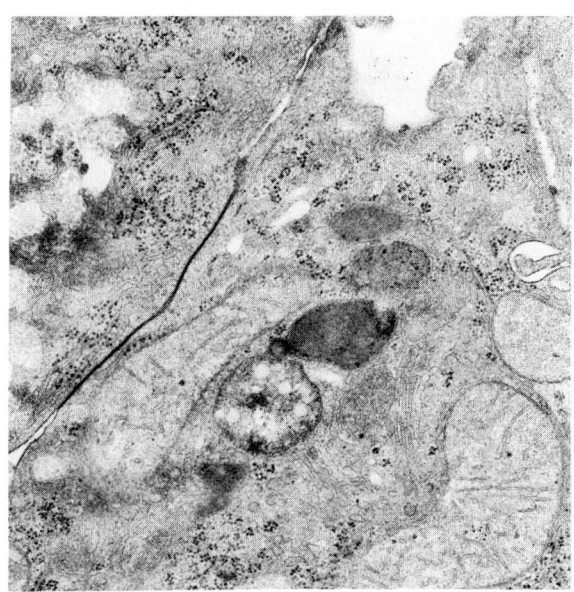

Most cells have one or more hollow structures in their cytoplasm called *vacuoles* (VAK-yoo-ohls). You might call these the cell's storage bins. Vacuoles may contain almost any stored material that the cell needs to keep for a while. Water, salts, foods, and waste materials are some of the things that are often stored in vacuoles. Usually plants need to store more water than animals, so plant cells have larger vacuoles than animal cells.

Around the outside of the cell is a very thin layer of material. It is called the **cell membrane.** The word membrane means a thin, sheetlike structure. Other membranes surround the nucleus, the vacuoles, and the mitochondria. The cell membrane separates the living material inside the cell from the nonliving substances around it. Anything that enters or leaves the cell must pass through this cell membrane. Some things can cross the cell membrane. Other materials cannot. The membrane has quite a bit of control over what enters and leaves the cell.

For now, it might be helpful to compare the cell to a factory. The nucleus is the cell's front office, controlling all of its activities. The cytoplasm is the workshop where some substances are built up and others are taken apart. The cell membrane is the gatekeeper, directing traffic in to and out of the cell.

Plant cells usually produce a layer of strong nonliving material around themselves, outside the cell membrane. This structure is called the **cell wall,** and it gives strength and protection to the cell. Wood is merely a mass of old empty cell walls. They were formed by the cells of the tree when that part of the tree was alive.

Fig. 2-3
A diagram (left) and a photograph (right) of an animal cell as seen through the electron microscope. (Dr. Guido Zampighi)

cell membrane

Reference: For excellent photographs see *National Geographic,* Sept. 1976 "Exploring the New Biology: Awesome Worlds Within a Cell."

Enrichment: The cells first observed by Hooke were actually the cell walls of dead plant cells.

cell wall

Animal cells are more flexible than plant cells because animal cells do not produce cell walls. The cytoplasm in animal cells is somewhat different from the cytoplasm in plant cells. Animal cytoplasm cannot manufacture the same kind of material that forms cell walls in plants. All animal cells are covered by a cell membrane, just as are all plant cells. Some animal cells form strong nonliving layers around themselves that give strength to the mass of cells.

Now you should know that protoplasm is not a single material that is always the same. The term *protoplasm* is simply a handy word for all living material. This living material is really a great many different materials organized to work together. The nucleus, cytoplasm, and membranes are all protoplasm. Structures such as cell walls are not living parts of the cell. This means that cell walls are not part of the protoplasm. Cells are the organized working units of protoplasm, so the study of protoplasm must always include the study of cells.

The life activities of organisms are caused by the activities of protoplasm. These activities are carried out by chemical action. You will need to know some chemistry to understand them. The next chapter will give you some of this simple chemical knowledge.

SUMMARY

Living things are able to carry on certain activities that nonliving objects cannot. These life activities make it possible for plants and animals to exist. Response to changes in the environment is one such requirement for life. Animals may respond by going from place to place. Plants respond by movement also. Their stems and leaves turn toward the sun.

All living things need nutrition, which supplies them with energy and the materials for growth. To continue in existence, plants and animals must have the means for reproduction. They need to produce more of their kind. Energy for life activities is obtained by a special and rather complicated process known as respiration. The wastes of energy production are released from the body through excretion. Living things are made of cells that are packages of protoplasm bounded by cell membranes. Within each cell are smaller structures that control certain life activities. The nucleus directs the activities of the cell. Mitochondria control respiration. Ribosomes help make proteins. All of the living parts of the cell together are known as protoplasm.

ACTIVITY I

Using the Microscope

Before you try to use a microscope learn the names of its parts shown in Fig. 2-4. This will help you to understand directions when your teacher tells you how to use it. Your microscope may look a bit different from the one in the picture but it will have mostly the same parts. Follow all of the directions on pages 17 to 18 very carefully. A microscope is an expensive instrument. *It must be handled with care.*

Fig. 2-4

The parts of the light microscope.

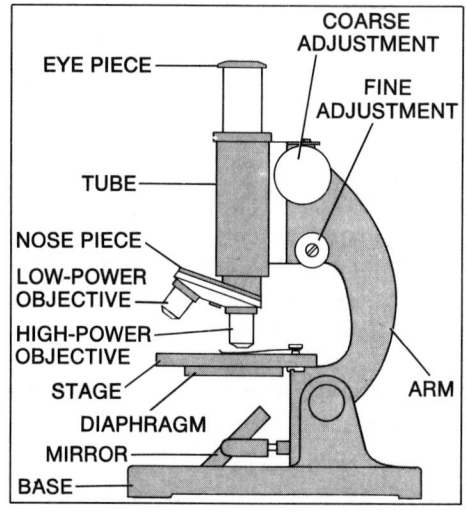

A. Carrying. Always carry a microscope in an upright position. Otherwise the eyepiece may fall out and break. Use a firm grip on the arm to carry it. Place your other hand under the base. Hold the microscope close to your body. **Do not place the microscope on the corner of a table where it may get knocked off.**

B. Mounting. Only very thin specimens can be viewed under the microscope. Most school microscopes give magnification of 100 times on low power and 430 times on high power. This is too much magnification for looking at fingers, pencils, or insects. For such work you need a lower power, such as you might get from a good magnifying lens.

1. What is the magnification given by your microscope on low power? On high power?

Objects viewed under the microscope must be placed on a small piece of glass called a *microscope slide*. These slides are nearly always covered with a still smaller piece of thin glass called a *coverslip*. At times your teacher may have you look at *prepared slides*. In these, the object that is to be studied has been permanently mounted on the slide with the coverslip cemented over the top.

Often you will use *temporary mounts*. To make a temporary mount you simply place the specimen in a drop of water on the slide and cover it with a coverslip. These are also known as *wet mounts* because water is used. In the activity on page 19 wet mounts are used for viewing the onion cells and cheek cells. When you place a coverslip on the drop of water, lower it slowly from the side like closing a trap door. See Fig. 2-5. This gives the air a chance to escape and there will not be so many bubbles caught under the coverslip.

Fig. 2-5

The coverslip should be lowered at an angle.

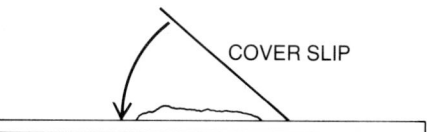

COVER SLIP

Activity I (Continued)

C. Focusing. When your slide is ready, place it on the *stage* of the microscope. The part you wish to see should be centered over the hole in the middle of the stage. Swing the *nosepiece* around so that the *low power* objective is in line with the tube. (This is the shorter objective.) Look through the *eyepiece* and adjust the mirror so that it reflects light up through the microscope.

Next, move your head to one side so that you can see the objective. Turn the *coarse adjustment wheel* until the low power objective nearly touches the coverslip. Then look through the eyepiece and slowly raise the objective until the object on the slide comes into focus. You may have to move the slide to get the exact view you wish to see. Only a small part of the slide can be seen at once, and moving the slide will enable you to see all parts of it.

2. What do you notice about the direction of this movement as seen through the microscope?

To get the specimen into perfect focus, use the fine adjustment. Turn the fine adjustment wheel very slightly to sharpen the focus.

If the object is very small you may wish to see it under high power. First focus in low power, as before. Then swing the nosepiece around slowly to bring the high power objective into line (it is the longer one). Only a small turn of the fine adjustment will then be needed to bring the object into focus. **Be very careful.** The high power objective is close to the cover glass, and **too much adjustment can break something.**

D. Problems. In general you will find that low power is much easier to use than high. It is easier to focus under low power, and it is much easier to get enough light. You can control the light on the stage by using the mirror. Move the mirror into the position that best lights up the stage. Use the low power to find the specimen on your slide. Switch into high power when you want greater magnification.

If the view through the microscope looks foggy, there are two possible causes. One cause is dirty lenses. Never clean a microscope lens with a handkerchief. Use special lens paper. It is soft and very clean. It has no dirt in it to scratch the lenses.

To remove dust and other particles from a microscope lens, wipe gently with lens paper. Stubborn spots are best removed with lens paper that has been moistened with water. Before using the microscope, clean the eyepiece and the objective lenses.

Another cause of a foggy view is too much light. The *diaphragm* can be turned to shut out some of this light. Often this gives a clearer view of the object, even if it is not as bright.

As soon as you finish observing a temporary mount, clean your slide and coverslip. Wash them in water and dry them with a cloth or a cleansing tissue. Lens paper is not necessary for drying slides, and it would be too expensive to use all the time for this purpose.

It will probably take a little time for you to get used to your microscope, but practice will soon enable you to become quite skilled in handling it.

Activity II

Observing Cells

A. Cut out a small piece from one layer of an onion. From its surface you can peel off a very thin sheet, which is actually a single layer of cells. Place the onion membrane in a drop of water on a slide. Cover with a coverslip. Focus the slide under low power.

1. How do the cells look?

2. What part of the cells can you see? Look for the cell walls, cytoplasm, and nuclei. Your teacher will show you how to stain the cells with iodine solution.

B. Now focus a cell under high power. Look at the nucleus. What do you see?

3. In addition to the nucleus, what other cell structure absorbed the iodine stain? Try to make a drawing of what you see under low power and under high power. Biologists often make drawings to record their observations.

C. Use the flat side of a toothpick to gently scrape a few cells from the inside of your cheek. Spread this material in the center of a microscope slide. Stain with iodine solution. Cover it with a coverslip. Cells are always coming loose from the lining of your mouth, and you should be able to see them under the microscope.

4. Do these cheek cells have cell walls, like the onion cells?

5. Can you see the nuclei?

6. Do your cells look like those in Fig. 2-6?

Suggestion: Peel the onion very close to the slide so that the layer doesn't roll up.

Enrichment: The cells lining the mouth are called squamous epithelium, a type of cell that covers other cells.

Fig. 2-6
Cheek cell. (Carolina Biological Supply Company)

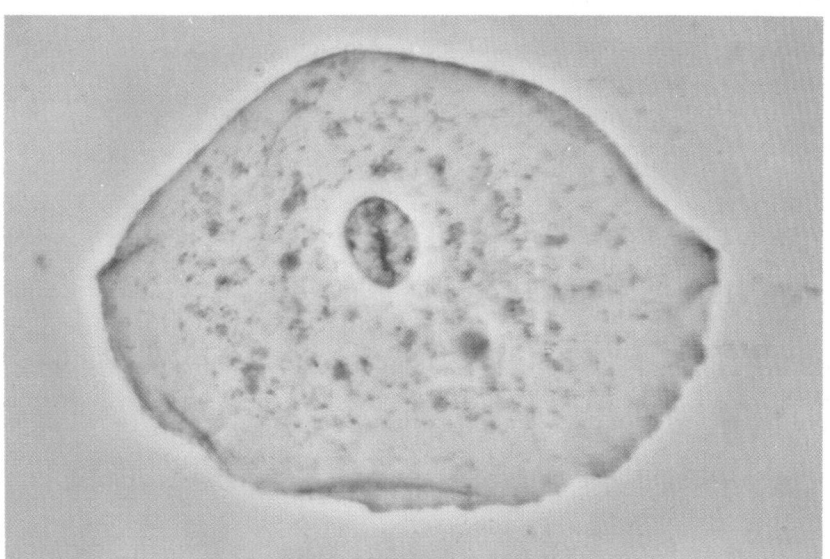

Word Quiz

1. Each of the words that follow are related to life activities. Do you know their meanings?

   movement   reproduction   respiration   excretion
   response   nutrition   growth

2. Each of the words that follow are related to cell structure. Can you identify these parts of the cell?

   protoplasm   cytoplasm   ribosome   cell wall
   cell   mitochondrion   vacuole

Check Your Facts

*Science
Reading Skills
2. Compare and
Contrast
7. Compare and
Contrast*

1. Read page 11 and the first five paragraphs on page 12. What is the main idea of this section?

2. List five examples of each group listed below.
   a. nonliving things   **b.** living things   **c.** dead things

3. Why should you learn about cells? Of what use is this information?

4. Draw a diagram of an animal cell. Label these parts: cell membrane, cytoplasm, nucleus, vacuole.

5. Draw a diagram of a plant cell. Label these parts: cell wall, cell membrane, cytoplasm, nucleus, vacuole.

6. Fig. 2-3 on page 15 is a drawing of a cell as it appears through the electron microscope. Name three parts of this cell that are not seen clearly through the light microscope.

7. Describe two differences between plant and animal cells.

Thought Questions

1. A ball rolls down a hill. A log floats in the river. An automobile travels on an open road. All of these are examples of movement. How do the movements of living things differ from the movements of these examples?

2. Explain how the structure of mitochondria is shown by the electron microscope and not by the light microscope.

3. How would you describe the appearance of protoplasm to a friend?

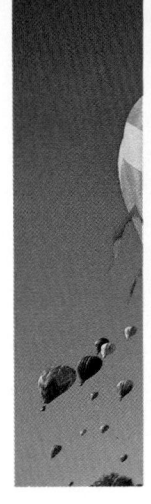
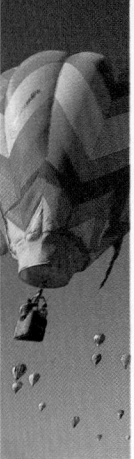

CHAPTER 3 Some Simple Chemistry

These balloons float in air because they are filled with gas molecules. What is a gas? What is a molecule? What is air? In this chapter you will find out the answers to these questions.

THE CHEMICAL MAKEUP OF MATTER

Chemistry is the study of matter. *Matter* is any kind of material or substance. It is anything that takes up space and has weight. There are thousands of different kinds of matter around you. Matter may be a gas, a liquid, or a solid.

There are only a limited number of basic substances. These substances are called *elements*. An element is a form of matter that cannot be divided into a simpler form of matter by ordinary chemical means. There are 107 elements. Of these elements, 90 can be found in nature. That is, they are in air, in soil, in rocks, and so on. These are called *natural elements*. Not all of these 90 natural elements are commonly used. The other elements have been made artificially by scientists in laboratories. All of the thousands of materials around us are made up of elements.

Two common elements. *Oxygen* is an element. It is a gas that looks and feels like air. In fact, air is about one-fifth oxygen. Another common element is *hydrogen*. It also is a gas and has no color, taste, or odor. Hydrogen is very light in weight. A balloon filled with hydrogen will float in air. Hydrogen gas is not usually found in air.

If we put hydrogen and oxygen together, nothing would happen. We simply have a *mixture* of hydrogen and oxygen. A mixture contains two or more substances that have been mixed together without a *chemical change*. Each substance keeps its own identity. This is like mixing red marbles and blue marbles together in a box. All you have is red marbles and blue marbles. You have not made anything new. But if we put an electric

objectives

After you read this chapter, you should be able to:

__Define the word matter

__Explain the difference between an element and a compound

__Explain the difference between an atom and a molecule

__Define chemical change

__List examples of different kinds of energy

__Explain the meaning of oxidation

Enrichment: Bring in examples of elements. Objects composed entirely of gold, silver, aluminum, copper, or iron would be good examples of pure elements.

matter

elements

mixture

Activity: Generate hydrogen and oxygen as a demonstration.

vapor

compound

Fig. 3-1
The formation of carbon dioxide during the burning of carbon. How many atoms are there in each molecule? What holds them together?

spark into a mixture of hydrogen and oxygen, something would happen. There would be an explosion. The hydrogen and oxygen would react chemically with one another to produce a new material—water. This is an example of a chemical change. Because of the heat from the explosion, this water would not be a liquid. It would be in the gas or *vapor* form. If we could cool this water vapor, it would become ordinary liquid water. So we see that under certain conditions it is possible for elements to unite with each other to form some new material.

Compounds are made up of elements. Water is a *compound.* When two or more elements unite and form a new substance, we call that substance a compound. Most of the common materials around you are made up of compounds. You are made up of compounds. A compound always contains two or more elements united in a very definite way. Water for instance, is made of two parts of hydrogen for every one part of oxygen. We often write this in a kind of shorthand, called a *formula,* such as H_2O. The H stands for hydrogen. The number 2 shows there are two units of hydrogen. The O stands for oxygen. There is no number next to the O. This means that there is only one unit of oxygen. Water is entirely different from either hydrogen or oxygen. It is not a gas at ordinary temperatures. It will not burn like hydrogen. It will not help things burn like oxygen. Water has its own characteristics, or *properties*. This is true of all compounds. Each compound has its own properties. These properties are entirely different from the properties of the elements that are in the compound.

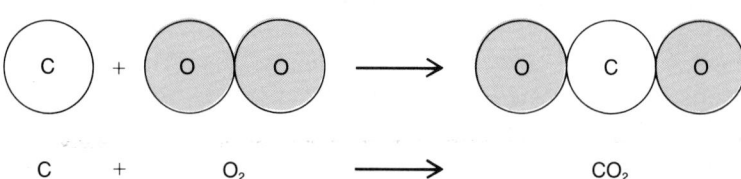

C + O_2 $\longrightarrow$ CO_2

Another common compound is carbon dioxide. This is the gas that bubbles out of soda pop. We also breathe carbon dioxide out of our lungs. Its formula is CO_2. This means that one unit of the element carbon (C) is united with two units of oxygen (O_2). See Fig. 3-1. Carbon is usually a black, solid material. It is familiar to you in the form of soot and charcoal, which are mostly carbon. Notice here, again, that the compound is different from either of the elements that combine to make it. Carbon dioxide looks like oxygen, but it does not act like it. Although carbon dioxide contains carbon, it has no resemblance to this solid black element.

So far, we have mentioned three elements. Some other common elements are nitrogen (the main gas in air), sulphur, iron, copper, phosphorus, aluminum, calcium, chlorine, and sodium. Most of these are never found naturally as pure elements. They are nearly always united in the many kinds of compounds that make up the world around us.

The structure of matter. Matter is made up of many small particles. Such particles are called *atoms*. An atom is the smallest unit of an element. Each element is made up of a different kind of atom. Hydrogen atoms are the lightest of all. Uranium atoms are the heaviest of all the *natural* elements. Each element has its own particular characteristics because each is made up of its own kind of atom. When elements unite and form compounds, the union takes place between the atoms which make up these elements. When two or more atoms are joined together chemically, a *molecule* is formed. A molecule is the smallest unit of a compound. The atoms in a molecule are held together by a force called a *chemical bond*. The atoms of some elements have an attraction for one another. These atoms can form chemical bonds with each other. In this way molecules are formed. Other elements have no attraction for one another. They cannot combine and form compounds. The ability of two elements to form chemical bonds depends on the structure of their particular atoms. Some molecules contain only two atoms. Some contain several atoms. Other molecules contain thousands of atoms. These giant molecules are especially important in the forming of protoplasm.

Some elements are usually found in the form of molecules. Oxygen atoms, for instance, are generally in pairs. Two oxygen atoms are held together by chemical bonds. This pair of oxygen atoms is an oxygen molecule. Its symbol is O_2.

The smallest unit of water is made up of two hydrogen atoms united with one oxygen atom. This is a molecule of water. Remember that its symbol is H_2O. The molecule of a compound is always made up of at least two kinds of atoms that are held together by their chemical bonds. Sometimes several different elements combine to form a compound.

CHEMICAL CHANGES

We have already mentioned that oxygen is one element that is usually found in the form of a molecule. This is written O_2. Hydrogen is another element that is usually found in the form of a molecule. This is written H_2. So, when hydrogen and oxygen unite to form water, two molecules of hydrogen unite with

atom

Enrichment: Emphasize the smallness of an atom. One drop of water contains 100 billion atoms.

molecule

chemical bond

Enrichment: Stress that atoms are neither lost nor gained in a chemical reaction. They are only rearranged.

one molecule of oxygen to form two molecules of water. See Fig. 3-2. We write the reaction this way:

$$2H_2 + O_2 \rightarrow 2H_2O$$

The arrow means that a chemical change has taken place. A new compound that has its own properties has been formed. **Energy** is required to produce some chemical changes. Other chemical changes result in the release of energy. In these chemical changes that release energy, the energy is stored in the bonds that hold the atoms together. When these bonds are broken, energy is released

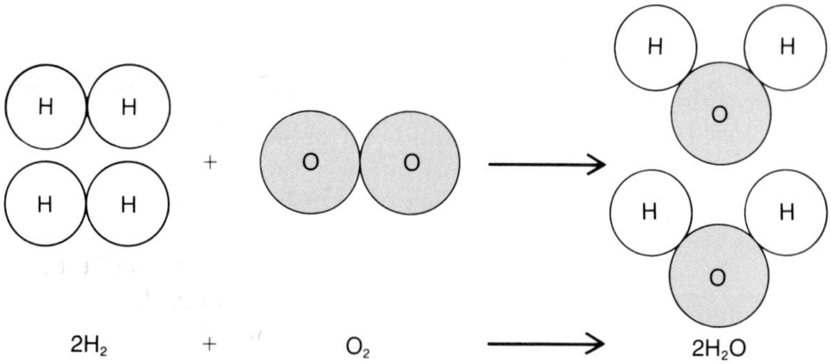

energy

Fig. 3-2

The formation of water. The larger circles represent oxygen atoms. The smaller ones are hydrogen atoms. See how the letters below tell the same story. How many molecules of each material are shown?

Enrichment: The burning of charcoal uses up oxygen in a room while producing carbon dioxide. If people are in the room, this can cause asphyxiation.

Fig. 3-3

A chemical reaction takes place during the burning of charcoal. Charcoal should not be burned indoors. Can you tell why? (HRW photo by Russell Dian)

There is more energy stored in molecules of hydrogen and oxygen than there is in the molecules of water that form when hydrogen and oxygen combine. This energy is released from the hydrogen and oxygen molecules when the water molecules are produced, and whenever any hydrogen gas burns. When this gas burns, oxygen in the air unites with the hydrogen to form water molecules. The water molecules appear as water vapor. While hydrogen is burning, light and heat are given off. Light and heat are forms of energy that are released when hydrogen burns. A chemical change takes place during this burning.

Another example of when a chemical change takes place is during the burning of charcoal. See Fig. 3-3. Charcoal is made of carbon. One atom of carbon (C) unites with one molecule of oxygen (O_2) to produce one molecule of carbon dioxide (CO_2). This reaction is shown in Fig. 3-1 on page 22, and in the following equation:

$$C + O_2 \rightarrow CO_2$$

Carbon and hydrogen are in all of our common fuels: wood, coal, oil, and gas. When oxygen unites with them, carbon dioxide and water vapor are formed and energy is released. Our foods also contain carbon and hydrogen. When oxygen unites with these elements in our bodies, energy is released and carbon dioxide and water are produced.

In the chemical changes we have mentioned so far, oxygen united with some other material. This process is called *oxidation* (ox-suh-DAY-shun). Oxidation is one very common type of chemical change, but there are many others. One important thing to know about chemical changes is that they always either use up or release energy.

The need for energy. Energy exists in several forms such as electricity, light, heat, and mechanical energy. The energy that is used to form chemical bonds or to break them is called *chemical energy*. Energy is easily changed from one form to another. For instance, the chemical energy stored in coal molecules becomes heat when the coal is burned. This heat energy can be used to make the steam that drives steam engines and generators. It has been changed to mechanical energy. Generators change mechanical energy into electricity. What forms of energy can you think of that are produced by electricity? Can you name the different types of energy shown in Fig. 3-4?

In somewhat the same way as mentioned above, the chemical energy stored in food molecules can be changed to do the work of the living cell. You can see evidence of this in your own body. Your body is warm. This heat is a form of energy being produced in your body. You can move. Movement also requires energy. The chemical energy stored in your food is changed into different forms of energy. This energy allows your body to perform all of its life activities. Without energy, life could not possibly continue.

Word Study: "oxid" means containing oxygen

oxidation

Enrichment: Give other examples of oxidation, such as iron rusting and silver tarnishing.

Enrichment: You may wish to mention the Law of Conservation of Energy that states: Energy may neither be created nor destroyed, but may pass from one form to another.

Enrichment: Another example is the chemical energy stored in trees. When the trees are cut and burned for fuel, chemical energy is released in the form of heat and light. Ask students to suggest other forms of chemical energy that are stored for future use.

Fig. 3-4
What different types of energy are being used here? (Steven Green/Photo Researchers)

SUMMARY

Matter is anything that has weight and occupies space. Air, water, protoplasm, and carbon are examples of matter. The basic substances from which matter is made are elements. Elements cannot be broken down into simpler forms of matter by ordinary chemical means. The smallest part of an element that has the properties of the element is an atom. An element is made of one kind of atom. Each element has its own atoms. When atoms of different elements combine, a chemical change takes place. Atoms are held together by the force of chemical bonds. When atoms of different elements are bonded together, a compound is formed. The smallest part of a compound that keeps the characteristics of that compound is known as a molecule. Sometimes substances are put together without chemical bonding to form a mixture. Chemical energy is stored in chemical bonds. The energy stored in food molecules may be released by oxidation and changed to the energy of heat or motion.

ACTIVITY

Caution: Warn students about the danger of using a flame. Warn students about long hair, loose clothing, discarding matches, and allowing for sufficient space between students.

Formation of Compounds

A. Pour a little limewater into a test tube. Blow your breath through a drinking straw into the limewater. Your breath contains carbon dioxide.

1. What change takes place in the limewater?

B. Place a short candle inside a large beaker. Your teacher will show you how to fix the candle to the bottom by using melted wax. Light the candle. Cover the top of the beaker with a glass plate. When the candle flame goes out, lift the lid of the beaker just enough to pour in a little limewater. Gently swirl the limewater in the beaker while carefully holding down the lid.

2. What change do you notice in the appearance of the limewater?

3. Where have you seen this change before?

4. What gas turns limewater milky?

5. What gas was produced by the burning candle?

C. Now dry out the beaker. Fix a candle to the bottom of the beaker as you did before. Hold a cold surface just above the flame for a moment. **CAUTION: Be careful not to burn your fingers.** A small jar containing an ice cube or *very cold water* will work very well. **Do not hold the jar low enough in the flame to deposit black soot on it.** Notice the moisture that collects on the glass.

6. Where did the moisture come from? Candle wax is made of the two elements carbon and hydrogen. Air in the beaker contains oxygen.

7. What gas is produced when 1 atom of carbon is joined chemically with 2 atoms of oxygen?

8. What gas is produced when 2 atoms of hydrogen are bonded chemically to 1 atom of oxygen?

Word Quiz

From the new words you learned in this chapter, write the word that correctly completes each statement. Do not write in this book.

1. Anything that takes up space and has weight is known as _____.
2. A substance that is made of the same kind of atoms is a (an) _____.
3. When salt and pepper are mixed together a (an) _____ is produced.
4. When water is in the form of a gas, it is described as water _____.
5. When two or more elements are joined together in chemical combination, a (an) _____ is formed.
6. Elements are made of tiny units known as _____.
7. The smallest part of a compound that has the properties of the compound is a (an) _____.
8. The atoms in a molecule are held together by a force known as a (an) _____.
9. Light and heat are forms of _____ given off when a candle burns.
10. When oxygen unites with another substance, _____ takes place.

Check Your Facts

Fig. 3-5

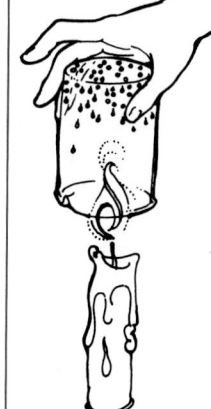

1. What is chemistry?
2. Name some natural elements.
3. What information is given by the equation that follows:

$$2H_2 + O_2 \rightarrow 2H_2O$$

4. Study Fig. 3-5, then answer the questions that follow. Use complete sentences for each answer.
   a. Name the compound that has collected on the inside of the beaker.
   b. Why is the substance named in answer "a" considered to be a compound?
   c. What other compound is formed by a burning candle?
   d. How do you know that energy is given off by a burning candle?
5. Explain the meaning of oxidation.

Thought Questions

1. Air is a mixture of these gases: nitrogen, oxygen, carbon dioxide, water vapor, and some other gases. Explain why air is not called a compound.
2. What might happen if hydrogen gas molecules were part of the air?
3. Of what use to the cell is the chemical energy that is stored in food?

CHAPTER 4 The Chemistry of Protoplasm

objectives

After you read this chapter, you should be able to:

___**List** the important compounds found in protoplasm

___**Describe** the functions of water in cell activity

___**Explain** the importance of carbohydrates to the cell

___**Discuss** the importance of proteins to the cell

___**Explain** why the cell needs energy for chemical activity

___**Describe** the function of ATP

Enrichment: The composition of protoplasm is constantly changing. The cells are continuously manufacturing different substances and breaking down other substances. Protoplasm is a special type of mixture in which large molecules of compounds are suspended.

All living things need water. Without it life would not be possible. Water is a compound that is quite familiar to us. Animals drink it to satisfy thirst. Plants take in water, too. But living things have need of water for many different kinds of activities. As you read this chapter, you will learn why water and several other substances are necessary for life.

THE COMPOSITION OF PROTOPLASM

You learned from Chapter 2 that living things are made up of cells. The living material of cells is protoplasm. Like any other matter, protoplasm is made up of elements. Some of the important elements present in living cells are shown in Table 4-1. As you look at this table, notice that each element has a chemical symbol of either one or two letters. The symbol is used in place of writing out the entire name of the element.

Protoplasm contains these elements in unequal amounts. Let us suppose that you weigh 54 kilograms. Your body is made up of about 35 kilograms of oxygen, 10 kilograms of carbon, 6 kilograms of hydrogen, 1 kilogram of nitrogen, and 1 kilogram of calcium. There are nearly 20 other elements in your body, but all together they weigh only about 1 kilogram.

Protoplasm is very special. This is so not simply because it contains the elements listed in Table 4-1. It is the way in which these elements are joined together that makes protoplasm so special. The elements are combined into very complicated compounds inside the cell. It is these large compounds that set protoplasm apart from nonliving substances. After the cell makes these compounds, it is able to use them for its life activities. Protoplasm contains many chemical compounds. Table 4-2 shows the main compounds in a typical sample of protoplasm. Let us now examine these compounds more closely.

Table 4-1 THE MOST COMMON ELEMENTS IN PROTOPLASM

Element	Symbol	Percent in Protoplasm
Oxygen	O	65.0
Carbon	C	18.5
Hydrogen	H	9.4
Nitrogen	N	3.1
Phosphorus	P	1.0
Sulfur	S	0.3
Potassium	K	0.4
Magnesium	Mg	0.1
Calcium	Ca	1.5
Iron	Fe	trace
Sodium	Na	0.1
Chlorine	Cl	0.1
Iodine	I	trace

Table 4-2 SOME COMPOUNDS FOUND IN PROTOPLASM

Compound	Percent
Water	80
Proteins	12
Carbohydrates, Nucleic acids, Vitamins, and Other compounds	3
Fats	3
Mineral salts	2

Water. The compound that makes up most of the protoplasm is water. Although water is a very common substance, it has most unusual properties. One of its properties is the ability of water to *dissolve* other materials. This means that water can separate other molecules and surround them. If you stir a teaspoon of sugar into a glass of water, can you see the sugar? You know that the sugar has not disappeared because you can taste the sweetened water. The sugar has merely been separated and surrounded by water molecules. When materials dissolve in water, a *solution* is formed. A solution is a mixture in which a substance is spread through a liquid such as water.

Another property of water depends upon its ability to form solutions. While many substances are in solution chemical activity takes place readily. For example, when you sprinkle dry scouring powder onto a dry sink surface, nothing happens. When water is added, the chemical activity of the powder in solution begins. You can now use the foaming powder to remove stains.

Sometimes the water in protoplasm breaks apart large food molecules forming smaller units. These small molecules can

dissolve

Word Study: "dis" means apart

solution

Word Study: "solu" means dissolve

diffusion

Activity: Demonstrate diffusion by opening a container of ammonia or peppermint oil or by dropping a crystal of copper sulfate into a container of water. CAUTION: Copper sulfate is poisonous. You may also wish to use diagrams to explain this concept.

carbohydrates

Word Study: "carbo" means coal, "hydro" means water

glucose

Word Study: "glu" means sweet

Enrichment: Show students a diagram or model of the glucose molecule.

Enrichment: Students should not think that they have little fires inside their cells.

Enrichment: ATP stands for adenosine triphosphate. ATP is put together in mitochondria and released to all parts of the cell.

ATP

be moved across cell membranes. The movement of molecules from a place in which there are many of them to a place where these molecules are few in number is called *diffusion* (di-FEW-shun). Dissolved mineral salts move from soil water into roots of plants by diffusion through water. You can now understand that water is a means of transportation for compounds through protoplasm. Some compounds move into cells. Other substances move out of cells.

Carbohydrates. The basic energy-making food molecules of the cell are called **carbohydrates** (kar-boh-HY-drayts). All carbohydrates contain carbon, hydrogen, and oxygen. There are always twice as many hydrogen atoms as oxygen atoms. This is the same as in water. Carbohydrates come in several forms, such as starches and sugars. One form can be changed to another rather easily. The particular form of carbohydrate we are most interested in right now is the simple sugar called *glucose*. This sugar is a little different from the sugar we use in coffee or tea. The formula for glucose is $C_6H_{12}O_6$. Living things can change large starch and sugar molecules into simple glucose. Both plant and animal cells need glucose. Plants can make their own glucose through a special process. Animals cannot do this. They get glucose from the carbohydrates that are found in the plants and animals that serve as food.

Protoplasm, as we have said, does its work by chemical action. We have also said that the power to carry on this work comes from chemical energy. The most important material which supplies this chemical energy to cells is glucose. Glucose is taken apart atom by atom in the process of cell respiration. This respiration occurs in the mitochondria of cells.

Cell respiration can be compared to burning. When something burns, it unites with oxygen and energy is released. When glucose unites with oxygen, the glucose molecules release energy also. We can say that respiration is the oxidation of the chemicals we get from food. But if sugar were simply burned, it would give off all of its energy in the form of heat. This heat could not be used to do the cell's work. Cell respiration releases most of the energy of the sugar molecule in a different way. In the mitochondrion the glucose molecule is taken apart one step at a time. The glucose molecule has six carbon atoms in it. During cell respiration, it is changed to one kind of molecule after another. It ends up as several molecules with only one carbon atom in each. These are carbon dioxide molecules (CO_2).

Each time one of these glucose molecules is broken down, energy is released. Most of this energy is not allowed to escape as heat. Instead, the energy is used to build up a special phosphorus compound called by its initials *ATP*. Energy stored in ATP can be released later when it is needed to do the work of the cell. See Fig. 4-1. Think of ATP as the energy carrier of the cell. The carbon and hydrogen in the glucose are removed by combining them with oxygen. This means that the final products of glucose oxidation are carbon dioxide and water. Some of the water formed during respiration is used by the cell. The excess water is excreted by the cell along with the carbon dioxide waste.

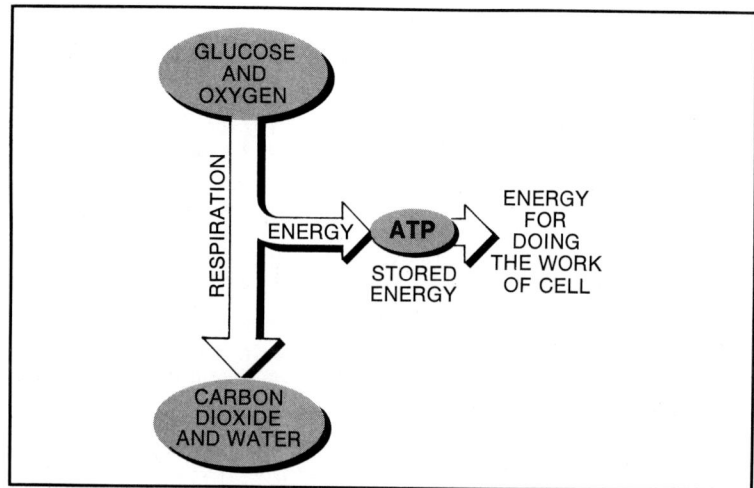

Fig. 4-1
Respiration releases energy to do the work of the cell. ATP carries this energy. The sugar is broken down into carbon dioxide and water.

Some cells do not use oxygen at all. These cells can only break down the sugar part of the way. Thus they get only part of the energy from it. Many bacteria and some other living things are like this.

Carbohydrates are not only used to supply energy. They can also be used as building materials. For example, cell walls of plants are made of carbohydrates. This is one of the things that gives plant structures like wood their strength.

Enrichment: This type of plant carbohydrate is called cellulose. Some examples are paper, wood, and cotton.

Fats and oils. *Fats* and *oils* are made of carbon, hydrogen, and a little oxygen. They are used by the cell in the building of its membranes. These include the cell membrane itself as well as those membranes covering the nucleus, vacuoles, and mitochondria. Fats and oils are also used as a reserve energy supply. Excess fat or oil is stored in some cell vacuoles. Later it can be broken down by respiration to release energy. The solid fats are most common in animal cells. The liquid oils are more common in plants.

fats and oils

Enrichment: More energy can be released from a given amount of fat than from the same amount of carbohydrate.

Assignment: Have students investigate terms such as cholesterol and saturated, unsaturated and polyunsaturated fats. Have students report on these terms which are commonly mentioned in advertising.

When cells build new materials, the simple compounds formed during glucose breakdown are first built up into fairly large molecules called *fatty acids*. Then these are put together to make the still bigger fat molecules. Energy needed for this building-up process is supplied by ATP.

fatty acids

Other compounds that are very important to the cell are *pro-teins*. Protein molecules often contain thousands of atoms.

proteins

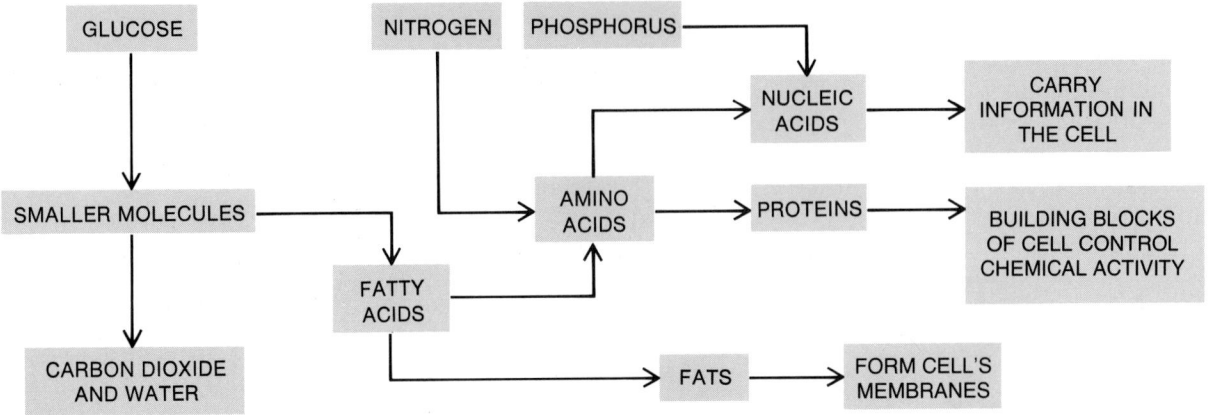

Fig. 4-2
Some of the compounds formed in the cell and their uses. What supplies the energy used by the cell to do this work?

amino acids

Enrichment: Proteins differ from one kind of plant or animal to another. They also differ in individuals of the same kind.

enzymes

Besides oxygen, carbon, and hydrogen, *proteins always contain nitrogen*. Other elements are sometimes present.

There are millions of proteins in the world. Your own body contains many thousands. Yet all of these different kinds of protein molecules are formed from a limited number of fairly simple compounds called **amino** (a-ME-no) **acids**. There are about twenty amino acids found in living cells. Chemical bonds can easily form to fasten amino acids together. A protein molecule is simply a large number of amino acid molecules which are bonded together in this way. The building of proteins takes place in the ribosomes. The energy for protein building is supplied by ATP.

The amino acids themselves are formed by adding a special form of nitrogen to fatty acids. Plants can get their nitrogen from mineral salts in the soil. Animals cannot do this. They get some amino acids ready-made from proteins in food. Animals can also break down amino acids or other nitrogen compounds. They use these broken-down substances to build up other amino acids they need. So we can see that animals make some amino acids and others have to be ready-made for them.

Some proteins are used to form parts of living things. The cell's membranes contain protein molecules that help to absorb needed materials into the cell. Animals produce tough protein fibers between the cells to bind their bodies together. Leather is strong because it contains large amounts of such protein fibers. Fingernails and hair are made of protein.

The use of proteins for building purposes is important. But many of the proteins in the cells have a more important use. They control the chemical changes that take place in the cell. These proteins which control chemical activity are called **enzymes** (EN-zimes). Each enzyme is a different kind of protein molecule. Each controls one particular chemical change. Each step in the breakdown of glucose is controlled by a different

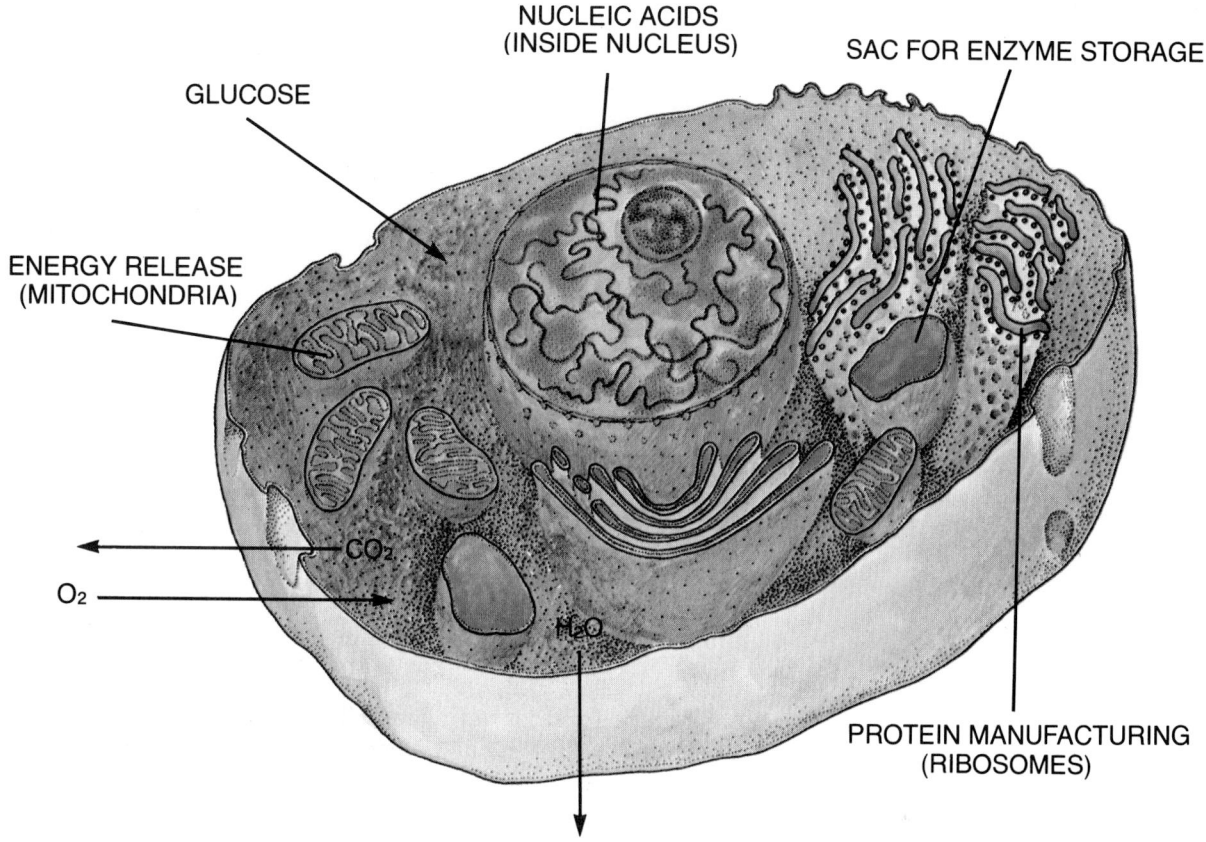

NUCLEIC ACIDS
(INSIDE NUCLEUS)

SAC FOR ENZYME STORAGE

GLUCOSE

ENERGY RELEASE
(MITOCHONDRIA)

CO_2

O_2

H_2O

PROTEIN MANUFACTURING
(RIBOSOMES)

enzyme. This is why mitochondria are able to carry on respiration. They contain the enzymes needed to do this job. Each step in the buildup of new compounds is also controlled by a particular enzyme. It takes several thousand different enzymes to carry on all of the different chemical activities that are going on in your cells right now. We might say that cytoplasm is a workshop in which enzymes are the experts directing the cell's work.

Nucleic acids. Some special compounds found in the cell nucleus are *nucleic* (new-CLAY-ik) *acids*. These are involved in three important activities of cells. Nucleic acids play a basic role in carrying information from one cell generation to another. These compounds are involved in the manufacture of proteins, and in the control and function of living cells.

If you have a general picture of the activities going on in protoplasm, that is enough for now. In the next chapter, we shall learn how plant and animal cells get the food materials they need in order to carry on these various chemical processes.

Fig. 4-3
Protoplasm is a chemical factory that performs many jobs. The arrows show the diffusion of some substances across the membrane.

nucleic acids

Enrichment: You may wish to introduce the terms DNA and RNA.

SUMMARY

Protoplasm is able to combine carbon, oxygen, hydrogen, and nitrogen into special compounds that do special work. These compounds include carbohydrates, fats, amino acids, proteins, and water.

Glucose is a simple sugar. When many glucose molecules are joined chemically, larger carbohydrates are formed. Glucose releases energy during cellular respiration, which takes place in the mitochondria.

All of the work of the cell depends upon energy obtained from carbohydrates and fats. Energy is stored in ATP molecules.

Amino acids are the building blocks of proteins. Enzymes are special types of protein molecules that control chemical activity. Other kinds of proteins form hair and nails. Cell membranes are made partly of proteins, also.

Fats are made from fatty acids. Sometimes fatty acids are used by the cell for energy. Excess fatty acids are changed into fat and stored in the cell. Cell membranes contain fatty acid molecules.

Nucleic acids are present in the nucleus of the cell. These molecules control the manufacture of proteins. They help also in controlling the activities of the cell.

ACTIVITY

Diffusion

A. Fill a beaker two-thirds full with water. Place the beaker on your desk. Wait until the water in the beaker stops moving. Now carefully drop one drop of red or black ink into the center of the water in the beaker. Watch the drop of ink closely.

1. As soon as the drop of ink enters the water, where does it go? Why?

B. Look at the beaker from the side. You will have to bend down to do this.

2. How can you tell that the ink is moving through the water? Describe what you see.

C. Wait for three minutes. Now look at the water in the beaker.

3. What is the general color of the water now?

Diffusion is the movement of molecules from a place where there are many of these molecules to a place where there are few of them.

4. To which place in the beaker of water does the ink move—where there are few ink molecules or many? Explain your answer using information given in the definition of diffusion.

Word Quiz

From the new words you learned in this chapter, write the word that correctly completes each statement. Do not write in this book.
A teaspoon of sugar is added to a glass of water.

1. The sugar is _____ in the water.

2. The water and sugar together form a _____.

3. The sugar molecules move through the water by means of _____.

4. Carbohydrate molecules can be broken down into the simple sugar _____.

5. Cellular energy is used to build up the compound _____.

6. Fat molecules are made from smaller units known as _____.

7. Protein molecules are built from units known as _____.

8. Proteins that control chemical activities are called _____.

9. Compounds that are involved in the manufacturing of proteins are called _____.

Check Your Facts

1. Name five important compounds found in protoplasm.

2. Why is water necessary to the chemical activities of cells?

3. How does the information in Table 4-2 differ from that given in Table 4-1?

4. How are carbohydrates used by cells?

5. Protoplasm is able to do things with elements that nonliving material cannot. Explain.

6. How does the cell make use of: a. amino acids; b. fatty acids?

7. How is energy used by cells?

8. What is the function of ATP?

9. How do enzymes help with the chemical work of protoplasm?

Science
Reading Skills
2. *Cause and Effect*
3. *Reading Illustrations*
9. *Cause and Effect*

Thought Questions

1. How does a knowledge of chemistry help us to understand the work of the cell?

2. Why is water needed by protoplasm?

3. How can a chemist tell if a compound in protoplasm is a protein?

CHAPTER 5 Energy and Photosynthesis

Each cell in the body of a living thing can be thought of as a factory. In an industrial factory something is made. Energy is used in the process. In cells, many compounds are manufactured. These are used to build new protoplasm. The cell gets the energy to do this work from glucose molecules. Where does this glucose come from? The glucose is made in cells of green plants like the one in the photograph above. This food-making process is the most important manufacturing process on earth. It is called *photosynthesis* (foe-toe-SIN-the-sis).

HOW PLANTS SURVIVE

Green plants make glucose. To carry on photosynthesis the plant must have four things. These are water, carbon dioxide, sunlight, and *chlorophyll* (KLOR-oh-fil). Chlorophyll is the green material in plants. In some plants the green material is hidden by another color. The red leaves of *Coleus* hide the green chlorophyll. See Fig. 5-1. Nevertheless, its chlorophyll still works as well as in plants where the leaves are green.

Glucose is made in the cells of green plants from carbon dioxide and water. The formula for glucose is $C_6H_{12}O_6$. This means that it contains carbon, hydrogen, and oxygen. Water (H_2O) is made up of hydrogen and oxygen. Carbon dioxide (CO_2) is made up of carbon and oxygen. Therefore, water and carbon dioxide have all the elements needed to make glucose.

Photosynthesis is really a series of chemical changes. The first of these changes depends directly upon light energy to make it happen. **Energy stored in ATP is used to carry on the later steps in food-making.**

Light energy enters the plant cells. Chlorophyll absorbs the light and changes it to chemical energy. In this form the energy can now be used to do its work in food-making. Part of the energy from the chlorophyll molecule is stored in ATP for later use. The other part of the energy splits water into hydrogen atoms and oxygen atoms. The oxygen escapes from the plant.

12 MOLECULES OF WATER	→	24 ATOMS OF HYDROGEN	+	6 MOLECULES OF WATER
$12H_2O$	→	$24H$	+	$6O_2$

REACTIONS USING LIGHT FOR ENERGY   **Fig. 5-2a**

Carbon dioxide enters the plant cells by diffusion. With the help of the energy stored in ATP, the second stage of photosynthesis begins. In a series of steps, carbon dioxide is joined with hydrogen to make glucose. Left over atoms of hydrogen and oxygen form water. Fig. 5-2a and Fig. 5-2b summarize the two stages of photosynthesis.

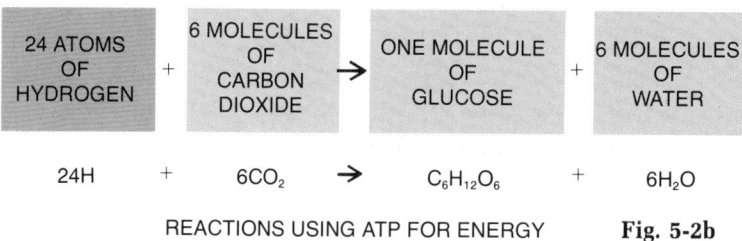

24 ATOMS OF HYDROGEN	+	6 MOLECULES OF CARBON DIOXIDE	→	ONE MOLECULE OF GLUCOSE	+	6 MOLECULES OF WATER
$24H$	+	$6CO_2$	→	$C_6H_{12}O_6$	+	$6H_2O$

REACTIONS USING ATP FOR ENERGY   **Fig. 5-2b**

It is important to remember that photosynthesis is really a series of chemical changes. Light energy starts the process going. ATP supplies the energy needed to bring about the joining of carbon dioxide and hydrogen. The main idea of photosynthesis is not hard to understand. Photosynthesis is a food-making process. It takes place in the cells of green plants. Energy for photosynthesis comes from sunlight. Light energy is absorbed and put to work by chlorophyll, the green material in the plant. The most common food produced is glucose. If we wish to sum up the process in a single chemical equation, we can do it this way:

$$12H_2O + 6CO_2 \rightarrow C_6H_{12}O_6 + 6O_2 + 6H_2O$$

Actually, the ATP formed during photosynthesis may be used to produce other compounds, such as fats or plant oils and proteins. ATP is also found in animal cells.

The energy connections. You will notice that photosynthesis and respiration depend on each other. In fact, photosynthesis is just the opposite of respiration. Photosynthesis builds up

Fig. 5-1
(Above) The leaves of green plants are not always all green.

Fig. 5-2a
Chemical changes using light energy. Water is split into hydrogen and oxygen.

Fig. 5-2b
Chemical changes using ATP for energy. The hydrogen from the splitting of water combines with carbon dioxide to form glucose.

Enrichment: After energy from the chlorophyll is used to do work, the chlorophyll can be used to absorb more light.

5

glucose. Respiration tears it down. Photosynthesis uses energy. Respiration supplies energy. Photosynthesis uses carbon dioxide and produces oxygen. Respiration uses oxygen and produces carbon dioxide. Photosynthesis takes place only in the presence of light. Respiration goes on all of the time.

How does photosynthesis help the plant? Photosynthesis gives the plant a supply of glucose for its cells. Remember that glucose supplies both energy and building materials. You can think of glucose as a sort of storage battery for the cell. Energy stored in the glucose molecules can be used by the cell whenever it is needed.

6

Photosynthesis must make enough glucose to keep the plant supplied all the time. Extra glucose is often stored in plants by changing it into starch. It may also be changed into other kinds of sugar. Common table sugar is formed by combining two simple sugars. Glucose, table sugar, and starch are all examples of carbohydrates.

Animals, including humans, cannot make glucose. They must get it from green plants. Animals carry on respiration but not photosynthesis. During respiration both animals and plants give off carbon dioxide. This carbon dioxide goes into the air. From the air, land plants get the carbon dioxide needed to make glucose. Animals can then eat the plants. See Fig. 5-3.

Fig. 5-3

Animals rely on plants for food and oxygen. During respiration animals give off carbon dioxide, which is used by plants for photosynthesis. During photosynthesis plants give off oxygen, which is used by animals for respiration.

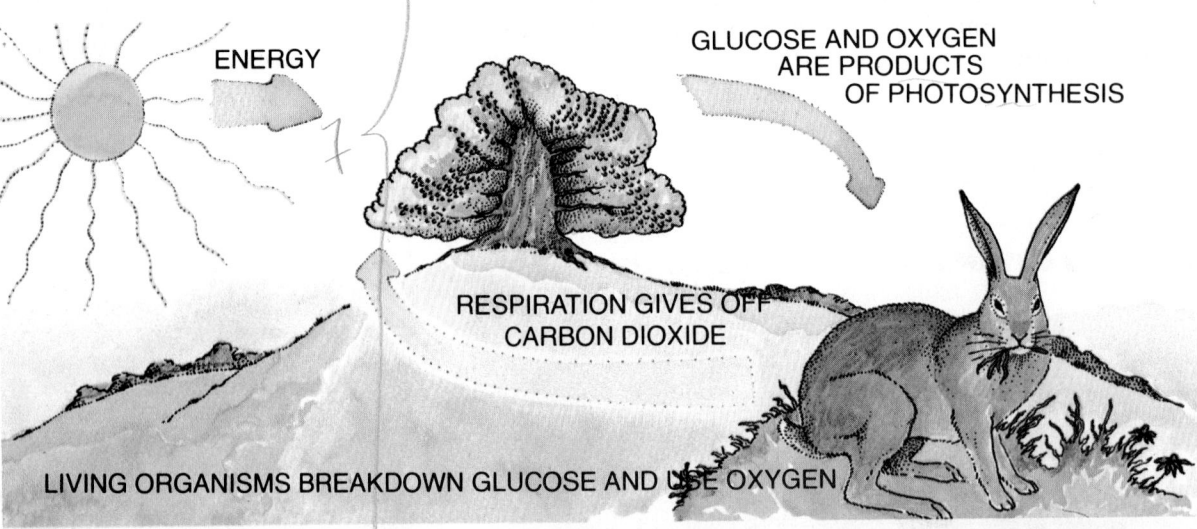

ENERGY

GLUCOSE AND OXYGEN
ARE PRODUCTS
OF PHOTOSYNTHESIS

RESPIRATION GIVES OFF
CARBON DIOXIDE

LIVING ORGANISMS BREAKDOWN GLUCOSE AND USE OXYGEN

You can see that the same carbon atoms that are now part of you have been used many times before. You are made of second-hand materials! The carbon dioxide you are breathing out right now will be used by plants to make food again and will become part of other living things. Atoms that are now part of you have, in the past, been part of other people, fish, dinosaurs, trees, seaweed, and all other kinds of plants and animals.

There is something else to notice about photosynthesis. Not only does it supply the food used by all living things; it also supplies the oxygen of the air. Respiration keeps using up this oxygen, but photosynthesis puts it right back again. The food you eat and the oxygen you breathe are both supplied by the green plants!

Green plants need mineral salts. Green plants not only make glucose; they make all the compounds needed by their cells. They build up plant oils, proteins, chlorophyll, vitamins, and many other substances. To do this they need a supply of many different elements. We have seen how carbon, hydrogen, and oxygen are supplied by water and carbon dioxide. The other elements are usually supplied by simple compounds, called *mineral salts*. These mineral salts are commonly found dissolved in water. Soil water supplies minerals to land plants. Dissolved mineral salts are absorbed by water plants directly from the water they live in. Nitrogen, for instance, often comes to plants in the form of mineral salts called *nitrates* (NY-trates). Nitrogen is used in making proteins, ATP, and several other compounds. Salts called *phosphates* (FOSS-fates) often supply the phosphorus needed for making ATP, nucleic acids, and other materials.

Animals have some of this same ability to use simple compounds for building their protoplasm. You get some useful elements like iron and calcium from salts in the water you drink. But animals cannot build all of the compounds needed in their protoplasm. There are many materials which they must get ready-made in the food they eat. Besides glucose, these include several amino acids and a long list of vitamins.

HOW DO MATERIALS ENTER AND LEAVE THE CELL?

Now you know of things which must get into and out of cells. These include food, oxygen, carbon dioxide, water, several kinds of wastes, and several kinds of mineral salts. Every cell is completely surrounded by a cell membrane, so how can anything enter or leave? The answer is not always the same.

Most materials must be dissolved before they can enter a cell. You have seen things dissolve. When you stir sugar or salt into water, it disappears. You know it is still there because you can taste it in the water. When something dissolves like this, its molecules have separated from one another. The separated molecules become scattered all through the water. It is in this form that most molecules enter a cell.

Molecules in a liquid are always moving. Each molecule shoots along at high speed until it hits another molecule. When two molecules hit each other, they bounce off and keep going.

Fig. 5-4
Sun, water, and carbon dioxide help to produce these vegetables. How do nitrogen and phosphorus help also? (HRW photo by Russell Dian)

mineral salts

nitrates

phosphates

Fig. 5-5
An electron micrograph of a cell membrane. You can see that the membrane has two layers. (Dr. J. David Robertson, M.D., Ph.D.)

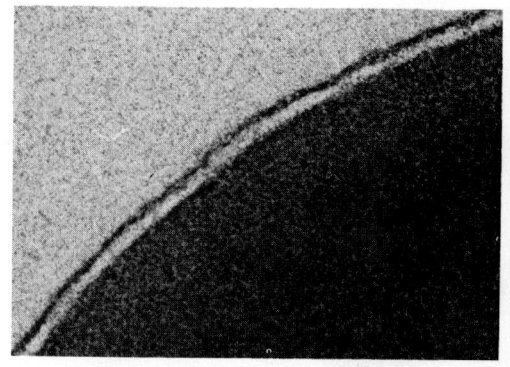

Cell walls of plant cells are no problem for the moving molecules. Since cell walls are made of nonliving carbohydrates, liquids pass easily through them. The living cell membrane is the real barrier between the inside of a cell and the outside surroundings. The cell membrane is very thin. It is made of two layers of fat and protein molecules. The fats form a very tight barrier which you might expect would keep everything out of the cytoplasm. Actually, it is thought that the cell membrane has tiny holes, or pores. These may be openings through the centers of the protein molecules. These openings are so small that only the smaller sized molecules can pass through, but the larger molecules are too big to enter. The natural motion of small dissolved molecules may send them into the cell through the pores in the membrane.

Diffusion in a living cell. Now let us picture what happens in a living cell. As an example, we shall use *Chlorella* (klo-REL-ah). *Chlorella* is one of the many kinds of simple cells that live unattached to other cells. *Chlorella* lives in water. A cell that lives alone carries on all of the life activities even though it is small and simple.

Fig. 5-6 shows what a *Chlorella* cell looks like. This one cell can live all by itself. It is a simple green plant-like cell. If there are a great many *Chlorella* cells drifting in a pond all at once, they make the water look green. The green, of course, is chlorophyll. Notice that the chlorophyll is contained in a special structure in the cytoplasm. This is called the *chloroplast* (KLOR-oh-plast). Besides chlorophyll, the chloroplast contains the many enzymes needed to control the food-making process. *Chlorella* has one big chloroplast in its cell. Most of our larger plants have many small chloroplasts in each cell. Chloroplasts are the food-making units of the cell.

Chlorella

Word Study: "plast" means formed or molded

chloroplast

Fig. 5-6
A diagram (right) and a photograph (left) of Chlorella, *a single-celled organism. (Walter Dawn)*

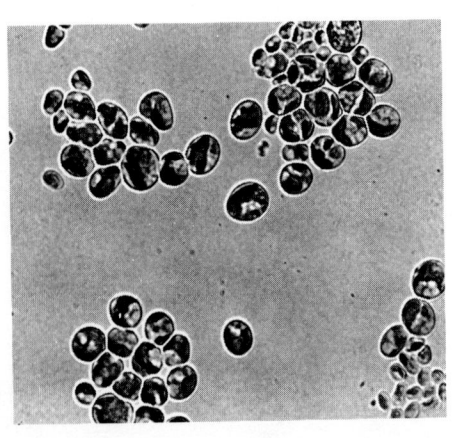

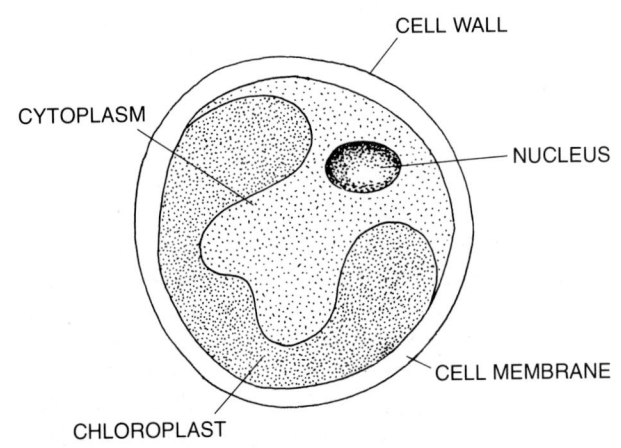

Now look at Fig. 5-7. It is an enlarged diagram that shows the conditions at the cell membrane of a *Chlorella*. Notice that the cell membrane has pores in it. Outside the cell is the water of the pond in which the *Chlorella* lives. Inside the membrane are molecules of the living protoplasm. Remember that the molecules are in motion all the time and that they are much smaller than shown in the figure. What happens when a water molecule hits a pore in the membrane? What about the other kinds of molecules?

As you can see, some molecules can pass right through the membrane. Others may be too large to get through the pores. Big molecules like those of sugars and fats and proteins stay in the cell. Undissolved particles outside the cell do not enter. Dissolved mineral substances, such as nitrates and phosphates enter and are used by the cell to build up the compounds in its protoplasm. There are water molecules on both sides of the membrane. Water molecules pass both into and out of the cell.

You learned in Chapter 4 that dissolved particles move through water by means of diffusion. The passage of materials through the cell membrane is another example of diffusion.

Passage of materials through the membrane is not all simple diffusion. Protein molecules in the membrane can form bonds with particularly large molecules that are much too big to pass through pores. These proteins then use energy from ATP molecules to draw the materials through the membrane into the cell. Each protein is able to draw in some particular material. In this way, the membrane controls what enters the cell.

If you watch a living *Chlorella* cell under the microscope you are not able to see anything going on. You simply see a little plant cell in the water. Yet, there is actually a great deal of activity taking place in that simple little cell. Let us examine this activity.

Dissolved in the water outside the cell are molecules of oxygen, carbon dioxide, and mineral salts. All of these molecules enter the cell by being absorbed through the cell membrane. Sunlight shines on the chloroplast. This gives the chlorophyll the energy to split water molecules and to make ATP. Energy carried in the ATP molecules is used to carry on the cell's activities. Glucose is produced. As it accumulates, glucose is saved for future use by being changed into solid starch grains, which are stored in vacuoles. Some food may be changed into plant oils and also stored. Proteins, vitamins, chlorophyll, nucleic acids, and many other useful materials are being manufactured. The cell is growing. Night comes, and photosynthesis stops because there is no more sunlight. Now the cell gets its energy by using the glucose and other foods it made during the daytime. Respiration breaks them down, forming ATP. Energy

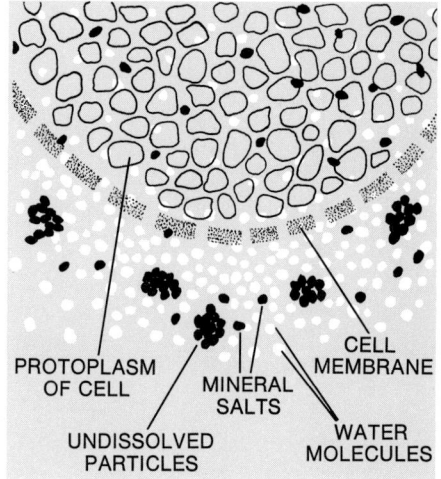

PROTOPLASM OF CELL

MINERAL SALTS

CELL MEMBRANE

UNDISSOLVED PARTICLES

WATER MOLECULES

Fig. 5-7
A diagram of the molecules on the inside and the outside of a cell membrane. Remember that molecules are in motion. Which ones will get through the membrane?

carried in the ATP molecules is used to manufacture proteins, vitamins, and other useful materials in the cytoplasm. The cell goes on growing. Both night and day the cell membrane uses energy to carry on its work of absorbing needed molecules from the pond water.

Molecules of waste produced by the cell pass through the cell membrane by diffusion. In the daytime, photosynthesis is going on very rapidly. Thus oxygen leaves the cell as a waste. At night, photosynthesis does not take place. It is at this time that oxygen enters the cell to be used in respiration. Carbon dioxide now leaves the cell as a waste. **Remember that the process of respiration goes on at all times in both plant and animal cells.**

DIFFERENT WAYS OF GETTING FOOD

Notice that the little *Chlorella* cell does not need any materials from other living things. It makes its own food and builds its protoplasm entirely from simple compounds found in the water where it lives. Any living thing that can do this can be called a *producer*. Cells with chlorophyll are producers. Animals, including humans, must take in food materials which have already been manufactured by other living things. We call animals *consumers*. Some living things which are not animals are consumers. For instance, a mushroom has no chlorophyll. It must get its food ready-made from decaying wood or dead leaves.

Large plants such as trees carry on photosynthesis. They are made of many cells, but the same processes go on in them that go on in the simple little *Chlorella*. These large plants carry on photosynthesis in their leaves.

Materials must pass in and out through the membranes of all cells. It does not matter whether they are plant cells or animal cells. It does not matter if they are single cells or if they come in groups. All living cells absorb the molecules they need through their cell membranes. All of them get rid of waste molecules through their cell membranes.

Animals do not make foods. *Chlorella* is an example of a plant-like cell in action. An example of an animal-like cell is *ameba* (uh-MEE-bah). Like *Chlorella*, an ameba is a single cell, but it has no chlorophyll so it cannot make foods. Fig. 5-8 shows what an ameba looks like. The shape of the cell can change as it moves about. The almost colorless cytoplasm is surrounded only by a thin cell membrane. A nucleus and vacuoles are present.

Since an ameba cannot make food, it must find food that is already made. The ameba cell is able to move by a flowing

ameba

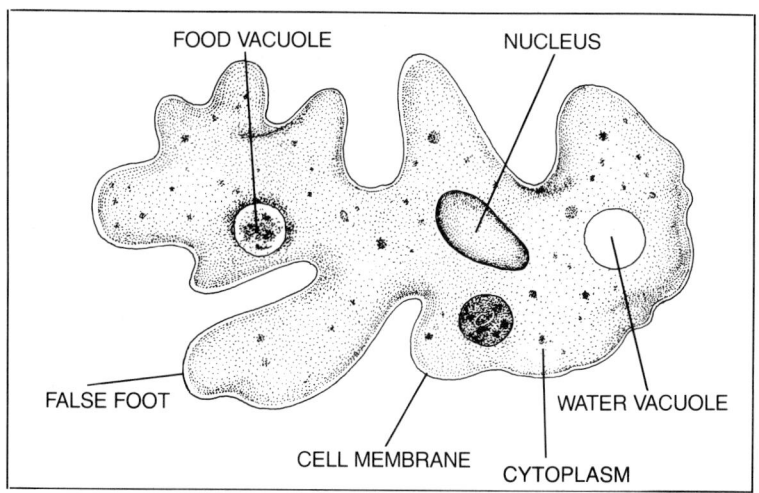

FOOD VACUOLE   NUCLEUS

FALSE FOOT

CELL MEMBRANE   CYTOPLASM

WATER VACUOLE

Fig. 5-8
Ameba, an animal type of cell.

Fig. 5-9
An ameba surrounding its food.

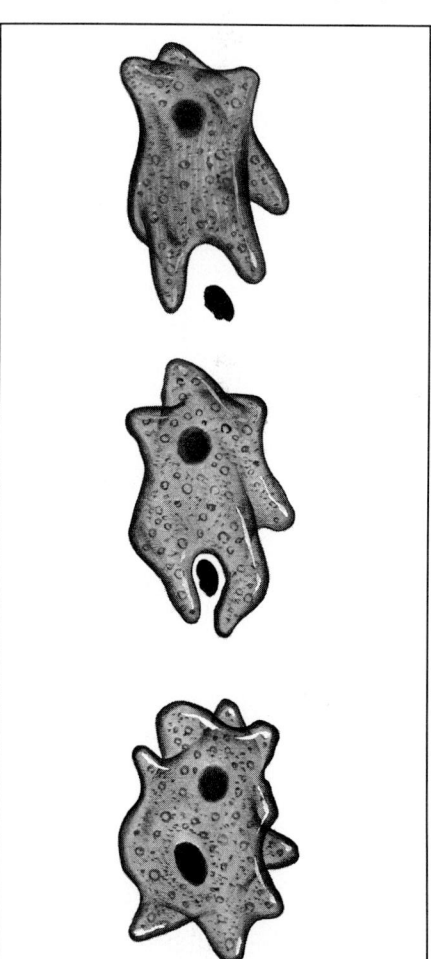

motion of its protoplasm. The cell bulges out on one side, and then the rest of the protoplasm flows into the bulge. Places where the cytoplasm bulges are known as *false-feet*. If it bumps into a piece of food, the ameba simply flows around the food on all sides. Finally the food is completely surrounded by the ameba's cytoplasm. There, inside the ameba, the food is slowly digested. See Fig. 5-9. The food might be any small piece of dead material in the water, or it might be a living cell. It could even be a *Chlorella* cell.

In the process of digestion an ameba uses enzymes to change the solid food into simpler, dissolved forms, such as glucose and amino acids. These can be used to supply energy through respiration or to build protoplasm. There may be parts of the food particle that cannot be digested. For instance, an ameba cannot digest the cell wall of a *Chlorella*. The ameba simply forces this undigested material out through its cell membrane. When the ameba moves on the waste material is left behind. Dissolved wastes like carbon dioxide pass out through the thin cell membrane, just as in *Chlorella*. Oxygen and mineral salts enter through the cell membrane.

The ability of an ameba to take in rather large solid particles is found also in certain cells of large animals. Your white blood cells and some liver cells can do this. Most cells cannot. However, it has recently been found that many animal cells are able to take in tiny particles, which are too big to go through the pores of the cell membrane. These include small undissolved food particles, and also some giant molecules, like protein molecules. A small part of the cell membrane folds out around the particle to draw it into the cell. The particle is surrounded by a bit of cell membrane that carries it to a vacuole or to some other part of the cytoplasm. We do not know yet just how many types of animal cells can do this. Probably plant cells cannot. Their cell walls would be in the way. Certainly the story of how materials pass through cell membranes is not entirely understood as yet.

We have used the ameba and *Chlorella* as examples of cells that get their food in different ways. Like the ameba, there are many consumer cells that need ready-made food. Other types of cells, like *Chlorella*, are able to carry on photosynthesis to make food. Aside from the ways in which cells get food, all cells carry on much the same life activities.

SUMMARY

Photosynthesis is the food-making process in green plants. Light is absorbed by chlorophyll and changed to chemical energy. Some of this energy is stored in ATP. The stored energy makes possible the manufacture of glucose from carbon dioxide and water.

During respiration, energy is stripped from glucose. The waste products produced are carbon dioxide and water. Respiration is carried on by both plants and animals at all times.

ACTIVITY

The Conditions Necessary for Photosynthesis

You will need three test tubes that are fitted with stoppers. You must also have sprigs of the water plant, *Elodea*, some bromthymol blue solution, a beaker, and a soda straw. Measure 1 teaspoon of bromthymol blue into the beaker. Add enough tap water so that you can fill the three test tubes. Use the drinking straw to blow your breath into the bromthymol blue.

1. How does the color change? The carbon dioxide from your breath causes the change in color. Into two of the test tubes place a sprig of *Elodea*. Fill with the bromthymol blue solution

and stopper. Fill the third test tube with the bromthymol blue solution and stopper. Now you have two test tubes with *Elodea* and one without. Place a test tube with *Elodea* and one without on the windowsill where they will get good sunlight. Put the other test tube in a dark place. Examine each in half an hour.

2. Do you see any changes?

3. In which tube do you see a change?

4. How do you explain the changes?

5. Why was it necessary to put the test tube without the *Elodea* on the windowsill?

Word Quiz

For each of the following words write a sentence that shows that you understand the meaning of the word. Do not write in this book.

1. photosynthesis

2. chlorophyll

3. mineral salts

4. nitrates

5. phosphates

6. *Chlorella*

7. chloroplast

8. ameba

Check Your Facts

1. Why is photosynthesis an accurate name for food-making in plants?

2. What is the relationship between each of the following pairs?   a. light and chlorophyll   b. ATP and energy   c. food and glucose

3. Tell how each of the following is used during photosynthesis and during respiration.
   a. glucose   b. ATP   c. carbon dioxide   d. oxygen   e. light

4. Why are green plants called producers?

5. What is the function of the cell membrane?

6. What is the difference between: a. chlorophyll and chloroplast; b. *Chlorella* and chloroplast?

7. Draw a diagram of an ameba. Name as many parts as you can.

8. Draw a diagram of *Chlorella*. Name its parts.

Science
Reading Skills
2. Cause and Effect
6. Compare and Contrast

Thought Questions

1. Draw a diagram showing how carbon dioxide is recycled.

2. Why do farmers add fertilizer to their soil?

3. Compare the green plant cell to a factory. Discuss: a. raw materials;   b. source of energy;   c. final products.

Science
Reading Skill
3. *Compare and Contrast*

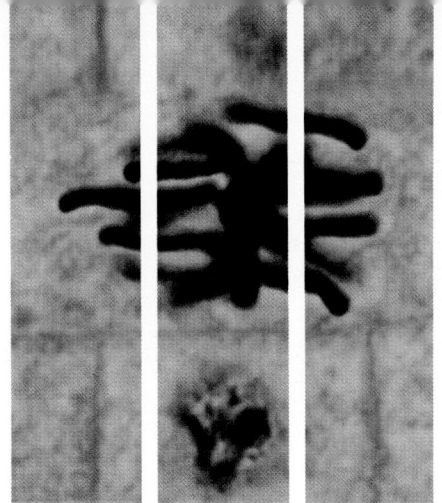

CHAPTER 6 Cells in Groups

The photograph above shows some cells of the onion root tip. Look closely at this picture. How can you tell that changes are taking place in the cells?

CELL DIVISION

In Chapter 2 you learned that protoplasm is the living material of cells. Living cells are able to carry out certain activities that nonliving things cannot. There are some activities that are common to all cells. One such life activity is the ability of a cell to reproduce itself.

A cell grows larger by an increase in the amount of its cytoplasm. The other cell parts also grow larger. But there is a limit to how large a cell can get. Instead of growing larger and larger, a cell will divide. The dividing of cells is known as *cell division.*

A complete living thing is called an *organism.* An ameba is an organism. So is *Chlorella.* When a single-celled organism divides, two new individual cells are formed. An onion plant is an organism with many cells. When a cell in a multi-cellular organism divides, the new cells stay together and grow to full size. When most cells reach full size, they divide again. You can see that there are two possible outcomes of cell division. Either new individuals are formed, as happens when *Chlorella* cells divide, new cells are formed that add to the size of the organism, or replace dead or damaged cells. This happens in the onion plant.

Cell division is the means by which new cells are formed. First the nucleus divides. Then the cytoplasm and the cell membrane divide.

Inside the nucleus. Let us now consider what takes place inside the nucleus. The process of cell division is not a simple

objectives

After you read this chapter, you should be able to:

___**Describe** the process of mitosis

___**Discuss** the function of genes

___**Define** tissue, organ, and system

___**List** five types of animal tissues

___**List** three types of plant tissues

___**List** eight body systems.

cell division

organism

Enrichment: An organism's size depends on the number, not the size, of its cells. In general the cells of an elephant are about the same size as an ant's cells. The elephant simply has more cells.

Enrichment: Discuss the meaning of organism.

one. If you stain the cells of an onion membrane with iodine, the nucleus absorbs the yellow-brown stain. This will make the nucleus show up more clearly.

Certain other dyes can be used to show detail within a nucleus. In most cells, the nucleus seems to have a grainy texture. When the cell is ready to divide, the appearance becomes different. Then, strands of nuclear material become clearly visible. These are called **chromosomes.** They are present all the time but are so long and thin that they cannot be seen until the cell is ready to divide. When the cell is preparing to divide, the chromosomes coil up into the shorter, thicker form that can be seen when properly stained. Chromosomes contain tiny units called **genes.** These genes control the cell's activities. They also control the heredity of the individual.

When cells divide, the two new cells must be just like the old one. They must be able to do the same things. Genes control the activities of the cell, so each new cell must receive all of the same genes the old one had. As you study cell division, watch for the way in which this happens.

The nucleus controls the cell's activities. Nucleic acids in the nucleus help to direct cell processes. Genes are made of nucleic acids. Genes in the nucleus control the production of enzymes in the cytoplasm.

Think of one gene as controlling the production of one kind of enzyme. Remember that each enzyme controls some chemical activity in the cell. It is the genes, then, that control what a cell can do. Organisms are different from each other because they contain different kinds of genes. Each living cell gets a particular collection of genes from its parent cell. Let's see what happens to genes when a nucleus divides.

An important characteristic of a gene is its ability to make an exact copy of itself. When a cell is about to divide, each gene becomes two genes. There are then twice as many genes as there were before. There are enough for each of the two cells to receive a complete set of genes. The genes of one cell are identical to those of the other cell.

Genes are made of a form of nucleic acid called **DNA.** DNA directs all of the activities of the cell. This form of nucleic acid passes information from one cell generation to another. Another form of nucleic acid, called **RNA,** carries directions from the genes to the cytoplasm. It is in the cytoplasm that enzymes and other proteins are made.

Nuclear division. There are a great many genes in a nucleus, perhaps as many as 100 thousand, yet the cell is able to keep these thousands of genes in order. This is so because many genes are linked together end to end. Each long string of genes (DNA) is bound chemically to protein to form chromosomes.

Enrichment: You may wish to introduce the term chromatin.

chromosome

Word Study: "chromo" means color

genes

DNA

RNA

Science Reading Skill: *Sequencing—* The sections on nuclear division reinforce this skill.

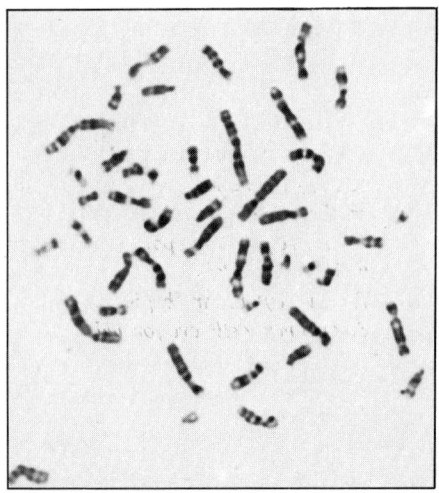

Fig. 6-1
This photograph was taken through a microscope. It shows the chromosomes found in a normal human body cell. Notice that they are already doubled.

mitosis

Activity: Use pop beads to demonstrate the relationship of genes to chromosomes.

A chromosome is really a package of genes. The cells in different kinds of organisms have their genes packaged into different numbers of chromosomes. In like organisms, the number of chromosomes in the cells is the same. Many bacteria have only one chromosome. Each cell in the body of a crayfish has more than 200 chromosomes. Each cell in your body has 46 chromosomes.

Before the nucleus begins to divide, each chromosome makes an exact copy of itself. This means that each gene in the chromosome has been duplicated also. Each chromosome now has a double like itself. These identical chromosomes remain paired until the nucleus is ready to divide. The chromosome pairs then separate. Each new nucleus gets a complete set of chromosomes having a complete set of genes. See Fig. 6-1.

This kind of nuclear division is called **mitosis** (my-TOE-sis). As a result of mitosis, one fully developed nucleus divides into two new nuclei. Each new nucleus has a complete set of chromosomes, and so it is able to produce the same enzymes and carry on the same activities as the old cell.

Stages in cell division. Genes are too small for us to see with a microscope. But we can see the chromosomes if we add stain to them. Even then, the only time we can see them well is during mitosis. At other times, the chromosomes are strung out so long and thin and are bunched so close together that they do not show clearly. When the nucleus gets ready to divide, the chromosomes become much shorter and thicker than usual, and then we can see them. They look like little rod-shaped objects in the nucleus.

Imagine that you are watching the nucleus of a cell during mitosis. See Fig. 6-2. First you see the chromosomes begin to appear. They become shorter and thicker until you can see them easily. If you look closely, you can see that each chromosome is double. We already know why this is so. At this time the membranes around the nucleus begin to disappear so the chromosomes can move to the middle of the cell. Now fibers of clear protoplasm form from two opposite ends of the cell. Each chromosome becomes attached to one of these fibers. Next, the pairs of chromosomes separate and move apart in opposite directions. They do this because they are being pulled by the clear fibers. All of the chromosome pairs separate at the same time. In this way, one of each kind of chromosome reaches one end of the cell. Each of these groups of chromosomes becomes enclosed by a nuclear membrane. The chromosomes again become long, thin, and tangled together. Finally the chromosomes cannot be seen any more. They have become parts of two new nuclei.

Now the cytoplasm can divide. In animal cells, the cell membrane pinches in around the middle. Finally it pinches all the

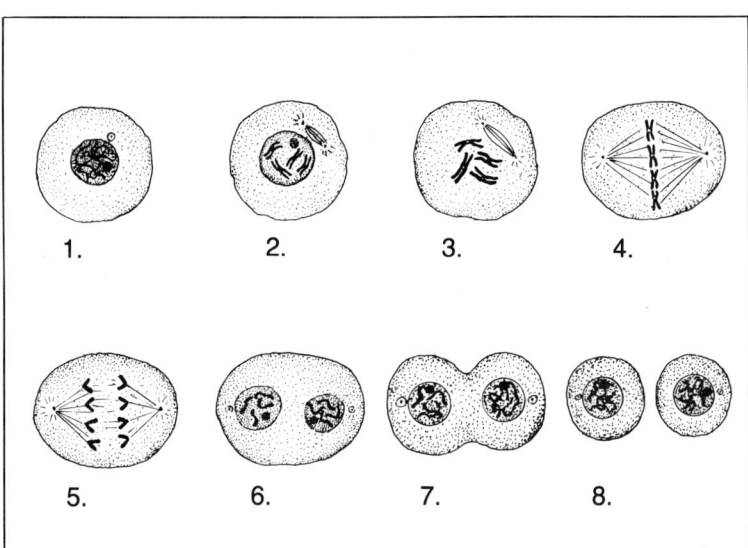

Fig. 6-2
Cell division in an animal cell. 1. The cell is about to begin division; 2. Chromosomes become thicker and can be seen more clearly—they are already doubled; 3. Nuclear membrane disappears; 4. Fibers of protoplasm attach to the centrosomes; 5. The chromosomes are pulled in opposite directions; 6. Two new nuclear membranes form; 7. The cytoplasm begins to divide; 8. Two new cells are formed.

way through, forming two new cells. Each has a nucleus just like the old one. Of course, the new cells are only half as large as the old one but they keep growing until they are full size. Then it may be the new cell's turn to divide. Plant cells cannot pinch in two because the cell wall is too stiff. Instead a new wall forms across the middle of the old cell, dividing it into two new cells.

CELLS ARE NOT ALL ALIKE

When an organism is made of many cells, its cells are not all alike. The cells in a multi-cellular organism do not all develop the same kind of cytoplasm. They grow into different sizes and shapes. Different kinds of cells develop which have special jobs to do. The cells in your body are specialists. This is true of the cells in any of the multi-cellular plants or animals.

The single cell of an ameba obtains everything it needs from its surroundings. The cell membrane of the ameba is in contact with the water around it. Water, food, and oxygen can enter such a cell directly. Wastes pass from the membrane into the surrounding water.

Most cells in humans, trees, or whales have no contact with the outside surroundings. They are buried thousands of cells deep, with nothing but other cells around them. These cells need oxygen and food also. They cannot get these things for themselves. So it is important that cells in larger organisms be arranged to help one another. Each kind of cell has a special job to do. And so cells of the same kind work together and carry out a specific function.

Enrichment: Stress the relationship between structure and function of different kinds of cells.

TYPES OF TISSUE

tissue

covering tissue

nerve tissue

muscle tissue

blood

Fig. 6-3

Diagrams and actual photomicrographs of, from left to right, covering tissue (Wards Natural Science Estab.), skeletal muscle tissue, cartilage, red and white human blood cells, nerve cell. (Carolina Biological Supply Co.)

A group of similar cells that do the same type of work is called a **tissue.** There are several types of tissues in the body of a multi-cellular animal. One type forms the body covering. The outer skin is made of this **covering tissue.** It also covers and lines structures that are inside of the body. The heart, stomach, and intestines are lined with this tissue.

A second type of tissue is specialized for carrying messages through the body. This is **nerve tissue.** Nerve cells have long fibers that extend to all parts of the body. Nerve tissue helps organisms to adjust to their surroundings.

Muscles in your arms, legs, or heart are made up of **muscle tissue.** The cells that make up this tissue have fibers inside of them that are able to get shorter. When these fibers receive messages from nerve cells, they shorten causing muscles to move. Of course, ATP supplies muscle cells with energy to do such work as movement.

A fourth type of tissue is **blood.** Although blood is a liquid, it is a tissue. It contains three types of cells that do their work while traveling through the body in a liquid. Blood transports digested food and oxygen to all body cells. It removes carbon dioxide from the cells. Some blood cells help to fight off certain diseases. Others help in the healing of wounds.

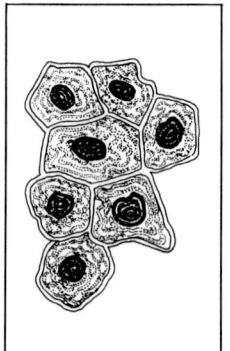 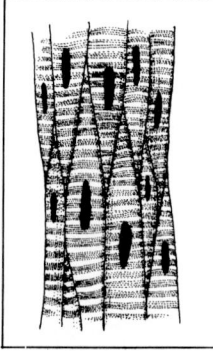 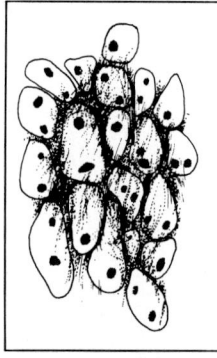 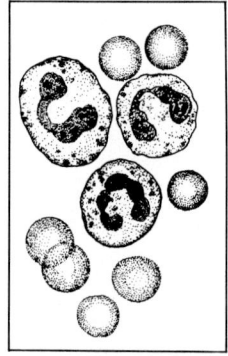 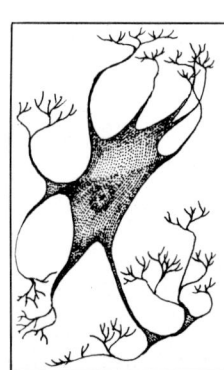

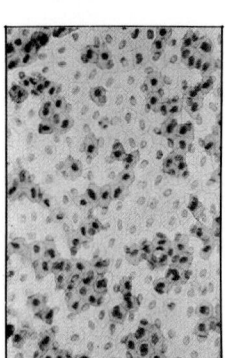 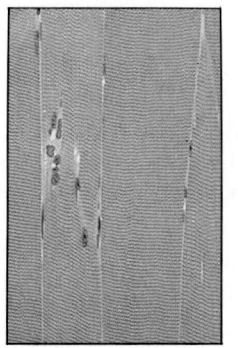 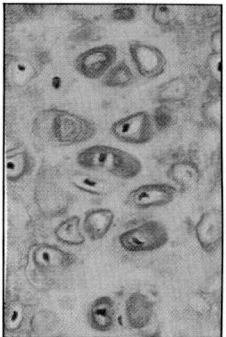 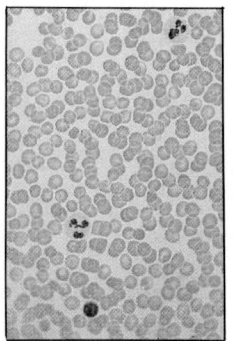 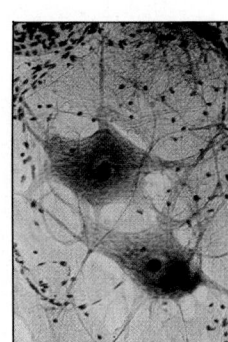

Connective tissue forms strong, tough fibers that bind body parts together. *Bone* is one kind of connective tissue. This forms the skeleton and gives shape and support to the animal body. Hard mineral matter is deposited between bone cells. This is why bone is so strong.

Cartilage (KART-ah-lij), which we sometimes call gristle, is another form of connective tissue. It contains scattered cells and a large amount of nonliving material between the cells. Other types of connective tissue attach bones to muscles and muscles to other muscles.

The tissues we have mentioned so far are a few of those found in animals. Examples of these are shown in Fig. 6-3. Plants also have tissues. One important type is called *vascular* (VAS-cue-lar) *tissue*. Vascular tissue is composed of tube-like cells. They carry liquids inside the plant. Water reaches the leaves from the roots by way of the vascular tissues. Plants also contain several other kinds of tissue. One of these is a covering tissue like the membrane in onions. Another type is supporting tissue. It is made of thick-walled, strong cells which hold up or protect the plants. Bark is a supporting tissue which is stiff and hard.

ORGANS AND SYSTEMS

Several tissues may be combined into a single working unit called an *organ*. Your heart is an organ. It has covering tissue, muscle tissue, connective tissue, and nerve tissue. Your skin is an organ too. It is made up of different types of cells. Lungs, eyes, and ears are examples of other organs. See Fig. 6-4 on page 52.

Usually, several organs work together forming a *system* of the body. A system gets a major job done. The higher animals have the following systems:

1. *Digestive system.* This system usually includes a tube which leads through the body. Food is dissolved as it passes through the tube. Enzymes break down the food into simpler substances. The dissolved food molecules pass through cell membranes. Parts of the tube may be enlarged to form special organs, such as a stomach or intestines.
2. *Circulatory system.* This is the blood system. Blood moves through the entire body. It absorbs digested food from the intestine and oxygen from the lungs. Blood supplies the body cells with food and oxygen. It carries away wastes given off by the cells. The body cells exchange molecules with the blood just as an ameba exchanges molecules with the water it lives in.
3. *Respiratory system.* This is the system that supplies oxygen. Fish have gills. Large land animals have lungs. In both

connective tissue

cartilage

Enrichment: You may wish to mention fat as a storage and protective tissue.

vascular tissue

organ

Activity: Use charts and models of various organs and systems that students can examine.

system

Science Reading Skills:
Generalization—Mention the systems only in terms of an overview at this time.

Reference: "I am Joe's Body" series in *Reader's Digest*

Fig. 6-4
These are examples of organs. (Left) A human heart; (right) a human brain. (Martin M. Rotker/Taurus Photos)

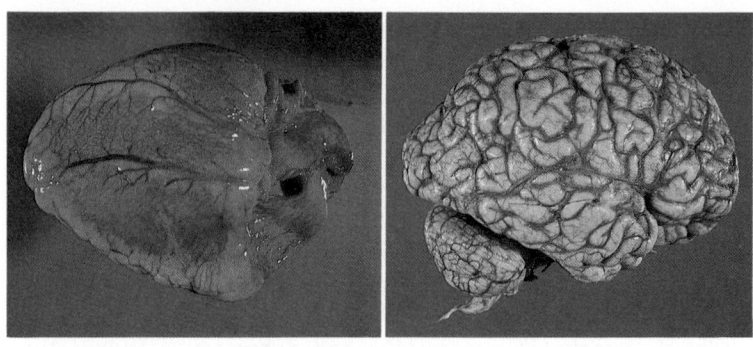

types of organs there is an arrangement of thin membranes. Oxygen diffuses through these membranes into the blood. Carbon dioxide leaves the blood through these same membranes.

4. *Excretory system*. Different animals have different types of excretory systems. But all excretory systems are arranged so that the body can get rid of its dissolved wastes. Nitrogen compounds and carbon dioxide are two types of wastes produced by cells.

5. *Reproductive system*. New individuals are produced by this system. The methods of reproduction may vary in different types of animals, but all organisms produce those of their own kind.

6. *Nervous system*. This system enables animals to respond to changes in the environment. Some nerve cells receive information. Others carry the information to muscles. The more complex an animal's nervous system, the more varied the responses it can cause.

7. *Muscular system*. Muscles are used for movement. You walk by using muscles in your legs and back. A fish swims by using muscles that move its fins and tail.

8. *Skeletal system*. A skeleton is any stiff material that helps support the soft part of the animal's body. In many animal groups the skeleton is a shell covering the body. A jointed skeleton with movable parts is found in only two of the animal groups. One of these groups has the skeleton on the inside, as you do. The other group, including insects, has the skeleton on the outside.

The simple and the complex. When we talk about "higher" animals or plants, we mean those with highly organized bodies. People, fish, and insects are examples of higher animals. Trees, grasses, and ferns are higher plants. "Lower" animals and plants are the simple ones without complex body structures. *Chlorella* and *Ameba* are two examples of lower forms of living things.

The simplest living things have no body systems. Each cell does everything for itself. The highest animals have all the eight systems which we have described. Other groups fall somewhere in between.

SUMMARY

Cell division includes nuclear division and division of the cytoplasm. The nucleus contains chromosomes made of DNA and organized into genes. Chromosomes duplicate themselves before cell division. Nuclear division is known as mitosis.

All of the cells in a multi-cellular organism are not alike. Groups of similar cells form tissues. Tissues form organs. Groups of organs make up special systems. Each body system controls a major body function. How living things carry out those functions depends upon the structure of the organism.

ACTIVITY

A Study of Tissues
A. Place the chicken wing in the pan or on the board. Either of these will give you a proper cutting surface. Use the tweezers to lift up the loose skin that is in the middle of the wing. Cut through the skin while holding it with the tweezers. Carefully remove as much of the skin as possible. CAUTION: **When using a pointed scissors, cut downward and outward—away from yourself.** As you remove the skin, notice the fine membranes that hold it in place.

1. Describe the color and the appearance of the outer skin.

2. How does the inner surface of the skin differ from its outer surface?

B. Look for layers of yellow fat. Fat is stored under the skin in small lumps.

3. How does the fat feel?

C. Locate the muscles. The muscles are pink in color and arranged in bundles.

4. What happens to the shape of the muscles as you bend the wing bones?

D. Carefully separate the muscles from the bones. As you do so, look for a vessel filled with blood. Look closely. You should be able to see a nerve. It is white and as wide as a thread.

E. Free the bone from all soft tissues. Examine the ends of the long bones. The whitish coverings are made of cartilage. Feel the cartilage.

5. How can you tell that it is different from bone?

Science Reading Skill: *Compare and Contrast*—Students are asked to use this skill in examining the differences and similarities in the tissues of a chicken wing.

Word Quiz

Match the item in **Column A** with its function in **Column B.**

Column A	**Column B**
1. cell division	**a.** a group of similar cells
2. organism	**b.** tissues that work together
3. chromosome	**c.** a complete living thing
4. genes	**d.** equal division of chromosomes
5. DNA	**e.** controls activities of cells
6. RNA	**f.** rod-like structures
7. mitosis	**g.** resemble beads on a string
8. tissue	**h.** control enzyme manufacturing
9. organ	**i.** one cell becomes two cells

Name the types of tissues described in each of the following:

10. carries messages   **11.** lines some organs
12. water-carrying plant tissue   **13.** controls movement
14. transports food and oxygen   **15.** joins bone to muscle

Check Your Facts

Science
Reading Skills
1. *Reading
Illustrations*
3. *Reading
Illustrations*
6. *Compare and
Contrast*

1. Look at Fig. 6-1. How do you know that this cell is not dividing?
2. Why does cell division take place?
3. Study Fig. 6-3. Define mitosis. What changes in the nucleus do you see at each stage of mitosis?
4. Explain the relationship of each of the following to each other: genes, DNA, chromosomes.
5. How do enzymes help the work of cells?
6. Explain how each of the following are related to each other: cell, tissue, organ, system.
7. From where do muscle cells get the energy for movement?
8. A *Chlorella* cell has no vascular system. How does it obtain water?
9. An ameba has no excretory system. How does it get rid of carbon dioxide?
10. What is meant by a "higher" animal?

Thought Questions

1. Genes cannot be seen. Yet scientists know that they exist. Explain.
2. Why is skin considered to be an organ?
3. How is it that an ameba can live alone and a skin cell cannot?

Scientific research and investigation are aided by the skills of many technical assistants.

Scientists work together with technical staff who have an interest in the subject and whose skills are needed (1) to produce and care for the instruments of research and (2) to carry out many of the testing procedures necessary to careful experimentation and reliable data.

Microscopes are essential to the work of medical personnel, research scientists, and industrial development scientists. While these instruments are mass produced, often overseas, they are changed or adapted to the special needs of the task at hand. This is particularly true in industry.

A series of jobs has grown around these purposes. The necessary skills are usually learned on the job. There are, for example, people who repair microscopes. There are also those who design accessories for them, that is, additions which make them more powerful or better suited for what they have to do. Designing as well as creating these changes is essential to research. Mechanical ability and the willingness to do precision work

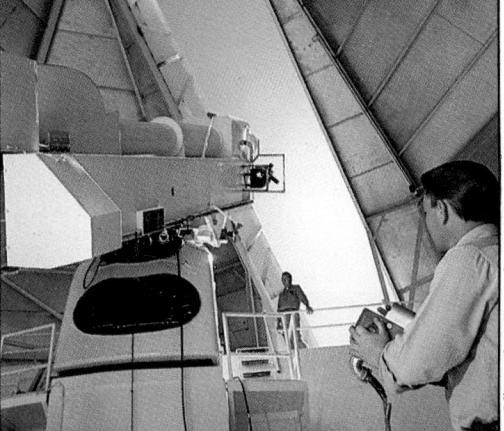

Guy Gillette/Photo Researchers, Inc.

are the basic requirements. A math and science background in high school will be especially helpful.

Similar support occupations surround astronomers who have, of course, an advanced degree in astronomy, usually a Ph.D. There are about 2,000 astronomers in the United States today and only about 8 openings in this field a year, but the research itself supports other personnel. There are telescope operators, instrument makers, opticians who work with the lenses, electronics technicians, and photographers. The photo of the space mirror under construction demonstrates the immensity of

Hank Morgan/Photo Researchers, Inc.

the operation. There are also management and public relations people who work with astronomical research organizations and planetariums. Advanced degrees are not necessary for these jobs, but interest in the subject is, as well as good interpersonal skills.

A variety of biological and chemical research and lab technicians are necessary, not only in medicine but in industry. The business of food processing provides only one example. The food processing technician buys, processes, preserves, packages, inspects, grades, and markets every type of food product processed for public consumption. In most large food companies there is a

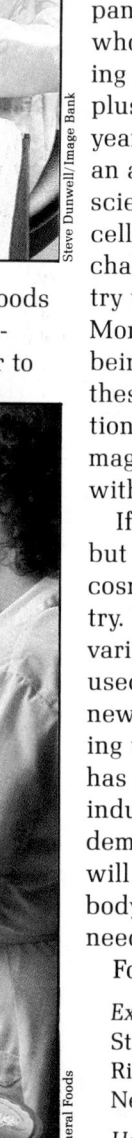

Steve Dunwell/ Image Bank

consumer division where foods are tested, tasted, and combined in new ways in order to

provide new marketing opportunities. Mixtures of ingredients must taste good. Methods of cooking and preserving must be appropriate to the varying conditions across the country. Baking a cake in Montana presents different problems from baking in Texas simply because of differences in altitude. Companies look to recruit persons who have a basic understanding of biology and chemistry, plus public relations skills. Two years of technical school with an associate degree in applied science would provide an excellent background. Consumers challenge the claims of industry to an ever greater extent. More persons are, therefore, being employed to deal with these challenges. Similar positions are to be found with food magazines and publications with cooking sections.

If chemistry is an interest, but food is not, consider the cosmetics and fragrance industry. Many lab technicians with various levels of expertise are used in developing and testing new products. Men's increasing use of associated products has added to the growth of this industry. Consumers' greater demands for ingredients that will not be damaging to the body have also increased the need for lab technicians.

For further information, read:

Exploring Careers in Science
Stanley Jay Shapiro
Richard Rose Press, Inc.
New York, NY   1981

How Scientists Find Out
Herman Schneider
McGraw-Hill Book Company
New York, NY   1976

Courtesy of General Foods

**Learning how to find information and
sort it out is a skill particularly useful in
your consumer and future voting decisions.**

Research is simply another name for "methods of finding out." The events of daily life in our society are complicated to understand. At the same time, public media—television, newspapers, and magazines—throw out a confusing mixture of information. Learning how to find information and sort it out is a skill particularly useful in your consumer and future voting decisions.

Libraries contain many research tools. Students sometimes associate a library only with literature and history books. However, there are books on subjects ranging from how to rewire your guitar to how to get the most out of your visit to Oklahoma City.

Newspapers also have libraries where back issues may be read, and industries may have libraries which stock information relevant to their history and methods of production. Hospitals have libraries that may be open to the public. And your local city hall is also a kind of library, with information on land ownership and municipal regulations, among other things.

HRW: Ken Karp

Learning to use instruments of research such as microscopes, telescopes, personal computers, word processors, and calculators could prove especially important in your future job search. Many types of work are now done through means of a computer. Keep alert to all available sources of research tools. Your library may have a computer for public use and your librarian will be a source of computer skills and information. Increasingly, books will be stored in computers, not on shelves. Sometimes community organizations and agencies hold special programs open to the public, which teach computer skills or which make these machines available for use. You don't need to have an immediate use for the skill. It may prove valuable later on.

Observations and listening provide information which can be used in problem-solving. With those skills, almost anything we do and anyone we speak with can be a source of information at some time. Business analysts say that sales and marketing are areas where there will be many available positions in the future and where incomes can be high. Developing sales techniques and marketing strategies depends on knowing the potential consumer, a knowledge that comes with good information-gathering skills.

Townsend P. Dickinson/Photo Researchers, Inc.

CONSUMER SCIENCE IN...

NEW MEXICO'S LEADING NEWSPAPER

ALBUQUERQUE JOURNAL

Mystery Find Stuns Astronomy
Orbiting Telescope May Have Discovered Elusive Member of Solar System

Los Angeles Times
Washington Post Service

WASHINGTON — A heavenly body possibly as large as the giant planet Jupiter and possibly so close to Earth that it would be part of this solar system has been found in the direction of the constellation Orion by an orbiting telescope called the Infrared Astronomical Observatory.

So mysterious is the object that astronomers do not know if it is a planet, a giant comet, a nearby "protostar" that never got hot enough to become a star, a distant galaxy so young that it is still in the process of forming its first stars or a galaxy so shrouded in dust that none of the light cast by its stars ever gets through.

"All I can tell you is that we don't know what it is," Dr. Gerry Neugebauer, IRAS chief scientist for California's Jet Propulsion Laboratory and director of the Palomar Observatory for the California Institute of Technology, said in an interview.

The most fascinating explanation of this mystery body, which is so cold it casts no light and has never been seen by optical telescopes on Earth or in space, is that it is a giant gaseous planet as large as Jupiter and as close to Earth as 50 trillion miles. While that may seem like a great distance in earthbound terms, it is a stone's throw in cosmological terms, so close in fact that it would be the nearest heavenly body to Earth beyond the outermost planet Pluto.

"If it is really that close, it would be part of our solar system," said Dr. James Houck of Cornell University's Center for Radio Physics and Space Research and a member of the IRAS science team. "If it is that close, I don't know how the world's planetary scientists would even begin to classify it."

Then, what is it? What if it is as large as Jupiter and so close to the sun it would be part of the solar system? Conceivably, it could be the 10th planet astronomers have searched for in vain. It also might be a Jupiter-like star that started out to become a star eons ago but never got hot enough, like the sun, to become a star.

While they cannot disprove that notion, Neugebauer and Houck are so bedeviled by it that they do not want to accept it. Neugebauer and Houck "hope" the mystery body is a distant galaxy either so young that its stars have not begun to shine or so surrounded by dust that its starlight cannot penetrate the shroud.

The mystery body was seen twice by the IRAS satellite as it scanned the northern sky from last January to November, when the satellite ran out of the supercold helium that allowed its telescope to see the coldest bodies in the heavens. The second observation took place six months after the first and suggested the mystery body had not moved from its spot in the sky near the western edge of the constellation Orion in that time.

"This suggests it's not a comet because a comet would not be as large as the one we've observed and a comet would probably have moved," Houck said. "A planet may have moved if it were as close as 50 trillion miles, but it could still be a more distant planet and not have moved in six months time."

Research is a continuing process of questioning and of observing. Nothing can be taken for granted. This curiosity, combined with the willingness to pursue an investigation thoroughly, is the kind of motivation necessary to do research.

The elements necessary for discovery include the technological developments that make them possible. In this case, a telescope orbits the earth using helium to make it possible for scientists to see the coldest bodies in space. Even with exceptional techniques and training, however, answers are not clear.

Of what importance could this new "find" be? Would it make a difference if it were the tenth planet in our solar system or so far away it is only what we can see of a completely new galaxy? Try to imagine how this information might prove useful in some now unknown way, for example, in space travel. Do you think it's possible that today's new find could eventually become a stop on a public transportation system 1,000 years from now?

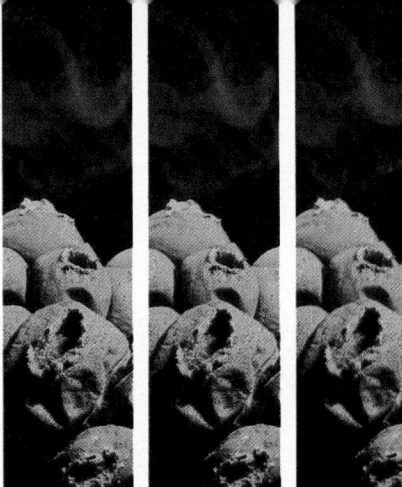

CHAPTER 7 Reproduction in Simple Organisms

objectives

After you read this chapter, you should be able to:

Analysis

___**Explain** the difference between sexual and asexual reproduction

___**List** and **Describe** several different methods of sexual reproduction

___**List** some of the organisms that reproduce by asexual methods

___**List** and **Describe** several different methods of asexual reproduction

___**List** some of the organisms that reproduce by sexual methods

___**Explain** the advantages of a-sexual reproduction and the advantages of sexual reproduction

sexual reproduction
asexual reproduction

Word Study: "a-" means without
Science Reading Skill: *Compare and contrast* asexual and sexual reproduction.

Sooner or later plants and animals die and new ones replace them. If each kind of living thing is to survive, it must have some way to reproduce. There are millions of different kinds of living things. They have many ways of reproducing, but there are only a few *basic* methods. In the photograph above some puffballs are giving off reproductive spores. In this chapter you will learn about this method of reproduction and others.

KINDS OF REPRODUCTION

There are two main types of reproduction. One is called *sexual reproduction* and the other is called *asexual reproduction.* Sexual reproduction happens in many different ways, but it always includes the **union of two cells.** These cells unite to form a single cell. This new cell is the beginning of a new individual. It can divide again and again until it forms all the cells of the new plant or animal. In sexual reproduction there are two parents. They produce the two cells that unite with each other. These cells contain genes, so the new individual is something like both parents. Do you begin to see how heredity works?

Asexual reproduction is a kind of reproduction in which **no union of cells takes place.** There is only one parent. A new individual is formed by division from a cell or group of cells coming from that one parent. The offspring have the same genes as the parent.

EXAMPLES OF ASEXUAL REPRODUCTION

Cell division. In Chapter 6, you learned how cells divide. In plants or animals having many cells in their bodies, such division results in growth. The individual becomes bigger as the cells multiply. When a one-celled living thing divides, the two cells formed separate. Each of them becomes a new individual. A single cell grows to full size. Then it divides to form two cells. This is the simplest form of reproduction. Each of the new cells is just like the parent cell because mitosis gives each the same genes. Single-celled living things usually reproduce in this way.

The ameba is a good example of an organism that reproduces by cell division. First the nucleus divides by mitosis. Then the cytoplasm pinches in two between the new nuclei. See Fig. 7-1. The parent ameba becomes two amebas. Each of them is only half the original size, but they take in food and grow until they too are large enough to divide again.

Reproduction by spores. Another form of asexual reproduction is by *spore* formation. A spore is a single cell that is able to grow into a new organism.

A mushroom produces spores. If you look on the underside of a mushroom cap you will see flat layers of tissue that produce spores all over their surfaces. See Fig. 7-2. One mushroom may form millions of spores. The tiny spore cells fall off and blow around in the wind like dust. A spore may be carried many kilometers before it finally lands. If it happens to land in a place where mushrooms can grow, it will grow into a new mushroom plant.

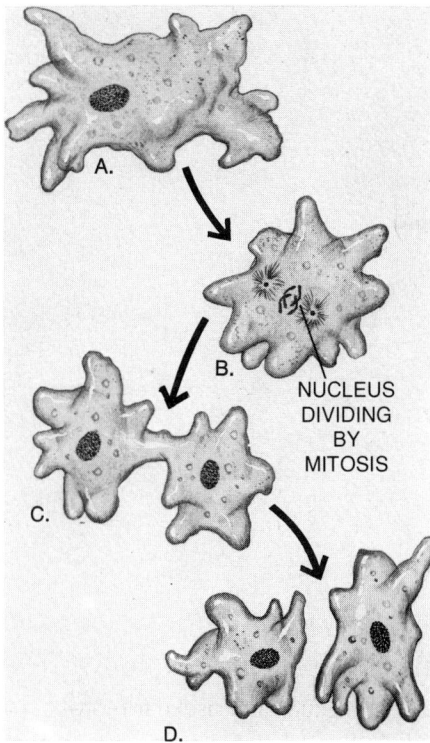

A.

B.   NUCLEUS DIVIDING BY MITOSIS

C.

D.

Fig. 7-1
Dividing ameba. In single-celled organisms cell division is a method of reproduction.

spore

Fig. 7-2
Mushrooms produce tiny spores on their lower surfaces. The diagram shows how these spores would look. (Earth Scenes/© F. E. Unverhau)

Fig. 7-3
The brown structures on the underside of this fern leaf are really groups of spore cases. (Manuel Rodriguez)

spore cases

vegetative reproduction

Other groups that reproduce by means of spores include the molds, mosses, and ferns. Perhaps you have seen small spots on the undersides of fern leaves. See Fig. 7-3. These are clusters of *spore cases*. Spores form in these. When they are ripe, the spore cases split open, and the spores fall out.

The big advantage in the spore method of reproduction is that millions of spores can be carried long distances by the wind. Spores spread the organisms far and wide. The disadvantage of spores is that they are so tiny. They cannot carry enough stored food in their cytoplasm to give the new organism a good start. If a fern spore lands on moist soil, it may grow into a tiny fern plant. But if the surface of the soil dries out too soon, this plant will die. The underground parts will not be long enough to reach down to water. How do seed plants have an advantage in this respect?

You can see that conditions must be just about perfect if a spore is to grow. Because most spores do not land where there are such perfect conditions and they die, the organism must produce spores by the millions. Suppose all the spores coming from one fair-sized puffball, a fungus something like a mushroom, were to grow into new puffballs. These new puffballs, if put together, would make a pile bigger than the entire earth! Obviously, the spores cannot all grow. With so many spores being produced by simple living things, the air is always loaded with them. They are a part of the dust that is always present in the air. This very minute, as you are reading this page, you are probably breathing in many different kinds of spores.

Reproduction by vegetative means. *Vegetative reproduction* is another form of asexual reproduction. In higher plants, the flower is the reproductive structure. It produces seeds by a sexual process. The stems, roots, and leaves are called the *vegetative* parts of the plant. They have to do with carrying on the ordinary activities of the plant. If one of these parts grows into a new plant, we call it vegetative reproduction.

There are many examples of vegetative reproduction. Strawberry plants send out long, slender stems called *runners*. These runners take root and form new plants at their tips. Iris plants spread by means of thick, branching stems that grow just under the surface of the soil. New plants form at the ends of these branches. Tulips, onions, and other plants form new *bulbs* on the sides of the old ones. These new bulbs grow into new plants. Raspberry bushes may form new plants when the tips of their stems touch the ground, take root, and grow.

It is often an advantage for people to use vegetative reproduction when starting new plants. It usually saves time. The growing of potatoes is an example. The potatoes you eat are

special underground stems that store food. When we wish to grow more potatoes, we do not plant seeds. Instead, we plant pieces of potato containing several buds or "eyes." These "eyes" send out sprouts that grow into new plants. It would take two years to produce a crop of potatoes that were started from seed. By planting pieces of potato, a crop is produced in one season. Another advantage is that the new plant is sure to be just like the old one. In vegetative reproduction the cells of the new plant have the same chromosomes as those in the parent plant. These chromosomes are passed on each time cells divide by mitosis. Fig. 7-4 shows several methods of vegetative reproduction.

Plants grown from seeds are more of a problem. The seeds are produced sexually. Each seed contains genes from two parents, so a seed has a new gene combination. You cannot be sure that it will grow into the exact variety of plant you want.

Now suppose you have one plant of the type you want and it is a plant that can be grown by vegetative means. When you grow it vegetatively you know exactly what you will get.

Some of the plants raised at home can be started by vegetative means. If you want a new geranium plant you can get it by

cutting

grafting

Reference: For his research on plants and breeding see *Luther Burbank: The Wizard and the Man* by Ken & Pat Kraft.

cambium layer

using a *cutting.* Cuttings are also called slips. Here is what you do. Get someone who has a geranium plant to allow you to cut off 10 or 15 centimeters from the end of a stem. Next, place the cut end of this slip in a glass of wet sand. When roots begin to grow, it is time to plant the cutting in a pot of good soil. African violets can be started from leaves instead of stems. A leaf half buried in soil will grow into a new plant. Many plants grown in greenhouses and nurseries are multiplied by cuttings and other vegetative means. Unfortunately, a good many plants will not grow from vegetative parts and must be produced from seeds.

Grafting. *Grafting* is a very special way of using cuttings. The stem, or bud, cut from one plant is made to grow on another closely related plant. Suppose the people in a plant nursery wish to have more Jonathan apple trees. First they plant seeds of some strong kind of crab apple. When these seeds have grown into young trees, they cut off their tops and graft on cuttings from a Jonathan apple tree. The Jonathan top and the crab apple base grow together. The entire top of the new tree will grow from the Jonathan cutting. The entire root system will grow from the crab apple seedling. Since the top produces fruit, this tree will yield Jonathan apples when it is big enough. Notice that **grafting produces nothing new.** It is simply a way to preserve and multiply a good variety of plant that we already have. When you buy fruit trees or rose bushes, the salespeople can guarantee what kind of fruit or flowers they will bear because they know what has been grafted on these plants. Only closely related plants can be grafted together. Apples and pears are related, so they could be grafted. So could plums and cherries. Actually this is not done very often. Most grafting is done between different varieties of the same basic types of plant.

It is possible for a single tree to produce several varieties of fruit. Such trees never grow naturally. They are produced by grafting. Cuttings from several varieties of tree are grafted onto a single trunk. Each cutting becomes a whole branch on the new tree, and each branch forms its own kind of fruit. In this way, early apples, late apples, eating apples, cooking apples, and so on can all grow on a single tree. Nurseries will sell you young trees that will produce four or five kinds of apple. This is of no use to a fruit grower who has many trees, but it is nice for a family that has room for only one tree in the back yard.

There are several ways of making grafts, but they are all based on the same idea. In trees and shrubs there is a special layer of cells between the bark and the wood. It is called the *cambium* (KAM-bee-um) *layer.* Cells in the cambium multiply and produce new growth in the thickness of the stem. In grafting, the cambium layers of the two plants must be brought together. The two plant parts are fastened together. Then wax is smeared on the outside to prevent the living tissues from drying up. Fig. 7-5 shows some of the ways grafts are made.

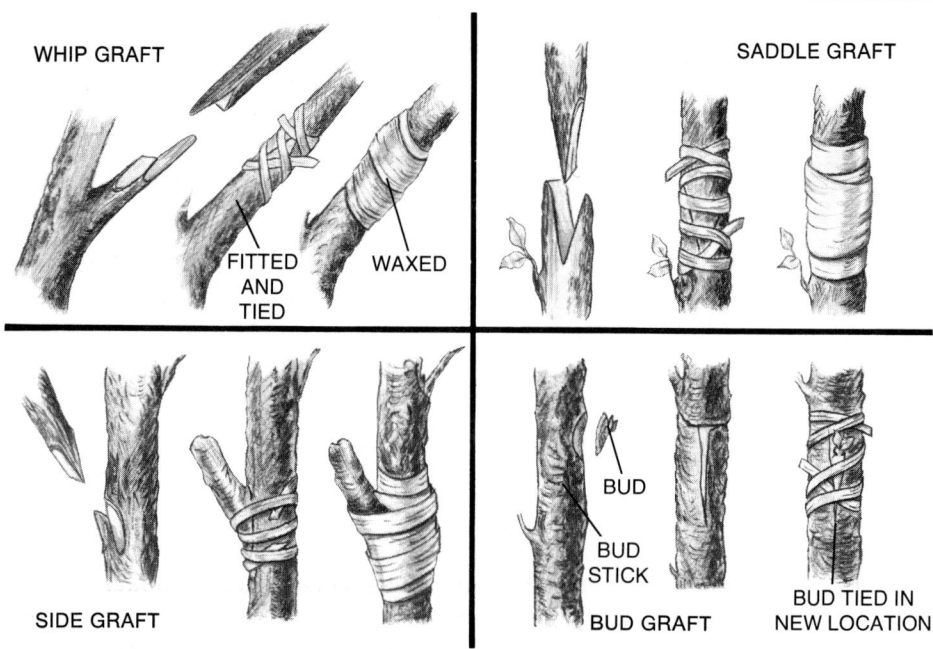

WHIP GRAFT

FITTED
AND
TIED

WAXED

SADDLE GRAFT

SIDE GRAFT

BUD

BUD
STICK

BUD GRAFT

BUD TIED IN
NEW LOCATION

Fig. 7-5
Several forms of grafts. What parts of the plants are brought together in a graft? Why are the parts tied? Why are they coated with wax?

budding

Budding. Some primitive animals reproduce by a process called *budding*. Budding in animals is the same as vegetative reproduction in plants. A part of the body grows outward and becomes a new individual. A good example of budding is found in *Hydra*. A *Hydra* is a tiny animal that lives in fresh-water ponds. It is about two millimeters long, and it has a tube-like body with a mouth at the top. Around the mouth is a ring of arm-like structures called *tentacles*. Most of the time *Hydra* reproduces asexually by budding. A bulge forms in the body wall. This is a bud. Its outer end develops tentacles and a mouth. When it is fully grown the bud breaks loose and becomes a new *Hydra*. See Fig. 7-7 on page 63.

EXAMPLES OF SEXUAL REPRODUCTION

Reproduction in *Spirogyra*. You will remember that in sexual reproduction two cells unite to form one cell. A simple example of this is found in an organism called *Spirogyra* (spy-roh-JY-ruh). *Spirogyra* is simply a long row of cells fastened end to end. Each cell has one or more spiral-shaped chloroplasts for making food, as shown in Fig. 7-6. In fact, this organism lives in much the same way as *Chlorella*, except that it is made up of rows of cells instead of single cells. *Spirogyra* growing in the water looks like a mass of green threads. These threads are merely rows of *Spirogyra* cells. The threads become longer by ordinary cell division and growth. When waves or fish break them, it results in a sort of vegetative reproduction. The cells keep on living and dividing. They can become long threads again.

Now and then, a mass of *Spirogyra* threads will start a process that is a simple type of sexual reproduction. Bulges will form along the sides of two threads that happen to lie side by side. A bulge will form in the wall of each cell where it touches a cell of the other organism. As the bulges grow longer they push the threads apart, so the entire structure looks like a ladder. The tips of the bulges open and thus produce a tube from each cell in one thread to each cell in the other. See Fig. 7-6.

While these tubes are developing, changes are taking place in the protoplasm of each cell. The protoplasm shrinks down to less than half its original size. It does this by getting rid of the water in its large vacuole. By this time the mass of protoplasm does not nearly fill the old cell wall. These shrunken cells in one *Spirogyra* move across, through the tubes, to join the cells in the other. Each of the pairs of cells flows together, forming a single cell. Now one *Spirogyra* has lost all of its living material. But each cell in the other one contains the material from two cells.

Fig. 7-6

Sexual reproduction in Spirogyra. *Different steps are pictured in the diagram on the left. An actual photograph of magnified* Spirogyra *is on the right. Note how the living parts of the two cells are uniting to form a single new cell. These new cells later grow into new* Spirogyra *threads (Photo, Carolina Biological Supply Company)*

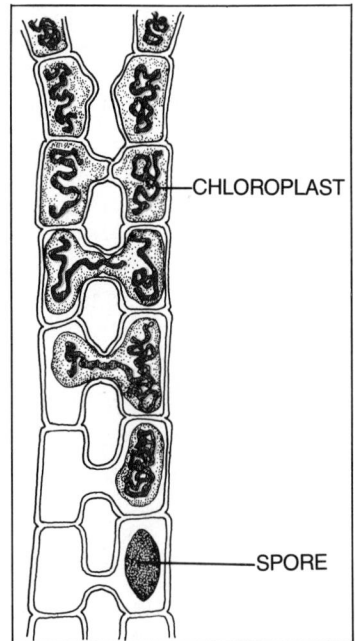

CHLOROPLAST

SPORE

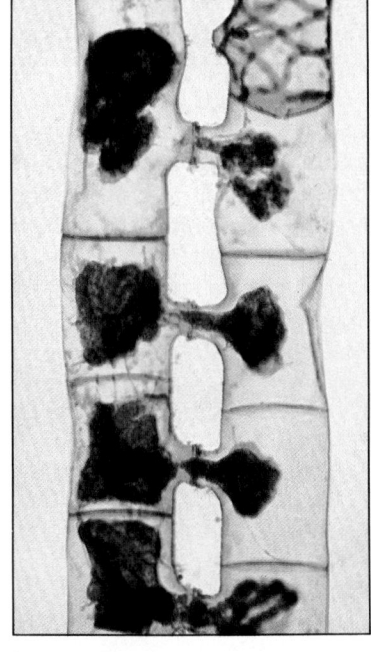

Each of these new cells can grow into a new *Spirogyra*, but they do not do it right away. They form heavy cell walls around themselves and become what are called *resting spores*. A resting spore can go without water for a long time, so the organisms are able to survive when the pond they live in dries up. When fresh, cool water floods the pond again, the walls of the resting spores split open. The cell in each spore grows and divides over and over again. A new *Spirogyra* is formed.

Reproduction by sperm and eggs. The *Spirogyra* method is the simplest kind of sexual reproduction because the ordinary body cells act as sex cells. Most living things produce special cells for this purpose. Usually there are two kinds of cells. One type is tiny and has a hair-like tail with which it can swim. The other type is larger and cannot move on its own. This larger kind of sex cell is called an **egg**. The smaller, swimming one is called a **sperm**. A sperm can swim to an egg and unite with it. Their nuclei join and form a single nucleus. This process is called *fertilization.*

egg
sperm

fertilization

We have already described how the little pond water animal called *Hydra* reproduces by budding. *Hydra* also reproduces sexually by forming sperm and eggs. See Fig. 7-7.

When *Hydra* reproduce sexually two kinds of bulges appear on the sides of their bodies. In one kind of bulge, a single large cell develops. This is an egg. In other bulges, large numbers of sperm are produced. When these sperm break out into the water, they swim about until they reach the egg cells. A single

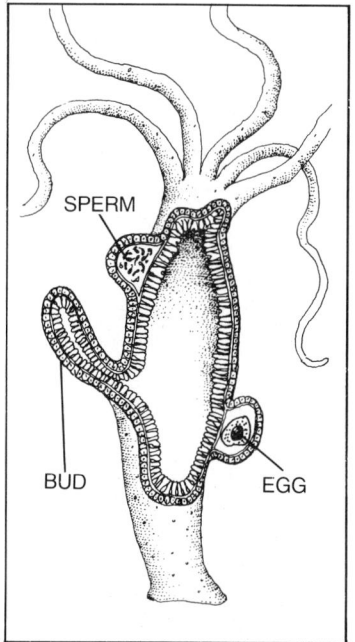
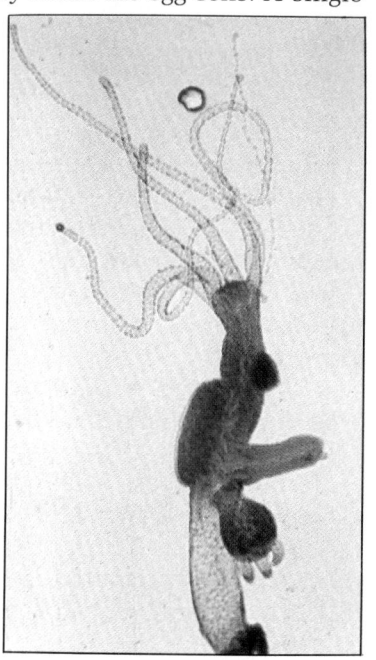

Fig. 7-7
Left: A diagram showing all of the possible structures a Hydra *may use for reproduction. These are not always all found on the same organism; Right: A photograph of a* Hydra *with a bud growing out from the body wall. (Photo, Carolina Biological Supply Company)*

63

sperm unites with each egg cell. This fertilized egg cell can then begin to divide. It divides many times to form the cells of a new *Hydra*. If a sperm does not happen to meet an egg, no new *Hydra* is produced.

All but the very simplest living things reproduce by means of sperm and egg cells. A sperm is a male cell. An egg is a female cell. This is a difference between males and females. A male produces sperm. A female produces eggs. In simple animals like the *Hydra* the sexes may be separate or one animal may produce both sperm and eggs. In most animals the sexes are separate.

You might think that because the egg cell is bigger, it would have more influence on heredity than the sperm. This is not true. Remember, the nucleus controls heredity. The nucleus of the sperm contains just as many chromosomes as the nucleus of the egg cell does. Both cells have equal effect on the heredity of the new *Hydra* they produce. The large cytoplasm of the egg cell contains stored food. This food gives the new animal the materials needed for rapid growth.

ADVANTAGES OF SEXUAL REPRODUCTION

Imagine a situation where *Hydra* never produced sperm and eggs, but always reproduced by budding. Every new *Hydra* would be exactly like the old *Hydra* it came from. Some members of the entire *Hydra* population might have a gene that made them especially successful at catching food. Others might have a gene that helped them resist disease. Could both of these genes be brought together in one *Hydra?* Obviously not. Asexual reproduction can only hand on the gene combination already present in a single parent.

Now let's see what happens when *Hydra* reproduce sexually. A sperm carrying the gene for one useful trait may fertilize an egg containing the gene for the other useful trait. The new *Hydra* growing from this fertilized egg will have both of these genes. It will be better than average at getting food and in its ability to resist disease. With these advantages it will probably live longer and it will reproduce more often. The useful combination of genes will be inherited by many of its offspring. This type of *Hydra* would then become more common in the total population.

Meanwhile, those *Hydra* that inherit neither of the useful genes will be less successful. On the average they will not live as long and they will not produce as many offspring. Their gene combination would become less and less common. In this way sexual reproduction helps to keep living things strong and able to survive.

SUMMARY

When two cells unite to form a new individual the process is called sexual reproduction. When a cell or cells form a new plant or animal without any such union of cells, it is an example of asexual reproduction.

Single-celled organisms divide to form two new individuals.

Spores are special cells that can grow into a new organism.

Vegetative reproduction is the production of a new individual from the stem, root, or leaf of a plant.

Grafting is the process of getting parts of two plants to grow together. It is a way of multiplying a good variety of fruit or flower.

In *Spirogyra*, cells from two plants may unite to form one cell. This cell grows into a new organism. The simple animal, *Hydra*, produces tiny swimming cells, called sperm. These unite with large stationary cells called eggs. These fertilized eggs grow into new *Hydra*.

Sexually produced organisms inherit a mixture of traits from their two parents. Asexually produced individuals will be just like the single parent from which they came.

Activity

Spores Produced on Land

Mushrooms spores are a good example of spores produced on land.

A. Examine some fresh mushrooms that have been bought at a grocery store. Older ones that are fully opened are best. Wild mushrooms from outdoors may be used, also, if any can be found.

CAUTION: **If you handle wild mushrooms be sure to wash your hands afterward. Some of them are poisonous.**

Notice the thin layers of tissue on the underside of the mushrooms. Spores grow on the surfaces of this tissue. Cut the cap off the stem. Lay this cap on a sheet of paper. Place it with the underside down. In a day or two, examine the paper where the mushroom cap has been lying. The colored design on the paper is called a spore print. It is made up of thousands of tiny spores that fell from the mushroom's underside. Place some of them in a wet mount and examine them under the microscope.

1. Why do you think these spores are so tiny?

2. Why does the mushroom produce so many spores?

3. Are the spores produced by the mushroom involved in sexual or a-sexual reproduction?

Word Quiz

Write the meaning of each of these terms. Do not write in this book.

1. sexual reproduction
2. asexual reproduction
3. spore
4. vegetative reproduction
5. cutting (slip)
6. grafting
7. cambium layer
8. egg
9. sperm
10. fertilization
11. budding

Check Your Facts

Science
Reading Skills
1. & 7. *Compare
and Contrast*
9. *Generaliza-
tion*

1. What is the difference between sexual and asexual reproduction?
2. When does cell division result in growth, and when does it result in reproduction?
3. What is a spore? Which organisms reproduce by means of spores?
4. What is vegetative reproduction? Give some examples.
5. Why are fruit trees usually produced by grafting?
6. In *Spirogyra,* how do cells unite in sexual reproduction?
7. What is the difference between sperm and eggs?
8. How does the sperm of a *Hydra* reach the egg?
9. Explain some advantages of asexual reproduction and sexual reproduction.

Thought Questions

Science
Reading Skills
1. & 2. *Problem
Solving*

1. Would you start a new apple orchard by planting seeds from the best apple you could find? If so, why?
2. It is often said that grapefruits were first formed by grafting orange branches onto lemon trees. Do you believe this? Why?

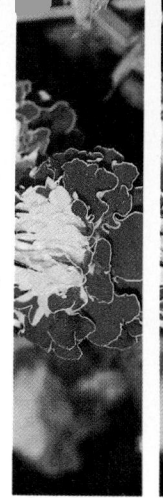
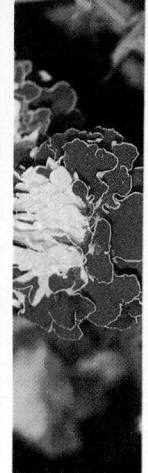
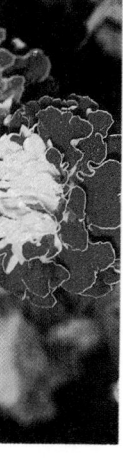

CHAPTER 8   Reproduction in Flowering Plants

Imagine that flowers and bees had the power to think and to talk. The bee in this picture would say, "I am stealing two valuable things from this flower." The flower would say, "I have fooled this bee into working for me." Both statements would be true, as you shall see in this chapter.

THE FLOWER

Flowers are the reproductive structures of the higher seed plants. The egg and sperm are produced by the flower. Fig. 8-1 on page 68 shows the main parts of a flower. In the center is the *pistil*. The lower part of the pistil is thicker than the upper part. This lower part is the *ovary* (OH-va-ree). The ovary is the part of the pistil in which the eggs are formed.

Around the pistil are a number of long, slender stalks with thicker structures at their upper ends. These are the *stamens* (STAY-mens). When the end of the stamen is ripe, it splits open and a large number of tiny particles come out. These are called pollen grains. If you rub your finger over the stamens of a flower, you will get a yellow dust on your skin. This is *pollen*. Each pollen grain is built like a cell. It has a cell wall, membrane, and cytoplasm. But it has at first two and later three nuclei. It is really a three-celled structure with no separate walls between the three nuclei.

Around the stamens is a circle of *petals*. These are the bright, showy parts that make flowers attractive to look at. Around the petals are the *sepals* (SEE-puls). In most flowers, the sepals are green and leaf-like. They cover the flower bud before it opens.

Flowers are not all alike. Some have only one pistil and others have more than one. Some flowers produce both pistils and stamens and others have only a pistil or only stamens. Some have no sepals or petals. Plants sometimes have many

objectives

After you read this chapter, you should be able to:

—**Name** the parts of a typical flower

—**Describe** the structure and function of the seed

—**Explain** several ways in which seeds are scattered

—**Compare** the three common types of life cycles found in plants

pistil

ovary

stamens

pollen

petals

sepals

Enrichment: You may wish to point out that the pistil is the female structure and the stamen is the male structure.

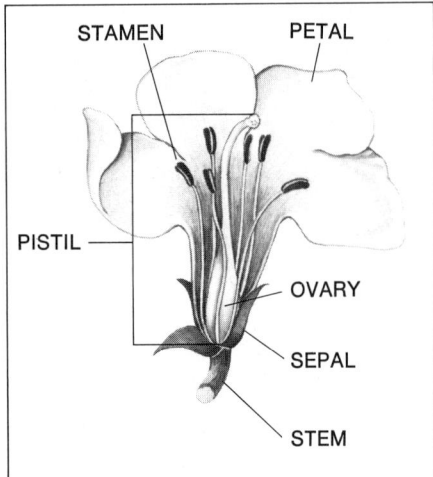

Fig. 8-1
The main parts of a flower.

pollination

pollen tube

nectar

Fig. 8-2
A future seed from the ovary of a flower.

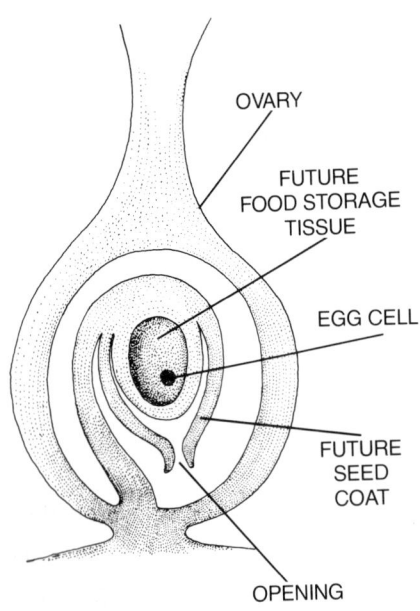

tiny flowers grouped together so they look like a single large flower. You may have seen the blooms of dandelions, asters, and daisies. If you looked at one of them closely you would see that it is made up of many tiny flowers.

POLLINATION AND FERTILIZATION IN FLOWERS

Inside the ovary of a flower are structures that are the future seeds. Each of these future seeds is made up of several cells. One of these cells is the egg cell. See Fig. 8-2.

Before fertilization can take place, a pollen grain must land on the upper end of a pistil. Of course, the pollen must come from that same kind of flower or the process will not continue. The arrival of the pollen on the pistil is called *pollination*. The end of the pistil is covered with a sticky juice that holds the pollen grain in place.

When a pollen grain gets on the upper end of a pistil, it begins to grow. As it absorbs water and food from the pistil, the pollen grain swells enough to crack its wall open. Then it grows downward in the form of a long slender tube called the *pollen tube.* This is shown in Fig. 8-3. The tube grows down through the pistil. It makes its way by digesting the tissues of the pistil. Notice in Fig. 8-3 that there are two sperm nuclei in the pollen tube. Either of these nuclei can act as a sperm and fertilize an egg. When the end of the pollen tube reaches the inside of the ovary, it grows through a tiny hole in the end of the future seed. Then the end of the pollen tube bursts open. The sperm nuclei escape from the tube and one of them unites with the egg. This is the act of fertilization.

METHODS OF POLLINATION

You may have wondered how the pollen arrives on the end of the right kind of pistil at just the right time. The pollen grains are usually carried either by wind or by insects.

Bees are insects that often carry pollen. Other insects such as wasps, moths, and butterflies sometimes carry pollen, also. When a bee crawls into a flower, its hairy surface becomes covered with pollen. When the same bee visits another flower, some of this pollen rubs off onto the pistil of that flower. This is an accident as far as the bee is concerned.

If you want people to work for you, you usually have to pay them. Do you know how the flower "pays" the bee? Bees get all of their food from flowers. They get two kinds of food. One is pollen. The flower produces enough pollen both for bee food and for pollination. The other food is *nectar.* This is a sweet juice which is produced by cells near the base of the petals.

Bees crawl into the flower to drink this nectar. It is carried back to the hive in a special part of the bee's stomach. Pollen is carried packed on the bees' hind legs. Pollen is stored in the hive. The nectar is made into honey. Look at the bee and flower at the beginning of this chapter. Can you understand their relationship better now?

Bees usually visit only one kind of flower on a single collecting trip. This means that the right kind of pollen is almost sure to land on each pistil. Flowers are often shaped in some special way that makes the bees' work easier. Certain petals are large and form "landing fields." In some flowers the end of the pistil is located just above this landing area. This causes the bees' backs to rub against it when they crawl in to get the nectar.

Many flowers have a sweet odor. Bees have a good sense of smell, so this perfume helps them to find these flowers. Bright, showy petals also help them to find flowers. Wind-pollinated flowers do not need any such display. They have no perfume, and their petals are either very small or missing entirely. You may never have noticed wind-pollinated flowers, yet they are very common. All of our grasses and most of our shade trees are pollinated by the wind. Their flowers are very small, and generally have just a pistil and stamens. The stamens produce large numbers of very small pollen grains. These are carried by the wind like spores. There are so many of them that some are almost sure to land on the right kind of pistils. Of course, most of them are just wasted, landing in many other places. In some plants, pollen falls from the stamens to the pistils in the same plant. In this case, the plants pollinate themselves.

THE FUNCTION OF FRUITS

A *fruit* is the ripened ovary of a flower. Fruit development starts as soon as pollination takes place. The walls of the ovary thicken and form the fruit. Some fruits are *fleshy fruits,* such as a peach or a berry. Others are *dry fruits,* such as a burr or a pea pod. You usually think a fruit is something like an apple or an orange that you can eat. You think it is the sort of thing one can buy in markets. These are examples of fleshy fruits. They can be eaten by animals. But to a biologist, a cockleburr is also a fruit. So are a lot of other dry structures that develop around seeds. They are the ripened ovaries of these plants. A fruit may simply protect the seed during its development, but it also may serve to scatter seeds. Suppose you were to walk past a tree covered with peaches. You could pick some, and, as you walked along, you could eat the fruit and throw away the seeds. You would be working for the plant by scattering its seeds. In the same way, wild animals scatter the seeds of many fruits.

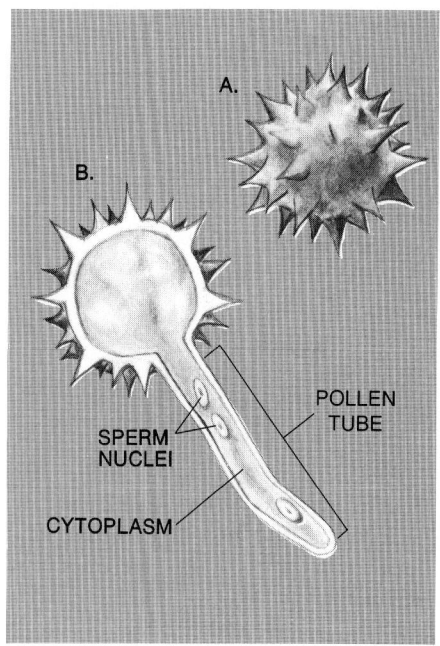

Fig. 8-3
A pollen grain before and after it has formed a pollen tube.

Science Reading Skills: Compare and contrast insect-pollinated and wind-pollinated flowers.

Activity: Have students look at flowers or pictures of flowers and classify them as wind-pollinated or insect-pollinated.

fruit

fleshy fruits

dry fruits

Enrichment: Stress to students that many so-called vegetables are really fruits: tomatoes, beans, and corn kernels. A discussion of the characteristics of a fruit should follow.
Enrichment: Mention water as an agent of seed dispersal.

Enrichment: Another group of plants, the gymnosperms produce "naked seeds" that are not enclosed within a fruit.

Fig. 8-4
A cutaway view of a bean seed (left) and a corn seed (right). Note that the corn has special tissue for food storage. In the bean, the embryo fills the entire space inside the seed coat. The bean stores food in its enlarged seed leaves. Which parts of these seeds are also parts of the embryo?

Some dry fruits are also carried about by animals. Have you ever discovered burrs sticking to your clothing? When you pull the burrs off they may land in new places where their seeds can grow. The hooked spines on the burrs can also stick to the fur of animals that brush past them. You may have seen dogs or other animals scratching burrs out of their fur.

When a pea pod is dry enough, it suddenly splits open throwing the seeds out in all directions. This scatters them at least a small distance away from the parent plant. The acorn is the seed of an oak tree. It may be carried off by a squirrel and buried in the ground. The squirrel is hiding the acorn to use later for food. Squirrels fail to find some of these hidden seeds. These will have a chance to grow into new trees. In this way squirrels help oak forests to spread outward. Many seeds are distributed by the wind. Either the seed or the fruit has some sort of wings or fuzz that catch in the wind to carry the seed long distances. See Fig. 8-5. The seeds of dandelions, maples, milkweeds, and cottonwoods are all scattered in this way.

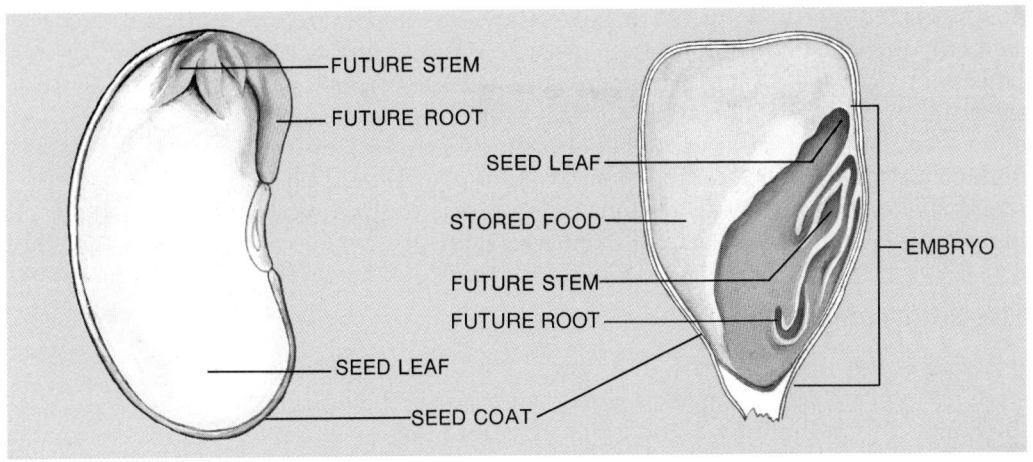

DEVELOPMENT OF THE SEED

embryo

As soon as the egg and sperm nucleus have united, the fertilized egg begins to divide, grow, and divide again, many times. A very small plant is formed in this way. It is called the *embryo* (EM-bree-oh). It has one or two seed leaves and parts that will become a root, a stem, and a stem tip. When it has formed all of these parts, the embryo stops growing. Meanwhile, the tissues around the embryo are growing to form the rest of the seed. Not all seeds are alike, but they always have an embryo, an outside seed coat, and a stored food supply. You can see

Activity: Have students collect and display numerous seeds and discuss their methods of dispersal.

this in Fig. 8-4. The stored food may be in cells between the embryo and the seed coat, as shown in the corn seed. But in some plants the food is stored in the seed leaves, as shown in the bean seed. These *seed leaves* are the first leaves to form on the embryo. The bean has two of them. In the bean the seed leaves are very thick because of the stored food they contain, and they do not look much like regular leaves.

Enrichment: You may want to introduce the terms monocotyledon and dicotyledon.

seed leaves

Fig. 8-5
Photographs of different kinds of seeds and fruits. Top left: raspberries; top right: pears; center left: beggar tick; center right: acorn; bottom left: dandelion; bottom right: maple. How are each of these scattered? (All photos by William E. Ferguson)

71

A seed will generally grow when it has moisture and the proper temperature. Water soaks through the seed coat, and the embryo begins to develop. Its root grows down into the soil, and its stem grows up into the air. New leaves spread out in the light. The stored food in the seed is used during this early growth process.

A seed may live for a long time before it begins to grow. Wheat or corn seeds are often able to grow after six or seven years. Seeds of the common pigweed can still grow after forty or even sixty years. A lotus is a kind of big water lily. Its seeds may grow after one hundred years or more.

As you can see, growth from a seed has several advantages. In the first place, the seed contains a multi-cellular plant, the embryo. It can grow quickly when the right time arrives. In the second place, the seed's supply of stored food gives the young plant the nourishment needed for growth. By the time this food is used up, the plant has roots in the ground to obtain water and leaves in the light to make its own food.

THE PLANT LIFE CYCLE

Some seed plants complete their entire life cycle in a single growing season. The seed sprouts in the spring. By autumn the plant produces seeds of its own. The old plant dies, and only the seeds remain. Zinnias, marigolds, corn, wheat, beans, and peas are examples of this.

Other seed plants take two growing seasons to complete their life cycles. In the first season they develop stems, roots, and leaves, but do not bear flowers. During this first year they store food. Often the parts above ground die during the next winter. But the plants grow up again from their underground parts during the second season. This time they produce flowers and seeds. Then the plants die. Beets, carrots, foxglove, and Canterbury bells are plants with this type of life cycle.

Finally, there are plants that live through several growing seasons producing seeds each year. Lilies, columbines, and daisies do this. So do trees, shrubs, and many of the vines and grasses. In fact some plants live for hundreds of years.

As you can see, all of these different types of life cycles have several things in common. In each case, the plant produces seeds from which new plants can grow. Also, in each case it is necessary for the plant to have its own food supply in order for it to be able to grow. Each type of plant has certain characteristics that make it suited to a particular type of life cycle. For example, if a certain plant is not able to decrease its activity in winter, it will have the type of life cycle that is completed in one season.

SUMMARY

Flowers are the organs for sexual reproduction in plants. The pistil is in the center. The ovary at its base contains the future seeds. Each future seed contains an egg. Stamens, surrounding the pistil, produce pollen grains. These pollen grains contain sperm nuclei. When pollen grains fall on a pistil they form long tubes. These tubes grow down through the pistil to the future seeds. Sperm nuclei fertilize the egg nuclei. The fertilized egg becomes the embryo. This embryo is the new young plant contained in the seed. Other parts of the future seed become the seed coat and the food storage tissue. Food may be stored in the seed leaves of the embryo.

Seeds are contained in a fruit. The fruit is the ripened ovary of the plant. Fleshy fruits have edible walls, which attract animals. In eating the fruits, animals scatter the seeds. Dry-walled fruits protect the seeds. Some dry fruits also help to distribute the seeds. They may have hooks that catch in the fur of animals, or they may have wings that help them to be scattered by the wind.

Some plants complete their life cycles in a single season. Others produce seeds in their second year and then die. Still others grow for many years.

Activity

A Study of Flowers
In this activity you will study the parts of a flower.

A. Obtain a flower and examine it carefully. Find the sepals, petals, stamens, and pistil.

   1. Describe what each of these parts looks like and explain what each does.

B. Rub your finger over a stamen of the flower. You should get a yellow dust on your skin.

   2. What is this yellow dust?

C. Remove the sepals and petals from the flower and closely examine the pistil. Notice that the top of the pistil is sticky.

   3. What is the advantage of the top of the pistil being sticky?

D. Make a temporary mount of the pollen. Observe the slide under low power then switch to high power.

   4. Describe what the pollen grains look like.

E. Use a needle to split the ovary of the flower in half lengthwise.

   5. What is inside the ovary?

Word Quiz

Choose the letter from **Column B** that best matches each number in **Column A**. Do not write in this book.

Column A	Column B
1. pistil	**a.** produces pollen
2. ovary	**b.** showy part of flower
3. stamen	**c.** young plant inside the seed
4. pollen	**d.** first leaves formed
5. petal	**e.** sweet juice that attracts bees
6. sepals	**f.** central part of flower
7. pollination	**g.** a ripened ovary that is good to eat
8. pollen tube	**h.** contains sperm
9. embryo	**i.** burrs, pods, etc.
10. seed leaves	**j.** contains future seed
11. nectar	**k.** carries sperm to egg
12. fleshy fruit	**l.** pollen arriving on pistil
13. dry fruits	**m.** covers flower bud

Check Your Facts

1. What is the function of the flower?
2. Draw a picture of a flower and label all of the important parts.
3. Name the parts of a flower and give the function of each part.
4. Where are the eggs of a flower located?
5. How does the sperm of a flower reach the eggs?
6. What is pollination? How is it different from fertilization?
7. What are the parts of a seed? What do they do?
8. What part of the flower becomes a fruit?
9. What are the two main ways that pollination takes place?
10. Explain the three common types of life cycles found in plants.

Thought Questions

Science
Reading Skills
1. & 2. *Problem Solving*

1. Would you expect the flowers in a bouquet to be wind-pollinated types or insect-pollinated types? Why?
2. Would a person with hay fever be affected more by a flower garden next door or a grassy field across the street? Why?

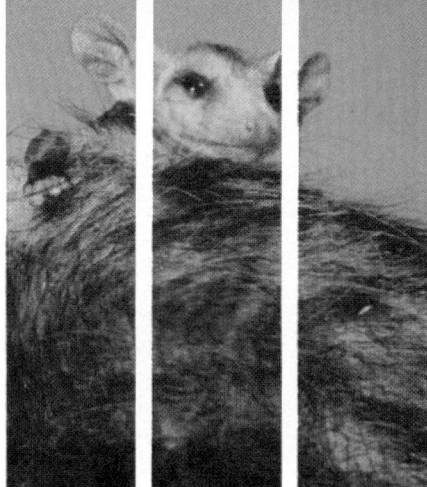

CHAPTER 9 Reproduction in Higher Animals

Obviously, the mother opossum in the photograph above has been able to reproduce. She has solved the baby-sitting problem by carrying her babies on her back. In Chapter 7 you learned how *Hydra*, a simple animal, reproduces. The sperm swims through the water to the egg. The fertilized egg grows into an adult *Hydra*. You are now about to study reproduction in higher, more complicated animals. The opossum is an example of such an animal. Reproduction in these animals is really not so very different from reproduction in *Hydra*. Eggs and sperm are produced, and the fertilized egg develops into a new animal.

EXTERNAL FERTILIZATION

The type of fertilization that takes place in the *Hydra* is called *external fertilization.* This means that the sperm fertilizes the egg *outside* the body of the female organism. This type of fertilization is used by many organisms that spend part or all of their time living in the water. This includes such higher organisms as fish and frogs.

Reproduction in fish. Female fish have organs called *ovaries,* which produce eggs. The eggs pass through tubes called *oviducts* (OH-ve-dukts), which lead from the ovaries to the outside. Male fish have organs called *testes* (TES-teez), which produce sperm cells. *Sperm ducts* lead from the testes to the outside. When the sperm are ready to leave the male, they are contained in a liquid called *semen* (SEE-men).

Breeding habits differ, but most fish lay and fertilize their eggs in the water. As the female lays her eggs the male fish swims close to her and sends out semen through his sperm ducts. Sperm from the semen swim through the water and fertilize the eggs.

Different kinds of fish lay their eggs in different places. Some fish lay their eggs on rocky reefs, some on gravel bottoms, and

After you read this chapter, you should be able to:

__**Compare** internal fertilization and external fertilization

__**Compare** the methods of reproduction in reptiles and birds

__**List** three types of reproduction in mammals

__**Explain** the reproductive advantage that mammals and birds have over most other groups

external fertilization

ovaries

oviducts

testes

sperm ducts

semen

Fig. 9-1

This clownfish is guarding the eggs near its mouth. Most fish do not do this, but when it does happen it is nearly always the male that gives the eggs this protection. (Zig Leszczynski/Animals Animals)

Science Reading Skill: *Sequencing* of events in the development of a frog. See Fig. 9-3.

Fig. 9-2

Note the clear jelly that surrounds each frog egg. (Runk/Rannels, Grant Heilman)

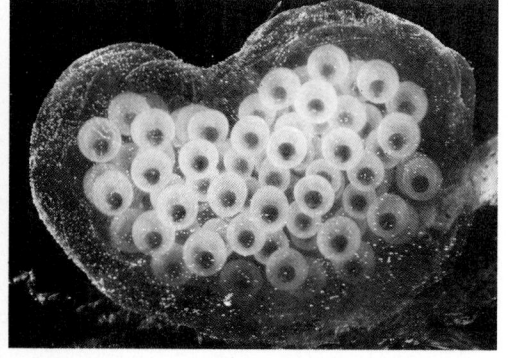

some scatter them among water plants. Usually the parents leave as soon as the eggs are laid and fertilized. Most fish give no care to their young. But in some kinds of fish the male stays near the "nest" and protects the eggs until they hatch. Fish of the bass family are one of the types that do this. See Fig. 9-1. One kind of male catfish carries the fertilized eggs in his mouth until they hatch!

Since most fish eggs are unprotected, many are eaten by other fish and water animals. Some species might be wiped out by these losses if it were not for the large number of eggs laid. A single fish may lay thousands or even millions of eggs each year. If only one in a thousand lives, a good many new fish are produced.

Reproduction in frogs. Adult frogs and toads are air breathing animals, but they lay their eggs in water. During spring, the male frogs in ponds make a certain sound. This "singing" attracts the female frogs. When the males and females come together they may mate.

A female frog carries many eggs. She cannot lay these eggs by herself. The male helps by squeezing her sides with his front feet. Together they force out some of the eggs. This process is repeated many times and may take a whole day. Each time eggs are laid, the male sends a cloud of sperm over them. The sperm swim to the eggs and fertilize them.

If you should find a mass of frog eggs in the water you might wonder how they could all have come out of one small frog. The outside layer of clear jelly does not swell up until after it is in the water. See Fig. 9-2. This makes the entire egg mass bigger than the frog it came from. Like most fish, frogs leave their eggs, giving them no protection. The young tadpoles that hatch from the eggs are like fish in many ways. They swim with their tails and get oxygen through their gills. Only after a period of growth do they develop legs and lungs. Finally, the tail is absorbed, and the tadpole becomes a frog. This life cycle is shown in Fig. 9-3.

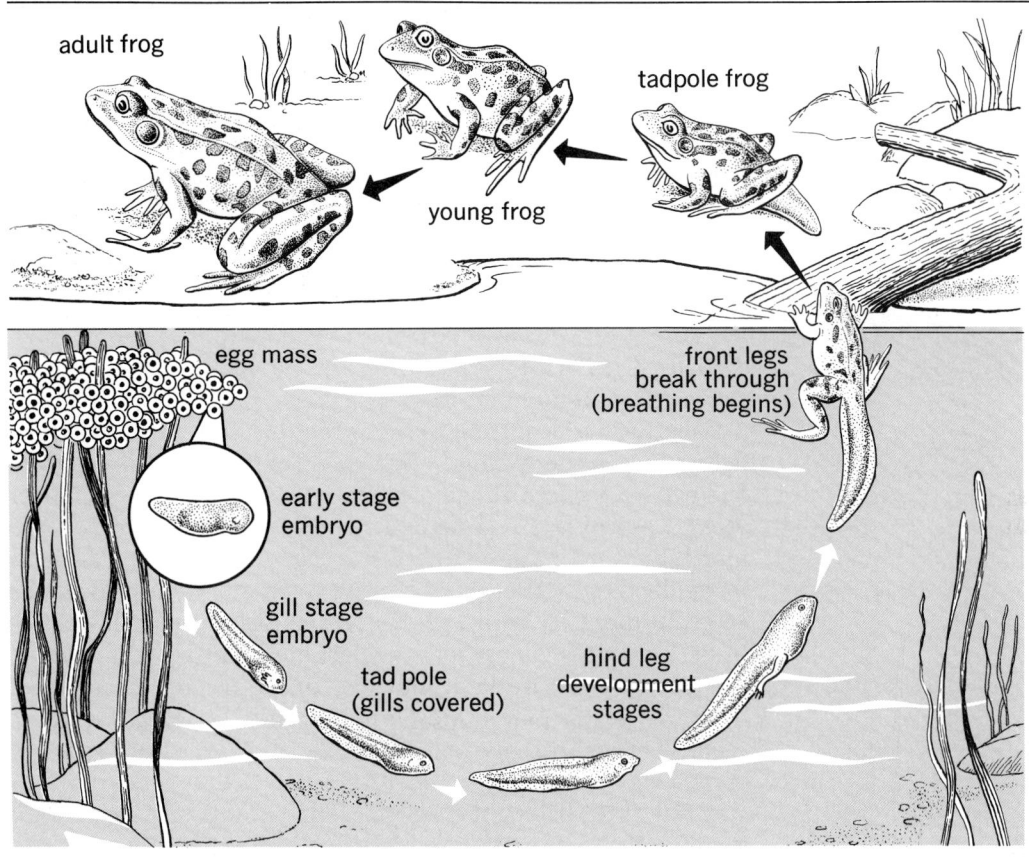

adult frog

young frog

tadpole frog

egg mass

early stage
embryo

gill stage
embryo

tad pole
(gills covered)

hind leg
development
stages

front legs
break through
(breathing begins)

Fig. 9-3
Stages in the life cycle of a frog.

INTERNAL FERTILIZATION

The higher land animals include *reptiles,* such as snakes, turtles, and lizards. They also include the *birds* and the *mammals.*
Mammals are warm-blooded, hairy animals, such as mice, dogs,
giraffes, and people. Each group has its own special way of
reproducing. In all three groups, the eggs are fertilized *within*
the body of the female. This is known as ***internal fertilization.***
In order for this to take place a male and female must come
together and mate. In this mating process, the male releases
semen into the oviduct of the female. The sperm swim upward
in the film of moisture that clings to the linings of the oviducts.
Hundreds of millions of sperm are released at a single mating.
One sperm may unite with one egg and fertilize it. This method
of fertilization makes sexual reproduction possible on land
where sperm would dry up and die if they were exposed to the
air. Actually, some kinds of fish also use internal fertilization.
The fertilized eggs may stay in the oviducts of the fish and
hatch there, in which case these fish produce living young.
Guppies are a good example of this.

internal fertilization

77

Fig. 9-4
Snake eggs hatching. Will the mother ever take care of these young reptiles? (Zig Leszczynski/Animals Animals)

Reproduction in reptiles and birds. The eggs of fish and frogs are small and delicate, like little balls of jelly. They would soon become dry and die if they were not in the water. Reptiles and birds have a type of egg that can be laid on land. You have seen birds' eggs; most likely, you have eaten them for breakfast. Reptile eggs look about the same as these. See Fig. 9-4. Most reptile eggs have a tough, leathery covering, but some have a hard shell. The only living part is the small white spot on the surface of the egg yolk. If you break an egg into a dish you will see this small bit of living protoplasm on the upper surface of the yolk. The yellow yolk material is a supply of stored food. This food from the yolk is used by the young bird or reptile during its development in the egg.

If mating has occurred, a bird or reptile egg is fertilized when it first enters the oviduct. This could not happen if the shell and egg white were already present. As the fertilized egg moves along, cells of the oviduct add the egg white. Then three thin membranes are added around the egg white, and the egg gets its shell. Finally the egg is laid. If no sperm were present the egg would not be fertilized. Unfertilized eggs can still be laid. In fact, the eggs we eat for breakfast are unfertilized.

Once the eggs are laid, the story of the reptiles is different from that of the birds. Most reptiles lay their eggs and leave them. Turtles, for instance, bury their eggs in the sand and crawl away. When the young turtles hatch, they must dig their way out and care for themselves. The parents take no interest in them. Birds, as you know, usually protect their eggs and feed their young. This means that each young bird has a much greater chance of surviving. Not so many eggs need to be laid to keep the number of birds about the same.

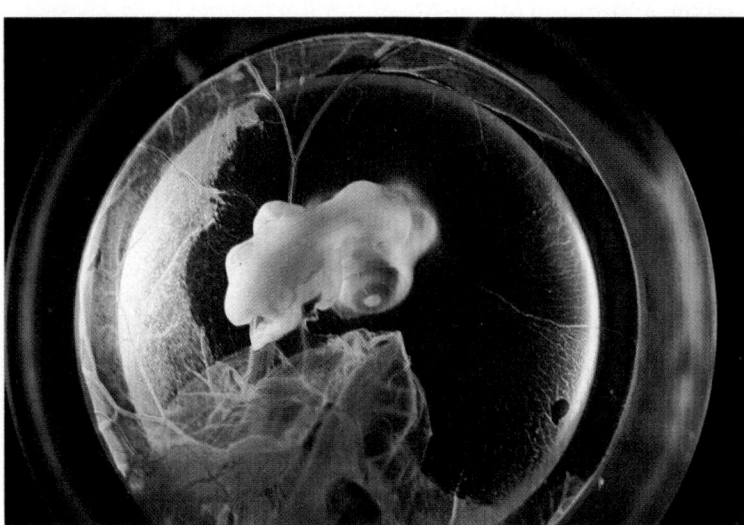

Fig. 9-5
A six-day-old chick embryo. It is contained in a sac of water. A stalk connects it to the yolk of the egg. (Animals Animals)

78

Fig. 9-6
What is happening in this photograph? Would a reptile do this? (F. J. Alsop III/Bruce Coleman)

Birds' eggs do not start to develop unless they are kept warm, or **incubated** (IN-cue-bate-ed). The heat they need is supplied by the bodies of adult birds when they sit on the eggs. They must continue to sit on them day and night until the eggs hatch. Then the young birds are fed and protected until they are grown. See Fig. 9-6. Some birds hatch quickly but are very helpless at first. Robins, for instance, hatch in just fourteen days, but they are blind and have no feathers. About all they are able to do is open their mouths. The parents work hard, gathering food and stuffing it into these open mouths. Other birds take longer to grow in the egg and are more fully developed when they hatch. Chickens take twenty-one days to hatch, and ducks take twenty-eight. But chicks and ducklings are able to follow their mothers and pick up their own food right away.

Reproduction in mammals. Most mammals bear living young. They are not the only animals that do this. Some female fish, like the guppies, and some female snakes, like garter snakes and rattlesnakes, never lay their eggs. They keep the eggs in their bodies until development has taken place. These young fish and snakes receive food for growth from the supply stored in the eggs. When the young are fully developed, they are born. This is not much different from the egg developing after it has been laid, but it means that the egg is protected in the body of the parent while development goes on. Just as a young plant developing inside a seed, a young animal developing in an egg or inside the body of the mother is called an embryo.

In mammals there are really three types of reproduction. Though most mammals give birth to live young, the duckbill (See Fig. 9-7.) and the spiny anteater lay eggs, just like most reptiles. But once the eggs are laid, the mother protects them. When they hatch, she feeds the young animals on her milk. Mammals are the only animals that produce milk to feed their young.

incubated

Fig. 9-7
The duckbilled platypus. This primitive mammal, which is a native of Australia, is a type that lays eggs. (Australian News and Information Bureau)

Assignment: Have students research the unusual marsupials of Australia.

placental mammals

placenta

The second type of mammal reproduction is found in kangaroos, opossums, and other pouched animals. A mother opossum is shown at the beginning of this chapter. These animals produce very small eggs that develop right in the oviduct. There is not much yolk material, so the embryo must absorb what food it can from the mother through the lining of the oviduct. The young are born when they are very small and only partly developed. A newborn opossum is smaller than a honeybee. A 200 kilogram kangaroo gives birth to young about the size of peanuts. The only well-developed parts of these young animals are their front legs. They use them to climb into the mother's pouch. This pouch is a warm, fur-lined pocket on the underside of the body. The nipples of the milk glands are located in the pouch. The young animal attaches its mouth to the nipple and does not let go until it is much older and more mature. So the young kangaroos live in the pouch where they are protected, warm, and well-fed. They leave the pouch when they are big enough to walk about on the ground. See Fig. 9-8.

In North America there are no egg-laying mammals and only one type of pouched mammal, the opossum. All of our other mammals use the third type of reproduction found in mammals. They are called *placental* (pla-SEN-tul) *mammals,* because their unborn young are nourished by a special structure called the *placenta.* Fig. 9-9 shows the female reproductive organs in an animal that usually has one young at a time. Some mammals produce many young at a time. See Fig. 9-10 on page 82. There are two ovaries in mammals, which produce tiny eggs. These eggs are smaller than the period at the end of this sentence. An oviduct leads away from each ovary, and the two oviducts join to form a "Y." The lower end of the Y is the birth canal leading to the outside. The region where the three parts of the Y come together is called the *uterus* (YOO-ter-us). The uterus is often called the *womb* (woom). It has thick, muscular walls and a rich blood supply. It holds the embryo during its development, which may take a fairly long time.

An egg leaves the ovary and enters the upper end of an oviduct. If mating takes place at this time, the sperm placed in the birth canal by the male will swim up through the uterus into the upper part of an oviduct. Here one of the sperm will fertilize the egg. Then the fertilized egg begins to divide. It moves down into the uterus and by this time it is already a little ball of cells. We can now call it an embryo.

The embryo becomes attached to the lining of the uterus. It absorbs nourishment from the blood vessels in the uterus. At first, the embryo is a ball of cells that are all alike. This ball has a layer of cells on the outside and another on the inside. Then a middle layer of cells forms between the first two. This gives the embryo three cell layers. All of the body parts develop from these three cell layers. Your outer skin and your nervous system developed from the outer layer of cells. Your stomach and intestinal linings came from the inner layer. The middle layer produced your muscles, bones, and blood vessels. Biologists have always marveled at how a fertilized egg can give rise to trillions of cells of many types, all perfectly arranged to form the adult body.

Fig. 9-8
A kangaroo with its young, which is about nine months old and almost ready to leave its mother's pouch. (Australian News and Information Bureau)

DEVELOPMENT AND BIRTH

In placental mammals, the placenta grows tightly against the wall of the uterus. The placenta is connected to the embryo by a tube called the *umbilical* (um-BIL-i-kul) *cord*. See Fig. 9-9. This

uterus

womb

umbilical cord

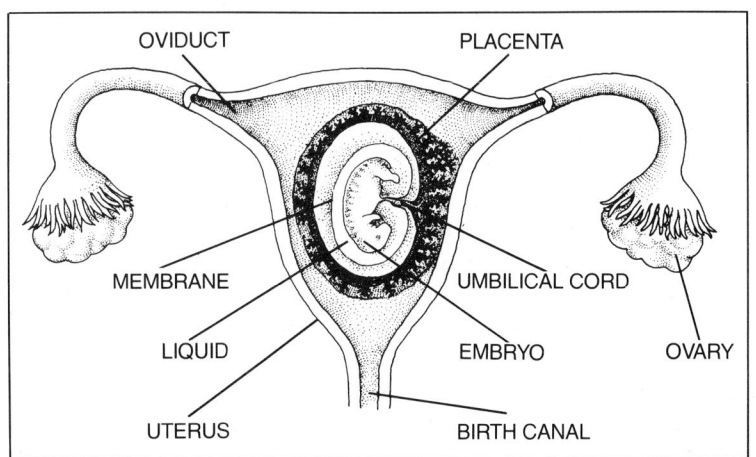

Fig. 9-9
The developing embryo in a mammal that produces one young at a time.

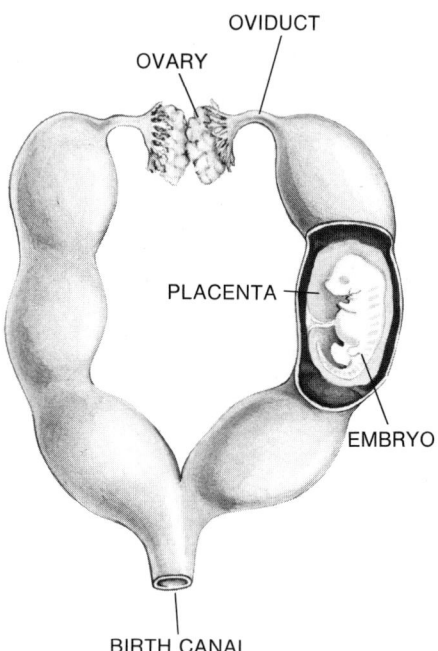

OVARY
OVIDUCT
PLACENTA
EMBRYO
BIRTH CANAL

Fig. 9-10
The uterus of a mammal that produces several young at one time. How many are developing here?

Enrichment: You may want to introduce the term fetus.

Enrichment: In humans, these uterine contractions signal the beginning of labor.

cord contains three blood vessels. Two of them carry blood from the embryo to the placenta. One carries blood to the embryo. Within the placenta, only thin membranes separate the blood of the mother from the blood of the embryo. Dissolved substances pass through these membranes. Food and oxygen pass from the mother's blood to the embryo's blood. Carbon dioxide and other wastes pass from the embryo's blood to the mother's blood. The mother's excretory system gets rid of these wastes.

The embryo has special membranes surrounding its body. These membranes contain a liquid so that the young mammal develops in water, just as a fish does. Bird embryos also have liquid-filled sacs around them as they develop inside the eggshell. A stalk connects the bird embryo with its food supply in the egg yolk in much the same way that the umbilical cord connects the mammal embryo with the placenta. This is shown in Fig. 9-5 on page 78.

Thus, the young mammal is nourished and grows. Tissues, organs, and systems are formed in its developing body. Finally, the time for birth arrives. The walls of the uterus begin to contract. The young animal is forced out through the birth canal. When this happens the umbilical cord is broken. The new-born mammal must do a very important thing—it must begin to breathe. Most female mammals lick their newborn young all over, starting with the nose. This clears the nostrils and helps the young animal to start breathing. Human babies sometimes have their bottoms slapped to shock them into breathing. Their breathing passages are carefully cleared of fluids, so the air can get through.

When the young mammal has been born there is no further use for the placenta. It passes out of the mother's body and is commonly called the *afterbirth*. A scar is left on the abdomen of each mammal where the umbilical cord was attached before birth. This is the navel.

CARE OF THE YOUNG

Like birds, some mammals develop more fully before birth than others do. Newborn puppies and kittens are blind and helpless. They must be kept in a protected place and fed on milk for quite a while before they can come out and follow their mother around. But a young calf gets up on its feet and follows its mother in about half an hour after it is born. Wild, grass-eating animals are always on the move. If their newborn young could not keep up with the herd, they would be killed by their natural enemies. These enemies are always on the lookout for stragglers.

Since both the birds and the mammals care for their young, we find similar kinds of family organizations in the two groups. There are some, for instance, in which the female alone does all the work of raising the young. Hummingbirds are an example of this. The male deserts the female soon after mating. The black bear is a mammal in which only the mother cares for her cubs.

Another arrangement is found among birds, such as chickens and pheasants, and among mammals, such as elk and musk oxen. In these groups one male mates with several females. In some cases he may stay with this group of females and protect them while they care for their growing young.

In still other birds and mammals the parents stay together and both of them help to raise the young. Most of our common songbirds are of this type. They include robins, blackbirds, and sparrows. Male and female share the work of building the nest, sitting on the eggs, and feeding the young. Wolves are mammals that form this sort of family group. Both parents prepare the den. Both protect and feed the young. Other wolves in the pack may also help.

Fertilization, growth of the embryo, and birth of the new baby are the same with humans as with other mammals. But when it comes to care of the young, people are very different from any other kind of living thing. Babies are born very helpless and knowing almost nothing. See Fig. 9-11. It takes them many years to grow up and to learn the things they must know in order to take care of themselves. In the meantime they must be cared for in order to survive.

This makes the choosing of a mate a much more serious matter for humans than it is for animals. A female deer, bear, or rabbit will mate with any male of her species. The female alone will care for the young, and it will take only a short time to raise them. They do not need to learn very much from older animals. Their inherited instincts give them the ability to survive.

Unlike these animals, human children can be influenced a great deal by the adults around them. Fathers, mothers, grandparents, aunts, uncles, and family friends can help to give valuable companionship and instruction to children as they grow up.

Care of the young child should start even before birth. As soon as a woman knows she is pregnant she should see a doctor. This is both for her own health and for the health of the child. It is important that a pregnant female eats the proper foods. Diet will affect the development of the baby. Certain drugs can also cause the baby not to develop properly or to be born sick. So, if a woman is pregnant she should not take any drugs unless

Fig. 9-11
Humans are born helpless and need to be cared for through the first years of life. (HRW photo by Richard Haynes)

Reference: See "Pregnancy: The Closest Human Relationship" by N. Newton and C. Modahl in *Human Nature*, March 1978.

Science Reading Skill: *Generalization*— Describe characteristics common among sexual reproduction in higher animals.

they are prescribed by a doctor. We also know that alcohol and nicotine can be absorbed through the placenta. These substances can have a bad effect on the embryo. Thus, it would be better for the baby if the mother could avoid drinking or smoking during the pregnancy.

SUMMARY

Higher animals produce eggs in the ovaries of the female, and sperm in the testes of the male. Eggs reach the outside through tubes called oviducts. Most fish eggs are fertilized in the water. Some fish guard the eggs, but most do not. Frog eggs are fertilized in the water and hatch into fish-like tadpoles.

In land animals, the eggs are fertilized inside the female by sperm from the male. Reptiles and birds lay large, well-covered eggs that contain much stored food. Birds incubate their eggs and care for the young. Reptiles do not.

In egg-laying mammals reproduction is similar to that of birds, except that the young are fed on milk. Pouched mammals give birth to tiny young, which finish their development in the pouch. Placental mammals keep the embryo in the uterus until it is well-developed. The embryo is nourished from the mother's blood through the placenta. All mammals care for their young. Human babies need even more care than the young of other mammals.

Activity

Suggestion: In spring frog eggs can be found in ponds. If possible you might have students gather these. Otherwise order them from a supply house.

Animal Development
In this activity you will see how frogs develop from eggs.

A. Place some frog eggs in an aquarium. Use plenty of water. If they are crowded they will die. Watch the changes that take place in the eggs as they develop. One or two at a time can be taken out and observed under a lens. Remove the clear jelly cases from the water as soon as the eggs have hatched. The tadpoles can be fed bits of canned spinach and boiled egg white. Keep track of their development. Keep the water fresh.

1. Make a drawing of the embryo at each stage of its development.

2. When can you first see eyes?

3. When does movement begin?

4. How well can tadpoles swim when they first hatch?

5. When do the hind legs first appear? front legs?

6. Does the tail drop off?

7. What change takes place in the mouth?

Word Quiz

Choose the letter from **Column B** that best matches each number in **Column A.** Do not write in this book.

Column A	Column B
1. ovary	**a.** contains the mammal embryo
2. oviduct	**b.** absorbs nourishment for the embryo
3. testes	**c.** a passageway for eggs
4. semen	**d.** produces eggs
5. incubation	**e.** produces sperm
6. placenta	**f.** connects embryo to placenta
7. uterus	**g.** fluid containing sperm
8. internal fertilization	**h.** keeping an egg warm
9. umbilical cord	**i.** when a sperm unites with an egg in the water
10. sperm ducts	**j.** lead sperm from the body of a male to the outside
11. external fertilization	**k.** takes place inside the body of a female

Check Your Facts

1. Describe how fish and frogs lay and fertilize their eggs.
2. How does fertilization take place in reptiles, birds, and mammals?
3. How are reptile and bird eggs better adapted to land conditions than the eggs of fish and frogs?
4. How does the reproduction of birds differ from the reproduction of reptiles?
5. What are the three types of reproduction in mammals?
6. How does care of the young by parents help birds and mammals to survive?
7. How do human young differ from those of other mammals?
8. How is internal fertilization different from external fertilization?

Science
Reading Skills
1.–8. *Compare and Contrast*

Thought Questions

1. Could we say that birds or mammals have a better method of reproduction? Why?
2. If rabbits gave birth to two young every two years, like bears, what would happen?

Science
Reading Skills
1. *Generalization*
2. *Problem Solving*

CHAPTER 10 Heredity

The twins in the photograph above look almost exactly alike. We know this is because the chromosomes in their cells are exactly alike. In this chapter, you will learn *why* their chromosomes are alike. You will also learn many other facts about heredity and how it works.

Early attempts to study heredity were not very successful. Then, in the middle 1800s, a man named Gregor Mendel made real progress.

Mendel was an Austrian monk. He lived in a monastery that had a large garden where he grew and studied the heredity of plants. His experiments were so simple that you can repeat them in a backyard garden, if you have one. Yet, simple as they were, he learned more about heredity than anyone else up to that time.

SOME EARLY STUDIES

Mendel's experiments. Mendel's best work was done with ordinary garden pea plants, which came in several varieties. The plants of one of these varieties are climbing vines that grow 1 or 2 meters high. Another variety of garden pea has short plants, less than ½ meter tall.

Seeds from Mendel's original tall plants always grew into tall plants. For the characteristic of height they inherited only information for tallness. We say that they were *pure* for tallness. The short plants were pure for shortness. Their offspring were always short.

We shall number the steps in Mendel's experiments with tall and short pea plants. The same numbers appear in Fig. 10-2. Study the diagram at the same time you read about the experiments.

1. Mendel *crossed* the two varieties of pea plants. This means that he took pollen from tall plants and put it on the pistils of short plants. He also put pollen from short plants on the pistils of tall plants. It made no difference which way the cross was made. He saved the seeds that were produced by this cross and planted them the next year. The embryos in each of these seeds had one parent that was tall and one that was short. When an organism has mixed heredity like this, we call it a *hybrid*. Every one of these hybrid peas grew into a tall plant. There were no short ones or in-between ones.

2. Mendel wondered if the trait for shortness had been lost. To find out, he pollinated the pistils of some of these hybrid peas with pollen from other hybrid peas, or he simply allowed them to pollinate themselves. When he planted the seeds produced by this cross, about three-fourths grew into tall plants, and about one-fourth became short plants. These were the average results when hundreds of seeds were planted.

3. Notice that one-fourth of the plants produced by pairs of hybrid parents were short. Mendel crossed pairs of these

Fig. 10-1
Mendel in his garden. His experiments led to the first scientific information about heredity. (The Bettman Archive)

hybrid

Science Reading Skill: Follow the *sequence* of steps in Mendel's research. See Fig. 10-2.

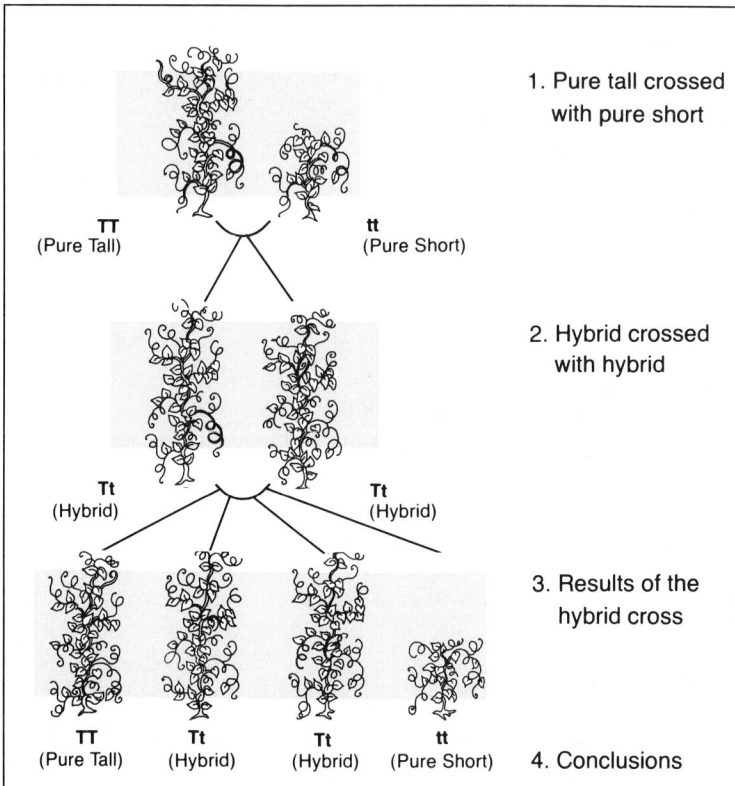

TT
(Pure Tall)

tt
(Pure Short)

1. Pure tall crossed with pure short

Tt
(Hybrid)

Tt
(Hybrid)

2. Hybrid crossed with hybrid

3. Results of the hybrid cross

TT
(Pure Tall)

Tt
(Hybrid)

Tt
(Hybrid)

tt
(Pure Short)

4. Conclusions

Fig. 10-2
Illustration of Mendel's experiments with pea plants. The letters (T and t) are symbols for the genes.

short plants. They always produced short plants. They carried no heredity for tallness. They were pure for shortness.

4. Next, Mendel experimented with the tall pea plants that had been produced by the hybrid parents. He wondered if they were pure for tallness or if they were hybrids. He allowed them to self-pollinate. One-third of these tall plants produced only tall offspring. They were pure for tallness.

The other two-thirds of the tall plants (with hybrid parents) produced two sizes of offspring. This tells us that the plants producing these offspring must have been hybrids, like their parent plants. Hybrid parents can be expected to have one-fourth of their offspring pure for one trait, one-fourth pure for the other trait, and two-fourths (½) hybrids.

The short plant, when self-pollinated, produced only short offspring. It was pure for shortness. Mendel reasoned that a trait for tallness inherited from one parent hid the trait for shortness that came from the other parent. The hybrid had both kinds of inheritance, but the trait for tallness was "stronger" so the hybrid plant grew tall. Mendel called the "strong" trait a *dominant* trait. He called the one that stayed hidden a *recessive* trait. There are many examples of dominant and recessive traits. In cattle, hornless is dominant over the normal horned condition. The trait for black fur is dominant over the trait for white fur in guinea pigs. White is dominant over black in hogs.

Understanding Mendel's results. Mendel did not know exactly why he got the results that he did. Now we can explain them, because we understand cell structure and sexual reproduction better than scientists did in his day. You have probably guessed that it is the genes passed on by parents that produce the tall or short pea plants or the white or black guinea pigs. This is why we now call the study of heredity *genetics*. You have already read in Chapter 6 how genes are carried in chromosomes and passed on each time the cell divides. What was not explained is the fact that each body cell in a plant or animal contains two of each kind of chromosome. A human body cell with its 46 chromosomes really has just 23 different kinds. There are two complete sets of 23 each. One of each kind of chromosome comes from one parent, and one of each kind comes from the other parent.

In Mendel's hybrid peas, there was a chromosome with a gene in it that made the peas grow tall. This same gene in the other chromosome of that pair was a little different. It was a gene for shortness. In this situation the gene for tallness was

Enrichment: Give examples of dominant and recessive traits in humans: Curly hair is dominant over straight hair; brown eye color is dominant over blue eye color; free ear lobes are dominant over attached ear lobes.

dominant
recessive

dominant and produced a tall pea plant. But the chromosome with the gene for shortness was still there, and could be passed on to the next generation. A recessive trait shows up only when both chromosomes of a pair have genes for that recessive trait.

Sometimes one gene of a pair is not entirely dominant over the other. If you cross a white snapdragon with a red snapdragon, the offspring will have pink flowers. See Fig. 10-3. A cross between red and white short-horn cattle will result in animals that have red and white patches of color. When a difference in genes results in a compromise like this we call it a case of *incomplete dominance.*

REDUCTION DIVISION

Remember that mitosis is the common type of cell division. In mitosis, each chromosome is duplicated. Each of the daughter cells gets a complete double set of chromosomes. Although a daughter cell is smaller than the parent cell at first, it has the same genes as the parent cell that produced it. And since all ordinary body cells divide by mitosis, they all have complete double sets of chromosomes. If sperm and egg cells were produced by mitosis, fertilization would result in a new individual with twice the original number of chromosomes. Do you see why? In the next generation, the number would double again, and so on. But this does not happen. Actually the sperm and eggs are produced by a special sort of cell division. It is called *reduction division.*

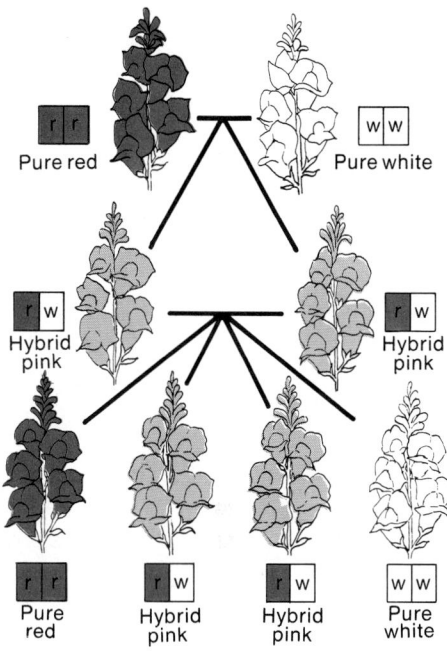

Fig. 10-3
A cross between red and white snapdragons results in pink flowers. The hybrid looks different from the pure types. This is a case of incomplete dominance.

incomplete dominance
reduction division

Fig. 10-4
A cross between red and white short-horned cattle produces offspring with patches of both colors. This is an example of incomplete dominance in animals. (Hans Reinhard/Bruce Coleman)

A reduction division looks very much like mitosis, but it is not quite the same. In a reduction division, each pair of chromosomes is separated. One chromosome of each pair goes to one cell. The other chromosome of this pair goes to the other cell. So, cells formed during a reduction division have only single sets of chromosomes. This means that each cell has only half of the usual chromosome number. See Fig. 10-5. Actually, the diagram shows only the second part of reduction division. The chromosomes are already doubled before this reduction division begins. An extra cell division is necessary to separate them. It is as if a division by mitosis is followed by a division that reduces the number of chromosomes. As a result four cells would be formed instead of two.

Both egg and sperm are produced by reduction division. Therefore each of them has a single set of chromosomes. When they unite, they produce a fertilized egg with two sets of chromosomes. Body cells of the developing embryo are produced from this egg by mitosis, so body cells also have the double set of chromosomes. See Fig. 10-5. In each of your body cells, with its 46 chromosomes, one set of 23 came from the egg produced by your mother and the other 23 came from the sperm produced by your father.

Of the 23 chromosomes that came from your mother, it is a matter of chance which of them were ones she got from her mother and which were ones she got from her father. In other words, it is a matter of chance just how much of your heredity came from each of your grandparents. Brothers or sisters are more or less alike depending on how many of the same chromosomes they happen to inherit.

Fig. 10-5

Reduction division. A. A nucleus with two sets of chromosomes. The first division has already taken place; B. The paired chromosomes line up; C. The chromosomes are pulled apart by fibers. One of each kind of chromosome goes to each side of the cell; D. Two new nuclear membranes form; E. the cytoplasm begins to pinch in so it can divide; F. Division of the cytoplasm is complete. Two new cells have been formed, each having only one set of chromosomes. After fertilization the chromosome number will be restored.

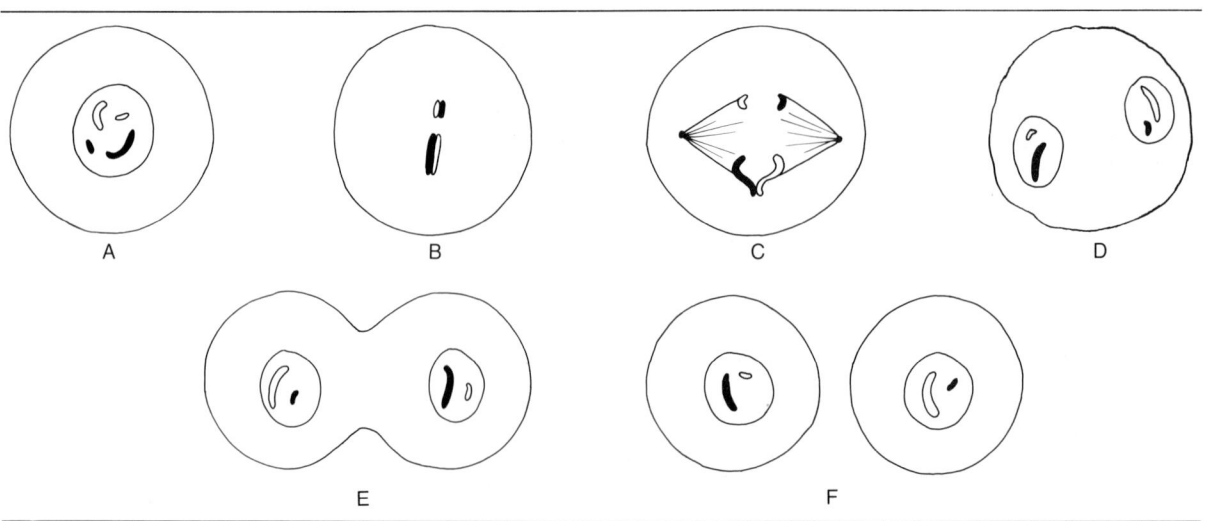

PREDICTING HEREDITY

Suppose you wish to predict what the results will be if you cross two similar plants or animals. Mendel's hybrid peas are a good example to use. If we use the capital letter T to represent the dominant gene for tallness and a small letter t to represent the recessive gene for shortness, then a hybrid with one of each gene would be Tt. This hybrid would be tall because it carried the dominant gene. The two genes would be separated during reduction division so that, in the sperm and eggs produced, there would be equal numbers carrying the gene for tallness (T) and the gene for shortness (t). There would be an equal chance that the gene for tallness (T) from one plant would unite with a gene for tallness (T) or with a gene for shortness (t) from the other plant. Likewise there would be equal chances for the gene for shortness (t) from the first plant to pair off with a gene for tallness (T) or a gene for shortness (t) during fertilization. It is simply a matter of chance which sperm happens to fertilize which egg cell. A handy way to show this is to use a chart. We list the kinds of genes present in the egg on one side and the kinds in the sperm on the other side, as shown below.

Genes in the sperm

	T	t
Genes in the egg cell T		
t		

Enrichment: This chart is known as a Punnett square.

Then we simply fill the spaces in the table, by pairing each gene from the sperm with each gene from the egg. The completed table is shown below.

Genes in the sperm

	T	t
Genes in the egg cell T	TT	Tt
t	Tt	tt

You will notice that these results are the same as those that Mendel had in his experiments. Three of the spaces have at least one dominant gene, T. These plants will all be tall. One of these is pure for tall (upper left). Two others are hybrids. One space has recessive genes only (lower right). This plant will be short. So there are three tall plants and one short plant. This is the three-to-one ratio that results when two hybrids are crossed. The same letters are shown in Fig. 10-2.

There are two things you should remember. First, **you cannot be sure what any offspring will be like. You only know what the chances are.** In the case of these hybrid peas, there is one *chance* in four of producing a short plant. Actually, you might get no short plants or you might get several of them from the next ten seeds you planted. It is like flipping a coin; you could get ten heads in a row, even though the chances are that you would get only five out of ten. Second, **the trait that you are interested in may depend upon more than one pair of genes.** Several pairs of genes influence some traits. Human skin color, for example, is affected by at least four pairs of genes, each pair showing incomplete dominance. As a result, many shades are possible from very dark to very light. Hair color is determined in a similar way, though in this case dark tends to be dominant over light.

Boy or girl? The sex of an individual is controlled by heredity. A pair of chromosomes in the cells of females are called *X chromosomes*. In males there is only one X chromosome. Paired with it is a shorter one called the *Y chromosome*. In other words,

X chromosomes

Y chromosome

Fig. 10-6
Left: A human egg cell surrounded by sperm. (Dr. Landrum B. Shettles, Columbia Presbyterian Medical Center); Right: A sperm and an egg cell each carry a set of chromosomes that come together during fertilization. This drawing represents a species having only four chromosomes in each body cell.

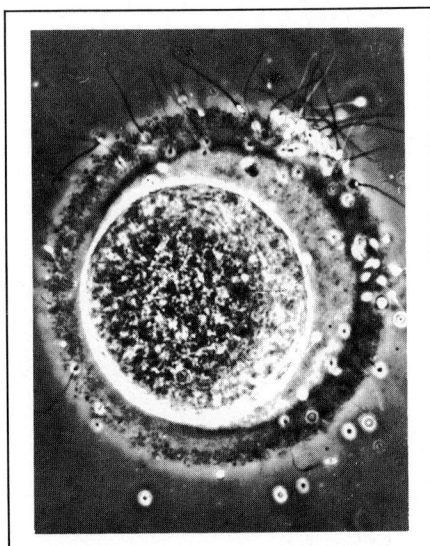

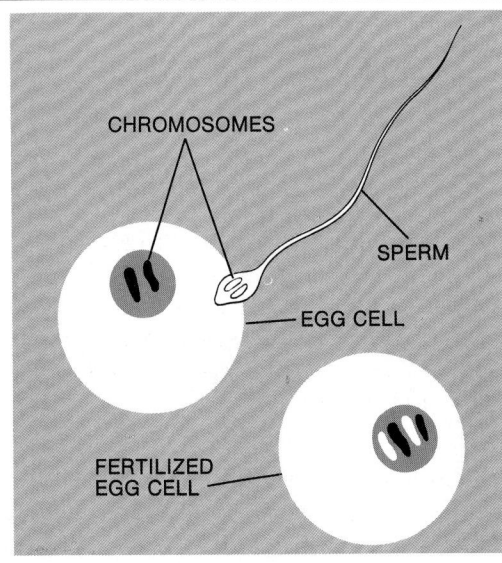

CHROMOSOMES

SPERM

EGG CELL

FERTILIZED EGG CELL

females have two X chromosomes while males have one X and one Y.

During reduction division, when eggs are being formed, the X chromosomes of the female are separated. Each egg receives one X chromosome. Reduction division in sperm formation causes one-half of the sperm to receive an X chromosome and one-half of the sperm to receive a Y chromosome. If a sperm carrying an X chromosome fertilizes an egg, then the fertilized egg will have two X chromosomes. It will develop into a female. If a sperm with a Y chromosome unites with the egg cell, the result is a male. The Y chromosome carries genes that produce male traits. See Fig. 10-6.

Thus, a male or female is produced depending on which of the two kinds of sperm fertilizes the egg cell. Since the two types of sperm are present in equal numbers, we can expect about equal numbers of males and females to be born. Actually, in the case of human beings, there are about 106 boys born to every 100 girls. It appears that Y-carrying sperm have a slightly better chance of fertilizing the egg.

Sex-linked traits. Any trait controlled by a gene in the X chromosome is called a *sex-linked* trait. We will use the gene for color blindness as an example. If a woman carries a gene for color blindness on one of her X chromosomes she will not be color blind. This gene is recessive. The normal gene in the other X chromosome is dominant, so she will have normal vision.

If a man's X chromosome carries a gene for color blindness the situation is different. There is no other X chromosome to carry a normal gene for that trait, so the abnormal gene is the only one present. This man will be color blind. All sex-linked traits act as dominant ones in males. A color-blind man has inherited a single gene for the trait from his mother. A color-blind woman must have inherited the gene from both parents. See Fig. 10-7. You can see why color blindness is far more common in men than in women.

There is a very serious sex-linked disease called *hemophilia* (he-mo-FEEL-ee-ah), or bleeder's disease. The victims of this disease bleed heavily from any small cut or bruise because the blood is unable to clot. The recessive gene for hemophilia is carried on the X chromosome, so the disease is far more common in men than in women.

Sex-influenced traits. Not all inherited differences between the sexes are due to genes in the X and Y chromosomes. Genes in other chromosomes may act differently in males than in females. The traits controlled by such genes are said to be *sex-influenced*. Common baldness is an example of this. The gene

Enrichment: Y-bearing sperm have a better chance of survival because they are lighter and swim faster.

sex-linked
sex-influenced

Fig. 10-7
Colored dots represent genes for color blindness. Black dots represent genes for normal vision. Normal is dominant. Can a normal gene hide the effects of a color blindness gene in a male? Could the sons of a female who is color blind have normal vision? Why?

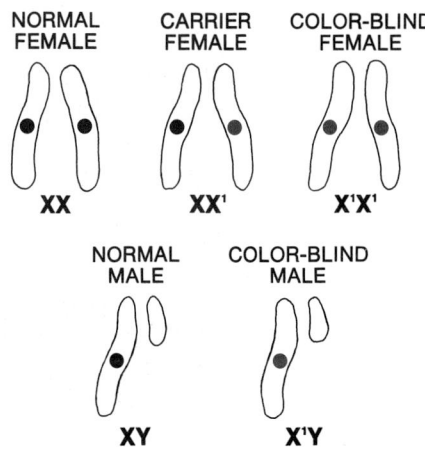

93

for baldness is not sex-linked. The general body chemistry in a man is a little different from that in a woman. This can cause the same gene to act differently in the two sexes. In a man the gene for baldness is dominant. One gene is enough to make him bald. In a woman it is a weak recessive. Even two of them will not usually cause baldness. They simply produce a growth of hair that is thinner than usual. There are several varieties of the baldness gene. Some cause the hairline to move back a little. Others cause partial baldness, and others complete baldness. Some produce baldness earlier than others. Different patterns of baldness run in different families.

So, boys, what are your chances of keeping your hair? If Dad is bald they are no better than 50–50. Why? If Mother's father and brothers showed baldness also, your chances are not very good. Can you ever be sure ahead of time?

How about twins? Twins may be produced in either of two ways. Remember, in the normal course of reproduction, an egg is released from an ovary, fertilized by a sperm in the oviduct, and passed down into the uterus. There it develops into an

Fig. 10-8
These are fraternal twins. Do they look exactly alike? Can you tell that they are related? How were these twins formed? (HRW photo by Ken Karp)

embryo. Many twins are formed by a simple change in this story—*two* eggs are released from the ovaries at the same time. Each is fertilized by a different sperm, and both develop in the uterus. Such twins are no more or less alike than any other two children of the same parents. They can be male and female, or they can be of the same sex, and are called *fraternal twins*. See Fig. 10-8.

Enrichment: Fraternal twins have separate placentas while identical twins usually share the same placenta.

fraternal twins

The other kind of twins start out as a single egg that is fertilized in the usual way. Then, at a very early stage in its development, the embryo splits in two, and each part develops into a complete individual. A pair of *identical twins* is the result. These twins look alike because they started out to be one person. Their heredity is exactly alike. They are always of the same sex. This is the type of twins shown at the beginning of this chapter.

identical twins

Of course many animals produce several young at a time. Twins seem unusual to us only because in our species one baby at a time is the usual number. When animals have entire litters of young at one time, they are usually of the fraternal type. The armadillo is one animal that commonly produces litters of identical young.

VARIATION WITHIN A SPECIES

What do we mean when we talk about a *species* (SPEE-sheez)? A species is a distinct kind of living thing. House cats all belong to one species. People all belong to one species. But, as you know, each cat or each person is different from all others in many ways. The individual members of each species of living thing vary in this way. We call these differences *individual variations*. You can recognize your friends because of individual variations. One person may be tall and blonde. Another may be short and brunette, and so on. There are small, long-haired cats and large, short-haired cats. Each person or each cat has its own special combination of characteristics. Look around the classroom. How many students have exactly the same shape of nose or ear or chin?

species

Enrichment: Members of different species usually cannot interbreed. However, some species are so closely related that mating may occur and viable offspring are born. These offspring are usually sterile. An example of this is a mule, which is the result of a mating between a horse and a donkey.

By now, you should be able to explain these variations. Remember that there are thousands of genes in the chromosomes. Many of these genes come in a variety of forms. For example, there is the gene that produces tallness in pea plants. Another variety of this gene produces short plants. There are genes that produce straight hair in people, but other forms of these same genes produce wavy hair, and still others produce curly hair. Some genes are known to come in as many as twenty different forms. Of course, no normal person has more than two of them.

Enrichment: The frequencies of blood types in humans vary in different populations. Certain diseases also affect particular groups of people, such as the occurrence of sickle cell anemia among Black people.

Reference: See *Heredity* by Jean C. Lipke.

Enrichment: Studies have been done on identical twins who were separated early in life and were brought up in different environments. Although twins remain closely alike in looks, basic personality, and learning ability, they are also influenced by different environments.

Just how likely is it that two individuals would inherit exactly the same genes? Two children in the same family have the same parents. How likely is it that both of them inherited the same chromosomes? That would depend on how many possible combinations of chromosomes their parents could produce. Think about it for a minute. Each of their parents has 46 chromosomes, but only 23 came to each child from each parent. It was a matter of chance which combination of these two sets each child has. It is also a matter of chance which sperm fertilizes the egg.

The chance of producing any one particular combination is about one in 70 trillion. This is the chance you, with your special combination of traits, had of being born. A child of your parents could just as easily have had one of the other 70 trillion possible combinations of chromosomes instead. As you see, it is not likely that there has ever been another person just like you. Nor is it likely that there ever will be again.

ENVIRONMENT

Of course the exception to individual variation is identical twins. They look alike because they have exactly the same genes. But even identical twins show slight differences. These differences cannot be due to heredity, so they must be due to environment. Heredity sets the pattern for growth and development of any living thing. This pattern can develop only if the environment allows it to. A tree seed has tree genes. But if it gets no water it will never be a tree. Water is a necessary part of its environment. Your genes may call for you to grow to a certain height, but unless you receive the right food you will never become that tall.

Some of your characteristics are not things you inherited. Suppose you cut yourself. The cut heals, but leaves a scar. This scar is a characteristic you acquired during your lifetime. You did not inherit it. You got it as a result of an accident. You may have it all your life, but it will not change your genes. You will not pass it on to your children. Such traits cannot be inherited.

Environment can also have an effect on traits that are inherited. Certain rabbits that have genes for white hair can end up growing black hair if their skin is kept at a temperature that is colder than usual.

As you see, both heredity and environment influence the development of any plant or animal. If you want to have an interesting class discussion, you might try to decide in what ways heredity and environment influence the following traits in people: height, build, appearance, intelligence, honesty, personality, health.

SUMMARY

Sperm and eggs are formed during reduction division. The combination of chromosomes and genes in each individual is determined by which sperm fertilizes which egg. A dominant gene on one of a pair of chromosomes can hide the effect of a recessive gene on the other.

In a cross between hybrids, an average of about three-fourths of the offspring will show the dominant trait and one-fourth will show the recessive one. Other crosses will show other percentages.

Females have two X chromosomes. Males have one X and one Y chromosome. Traits carried on the X chromosome are sex-linked. A trait controlled by a gene on some other chromosome is said to be sex-influenced if it acts differently in males and females.

Fraternal twins come from two eggs that were produced at the same time. Identical twins develop from a single fertilized egg that divides to form two individuals. Both heredity and environment influence the development of an individual.

ACTIVITY

A Study of Human Heredity
In this activity your class will investigate an actual example of a human hereditary trait.

Some people can roll their tongue up at the edges to form a roll, or tube. Others cannot. This is an inherited trait that is controlled by a single pair of genes.

A. Let the members of the class make a study of any families they know. Ask the parents and their children to stick out their tongues and roll them up at the edges. Some people do not like to admit they cannot do it. Do not take their word for it. Make sure you see them do it.

  1. Write down the results of this study for all of the families you were able to investigate.

B. On the blackboard add up all of the results brought in by the entire class.

  2. In all families where both parents could roll their tongues, how many children could roll theirs? How many could not?

  3. In all families where one parent could and one could not roll their tongues, how many children could and how many could not roll theirs?

  4. In all families where neither of the parents could roll their tongues, how many of the children could and how many could not roll theirs?

  5. Based on these results, which gene is dominant—the one for tongue rolling or the one for nontongue rolling? Why do you think so?

Word Quiz

From the new words you learned in this chapter, write the word that correctly completes each statement. Do not write in this book.

1. Because of _____ a sperm or egg has half as many chromosomes as a body cell.
2. A color-blind man has a defect in his _____ .
3. A boy and girl born at the same time have to be _____ .
4. A guinea pig that inherits one black and one white gene is a(an) _____ for that trait.
5. The guinea pig in question 4 will look black because black is _____ in guinea pigs.
6. When pure red crossed with pure white always produces pink or streaked flowers in the offspring this illustrates _____.
7. A trait that never shows up in hybrids is called a (an) _____.
8. The one sex chromosome that a man transmits to his son is the _____.
9. A trait like baldness that fails to show in females is _____.
10. A trait whose gene is proven to be carried in the X chromosome is called _____.
11. Only _____ have all 46 of their chromosomes exactly alike.

Check Your Facts

1. How did Mendel experiment with heredity?
2. What is meant by a dominant trait? Give examples.
3. What is meant by a recessive trait? Give examples.
4. What is meant by incomplete dominance?
5. What is reduction division? When does it take place?
6. What determines the sex of a new baby?
7. What is meant by a sex-linked trait?
8. What is meant by a sex-influenced trait?
9. Name the two kinds of twins and explain how each is produced.
10. How do heredity and environment influence the development of a living thing?

Thought Questions

Science Reading Skills 1. & 2. Problem Solving

1. If a German baby is adopted by an Irish family, will he speak with a German accent when he learns to talk? Why?
2. What percent of the offspring would show the recessive trait in each of the following crosses: pure dominant mated with pure dominant; pure dominant mated with pure recessive; pure dominant mated with hybrid; pure recessive mated with pure recessive; pure recessive mated with hybrid; hybrid mated with hybrid. Use the chart method shown on page 91.

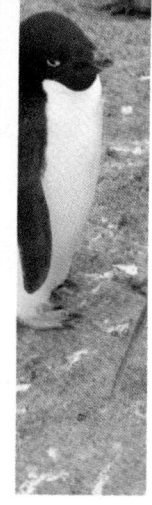

CHAPTER 11 Mutation and Adaptation

Remember that the nucleus of each cell contains DNA molecules that make up the heredity units called genes. Each gene controls the production of a particular enzyme. These enzymes control all the chemical activities of the cell, and these activities control the development of the plant or animal. The way a plant or animal looks depends on the structure of the DNA molecules in the fertilized egg from which its body cells developed.

We usually think of genes as things that remain unchanged. They merely turn up in new combinations in each generation. This is true most of the time, but genes may not always remain the same. Once in a while a gene does change to a new form and this results in a new hereditary trait. The penguin in the photograph above has white fur due to a change in a gene.

MUTATION

Sometimes we observe the sudden appearance of a new trait. Suppose, for instance, that a baby is born with two thumbs on each hand. Even though all of its ancestors had normal thumbs this baby was different. Suppose, also, that when this baby grows up some of its children and grandchildren also have double thumbs. It becomes clear that a new trait has appeared and that it is inherited. One or more genes must have changed. After all, it is genes that control the development of thumbs.

DNA molecules are long and complicated. They contain thousands of atoms. These atoms can be arranged in a great number of patterns. Every time a cell divides, the particular pattern in its DNA molecules must be duplicated exactly. It is really not surprising that something "goes wrong" once in a while, and the pattern of the atoms in a DNA molecule is changed. **It is no longer the same DNA molecule.** It causes a new type of enzyme to be produced. **A new trait appears due to the action of the new enzyme.**

objectives

After you read this chapter, you should be able to:

—**Explain** what is meant by mutation

—**List** some causes of mutation

—**Explain** what becomes of both helpful and harmful mutations

—**Compare** natural selection and artificial selection

—**Discuss** the ways in which fossils can be used to study the history of living things

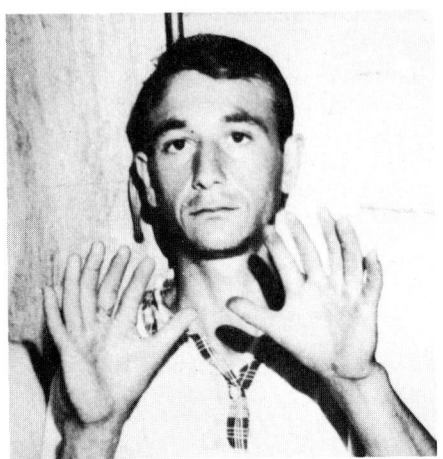

Fig. 11-1

The extra finger on each hand results from a gene mutation. Would this be helpful or harmful? Would it affect the length of a person's life? (UPI Photo)

mutation

Word Study: "muta-" means change

Enrichment: Some mutations are so lethal that the organism dies in the embryonic stage of development.

Enrichment: These naturally occurring mutations have been found in many plants and animals. Scientists have learned the rate of mutation for certain traits in many of these organisms.

Enrichment: Emphasize the difference between mutations in reproductive cells and in body cells.

Such a new trait is called *mutation* (mew-TAY-shun). The new type of gene keeps on duplicating itself and producing the new trait. There has been a change in heredity. The new trait may be dominant or recessive. In the case of the double thumbs we have described, it would be dominant.

When a gene mutates, the new form may be an improvement, but in most cases it is not. Some mutations are so harmful that they cause early death.

The body of a higher plant or animal is like a complicated machine. It works well just the way it is. This is why the plant or animal species has been able to survive. Mutations are changes that happen by chance. Such an accidental change will often be harmful. If it is very harmful, early death takes place. This is the end of the mutation and it is not passed on to another generation.

But many harmful mutations survive for a while. This can easily happen when they are inherited as recessive traits. Suppose, for instance, that a mutation causes some physical defect. The defect is serious, but it is inherited as a recessive trait. Any individual who gets two genes for the defect will develop the handicap. This individual may die early in life, or may live, but fail to produce offspring. But an individual who has only one of these genes does not develop that defect. This individual is likely to survive and to produce offspring. Some of these offspring inherit the mutated gene. So the gene is passed on from generation to generation. No one knows it is there. But suppose two people carrying this hidden gene should marry. There is then a chance that some of their children will inherit the gene from both parents. These children will develop the defect. This explains why some human defects show up only now and then in the population.

CAUSES OF MUTATION

Naturally we wonder what causes mutations. We cannot be sure about all the causes, but we do know about some of them. The natural vibrations of the molecules, for instance, may produce some mutations. Molecules are always in motion. This may cause a few atoms to line up out of order in the DNA molecules.

Chemicals. Certain chemicals can cause mutations. The chemical substance damages the cell, but not enough to kill it. If genes are changed, mutation is the result. Such things as formaldehyde (a tissue preservative), mustard gas (an irritant used as a war gas), and creosote (a wood preservative) can produce mutations. Many other industrial chemicals are now known to cause mutations. Still others are suspected of doing so. Many of these same chemicals can also cause cancer. Very likely there are chemicals in natural foods that cause some mutations. We should be especially careful when new chemicals are used. Insecticides, food preservatives, and drugs need to be thoroughly tested before they are sold to the general public.

Radiation. Mutations can also be caused by radiation, such as X-rays. There is always some radiation present on earth. It comes from outer space in the form of cosmic rays. Another form, called gamma rays, is produced by some of the elements found in rocks and soil. All of this *background radiation*, as it is called, is thought to be responsible for about 10 percent of the mutations that commonly happen in living things. In humans this percentage may be higher. Obviously, X-rays or nuclear energy should not be used carelessly or they may add to the number of mutations occurring in people. Of course, no lasting damage is done by taking an X-ray of a broken arm or leg. X-rays are helpful to the doctor and there is no reason to be afraid of them. However, the cells located in the testes or ovaries can be harmed by excessive exposure to X-rays. These cells produce the sperm and eggs that develop into new individuals. A mutation in one of these cells might be passed on to every cell in the body of the new individual. It is very important that these reproductive organs not be exposed to X-rays.

Enrichment: The angle of the X-rays is in a direct line with both the patient's teeth and reproductive organs. The lead apron is worn to screen out the X-rays from the reproductive organs.

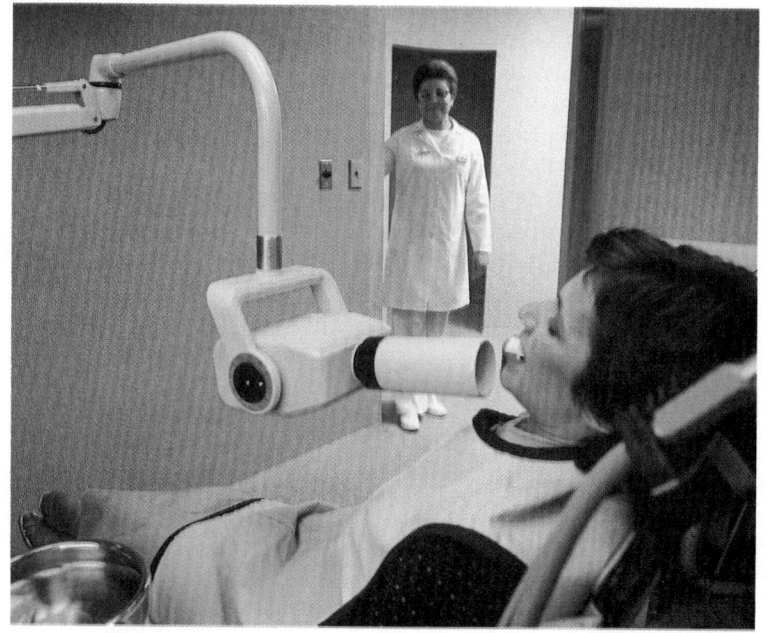

Fig. 11-2
Dental assistants frequently take X-rays of patients' teeth. To protect themselves from overexposure to these X-rays they stand away from the area. The patient wears a protective lead apron. (HRW photo by Ken Lax)

Assignment: Have students research the effects of radiation on survivors of Hiroshima and victims of radiation accidents.

When a nuclear bomb explodes, a very large amount of radiation is given off. People not quite close enough to be killed by a bomb blast may still receive a good deal of radiation. This causes many mutations, and it increases the number of defective genes in the population.

WHAT BECOMES OF MUTATIONS?

As we have said, most mutations are harmful. If new ones keep forming in each generation, you might think that finally every species of living thing would die out. To see why this does not happen remember how we explained the advantages of sexual reproduction in Chapter 7. Any animal or plant that inherits genes that greatly handicap it will probably die before it reproduces. The abnormal genes will not be passed on. If a living thing is handicapped just a little, it may not die, but it probably will not reproduce itself as well as the healthier members of its species. There will be fewer and fewer of that kind of harmful gene as time goes on.

In any species, then, a number of new mutations turn up in each generation. At the same time, about the same number of old mutations are lost through death. Thus, the number of harmful genes present in the living population remains about the same at all times. If anything should increase the number of mutations being formed there would be more harmful genes and the death rate would increase. This effect would not show up right away because many of the mutations would be recessives. It would take a very long time for them to be eliminated through death. But eventually the genes would be eliminated.

There are many examples of harmful mutations. We have already mentioned the recessive sex-linked gene for hemophilia, which causes bleeding. This gene appears as a new mutation every once in a while. About 10 percent of our population is nearsighted. This nearsightedness is hereditary, and no doubt it started by mutation. At the beginning of this chapter you saw a photo of an *albino* (al-BY-no). An albino animal has no coloring in its skin or eyes. White mice and white rabbits with pink eyes are albinos. The albino trait is due to recessive genes. Wild albinos are soon killed because they do not see very well, and they are easily seen by their enemies. Their genes die with them, but new genes for the albino trait keep appearing through mutation.

Reference: For excellent pictures on albinos, see *National Geographic World*, April, 1976, pp. 20-25.

ADAPTATION

Each kind of wild animal or plant is well suited to live in some particular kind of place. We say that each living thing is *adapted* to its way of life. A squirrel, for instance, is well adapted to a woodland environment. It has curved claws and muscular legs for climbing. It has a long tail for balancing. Its teeth and digestive system allow it to eat such foods as nuts, acorns, buds, and bark.

Think what a different set of adaptations a fish has! A fish could not live in the trees, nor could a squirrel live in a lake. Each is adapted to a particular kind of life. These adaptations are mostly inherited. The genes of a squirrel adapt it to life in the trees. The genes of a fish adapt it to live in the water. Any mutation that causes a living thing to be less well adapted to its surroundings is a harmful mutation. A mutation that makes the living thing better adapted is helpful. Such a helpful mutation aids the animal or plant to live longer and reproduce more often. Thus, new helpful genes become more and more common.

Now we can see why living things are so well adapted to their surroundings. Any mutation that makes them better able to survive becomes a standard gene for that species. See Fig. 11-3. Harmful genes tend to be lost. Plants or animals that have such genes are less likely to survive and produce offspring.

adapted

Activity: Collect pictures of plants and animals from different environments and identify traits that aid in adaptation.

Fig. 11-3
What adaptation has this mother giraffe passed along to her offspring for feeding off tree leaves? (N. Meyers/Bruce Coleman Inc.)

103

The squirrel we were just speaking of was the ordinary tree-climbing type. There are also ground squirrels, which live in open country and dig holes to hide in. Suppose a ground squirrel had a mutation that gave it curved, slender, pointed claws like those of the tree squirrel. Such claws are fine for climbing trees, but they are not good for digging holes. This would be a harmful mutation for the ground squirrel. It could not dig a good hole to hide in, so it might soon be eaten by a hawk or fox. If a tree squirrel inherited a mutation that gave it straight, strong, digging claws, this squirrel, also, might soon be caught and eaten. The same gene that is useful for one animal may be harmful for another. This depends on whether the gene makes the animal fit for life in its particular environment or not.

Natural selection.   Species of living things usually produce more young than can possibly survive. A pair of squirrels has several young squirrels each year. Mice have several young every few months. A pair of robins produces about eight young birds each year. At this rate of increase there would soon be too many squirrels, mice, robins, and other animals. There would not be enough food available for this increase in numbers. A single tree often produces thousands of seeds. There is not enough space for all of these to grow.

Many young plants and animals die before they grow up. They fail to obtain enough food or living space. Often they are killed by their natural enemies. Those that are best adapted to their environment are most likely to survive. They will be best able to avoid their enemies and to get the things they need. The individuals that are not so well adapted have less chance to survive.

The well-adapted plant or animal is one that inherits genes that enable it to survive. It will probably live longer than a poorly adapted individual. It reproduces and its genes become more common than they were before. These genes are handed down to new generations of the plant or animal. Meanwhile, harmful genes are lost from the species. This happens quickly if they are dominant. Can you explain why harmful, dominant genes are more quickly weeded out?

You can see, then, that the conditions under which a plant or animal lives select the genes that will survive. We call this *natural selection.* Natural selection works slowly, but it is very effective. It keeps the different species well adapted to their environment. It can even help them to become adapted to new environments.

If tree squirrels lived in a region where trees were slowly becoming less common, then any mutations that adapted them to life on the ground would be more useful. Imagine that the

natural selection

Fig. 11-4
Left: The tree squirrel is adapted to life in the trees; Right: The prairie dog is really a form of squirrel that lives on the ground. Each is adapted to its own environment. (HRW photo by Russell Dian; J.M. Burnley/Bruce Coleman Inc.)

change from woodland to open country came very slowly. Imagine, also, that the right mutations appeared. Natural selection, working over hundreds of thousands of years, could adapt the descendants of these squirrels to live on the ground all the time. They would become ground squirrels. See Fig. 11-4.

Each time the environment changes in a region, many animals and plants are no longer well adapted. Many species die out completely. Others become adapted to the new conditions through natural selection. They no longer look quite the way they did before.

The earth is over 4½ billion years old. There has probably been life on it for over 3 billion years. During this long period of time conditions on the earth have changed over and over again. Natural selection has taken place throughout this history.

During most of the earth's history, there has been a warm climate the world over. But about every 250 million years there has been a cold period. During these cold periods great layers of ice, called glaciers, have covered large areas of the earth's surface. The latest cold period started about 1½ million years ago and is usually called the *Ice Age*. Glaciers formed and melted several times. Ice covered much of the northern states and Canada until 12 thousand years ago. It finally disappeared from northern Canada about 6 thousand years ago.

Climate is not the only thing that changes. The surface of the land and the shape of the continents have also changed from

Enrichment: Present the case of light- and dark-peppered moths in England as an example of natural selection.

105

time to time. The rocks of the earth's surface may sink or rise very slowly, so that the sea comes in and covers some of the land, or land may rise out of the sea. Whole mountain ranges are pushed up and then completely worn away by the slow action of wind and rain. The sand and mud that form when the land wears down settle in the bottom of the sea and harden into rocks. Later, when the sea bottom is heaved up, these rocks become part of the land again. The land you are walking on today may have been covered by sea several times. Through all of these changes, living things have continued to exist on earth.

Artificial selection. When factors in the environment "select" which genes will survive we call it natural selection. When people do the selecting we call it *artificial selection.*

artificial selection

An interesting example of this can be seen in our farm animals. The ancestors of our cattle, for instance, were fierce, wild animals that only a skilled warrior dared to hunt. Then a few thousand years ago, brave men caught and tamed these dangerous beasts. Naturally, the wildest ones in each generation were killed and used as food. The tame ones were kept for producing calves. The ones that gave more milk were also kept. Those that gave less milk were slaughtered. Or if the cattle were being raised for beef, the ones having the most meat were kept for breeding. People selected the cattle having the genes that produced the kind of animals they wanted.

The result of this selection is the many breeds of cattle found on our farms today. They are of many color patterns. After all, we protect them from natural enemies, so a gene for coloring that can be seen easily does not matter, as it would in wild cattle. Milk cows have large udders and often produce over 25 liters of milk a day. This is much more than is needed for their own calves. Beef breeds have stocky, fat bodies that would slow them down under wild conditions. Selection has adapted these animals to live on the modern farm. See Fig. 11-5.

Fig. 11-5
Left: A Holstein cow, which is a good dairy type; Right: A Hereford, which is a good beef type. Both of these breeds are descended from the same ancestors. How did they become so different? (Webb Photo; American Hereford Association)

Artificial selection is really very similar to natural selection. Artificial selection has produced all of the breeds of plants and animals that we raise today. See Fig. 11-6. Since we now understand how heredity works, it is possible to develop new and better breeds more rapidly than they were produced in the past. Some new types are developed every year.

Fig. 11-6
Left: Irish setter; Right: Komondor. How did these breeds of dogs come to look so different from each other? (Alton Anderson)

THE CHANGING WORLD OF LIFE

One very interesting branch of biology deals with the study of past life on the earth. Mud and sand settling to the bottom of water eventually harden into rock. This process is something like the setting of cement, only it takes much longer. If some dead animal or plant lies buried in this mud, it may leave an imprint in the rock that we dig up and study millions of years later. Such records of past life are called *fossils*. See Fig. 11-7. Hard parts like bones or shells are most likely to become fossils. Softer parts are sometimes preserved. There may be a stone quarry near you in which you can find some fossils.

Scientists have been digging up fossils for many years. These tell us what kinds of plants or animals lived at various times in the past. Where the rock layers have not been folded and broken up, the lower layers are older than the upper layers. So fossils in the lower layers are generally older than fossils in the upper layers.

Reference: See *Genetic Engineering* by Carl Heintz.

fossils

Activity: Take students out on a field trip to look for fossils. See *Fossils* by Frank Rhodes on where to look and how to proceed when collecting fossils.

From fossil studies it appears that life began in the water. Throughout most of the earth's history, living things were small, simple forms found in the sea. By about 600 million years ago, complicated, multi-cellular forms of life had become common. Life became abundant on the land about 340 million years ago. In each period, the forms of life seem to have been somewhat different from what they were in the period before or after. Most scientists explain this as the result of natural selection, enabling plants and animals to become adapted to environments that have changed again and again.

Fig. 11-7

Left: A fossil of a fern leaf that lived about 300 million years ago; Right: A fossil of the shell of a sea animal that lived millions of years ago. How did these organisms become embedded in the rocks? (American Museum of Natural History; The Smithsonian Institute)

In other words, the forms of life upon the earth appear to have gradually changed. Simple plants and animals gave rise to more complicated ones. Certain water plants gave rise to land plants. Certain water animals gave rise to land animals. The living things of today are descended from ancient forms that often looked very different. No doubt you have seen pictures of dinosaurs. They died out because they failed to adapt to changing conditions. Other types of animals alive at that time survived to become the reptiles, birds, and mammals of today.

There are many scientists who spend their time studying fossils, volcanos, rock formations, and other evidence of the past happenings on the earth. It is like studying history with the earth itself as the history book. They have traced the record of the past climates. They have traced the changing forms of the continents. The history of living things is an interesting part of this study.

SUMMARY

When the structure of a gene is changed it will continue to duplicate itself in the new form. This changes the hereditary trait governed by that gene. Such mutations are caused by accidents during gene duplication, by certain chemicals, and by penetrating radiation. Most mutations are harmful and handicap the individuals carrying them. This can lead to the disappearance of the gene. Rarely, a good gene may arise by mutation. This gene can help the individual to live longer and thus to pass the gene on to descendants. By this process a population will accumulate the genes that adapt it to its environment. This is called natural selection.

The earth is very old, and living conditions have changed many times. Some species have become extinct, but others have been changed by natural selection. They became adapted to new environments. A study of fossils can show us what forms of life existed in the past.

ACTIVITY

Making a Fossil

A. Take a saucer-sized dish (The aluminum ones that frozen pies come in are good.), and place the leaf of any tree on the bottom of the dish. Next, mix some plaster of Paris with water until it is about as thick as cake batter.

B. Carefully pour this over the leaf in the dish. This is similar to the way mud might bury a leaf under water. It takes thousands of years for mud to harden into solid rock. But your plaster of Paris will harden in a few minutes.

C. After several minutes lift the plaster out of the dish and remove the leaf.

    1. What is left in the plaster?

    2. Does it look like a fossil?

3. Is it a fossil?

4. Are leaf prints and footprints in a cement sidewalk real fossils? Remember that a fossil is any record of past life.

D. You may want to try making fossils of other objects, such as shells, twigs, or coins. You could press your hand into the soft plaster to make a handprint. This would be like the fossil footprints of dinosaurs that have been found by scientists.

E. After your plaster fossils have dried for a day or two, they can be painted. You might use different colors for the fossil and for the background.

109

Word Quiz

Choose the letter from **Column B** that best matches each number in **Column A.** Do not write in this book.

Column A

1. mutation
2. adapted
3. natural selection
4. artificial selection
5. fossil

Column B

a. development of new crops and farm animals
b. survival of genes that adapt species to its environment
c. record of past life
d. change in a gene
e. the condition of being well-suited to the environment

Check Your Facts

Science
Reading Skills
5. &
6. Sequencing

1. What is mutation?
2. What causes mutation?
3. Are mutations usually helpful or harmful? Why?
4. What keeps harmful genes from becoming more and more common?
5. How does natural selection take place? How does artificial selection take place?
6. What are some of the changes that have taken place on the earth during the past 5 billion years? How do fossils help in the study of the earth's history?
7. What happens to living things when their environments change a great deal?
8. Why do our present-day catttle look different from their wild ancestors?

Thought Questions

Science
Reading Skills
1. & 2. Problem
Solving

1. If dark-haired parents have a blond-haired child, can we assume this is due to mutation? Why?
2. Suppose some of the calves born on a farm are dwarfs. This is a recessive trait. How could the farmer get rid of dwarfism in his herd?

THE WORKING WORLD

Breeding animals and plants is both a science and an industry involving thousands of persons.

The word "genetics" is most often associated with research involving elements such as DNA, and may seem related to the laboratory rather than to everyday concerns. "Genetic," though, refers to the origins and development of things. Many people have occupations that demand knowledge and interest in this aspect of life forms.

Breeding animals and plants is both a science and an industry involving thousands of persons. Interest and experience are the essentials. Training and education may be formal or informal, with many people having a mixture of both. One or two years in an agricultural college learning animal husbandry provides a good foundation for an animal breeder, if you are considering an entry-level position.

Businesses in this field range from small retail operations to large animal and plant producers. There are worm farmers, some of whom ship their products across the nation to both individual consumers and retail bait sellers, there are reptile farmers who breed reptiles for their skins and their venom, there are farmers who breed

Richard Hutchins/Photo Researchers, Inc.

Mark Sherman/Bruce Coleman, Inc.

fur-bearing animals such as mink and chinchillas, and people who keep bees for commercial honey production. Chicken hatcheries provide eggs and meat for consumption (as well as for seasonal gifts); fish "farming" has also grown as an industry and, therefore, as a source of jobs and food for the table. On a larger scale, there are breeding ranches where different varieties of cattle with particularly desirable traits are developed, usually to raise their market value.

Small nurseries need people who are familiar with the varieties of plants and the conditions necessary to their growth.

Eric Kroll/Taurus Photos

"Plant propagators," a more official title for this occupation, are needed in large commercial nurseries, where plants are bred for heartiness, color, and shape. Orchards also hire people skilled in the techniques of pollinating the trees they grow.

For more information, write:

The Society of American Florists
901 N. Washington Street
Alexandra, VA 22314

Or, for the pamphlet "Careers in Horticulture" write:

The American Society for
   Horticultural Science
National Center for American
   Horticulture
Mount Vernon, VA 22121

Commercial propagation extends to many areas, including the fungus known as the mushroom. Mushrooms require a very special set of growing conditions that allow them to thrive in places where green plants could not exist. Abandoned mines, caves, dark warehouses, and damp basements are good locations for growing mushrooms. This kind of farming, therefore, doesn't have to be done in a rural area.

Genetic information is also being used by genetic counselors to aid the healthy development of human beings. This field has developed in the last 15 years, and qualifications are not completely standardized. Therefore, not all genetic counselors have the same background, although a master's degree in science is now recommended. Some counselors are nurses, some are psychologists, some have a master's degree in genetics, and some are social workers. To date only two licensing exams have ever been given by the American Board of Medical Genetics (which is not currently affiliated with the A.M.A.). Many job openings in this field require that a person be board certified or board qualified. Genetic counselors act as guides to couples who are concerned about possibilities of disease or malformation in their future offspring. By asking the couple questions about their family histories, a counselor can determine the probability of certain defects showing up. Genetic counselors are usually affiliated with hospitals and clinics. For further information, contact:

The American Board of Medical
   Genetics
Harbor–UCLA Medical Center
Division of Medical Genetics
1000 W. Carson Street
Torrance, CA 90509

Paul Moylett/Taurus Photos

YOUR LEISURE TIME

Hobbies, leisure activities, and even chores give you valuable experience.

Keith Murakami/Tom Stack & Assoc.

Breeding and raising pets and plants is not an unusual hobby. However, many people have taken their knowledge in unique directions.

"Pet-sitting" for neighbors and friends who are on vacation has become a business for some people. Walking people's dogs on a routine basis, particularly in urban areas, has also been turned to profit. In spring, chicks, rabbits, and ducks are often raised for sale. Over the years, though, increasing concern has been expressed about the care of those seasonal pets. Consumers do not always know what is necessary for the animals' survival and sellers may simply be in the market for quick profit. A knowledgeable seller can be very helpful to a potential consumer.

Business projects are not the only way to gain experience.

Alec Duncan/Taurus Photos

Offering to take care of a sick neighbor's pet or walk an elderly person's dog provides a useful service and gives you experience as well. Of course, many organizations such as 4-H and Scouts provide opportunities for raising an animal of your own and showing it in competition.

Planting and tending a garden, no matter how small, can be a source of relaxation and learning, as well as food. Going from planting to harvest teaches you about seed varieties, soil conditions, weather problems, and stages of plant development. This knowledge can also be applied in helping others who may not have the time or energy for the job.

When something is learned in the course of your everyday life rather than at school, you may not realize the extent of your knowledge. You may not label what you know in a way that seems job-related. Review your work experience, whether it was done as a home chore, a community service, a leisure activity, or a paying job. What would that work be called if it were being discussed in school or in a book?

THE CHRISTIAN SCIENCE MONITOR

Genetic engineering of plants expected to yield 'supercrops' by year 2000

By SCOTT ARMSTRONG

Stumpy cornstalks laden with ears. Square tomatoes that can be crated and shipped easily. Orange trees capable of surviving subfreezing temperatures — without the need for smoldering smudge pots or gas heaters to stave off a deadly frost.

These are among the "supercrops" that farmers of the year 2000 may see in their furrowed fields or orchards. They will be the fruits of the genetic engineering of plants — a still-nascent technology that is expected to transform agriculture over the next two decades.

The green revolution of the past two decades was brought about largely by conventional plant breeding. But the big increases in agricultural productivity over the next two decades are expected to come through genetic manipulation of plants.

Recent advances in gene splicing have compressed the time scale that some experts believe it will take for this technology to move from the lab petri dish to the farm.

None of the changes, to be sure, will occur much before the late 1980s. But by the turn of the century, major crops should be genetically altered to resist pests and diseases, survive in salty soils and harsh climates, and perhaps grow without fertilizers.

The main obstacle to getting there remains technical: knowledge of the cell structure and behavior of plants lags behind that of animal and human cells. But ethical and political factors

are also coming more into play, as the field of genetic engineering moves from the lab toward the marketplace. The U.S. Environmental Protection Agency, for one, is grappling with how to regulate the release of genetically tailored organisms into the environment — partly as a result of safety concerns surrounding a California agricultural research project.

Politics notwithstanding, a glimpse of the changes on the farm and how long they will take to come about is mirrored in a recent study by L. William Teweles & Co., a Milwaukee, Wis.-based international consulting company specializing in seed and plant science. It projects that plant genetics will add $5 billion a year to the total crop value in major nations by the year 2000 and $20 billion annually shortly thereafter. The new technology is also expected to boost agricultural production by 15 to 20 percent in the industrial world by the turn of the century. Among its other findings:

— The biggest agricultural gains will be in the industrialized world, since that's where most of the know-how is being developed.

— Yield gains will be the most dramatic in 10 major crops, including corn, wheat, soybeans, barley, rice and tomatoes, as a result of the development of seeds resistant to diseases, insects and bad weather.

— Other than in some specialized areas, the United States is well ahead in research and development in the field.

It is expected that genetically engineered corn, tomato and wheat varieties will emerge by the late '80s.

Genetic research is often discussed in relation to humans. Similar research is being done with plant life as well. Creating a food supply more resistant to diseases and producing a crop more easily harvested by machines are immediate goals, but the future holds other interesting possibilities. Researchers are looking to the day when plants could be engineered for higher nutritional values or for the ability to produce their own nitrogen fertilizer. The entire world food supply could eventually be affected.

Of course, every new development raises problems and questions. People are concerned about how decisions will be made and who will make them. Look at the number and types of groups already involved: genetic researchers, seed and plant science consulting firms, food companies, gene-producing companies, the U.S. Environmental Protection Agency, and the U.S. Department of Agriculture. In what ways could these groups affect agribusiness and even our eating habits?

UNIT 3 Ecology: Living Things in the Environment

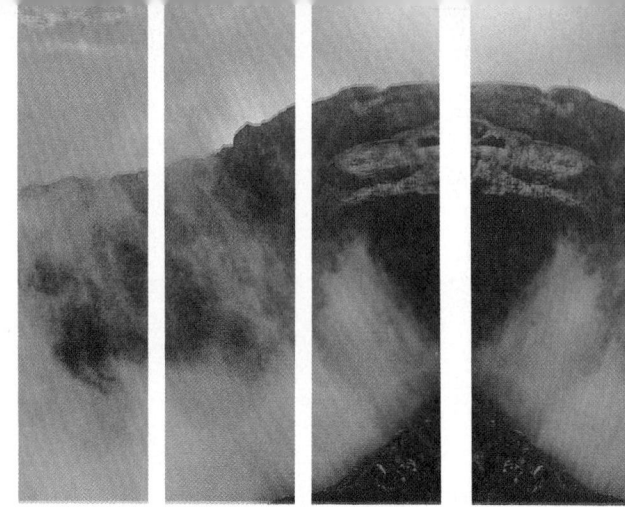

CHAPTER 12 The Nonliving Environment

ecology

The earth is not just made up of people, dogs, insects, trees, and other living organisms. It is also composed of nonliving things such as air, water, soil, and rocks. Each of the living organisms on earth is affected by the nonliving things around it and each particular living organism must be adapted to live in its own particular environment.

THE EARTH'S ENVIRONMENT

No living thing can exist, even for a few seconds, without being affected by its environment. You would realize this if you tried holding your breath while you read the rest of this chapter! Air is part of your environment. It affects you by supplying the oxygen you need. You affect it by removing some oxygen from it and adding some carbon dioxide to it. In what other ways do you and your environment affect one another?

The environment is everything that surrounds or affects the individual, including such nonliving things as light, heat, water, soil, and air. It also includes all the other living things in a region. In this chapter, we shall study the nonliving environment, and in the next, the living environment. This whole study of living things in relation to their environment is the science of *ecology.*

Let us think about the earth as a whole. The earth is the home of plants, animals, and people. What kind of environment must these living things have if they are to succeed on the earth?

First, the environment for living things must contain *water*. Protoplasm is made up largely of water. The other molecules in this living material can only meet and react with each other if they are dissolved in a liquid. There is probably no other liquid in the universe that can take the place of water in living things.

Second, the environment must contain a variety of *chemical elements* in order to form the complex compounds found in protoplasm. These chemical elements must be present if life is to exist. The most important of these is carbon. Carbon holds the other elements together in such important materials as carbohydrates, fats, proteins, and nucleic acids.

The elements for building protoplasm come from air, soil, and water. If the earth were a slightly smaller planet, it would not have enough gravity to hold onto its atmosphere. The air would be lost in outer space, and living things would not get elements like oxygen, hydrogen, and carbon.

Third, *temperature* is an important part of the environment. To support life, water must be in the liquid form, at least part of the time. If the earth were too near the sun, all of the water on it would boil and turn into steam. If it were too far from the sun, water would be frozen all the time.

Fourth, living things need a steady supply of *energy* in their environment. As you have learned, this comes to the earth in the form of sunlight. Organisms that contain chlorophyll absorb this energy during the process of photosynthesis. Plants and animals get their energy from the food made by this process.

So you see what a very special place the earth is. It is large enough. It has enough of the right elements. It is the right distance from the sun and receives a steady supply of sunlight. Do you think that life could exist on any other planets?

The water cycle. The materials and conditions for life are not present in equal amounts on all parts of the earth. It makes a difference to living things where they live. Suppose we take water and see how it is distributed over the earth. Most of the earth's water is in the oceans. When sunlight warms the surface of the sea it supplies the energy needed for evaporation. Heat from the sun also supplies the energy that makes winds blow. Cold air is heavier than the same volume of warm air. Whenever the sun heats the air in one place, heavier, cooler air comes flowing in under the lighter, warmer air, causing it to rise. The horizontal part of this air movement is what we feel as wind.

Warm, moist air from the sea may be blown over the land. It may move upward over mountains, it may move hundreds of kilometers northward to cooler regions, or it may be forced upward over a mass of cold air. In these cases, the air is cooled, clouds form, and rain falls. This happens because warm air can hold more water vapor than cold air. If it is cooled enough, some of its water vapor will turn back into liquid. You can see this happen when cold windows become misty in a warm room. Clouds are simply masses of very tiny drops of water drifting in the air. As the drops get bigger, they begin to fall as rain.

Science Reading Skill: *Sequencing—* Outline the steps in the water cycle.

Thus the sun supplies energy for moving water from the sea to the land. If this movement of water did not exist, there could be no land life. What happens to the water that falls on the land? Some of it runs off into rivers and back to the sea. Some soaks into the ground, where it may be taken in by plant roots. Some evaporates from soil and from leaves. This water may fall again as rain. Water sinking deeply into the ground will flow out into a valley and form a spring, lake, or stream. Eventually all water returns to the sea.

water cycle

Since water starts in the ocean and returns to the ocean, we call this circulation of water the ***water cycle***. A cycle is any situation in which something keeps repeating itself. The water cycle is very important. Without it there would be no water for the cells of organisms to use in their life activities. See Fig. 12-1.

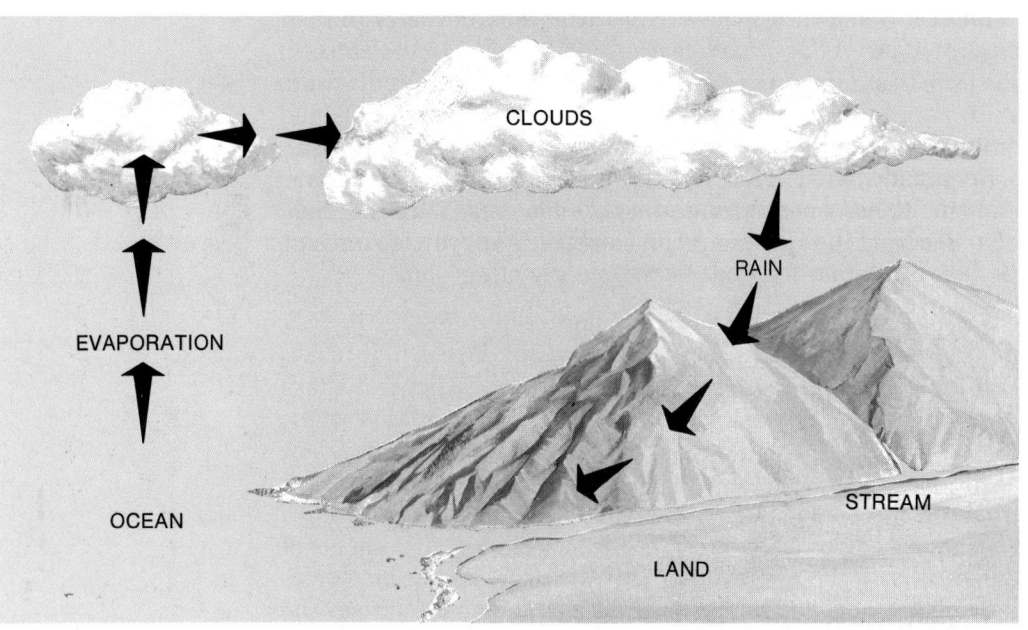

Fig. 12-1

The water cycle. How does water get to the land? How does it return to the sea?

The effects of water supply. Each kind of living thing is adapted to live in a particular kind of environment. We shall see how the water supply affects the kinds of life found in different environments of the world.

Many living things are adapted to make their homes in water. See Fig. 12-2. Some of them live at the bottom of the sea. Some live in more shallow water while some drift or swim near the surface of the sea. Others live in freshwater lakes, ponds, or streams. Each of these environments has different conditions of life, and each contains different kinds of living things. These organisms have become adapted to exist under the conditions of that environment.

On land there are many variations in the amount of water available. Some living things need more water than others and certain species can live in environments where others might die. If there is plenty of rain, trees will grow. *Forests* usually cover the land where rainfall is abundant. See Fig. 12-3.

Where there is little rainfall, the land becomes a *desert*. Plants and animals that live in the desert must be adapted to live where there is very little water. Some desert plants, like the cactus, store water in their thick stems. Others, like the aloe (burn plant), store water in their thick leaves. Such plants have wide-spreading roots that soak up the water quickly after a rain. The soil will soon dry up again, but the plant can live on its stored water for a long time.

Ordinary leaves are so thin and expose so much surface to the air that they lose water easily. Most cactuses have no such leaves. Other desert plants, like the paloverde of Arizona, have only tiny ones. Their stems carry on most of the photosynthesis.

Some desert plants have no special structures to adapt them to dry conditions. They are small plants that specialize in rapid growth. When a storm wets the ground, these plants sprout from seeds in the ground. They quickly put out leaves and make enough food to form flowers and seeds. Then they dry up and die. This new crop of seeds lies in the ground for months or years until the next rainstorm passes. The rain actually washes a certain chemical off the seeds. This chemical prevents the seeds from sprouting when there is no rain. When enough water is present the seeds sprout and the process is repeated. This is called a desert "bloom." See Fig. 12-4.

Fig. 12-2
What adaptations enable this fish to live in the water? What changes in structure would be needed if it were to live on land? (Bill Wood/Bruce Coleman, Inc.)

Fig. 12-3
Forests grow only where there is enough rainfall. (Renée Purse/Photo Researchers, Inc.)

Fig. 12-4
A desert "bloom" after a heavy rainfall. These flowers will die and fall off when the water is used up. (Linda Vartoogian)

115

Science Reading Skill: *Generalization* —Describe the general adaptations of desert organisms.

Desert animals get water from the food they eat and from dew that sometimes forms at night. Many of them spend the day in burrows, so they lose less moisture to the air. They come out at night to search for food.

What about regions that are not as dry as a desert but lack enough rain to grow trees? There was a large area like this between the Rocky Mountains and the eastern forests. This was an area of natural *grassland*—the Great Plains and the prairies of North America. Because of extensive farming, most of the natural grassland left is found only in national parks. In all continents there are similar grasslands in the fairly dry regions between the forests and the deserts. These include the *steppes* of Asia, the *veldt* of Africa, and the *pampas* of South America. Grasses grow well when there is water. Their tops die down during the dry spells, but the underground parts remain alive and send up new shoots when rains come again. Sheep and cattle can be raised on such lands. See Fig. 12-5.

Fig. 12-5
Grasslands like this get more rainfall than a desert, but not as much as a forest. Here buffalo are grazing. (Harold E. Teter)

Other cycles in nature. The water cycle is not the only cycle in nature. You already know how plants take in carbon dioxide and give off oxygen during photosynthesis. You also know how both plants and animals take in oxygen and give off carbon dioxide during respiration. This is another cycle. The same atoms that are in the air and sea part of the time are in protoplasm part of the time.

As you know, living things must take in mineral substances to supply them with the elements needed to build protoplasm.

Sea and fresh-water plants absorb these minerals from the water they live in. Land plants absorb them through their roots from the soil. Animals obtain minerals mostly from their food. Here again is a cycle. An element may be in water or soil as part of a dissolved mineral salt. Then it is taken in by a plant and becomes part of its protoplasm. When the plant dies, decay returns the element to water or soil as part of a mineral salt.

Air and soil as parts of the environment. *Soil* is important to land life as a source of water for plant roots and also as a source of minerals. Good soil is rich in minerals and is able to hold water well. Such soil will support more plants than poor soil. Poor soil contains few minerals and does not hold water well. Very sandy soils are an example of this.

Water sinks away quickly through sandy soils and is lost. Some pines and oaks are adapted to dry conditions, so you will often find them growing in sand. Maple and elm trees need more water, so they have to grow somewhere else. You see, then, that the kind of soil in a region will have a great deal to do with the plants and animals that can live there.

The *air* is an important part of the environment. It supplies organisms with oxygen and carbon dioxide. The air is also the source of the nitrogen used by all living things to build protoplasm. The air carries moisture that falls as rain. Air is important even to water life. Oxygen and carbon dioxide from the air dissolve in the waters of lakes, streams, and oceans. Oxygen and carbon dioxide can also pass from water into the air.

Temperature and living things. Water and soil are not the only things that vary over the earth. There is also a great range in temperature, and each living thing is adapted to a certain temperature range.

Land animals and plants have the greatest temperature problem. On land, temperatures are much more changeable than they are in the water. On land the winter readings in an area may be far below freezing, and the summer readings may be very high. Birds and mammals are the only animals that stay active in really cold weather. They are adapted to the cold by being *warm-blooded*. This means that they oxidize their food so rapidly that the energy produced keeps their bodies warm all the time. Even when the weather is very cold, a bird or a mammal keeps a warm body temperature, and its protoplasm remains active.

Other forms of northern land life must remain inactive during the winter. In trees and other higher plants, liquids flow slowly in the vascular tissue; photosynthesis may stop; ice crystals form among the cells. All life functions slow down. *Cold-blooded* animals have a similar problem. They oxidize food too

Activity: Plant one kind of seed in rich soil, sandy soil, and poor soil. Compare the differences in growth.

Science Reading Skill: *Compare and contrast* warm-blooded and cold-blooded animals.

warm-blooded

cold-blooded

slowly to keep their bodies warm in cold weather. They must find protected hiding places. Many insects die, but they leave eggs that hatch when warm weather returns. Some frogs burrow down in the mud on the bottom of ponds. There they lie through the cold winter months. Their blood flow slows down greatly. They get a little oxygen through their skins—enough to keep them alive. They use up the excess fat that was stored in their bodies during warm weather. This condition is called **hibernation** (hy-bur-NAY-shun).

Even some of the mammals hibernate. Woodchucks (also known as ground hogs) and ground squirrels are good examples of this. When they hibernate in burrows, mammals avoid the winter cold. In addition, they live through a period when food is not to be found above ground. What do birds do when their food supply disappears?

When a lake freezes over, fish still swim under the ice. There the temperature stays at about 4°C all winter long. Northern fish can withstand this temperature and remain somewhat active. So can many of the water insects and worms. But if dissolved oxygen is used up, ice prevents more oxygen from entering the lake from the air, and the fish die. This is called a *winter kill*. Winter kills are more common in shallow lakes because the smaller amount of water stores less dissolved oxygen. They are also more common in years when snow covers the ice all winter. Snow shuts out the light, so the water plants cannot produce oxygen by photosynthesis.

Most palm trees can stand no freezing, so they grow only in the warm tropics. Maples and oaks grow best where the temperatures range from fairly cold in winter to fairly warm in summer. Canadian spruce forests are able to stand freezing conditions for about half of the year. Notice that all varieties of forests grow only where there is ample rainfall. But the kind of forest depends on the temperature.

The regions in the far northern parts of the world are too cold for trees. These are the *tundra* areas of northern Canada, Alaska, and Siberia. The tundra is land that is covered by small, stunted shrubs, grasses, sedges, and lichens. It has the look of a wide, bleak, open plain. Only the surface layer of soil thaws in the summer. The deeper soil stays frozen the year around.

Altitude and living things. Higher altitudes have colder temperatures. In climbing a mountain you could see the same changes in climate as you would in going north, but the effect shows up much more rapidly. In an afternoon's drive up into the Rocky Mountains or the High Sierras you can pass from grassy plains or from broad-leaved forest up through pine forest, spruce forest, fir forest, and out on to the open tundra. Moving

Word Study: "hibern-" means winter

hibernation

Science Reading Skill: *Cause and effect* of winter kill.

tundra

Science Reading Skill: *Cause and effect* of altitude and temperature.

Fig. 12-6

A tundra is a region of low, stunted plant growth. Large areas of tundra cover the northern parts of Canada, Alaska, and Siberia. (Manuel Rodriguez)

1,000 meters higher has about the same effect as going 1,700 kilometers northward.

Light and living things. *Light* is necessary for nearly all of the life on earth because it supplies the energy needed for photosynthesis. Without light from the sun, there could be no food production. Organisms that contain chlorophyll must live where there is light. They cannot live underground or in the deep sea. Animals live in these environments, but their food comes from living things that live in the light. Enough light for

Fig. 12-7

The effects of altitude and latitude on the landscape.

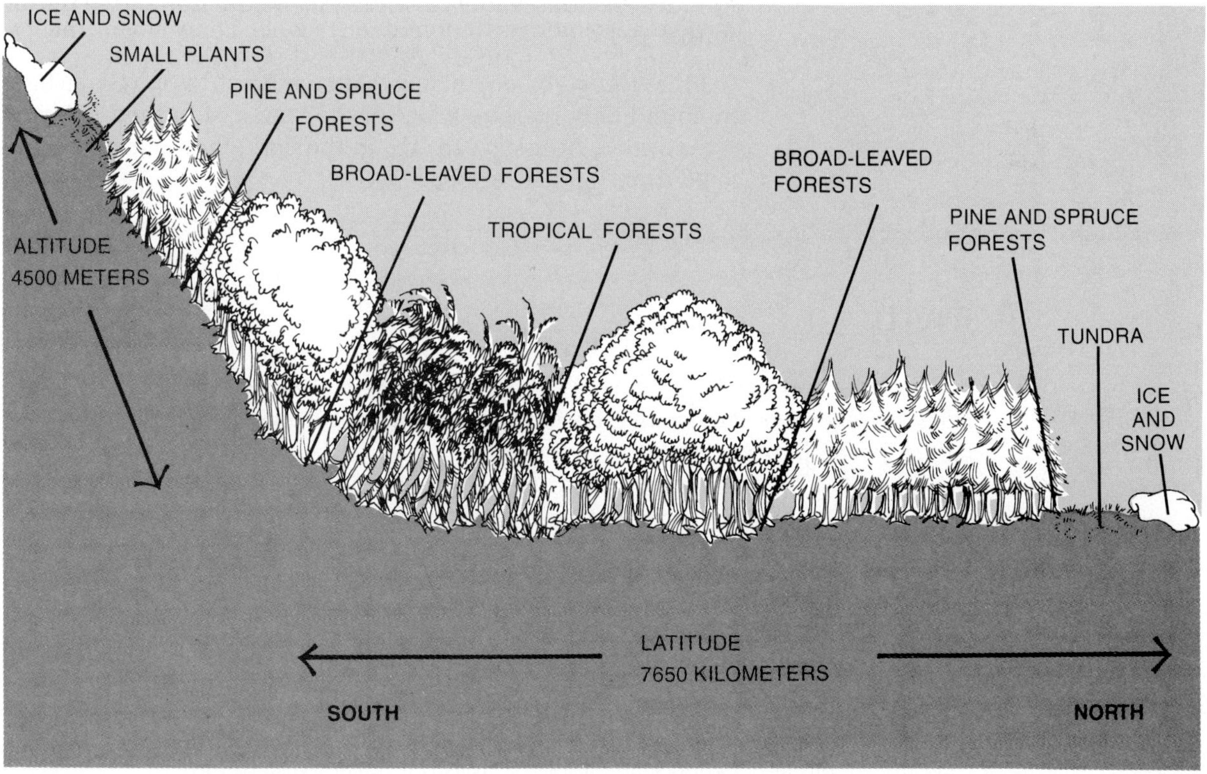

ICE AND SNOW

SMALL PLANTS

PINE AND SPRUCE FORESTS

BROAD-LEAVED FORESTS

TROPICAL FORESTS

BROAD-LEAVED FORESTS

PINE AND SPRUCE FORESTS

TUNDRA

ICE AND SNOW

ALTITUDE 4500 METERS

LATITUDE 7650 KILOMETERS

SOUTH

NORTH

119

photosynthesis may reach a little more than 60 meters down into clear water. Below that level, photosyntheis cannot take place. The food of deep sea animals comes from plant and animal materials that sink downward from the surface. Some of the deep sea animals live as much as 11 kilometers below the surface. There is no sunlight in such an environment.

SUMMARY

Ecology is the study of living things in relation to their environment. Conditions in the nonliving environment that affect living things include water, soil, temperature, and light.

The sun heats the ocean, causing evaporation. Moist sea air blows over the land, cools, and drops rain on the ground. We find forests where there is the most rainfall, grasslands where there is less rainfall, and deserts where there is still less rainfall.

Soils vary in their mineral content and in their ability to hold water. The differences in soils cause different plant life to grow in different regions.

Temperatures vary from south to north and from low to high altitudes. Similar plants grow on high mountains and in the far north.

Light is needed by organisms that contain chlorophyll, so they are found only on the surface of the land and in the top layers of the water. Animals can live in the dark, but their food must come from lighted regions.

ACTIVITY

Science
Reading Skill:
Cause and Effect

Temperature and Growth Rate
Study the effects of temperature on the rate of growth in plants in the following way:

A. Soak radish seeds in water for a few hours and then plant them in trays of moist sand or soil. Keep one tray in the freezer, one in the refrigerator, one in the room, and one in an incubator at about 38°C.

   1. Make a table showing when each group of plants first appears above ground and when each group has grown at least one centimeter tall.

B. Test other kinds of seeds in the same way. (Oats, squash, and beans would be good ones to try.)

   2. Did all seeds respond to all temperatures the same way?

C. See if you can plan similar experiments to show the effects of light and moisture upon living things.

Word Quiz

Choose the letter from **Column B** that best matches each number in **Column A**. Do not write in this book.

Column A	Column B
1. ecology	**a.** maintaining an even body temperature
2. water cycle	**b.** moisture moving from the ocean to the land and back again
3. warm-blooded	**c.** body temperature varies
4. cold-blooded	**d.** cold region with stunted plant growth
5. hibernation	**e.** study of environment
6. tundra	**f.** slowing down of all body processes in winter

Check Your Facts

1. What makes light very important to living things? Do all living things need the same amount of light?
2. Diagram the water cycle.
3. What general type of plant life covers the land in a wet region? In a fairly dry region? In a very dry one?
4. What does soil supply to living things? How can soil influence what grows in an area?
5. What is supplied to living things by the air?
6. How does temperature influence what grows in a region?
7. Why do animals hibernate?
8. What is meant by cold-blooded and warm-blooded? What advantage is there in being warm-blooded?
9. What causes winter kills in lakes?
10. What is the natural type of plant cover growing in the region where you live? What conditions are responsible for this?
11. "Changes in altitude have the same effect as changes in latitude." Look back at Fig. 12-7 and explain this statement.

Science
Reading Skills
1., 9., & 11.
Cause and Effect
2. *Sequencing*
3. *Generaliza-tion*
8. *Compare and Contrast*

Thought Questions

1. In what ways are the living conditions in a cave similar to those deep in the sea?
2. During the Ice Age, glaciers reached as far south as the Ohio River. If this happens again, what living things will grow in Michigan, in Kentucky, and in Mississippi? What changes would it make where you live?

Science
Reading Skills
1. *Compare and Contrast*
2. *Problem Solving*

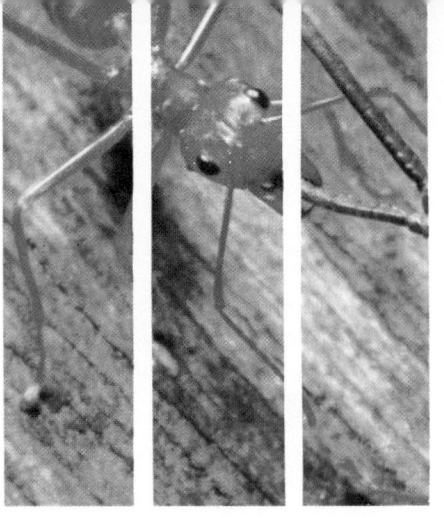

CHAPTER 13 THE LIVING ENVIRONMENT

food group

producers

consumers

plant eaters

Enrichment: You may want to introduce the terms herbivore, carnivore, and omnivore.

For each living thing, other living things are a part of the environment. Living things affect one another in many ways. One interesting way to discover how they affect one another is to see how they obtain their food. Like the ants carrying off the grasshopper in the photograph above, all living things need food, and many of their adaptations have to do with getting this food.

FOOD PATHWAYS

The food groups. We can group living things according to how they get their food. All living things that get their food in the same way make up a *food group.* The most important food group is made up of the *producers.*

These are mainly the plants with chlorophyll. When you are in a park or in the country on any summer day, look around you. Unless you live in a desert, you will see green wherever you turn—green grass, green trees, green crop plants on the plowed fields. This green plant tissue is using the energy of sunlight to make food. In the sea, seaweeds and tiny drifting sea plants are making food. **This food that the green plants make supplies the whole living world with the materials and energy it needs to live.**

All of the other food groups come under the general heading of *consumers.* They consume, or use, food that was first made by the producers. All animals are consumers, and so are the fungi, such as mushrooms and molds. Most bacteria are also consumers. These fungi and bacteria, like animals, do not have chlorophyll, so they must obtain their food ready-made.

The largest food group among the animals is made up of *plant eaters.* Antelope, mice, elephants, and grasshoppers are some types of plant eaters. Many fish and other sea animals are also plant eaters. Plant-eating mammals need the kind of teeth that can grind up coarse plant foods. They also need a digestive system that is able to handle such food. Nearly all of our common farm animals are plant eaters.

Another food group is made up of the *flesh eaters*. Tigers, weasels, ground beetles, robber flies, and trout are a few examples of flesh eaters. Flesh eaters need to have teeth, claws, or other weapons to catch and kill the animals they eat. Some flesh-eating animals do not kill their own meat. They feed on the bodies of any dead animals they find lying around. Vultures and blowflies are examples. Animals in this group are called the *scavengers*.

There are some animals that will eat many kinds of food, both plant and animal. Chickens, for instance, will eat green grass or clover. They will eat seeds, like corn or wheat. They will also eat insects, worms, and even meat, if they can get it. Such an animal can be called a *variety eater*. Other variety eaters include rats, bears, pigs, crows, goldfish, and people.

Parasites form another food group. A parasite feeds on another living thing, usually without killing it. The parasite lives right on or in the animal or plant it feeds from. This animal or plant is called its *host*. Lice are parasites. They are little insects that ride around on some larger animal and suck its blood. Fleas, ticks, and tapeworms are other examples of parasites. Not all parasites are animals. Fungi are sometimes parasitic, especially on plants. Wheat rust is a disease of wheat caused by a fungus growing on the wheat plant. When people become sick with diphtheria (dip-THIR-ee-uh) or pneumonia (new-MOE-nyuh) it is because bacteria are living as parasites in their bodies.

Many of the bacteria and fungi are not parasites but are the types of consumers that are called *decomposers*. Decomposers cause any dead plant or animal material to break down and decay. You have seen how wood, leather, cloth, meat, leaves, and many other things decay. They seem to disappear completely. Actually, they have been used as food by decomposers. Usually you do not see the decomposers that cause the decay. They are too small. Often they are single-celled bacteria. If large forms like mushrooms or molds are causing the decay, then you can see them. These decomposers digest the food first. Then this food is absorbed into their cells and used up. When they are finished, there is nothing left but water, carbon dioxide, and simple mineral salts. The energy that was contained in the original food material is also used up by the decomposers.

Food chains. Food that was first made by plants may be eaten by plant eaters. Plant eaters are eaten by flesh eaters. Flesh eaters may be eaten by other flesh eaters and these in turn will die and be used for food by the scavengers. Meanwhile, many of these forms are fed upon by parasites. Whenever food is

flesh eaters

scavengers

variety eaters

parasites

host

decomposers

food chain

Enrichment: Discuss the relationships between food chain, food web, and food pyramid.

passed along in this manner, we have what is called a *food chain*.

The following is an example of a food chain. Grass plants make food. Grasshoppers feed on grass. Toads eat grasshoppers, and garter snakes often eat toads. Hawks eat garter snakes. See Fig. 13-1. The hawk is at the top of this food chain. There is more grass than there are grasshoppers, more grasshoppers than toads, more toads than snakes, and more snakes than hawks.

Fig. 13-1
A food chain. What does the sun have to do with this? What do you think becomes of the hawk when it dies?

Activity: Make a food web using labeled cards as different organisms; connect producer to consumer, etc. with string. Then cut the string and have students identify the affected organisms.

This is true because each member of the chain uses up some of the food, so there is less and less as it moves along in the chain. If any plant or animal in the food chain dies, it will decay. Waste material from the living animals will also decay. So all of the food that is made by the producers is finally used up. It is used partly by the green plants and partly by the animals and decomposers in the food chain. Can you think of a food chain with people in it?

Cycles. Study Fig. 13-2. It shows the food relationships we have just been talking about. How does it also illustrate the cycles we studied in Chapter 12? Carbon dioxide from the air

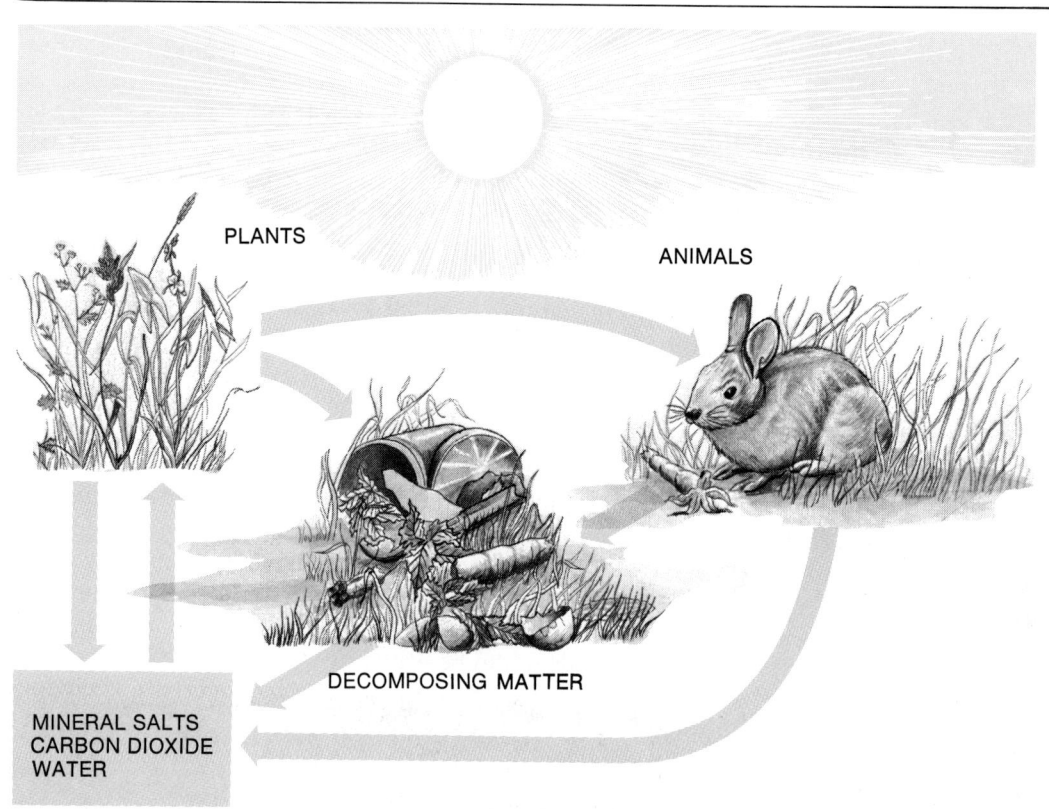

PLANTS

ANIMALS

DECOMPOSING MATTER

MINERAL SALTS
CARBON DIOXIDE
WATER

is combined with water and mineral salts from the soil to form food and living matter. This is the process of photosynthesis. Respiration and decay break down these materials and return them to the soil and air. They are used over and over again.

Energy from the sun does not go through cycles. When it comes into the living world it is changed into chemical energy by photosynthesis and stored in food molecules. Each member of the food chain gets some of this energy and uses it to carry on its life activities. In each living thing the energy is changed into heat, which passes off into the surrounding environment. This heat finally radiates out into outer space. It does not return to the sun to be used over again.

The greatest amount of food is used by the green plants themselves. They need it for energy and for growth. The rest may be used mainly by animals. In this case the decomposers obtain only what food value is left in dead plants, dead animals, and animal wastes. They may not even have a chance to decompose the dead animals. The scavengers get them first. A grassland is like this. In a forest the situation is different. Few animals are able to use tree leaves or wood as food. Decomposers are

Fig. 13-2
Food relationships. What materials are going in cycles? What source of energy keeps the cycles going? What becomes of this energy?

Science Reading Skill: *Sequencing*—Describe the stages in the recycling of matter and in the renewal of energy.

more important than animals in the food chains of forests. They cause decay of dead leaves and fallen tree trunks.

SURVIVAL IN THE ENVIRONMENT

Hunter and hunted. The process of eating and of being eaten takes place all the time. You can easily see, then, that living things do influence each other in many very important ways.

The *natural enemies* of a species are the animals that feed on it. Toads are natural enemies of insects. Snakes are natural enemies of toads and mice. You may think that natural enemies are harmful to their victims. Certainly they can be harmful to individual organisms. But flesh eaters may actually be of benefit to the *species* of animal they prey on. To see why this is true, let us suppose that a plant-eating animal, such as the deer, had no natural enemies. If the deer had no natural enemies, too many deer would live. Soon there would not be enough food for all of them. The deer would feed so heavily that they would kill their food plants, and many deer would die. The few that lived would be in a poor, half-starved condition. There are actually more deer and healthier deer in a region where natural enemies kill enough to keep them in balance with their food supply. See Fig. 13-3.

Enrichment: You may want to introduce the terms predator and prey.

natural enemies

Enrichment: In Isle Royale National Park in Lake Superior, the population of moose increased to over 2,000 within 20 years because no predators were on the island. A food shortage resulted in which all but a few hundred moose died within a few years.

Fig. 13-3
How do lions help the wildebeests living in a region? Where there are no lions what would probably take their place? (Y. Arthus-Bertrand/Peter Arnold, Inc.)

Adaptations for protection. Many animals and plants have adaptations to protect them from their natural enemies. Let us take coloring as an example. Animals that are hunted often have color patterns that make them difficult to see. You may walk within 2 meters of a rabbit crouched on the ground without seeing it. Its brown and gray fur matches the color of the ground and dead leaves so well that you can look right at the rabbit without realizing it is there. The rabbit may not move until you are very close. Then it suddenly runs away. Its running ability is, of course, another adaptation for escaping from flesh eaters. Most hunted animals have coloring that blends with their backgrounds. See Fig. 13-4. The giraffe blends in well with the light and shade under trees. Many moths rest all day out in the open on tree trunks. Their color pattern makes them look like part of the bark.

Animals that hunt also need color patterns that conceal them. After all, the hunter would go hungry if its prey could always see it coming. A lion matches the dead grass it hides in. A tiger blends with the sunlight and shade in the jungle. There are brightly colored spiders that wait on flowers for insects to come to them. The spiders match the flower petals so well that their victims do not see them in time.

Some animals not only match the color of their backgrounds, they also match the shapes of things around them. You may have seen a walking stick. This is an insect about 10 centimeters long. It is shaped like a twig and it moves very slowly as it feeds on tree leaves. Hungry birds may fail to see the walking stick because of its twig-like appearance. See Fig. 13-5.

Competition. Competition among living things does not mean that all living things are always fighting with one another. It means that they are after the same things, such as food and space to live in, and there may not be enough to go around. Trees, for instance, compete for space. Each tree must have room in the ground for its roots and room in the air to spread its leaves in the light. If another tree grows above it and shuts out the light, then the shorter tree may die. Competition is keenest among living things that are most alike. They are after the same things. The ones that are best adapted to their environment are the ones that survive.

Some animals have *territories,* which help to control competition. Some male birds, for instance, return north in the spring and each one selects his own piece of land for nest building. The bird sits on a high perch and sings, thus letting other male birds of the same species know he has claimed this particular piece of territory. The others keep out unless they want a fight. By the time the females arrive, the various territories are well established. This is something like a farmer who

Fig. 13-4
Coloring helps some animals to hide from predators. This ermine seems to blend in with its snowy environment. (Peter Arnold, Inc.)

Fig. 13-5
A walking stick. Would it be easy for a bird to see this insect? (G. Ziesler/Peter Arnold, Inc.)

territories

127

goes down to the courthouse to record the deed to his land. This is his way of telling other people, ''This is my land. I will earn my living here.'' He does not have to sit on a fence post and sing about it, like a bird, but the result is the same. Both bird and man have territories on which to obtain food.

Territories represent food supplies. Animals with territories are not so crowded that they cannot feed their young. Wolf packs hold large territories and will not allow other wolves to hunt there. Hyenas and lions also hunt in groups. They defend their hunting territories. Even a male cottontail rabbit defends its territory from other male rabbits.

Cooperation. Living things do not always compete. Sometimes the presence of one type of organism helps another type. Bees and flowers are a good example. The bee obtains its food from the flower and the flower is pollinated by the bee. Both benefit. Earthworms in the soil loosen the ground, letting in air and water. Then plant roots can grow better. When roots and leaves die, they become food for the worms.

The greatest degree of cooperation is found among the *social animals*. These animals form societies in which many individuals cooperate with each other for defense, food-getting, and care of the young. Bees, ants, hornets, and termites are social insects. They are all very successful. We also belong to a social species. Think of the many ways in which people work together. This ability to cooperate with each other has had a great deal to do with the survival of our species.

We are not always sure of how each living thing affects the others around it. Some duck hunters once found that skunks were eating eggs from the duck nests in a certain marsh. These hunters thought it might improve the supply of ducks if they eliminated this competition. They did their best to wipe out the skunks with guns, traps, and poisonous baits. The result was fewer ducks! This was not what they had expected.

Fig. 13-6
Members of human species must cooperate with each other also. These people are planning to build a new community. (HRW photo by Russell Dian)

Careful study showed the reason for this surprising result. Duck eggs were not the only things skunks ate. They also ate turtle eggs. With the skunks gone, snapping turtles multiplied rapidly. They ate young ducks and reduced their numbers far more than the skunks ever had. When skunks were allowed to increase in numbers again, the ducks also increased.

COMMUNITIES

All of the living things in a local environment make up what is called a *community*. Communities are separated from each other by some difference in living conditions. The underwater condition separates lake or ocean communities from land communities. Mountain ranges or deep canyons may separate communities. An oak forest may grow right next to a maple forest. Differences in soil explain this. Oaks can grow on sandy soil. Maples must have a richer, moister soil. A change from a ponderosa pine forest to one of spruce is due to climate, even though they grow in sight of one another. Ponderosa pines grow at a lower, warmer level, spruce at a higher, colder level on the same mountain.

The kinds of plants that grow in a community determine what animals can live there. Squirrels, porcupines, and bears can live in a forest. Antelope, prairie dogs, and bison can live on a grassland. There are several reasons for this. The animals must find enough of their kind of food and a place that offers escape from enemies. The climate must be right for the particular species. Some plants and animals can live in several different communities, but the members of two communities are never *all* the same.

Succession. Imagine an experiment that takes a thousand years to complete. Suppose we pick a large field of good soil in a region that has a moist climate. We plow this field and leave it bare. Then we watch for the next thousand years to see what happens. Perhaps you think that this is too long to wait, but remember that this is an *imaginary* experiment.

The very next year, the ground will be covered with plants, mostly weeds. These weeds come up quickly from seed. Some of this seed was already in the plowed soil and some was brought by wind. Here and there will be a few grass plants. In just a year or two we will see a change. The grasses do not have to start from seed each year. They live through the winter and spread out more each year. After a few years, the clumps of grass grow together. The entire field becomes a grassy meadow. Weeds no longer have a chance. Their seedlings cannot grow where grass plants are tall enough to shade them out. Tree seedlings do not have much chance either. They are also crowded out by the grasses. But certain plants that can grow

Fig. 13-7
Owls eat mice. Can you tell why owls are protected by game laws? (Lynn M. Stone)

community

Science Reading Skill: *Sequencing*— Outline the stages of succession. The activity in Chapter 19 covers succession.

with grass will be present. Wild carrot, wild aster, and golden-rod will find a place.

If we wait many years, we see another change taking place. Here and there, a young tree or bush manages to grow. Once it becomes taller than the grass, a young tree cannot be stopped. These trees keep the sunlight away from the grasses underneath. New grasses that can live in the shade crowd out the earlier types. When enough trees have grown so that their tops touch, nearly all ground plants die out. The field now is covered by a forest.

But this is not the end. The first trees are a mixture of different kinds that grow in the open. When each dies, it will be replaced by others better adapted to growing in a shady forest. Finally the forest is dominated by just a few kinds of trees, such as maple, beech, and basswood, which are the ones best adapted to the combination of soil and climate in this area. There are a few other kinds of trees, and underneath are a few small plants. These are types that can get along with little light. This final community is called a *climax community*. It will go on unchanged as long as conditions remain the same. See Fig. 13-8.

climax community

FIELD STAGE	GRASS WEED STAGE	SHRUB STAGE	PINE STAGE	HARDWOOD CLIMAX STAGE
	1-2 YEARS	3-20 YEARS	20-50 YEARS	100+ YEARS

Fig. 13-8

Succession from bare ground to a climax forest.

succession

When we have a series of different communities replacing one another in order, we call it a *succession*. In the succession we were just imagining, there was first a community dominated by weeds. Then came a grassland community, then a forest community, and finally the climax forest.

There would, of course, also be animals in these communities. In the earlier stages meadow mice, meadow larks, and grasshoppers would be among those present.

During the brush stage of the succession new animals would appear. These might include deer, foxes, robins, and towhees.

The final forest community would probably include raccoons, squirrels, deer mice, woodpeckers, and thrushes.

You may wonder why the climax community does not grow up at once. Why must there be a succession? In the case we have pictured, the climax community is a forest. The members of this community are adapted to forest conditions and cannot live in the open. The ferns and mosses on the forest floor would dry up and die in the open sunlight. Young maple and beech trees grow best in the shade. **Each community in the succession changes the living conditions so that the members of the next community are able to move in.** In our example the succession resulted in deeper shade. Shade-loving types finally win out and establish the climax community. The earlier species did not survive.

If our imaginary experiment were carried out in some other region where conditions were different, the succession would be different also. It would end in a different climax community. In Kansas or Nebraska, for instance, the climax community would be grassland. In parts of Arizona it would be desert. In northern Canada it would be tundra. The type of climax community established depends on the weather conditions in the area. Succession can even change an area from a water community to a land community. See Fig. 13-10.

Fig. 13-9
This photograph shows succession from a grassland to a forest community. (Harold E. Teter)

Fig. 13-10
Sometimes a succession can take place causing a water community to become a land community. (Dr. E. R. Degginger)

The city community is unusual because it is completely dominated by humans. Most plants in parks, yards, and along the streets were placed there by people. Cats, dogs, and other pets were also brought in by humans.

Even though the city and the country are communities in themselves, we can think of them together as a much larger

Fig. 13-11
Plants have the ability to overcome many obstacles. This tree is growing around a metal fence. (Karen Kennedy Gotimer)

balance

Fig. 13-12
Fire destroyed this forest. Can you see the new young plants that are the start of the new forest succession? (Standard Oil Co., NJ)

community. Food produced on farms in the country is sent to the people in the city. The city furnishes manufactured goods to the people in the country. We rely on cooperation between these two types of communities.

Some species live in cities in spite of the dominance of humans. Weeds sprout from cracks in the pavement and grow in neglected yards and vacant lots. Rats and mice may be numerous in old buildings and alleys. Squirrels survive in the parks the same way their relatives do in forests outside of town.

You can watch for places in your region where plant succession is actually going on. These could include vacant lots, abandoned fields, dumps, or places where fire has destroyed the former community. The longer these spots have been left alone, the further along they will be in their succession. Can you predict what the climax will be in a particular area? You can if you study nearby areas that have not been disturbed. They will be covered by the climax community that is normal for that region.

Balance in a community. There is constant change in a community during succession. New species come in and old ones die out. This kind of change stops when the climax is established.

Think of all the conditions that affect living things in any environment. These include climate, soil, food supply, hunting, competition, cooperation, parasitism, and decay. These conditions act upon each species to limit its numbers. The members of some species will be more abundant than others, but their numbers will remain about the same for long periods of time. We say that the species are in *balance* with one another. The climax community is a *balanced community*.

The trees in a certain climax forest might be 55 percent maple, 30 percent beech, 8 percent basswood, 5 percent yellow birch, and 2 percent other species. There would be many mice and squirrels, fewer small flesh eaters, and still fewer large flesh eaters. These percentages would remain about the same, because the chances for survival remain the same in a balanced community.

Upsetting the balance. As we have said, the climax community will be permanent as long as conditions remain the same. Sometimes conditions change. For instance, fire may destroy a forest community. Then a new succession will begin. This succession might include communities of weeds, of grass, of brush, and of small trees. Finally the original type of forest community would become established once more. It would return because it is the community best adapted to that particular soil and climate. See Fig. 13-12.

A *permanent change* in conditions will upset the balance so that the old climax community can never return. Then succession leads to some new climax. A change in *climate* can do this. About 15 thousand years ago there were communities of grassland and forest in parts of Arizona and New Mexico that today are desert. Ice and arctic tundra covered much of the land in the northern states where today there is rich farmland. These changes in the living communities resulted from changes in climate.

New kinds of plants or animals can enter a community and change it permanently. An example of this may be seen in some hemlock communities of the North. Hemlock trees formed heavy, dark stands of forest on the better soils. They were the most important trees in these climax communities. Then sugar maple trees entered the region. They are now slowly crowding out the hemlocks. In time the hemlocks may be entirely replaced by maple.

People may enter a community and upset the balance also. When the first settlers came to North America they found the whole eastern half of the continent covered with forests. They cut down most of these forests and made farms. Thus, a new kind of living thing entered the community and changed it. People established a new climax community dominated by themselves, their farm animals, and their crop plants.

In some other parts of the world people farmed the land only to start successions leading to semidesert conditions. Careless farming methods destroyed the soil. The community that resulted is an open growth of grass and weeds, with much bare ground. Today a few people wander across this poor land, seeking food for their sheep and goats. This is the new climax community.

Careful study is needed whenever people want to do something that might upset the balance in a natural community. Otherwise the results can be very different from what was expected.

Sometimes these results can begin to be seen in a short period of time. However it seems that most of the time these results do not begin to be seen until after a long period of time. In fact, upsetting the balance of nature is a process that may take place over a period of time as long as several hundreds of years. An example of this is the using up of our natural resources, which you will learn more about in Chapter 16. The fact that these changes may not be noticed for a long period of time adds to the problem. The longer it takes for people to realize that their actions are upsetting the natural community, the more difficult it may be to correct the situaton.

Enrichment: Explain the difference between primary and secondary succession.

SUMMARY

Organisms obtain food in several different ways. Those that make their own food are called producers. Organisms that depend on other organisms for food are called consumers.

Natural enemies help to keep the population of a species from getting too large. Some organisms have protective adaptations such as coloring, which blends into their surroundings.

Organisms often cooperate with each other for defense, food getting, or care of the young.

All living things in a particular area form a community. These living things react with each other in such a way that about the same numbers of each species will be present at all times.

When a community is destroyed, species adapted to the new conditions will grow and establish a new community.

ACTIVITY

Sealed Communities

You may wonder how small a community could be and still remain fairly well balanced. Try the following experiments to find out. Keep careful records of each experiment.

A. Place a small snail and a piece of water plant in a test tube three-quarters full of aquarium water. Seal the tube with a cork or rubber stopper. Set this simple "community" where it receives plenty of light but not direct sunlight.

B. Prepare another tube in the same way but without any plant life.

C. Prepare another test tube like the first one, but include only plant life.

D. Prepare a test tube exactly like the first one, but keep it in a dark closet.

E. Make a small aquarium in a liter mason jar. Put a little soil on the bottom. Cover this with sand. Fill three-quarters full with aquarium water. Plant several water weeds. Add three or four snails. Seal shut with a lid screwed tightly down against a jar ring. Place the jar in a light (not sunny) location.

F. Make a similar sealed aquarium using a 4 liter jar.

G. Examine each community frequently and answer the following questions:

   1. Did the plants in any of these communities die? If yes, which ones? Why do you think this happened?

   2. Are the snails still alive? Which ones?

   3. What do the snails need to stay alive?

   4. Where did the snails get these "life requirements" from?

   5. What evidence do you see that decomposers are present in the community?

Word Quiz

Choose the number from **Column B** that best matches each letter in **Column A.** Do not write in this book.

Column A	Column B
a. consumers	**1.** living things that get their food in the same way
b. flesh eaters	**2.** make their own food
c. host	**3.** all species that cannot make their own food
d. food group	**4.** feed upon producers
e. variety eaters	**5.** eat only other animals
f. scavengers	**6.** eat animals that were already dead
g. food chain	**7.** eat many kinds of food
h. decomposers	**8.** lives on or within the organism it feeds from
i. natural enemies	**9.** fed upon by a parasite
j. territory	**10.** their eating is the cause of decay
k. parasites	**11.** a series of events in which each of several living things eats and then gets eaten
l. plant eaters	**12.** all of the flesh eaters that normally feed upon a particular species
m. producers	**13.** an area of land claimed by a particular animal
n. balance	**14.** all of the plants and animals living in an area
o. community	**15.** the process by which one group of organisms is replaced by another
p. climax community	**16.** the final group of organisms to appear in an area
q. succession	**17.** when the number of organisms in a species remains about the same over a long period

Check Your Facts

1. Name each food group and list at least three organisms that belong in each.
2. Draw a cartoon showing a food chain of which you are a part.
3. List some ways in which organisms are adapted to protect themselves.
4. Use examples to show how competition and cooperation in nature are different.
5. What do we mean by a community in nature? Give some examples.
6. What are some of the conditions that cause certain communities to develop.
7. What is succession? Give an example.
8. What do we mean when we say that a climax community is in balance?

Science
Reading Skills
2. & 7.
Sequencing
4. *Compare and Contrast*
6., 9., & 10.
Cause and Effect

Thought Questions

1. Most states have deer, but no wolves. What keeps these deer in balance with their food supply?
2. What would happen if all of Canada became as warm as the southern United States?

Science
Reading Skills
1. & 3. *Problem Solving*

CHAPTER 14

objectives

After you read this chapter, you should be able to:

Explain how soil is formed

List some methods of preventing soil depletion

Describe several methods of preventing soil erosion

Explain why conservation of soil and conservation of water are closely related

conservation

Enrichment: Air spaces between soil particles determine the suitability of soil for plants. Sandy soils have large spaces that water passes through rapidly. Clay particles have small spaces making it difficult for water to pass through. Humus mixed with mineral particles loosens the soil and increases the number of spaces.

weathering

Conservation of Soil and Water

There is a limit to the number of living things that can exist in any community. This is just as true for us as it is for deer or wolves. There is a limit to the amount of food we can produce. Clean water may become scarce. The open spaces that make life so enjoyable can get too crowded. The more careless we are in managing our natural resources, the more likely we are to reach a time when the land can no longer support us.

Conservation is the wise management of natural resources. If we practice good conservation we can always have soil, water, forests, and a pleasant environment for ourselves. These resources can keep renewing themselves if we maintain balanced communities. In this chapter we shall deal with one aspect of conservation, the conservation of soil and water.

SOIL STRUCTURE

How soil is formed. Soil is made up mostly of small bits of rock. These tiny rock particles vary a great deal in size. In clay soils many of them are too small to be seen even under a microscope. Many sand grains are large enough to be seen without a lens. Pieces of gravel are even larger. Other soil particles can be anywhere between the sizes found in clay and in gravel. A *loam* soil is one that has a variety of particle sizes.

Soil begins to form when solid rock breaks down. Any rock exposed to the weather will break down in time. Heating in the hot sun and cooling at night will crack the rock. Water freezing in the cracks will break it down still more. Rainwater is slightly acidic. It will attack the rock chemically. Oxygen from the air and carbon dioxide from the plant roots also affect it. In time, the surface of the solid rock crumbles completely. It becomes soil. This breakdown process is called *weathering*. This weathered rock forms a layer of soil that sits on top of the solid, unweathered bedrock underneath.

Glaciers also break up rocks. During the last Ice Age, the moving ice scraped against the rocks it passed over, grinding them into soil. Some of these glacial soils are as much as 100 meters deep.

Chemical changes in the rock particles release dissolved mineral salts into soil water. Some soils contain more minerals than others. Sandy soils are often low in their supply of mineral salts. Clays are rich in minerals but they are very dense and hard to work with. In general, a farmer likes loam soil best. It is easier to cultivate than the dense, gummy clay soils and richer in minerals than the sandy soils.

Humus in the soil. Plants grow on the surface of the soil, and their roots grow down into it. When plants die, their leaves, stems, and roots are mixed into the soil and decay. The decaying bits of plant material are called *humus.* Humus is dark in color, so the upper layer of soil is darker than the soil underneath. This upper layer is called *topsoil. Subsoil* is the lighter colored soil underneath. See Fig. 14-1. Humus contains the minerals that were once a part of the plants it came from. These minerals are released during decay. They can be taken in by plant roots and used over again.

Humus also contains spongy fibers that hold moisture in the topsoil where the roots can reach it most easily. It also has the ability to hold the dissolved minerals that keep coming from the rock particles. These minerals would normally sink deeply into the ground along with the rainwater. Humus tends to keep the minerals in the topsoil where roots can absorb them.

Topsoil is a very complicated mixture. It contains rock particles, air, moisture, dissolved minerals, humus, and many forms of living things, both plant and animal. You will begin to see that a good rich topsoil takes a long time to form. It may take over 400 years to form 2 to 3 centimeters of topsoil. It is the richest part of the soil, and is what we are most concerned with in soil conservation.

Of course water must be present in the soil. Nothing will grow without it. *Groundwater* is rainwater that has soaked into the soil.

There are little spaces between soil particles. When rain falls on the land, some of the rainwater soaks into the ground and sinks downward through these spaces. When water reaches the solid rock layers beneath the soil, it can go no farther. The spaces in the lower levels of the soil become filled with water. The top level of this water-soaked part of the soil is called the *water table.* See Fig. 14-2. Above the water table the soil is moist, but not really water-filled. Here, the spaces between soil particles contain some air. The water forms a film around each soil

Fig. 14-1
Why is the top layer of soil so much darker than the deeper layers? (USDA—Soil Conservation Service)

humus

topsoil

subsoil

Enrichment: Soil also contains bacteria, fungi, algae, protozoa, worms, insects, and occasionally larger animals, such as moles, mice, and gophers.

groundwater

water table

137

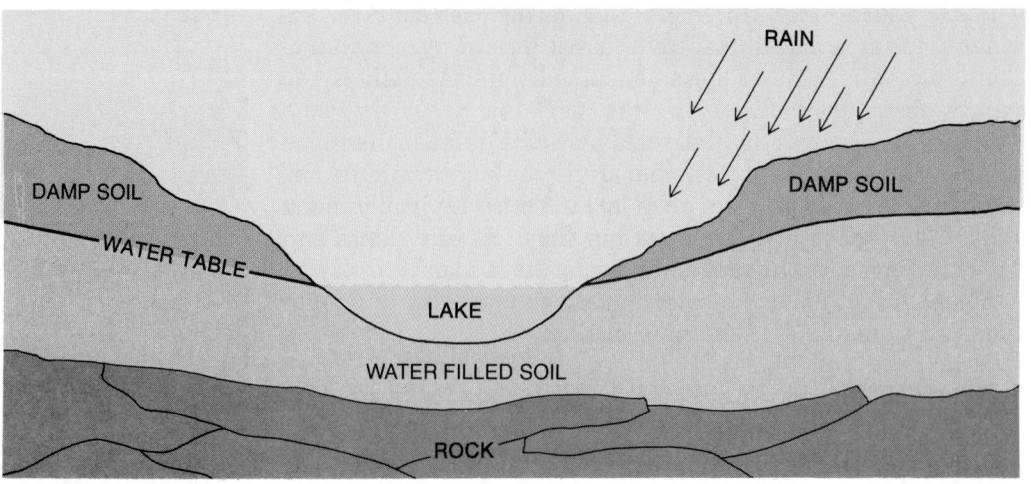

Fig. 14-2
Rainwater soaks into the ground to form groundwater. Lakes and streams are places where the water table stands above the surface of the ground.

soil depletion

Enrichment: Animal manures make excellent fertilizers. They contain minerals and add humus to the soil. Chemical fertilizers are manufactured to meet needs of different soils.

Enrichment: Alfalfa adds nitrogen to the soil and hay increases the humus content.

particle. Roots get their water from this damp soil. They do not usually go into the water-filled soil below the water table. There is no oxygen there for root respiration.

Minerals in the soil. Good topsoil supplies plants with all the minerals they need. But even with good soil, the farmers have problems. They grow crops year after year and send them to market. Minerals that came from the land go into town as part of meat, grain, vegetables, and other farm produce. These minerals do not return to the soil as they would in a natural, balanced community. New minerals may not form rapidly enough to make up for the loss. Such loss of fertility is called *soil depletion* (di-PLEE-shun). Depleted soils produce less and less as time goes on.

The old method of dealing with soil depletion was to let the land sit unused for a period of time. The action of weathering and of the grass roots upon the rock particles would build up a new supply of minerals. The soil's fertility would be restored. This method is not used much today because it takes too long. Most farmers use *fertilizers* to prevent soil depletion. A fertilizer is any material that adds useful minerals to the soil. About one-third of all the food produced in the United States could not be grown if fertilizers were not used.

The three elements most likely to be lacking in farm soils are nitrogen, phosphorus, and potassium. If a fertilizer has all three, it is called a *complete fertilizer*. For some soils other elements must be added. In certain southern areas cobalt is lacking. Plants do not need cobalt but animals do, so fertilizing with very small amounts of cobalt makes it possible to use these soils for cattle grazing.

138

Soil depletion can also be slowed down by ***crop rotation.*** This is the practice of not planting the same crop on the same ground every year. One common rotation is to plant corn for one or two years. The corn is followed by oats, which in turn is followed by hay. The hay is left in for about three years, then the rotation is repeated. Other crops are used in other places, but the idea is the same. Each crop takes a little different combination of minerals from the soil, so no one mineral is used up too rapidly. On some corn belt farms rotation is not used. Heavy use of fertilizers makes it possible to plant corn year after year and still have high yields.

THE WEARING AWAY OF SOIL

Water erosion. *Erosion* (i-ROE-zhun) is a more serious form of damage. Depleted soils can be improved by proper handling, but erosion carries away the soil itself. The fertile topsoil is lost.

Water erosion takes place in a number of ways. Raindrops striking bare ground splash the soil particles loose. As the water runs downhill, these loose particles are carried with it. Each storm carries more topsoil downhill. This process may go on so gradually that we do not notice it. The entire surface of the hillside loses its topsoil, which is the fertile part of the land. When the whole surface wears down evenly in this way the process is called *sheet erosion.*

Gully erosion is easier to see. Sometimes the water running downhill forms channels in the soil. These channels are cut deeper and deeper until gullies are formed. This is known as gully erosion. See Fig. 14-4.

Fig. 14-3
What can farmers do to prevent the minerals in their soil from becoming depleted? (Nicholas Foster/Image Bank)

crop rotation

erosion

Science Reading Skill: *Causes and effects of erosion and prevention of erosion.*

Enrichment: Once gullies form on a hillside, the land is abandoned since farm machinery cannot be used on it.

Fig. 14-4
Erosion has cut a gully through this vineyard. Where has the missing soil gone? Can the soil that is left ever be used for growing crops again? (USDA Soil Conservation Service)

Bad farming methods greatly increase the rate of water erosion. Some crops should not be planted on steep slopes. These are *row crops* such as corn, cotton, tobacco, potatoes, and beans. The bare ground between the rows is loose and very easily eroded. The damage is worse when the rows run up and down the slope. Each furrow, or groove, left by the cultivator becomes a channel leading water downhill rapidly. Gullies often begin to form in this way.

Soil erodes much more slowly when it is covered with plant life. Leaves break the force of raindrops striking the ground. Dead leaves and stems protect the surface from the main force of the running water. They hold back the water, giving it more time to soak into the ground instead of running downhill. This not only prevents erosion, but also puts more water into the ground, where it can be used by plants.

Ways to prevent water erosion. The best protection for soil is a forest cover. The next best is a dense sod of grass and other pasture plants. Small grains, such as wheat, oats, and rye are fairly good. Good conservation practice calls for keeping the steepest slopes in forests. Trees are useful here because their roots hold the soil and fallen leaves hold the water. Steep slopes would soon wash away if they were farmed.

Slopes that are a little less steep can be used as permanent pasture. The grasses provide food for cattle without exposing the soil. Of course it must not be overgrazed, or the soil will be exposed to wind and water erosion.

Hay can be planted on fairly steep slopes. The sod in a hayfield gives good protection to the soil, though not quite as good as that of a permanent pasture.

The more gentle slopes can be planted with small grains. These give fair protection while they are growing, though there is danger of erosion at plowing and planting time, when the soil lies bare.

contour plowing

For many years conservation experts have recommended *contour plowing*. This means plowing sideways, across the hillside, instead of up and down the slope. Each furrow becomes a little dam which slows the flow of the water. Slow-moving water does not do a great amount of damage. Also, slowing the rate of flow gives the water more time to soak into the ground. See Fig. 14-5.

Plowing up and down the slope is easier than plowing across it. As a result, the old harmful ways of plowing are still used more often than the contour method.

strip cropping

Strip Cropping is a conservation method that can be used along with contour plowing. Instead of laying out the land in square fields, each crop is planted in long, narrow strips that run across

the slope. Water running downhill across these strips is slowed down. A strip of hay, for instance, soaks up much of the water flowing across it. This protects a strip of some row crop that grows next to the hay. On the other side of the row crop another strip of hay also slows down the water flow.

Terracing can also be used. A permanent, low, broad ridge is formed by grading machinery at the lower edge of each strip of crop land. Farm tools can be operated right over the tops of these terraces, so they are not in the way. Water collects behind them and is led around the slope of specially constructed outlets. These outlets may have stone or cement at the ends of the terraces to prevent gullying. The water flows downhill through grassy runoff channels.

The use of terraces, contour plowing, and strip cropping makes it possible to grow row crops higher on the hillsides. Each strip of soil above a terrace is protected from erosion almost as if it were on level ground.

When topsoil has been lost by erosion it is possible to rebuild it from the exposed subsoil. This is done by a repeated process of fertilizing, planting certain plants, such as rye and clover, and then plowing them back into the soil. However, this process takes a long time and few farmers can afford to wait several years to start earning an income from their land again. It is very important to keep the topsoil in good condition in the first place.

Floods. Floods can produce severe water erosion. Swift currents from flooding streams can either wash away topsoil or

Fig. 14-5
Contour plowing. Notice that the fields are not plowed in straight rows. How does this help to prevent soil erosion? (USDA)

terracing

bury it under layers of gravel. Only the more violent floods do this. In general, floods improve the soil in the flat lands that cover the floors of the river valleys. Flood waters carry fine particles of topsoil that have eroded from the higher slopes of the valley. This material settles on the surface of the valley floor, deepening its layer of topsoil. Much of the best land in the world is found on these flood plains of river valleys. People who farm this kind of land expect floods and therefore place any buildings on high ground.

Floods become violent and destructive when too much water runs off the land instead of soaking in. Good soil conservation on the upland farms reduces flooding in the lowlands. Forests on the steep slopes also help. These lands should not be plowed. If the trees on them are destroyed, the soil erodes away. Water does not soak into the ground as it should, but runs rapidly downhill, causing floods. See Fig. 14-6. In a forest, the rainwater soaks into the dead leaf layer and on into the ground without causing erosion. This groundwater comes out later in springs that keep the streams flowing all summer.

Fig. 14-6
Good soil conservation helps to prevent this type of destruction from happening. (United Press International Photo)

Water that soaks into the ground is available for plants. Allowing water to run off the surface has the same effect as if the area had little rainfall. Less food can grow on the land because of poor water supply. Wells often go dry in regions where soil conservation is not practiced. Water runs off in streams instead of being added to the groundwater supply.

You will notice that in talking about soil conservation we have ended by talking about water conservation. The two are closely related. In the West, the conservation of water is important because of *irrigation*. Water from melting snow and from rainfall in the mountains is stored behind dams. This water can then be used all summer long to irrigate dry lands in the valleys and plains at the base of the mountains. River and well water are also used. In all of these irrigation projects, the use of water must be gauged so that the water supply lasts through the summer.

Wind erosion. *Wind* may cause erosion wherever soil becomes very dry. The topsoil blows away as dust. Wind erosion is a problem all through the West. Much of this land was originally dry grassland. The grass protected the dry soil from blowing away. Bison, elk, and antelope ate the grass. If it dried up, they moved on.

Then cattle and sheep were brought in to replace the bison. Often the land was over-grazed. The grass was eaten so close to the ground that it died out. Wind blew on the bare soil, and dust storms became common. Much of this land looks like a desert today. The obvious cure for this situation is to limit the animals to numbers the range can feed. Programs to restore a grass cover are also needed.

In other parts of the dry West the land was planted with wheat and other crops. In wet years there was enough rain to keep the crops growing. But the dry years came, there was little rain, and the crops failed. Wind blew the dry topsoil off the bare, plowed fields. The result was known as the American "dust bowl." See Fig. 14-7.

The worst dust storms blew great, dark clouds of dust all the way from Oklahoma and Kansas to the east coast and out to sea. The dust was so thick it was hard to see the sun at noon in places like Chicago and New York. This dust was actually good topsoil. It was lost forever to the land it came from.

Fig. 14-7

Dust storms, such as the one pictured here, are caused by poor soil management. Good conservation programs could prevent this from happening in the future. (USDA—Soil Conservation Service)

SOIL AND WATER CONSERVATION PROGRAMS

Our government helps in setting up local soil conservation programs all over the country. Agricultural engineers study the best methods for use in each particular area, give advice to the farmers in the region, and operate demonstration farms.

Actually a great deal still needs to be learned about soil conservation. Better methods need to be developed. Better plows, cultivators, and other equipment would serve to make the job easier and more effective. Many farmers still need to become conscious of the great importance of soil conservation to themselves and to the future of their country. People who are not farmers are affected by soil and water erosion also. These people should be concerned with the development of good conservation programs.

SUMMARY

Conservation is the wise management of natural resources. In the area of soil and water conservation, we should be concerned with loss of minerals from the soil and erosion caused by wind and rain.

Contour plowing, strip cropping, and terracing are three methods that can help stop erosion caused by water. Wind erosion can be lessened by not allowing land to become over-grazed.

Conservation is everyone's concern because it affects the food and water supply of the entire population. Therefore, everyone should be willing to support conservation programs.

ACTIVITY

Effect of Plant Cover on Soil

In this activity you will see if plant life can protect the soil it grows on from erosion.

A. Fill a large, shallow container with sod. Fill another container with soil. Place each container on a slant as if it were part of a hillside. The lower edge of each container should rest in a shallow pan.

B. Pour a pitcher of water down the slope of one container and then repeat with the other.

1. Which surface allowed more soil to wash into the pan?

2. How many pitchers of water must flow down each container before a small gully is formed in the surface?

3. What does this experiment show about the way steep hillsides can be protected from soil erosion?

Suggestion: Greenhouse flats can be used.

144

Word Quiz

Choose the number from **Column B** that best matches each letter in **Column A**. Do not write in this book.

Column A	Column B
a. conservation	**1.** water in soil
b. weathering	**2.** method of cutting ridges or shelves on hillside
c. humus	**3.** soil being carried away
d. topsoil	**4.** wise management of resources
e. subsoil	**5.** decaying materials in soil
f. groundwater	**6.** narrow bands of crop land
g. water table	**7.** loss of minerals from soil
h. depletion	**8.** different crops planted different years
i. rotation	**9.** cultivating across the hillside
j. erosion	**10.** breakdown of rock to form soil
k. contour plowing	**11.** dark colored, rich layer of soil
l. strip cropping	**12.** lighter colored, less fertile soil layer
m. terracing	**13.** upper level of the water soaked layer of soil

Check Your Facts

1. What is conservation? Why is soil conservation very important to all of us?
2. How is soil formed?
3. Name several ways in which topsoil is different from subsoil.
4. What is the water table? From what part of the soil do roots get their water?
5. What causes soil depletion? How can it be overcome?
6. What is erosion?
7. How does a growth of plants help protect soil?
8. Explain how each of the following helps to reduce soil erosion: contour plowing, strip cropping, terracing.
9. How may floods improve the soil? How may they harm it?
10. How can forests on watersheds serve to reduce river flooding?
11. Explain three or more ways in which land may be protected from wind erosion.

Science Reading Skills
3. Compare and Contrast
5.,7.,8.,9.,10., & 11. Cause and Effect

Thought Questions

1. Many people do not frequently see farms. How would they know when too much soil has been destroyed by erosion?
2. Some new methods of plowing leave dead stalks and leaves from the old crop all over the surface of the ground. Can you see any advantage in this?

Science Reading Skills
1. & 2. Problem Solving

CHAPTER 15 Conservation of Forests and Wildlife

objectives

After you read this chapter, you should be able to:

__**Describe** the seven forest regions of the United States and Canada

__**State** the importance of forests

__**Explain** several forest management practices

__**Discuss** the importance of environment to wildlife conservation

Nearly half of the United States and Canada were once covered with forests. See Fig. 15-1. A large part of these forests has been cleared for farming, but there are still many forest areas remaining. These forests provide a place to live for many different types of wildlife and they must supply us with our lumber now and in the future. Many of these plant and animal species could be destroyed by forest fires. In this chapter, we will see how fires and other such conservation failures can be avoided. We will study ways to insure the future survival of forests and wildlife.

PRESERVING OUR FORESTS

The forest regions. In the United States and Canada there are seven forest regions:

1. *Northern spruce and fir forest.* See Fig. 15-2. This forest reaches across Alaska and Canada. It is a huge area far to the north, just south of the Arctic tundra. The trees are not very big and growth is slow because of the cold climate.
2. *Central hardwood forest.* See Fig. 15-2. Once this was a very large forest region. It reached from the east coast to the prairies. Much of this forest grew on very good soil, so it was cleared away to make room for farmlands. Some forest remains on the slopes of the eastern mountains. Some also remains in woodlots. A *woodlot* is a part of a farm that is left in trees.

Maple, beech, ash, oak, and hickory are a few of the many trees growing in the Eastern hardwood region. Lumber from these trees is used mainly for making floors, furniture, woodwork, and tool handles.

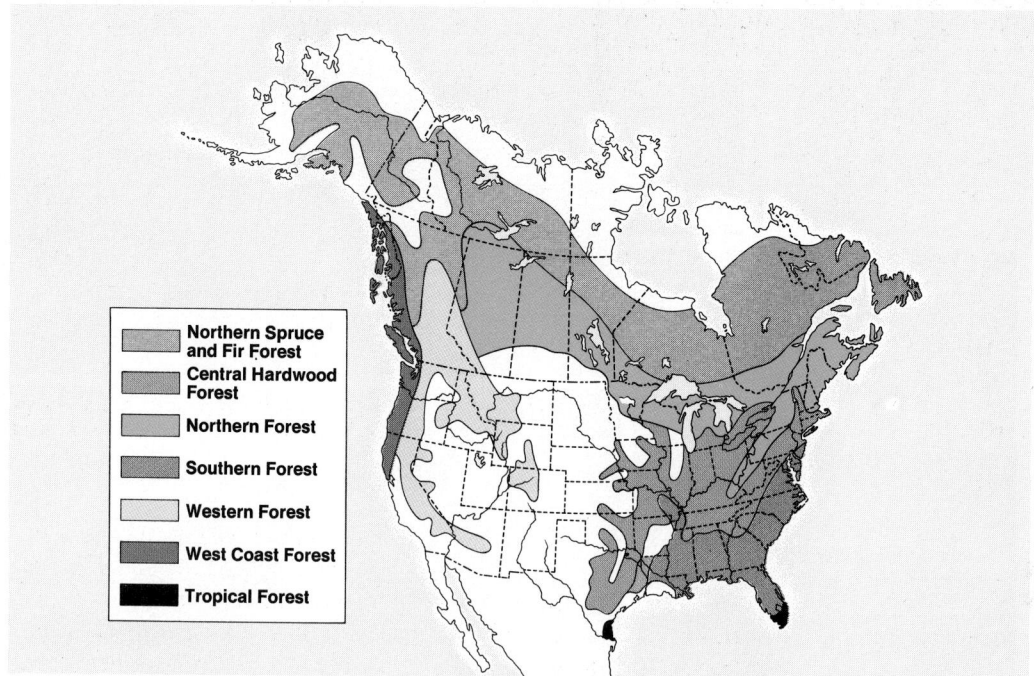

Northern Spruce and Fir Forest

Central Hardwood Forest

Northern Forest

Southern Forest

Western Forest

West Coast Forest

Tropical Forest

Fig. 15-1
This map shows the original forests of the United States and Canada. Are all these areas still in forest today?

3. *Southern forest*. See Fig. 15-2. The South has some large areas of sandy soil, much of which is in forest. Longleaf pine, loblolly pine, and slash pine are some of the trees that grow here. Much of the "yellow pine" lumber sold today comes from the South. These southern pines grow rapidly, sometimes becoming large enough to be cut in 50 years. This is considered a short time for a tree to grow that large. Southern forests produce over a third of the nation's lumber. On better soils the Southern forest contains oak, hickory, magnolia, bay, and other hardwood trees. In these places the Southern forest looks very much like the Central hardwood forest. It even contains a few of the same species. On low ground, cypress swamps are common.

4. *Northern forest*. See Fig. 15-2. This is a region where the change takes place between the Central hardwood and the Northern spruce and fir forests. A great deal of pine grows in the Northern forests along with stands of spruce, fir, and hardwoods. The Northern forest once led the nation in lumber production, but lumbering and fires destroyed the orginal forests. Much of its present production is in small logs used for paper pulp.

147

Fig. 15-2
Left side from top to bottom: Northern spruce forest (Harold E. Teter), Central hardwood forest (Karen Kennedy Gotimer), Tropical forest (Harold E. Teter). Right side from top to bottom: Redwood forest (Keith Gunnar/Bruce Coleman, Inc.), Southern pine forest (Lynn M. Stone), Northern forest (Harold E. Teter), Western mountain forest (Lynn M. Stone).

148

5. *Western mountain forests*. See Fig. 15-2. These forests grow on the slopes of the Rockies, Sierras, and other interior ranges of the West. Rain falls on the high slopes, providing the water necessary to grow trees. These are evergreen forests, producing important amounts of lumber for general construction.

6. *West Coast forest*. See Fig. 15-2. This forest grows along the Pacific coast and on the west slopes of the Sierras and Cascade Mountains. It extends from the middle of California all the way to Alaska. The rainfall here is very heavy. This forest includes beautiful stands of redwood, pine, Douglas fir, spruce, and other important evergreens. Away from the coast this forest blends into the drier Western mountain forests. This coastal forest region supplies more lumber than any of the others at the present time.

7. *Tropical forest*. See Fig. 15-2. Frost is unusual in the southern tips of Florida and Texas. In these areas the forests are more like those of the tropical Caribbean region than those farther north. Palms, mangroves, and mahogany grow there. If you were to travel north through Florida you would see a gradual change in the kinds of trees, from Tropical forest to Southern forest types.

Forest uses. Lumber is the most valuable product of the forests, but forests supply other important things also. Most paper is made from wood fiber. This *paper pulp* can come from small trees only 20 centimeters thick. It takes all the trees on 40 acres of land to produce the paper for just one Sunday edition of a large city newspaper.

Turpentine, pine tar, and resin all come from southern pines. These products have become less important than they used to be because man-made chemicals have taken their place in many paints and varnishes. Wood fiber from trees is now used to make wallboard, plastics, alcohol, camera films, and soil conditioners. From wood we also get rayon, dyes, oils, stains, sugars, and chemicals used in paints, soaps, and floor coverings. Even the sawdust that is formed when trees are sawed into boards has various uses.

The forests are also important as homes for wildlife and recreational areas for hunters and campers. Forests on steep slopes prevent erosion and reduce flooding in rivers.

Forest fires. Fire is the most serious threat to a forest. Fire not only kills the large trees, it kills the seeds in the ground and destroys much of the humus. This leaves bare soil lying open to erosion.

Assignment: Have students research articles pertaining to forest conservation and the use of forestry products. Some of the major lumber companies may be of help in this project.

When the trees were cut in the Northern forest region, fires were allowed to burn through the dead pine branches that were left on the ground. If these fires had been prevented, the young trees already present would have grown up to replace the old ones. By now the Northern forest would again be producing large amounts of lumber. As it is, it will be a long time before the slow succession of plant communities finally restores the climax forest.

In most of the forest regions of this country there is now good fire protection. Lookouts are stationed in fire towers all through these areas. If a fire is sighted from a tower, the forest ranger in charge of that district directs a crew of firefighters to put it out. Airplane patrols have replaced fire towers in many areas. Forest fires still do damage, but they do not spread unchecked over hundreds of square kilometers the way they once did.

Lightning can start forest fires, but most of them result from human carelessness. People fail to prevent the spread of the fire when they are burning brush, or they throw away lighted matches and cigarettes, or they fail to put out campfires. Some fires are started deliberately by "firebugs."

If you ever camp in a forest area, be sure you are careful with fire. Campfires should not be built on old roots or rotting logs. A rotting log can smoulder for days and finally break out into open flames that start a forest fire. Scrape the humus layer off the ground and build your fire on bare soil. Before you build any fire be sure to check with your local rangers or fire wardens. In dry weather it is sometimes against the law to build any fires outside the stoves found in the forest campgrounds. If you see a forest fire, go at once to the nearest telephone or ranger station and report it.

Fire used intentionally. Strange as it may seem, fire is sometimes used as a form of forest and grassland management. For the first several years of their lives longleaf pines grow only a few inches tall. If too much dead leaf material piles up on the ground and if bushes grow up around these small pine trees, their environment becomes too shady and damp and they are likely to be attacked and killed by fungus disease.

In the Southern forests it is a common practice to deliberately start ground fires every few years. Fire burns up the dead leaves and dries out the ground, so fungi cannot grow. It kills the bushes that would compete with the pines. There is not a great deal of material to burn, so the fire doesn't get hot enough to kill the tall trees. It does not kill the small longleaf pines either. Their thick masses of long needles are scorched at the outer ends, but these masses insulate the stems in the center. Therefore, the small trees are not killed by the fire or later on by the fungus. After some years the pines start growing rapidly and become tall trees, which can be cut for lumber.

It is common practice to burn the old tree tops and branches left on the ground after block cutting. This removes fuels from the area, so the danger of forest fires will be much less during the time the new young forest is growing.

Remember that these are special cases. In most forests, fire destroys the whole living community. Even where fire may be useful it must be used only by experts after they have made a careful study. It must be set only in the right place, at the right time. There must be a crew of fire fighters ready to prevent it from spreading to other areas where it would do harm.

Fig. 15-3
We can enjoy our forests for a long time if we are careful not to destroy them. (HRW photo by Ken Karp)

Careers: There are many careers that may be of interest to students who enjoy the outdoors. Ask students to report on careers such as a forest ranger, wildlife manager, and fish and game specialist. What other careers are open in these areas?

Forest management. Fire protection is the most important part of forest management, but it is not the only part. There are several other things that need to be done also. Where small trees grow too close together, none can grow well. Some of these trees may be cut out to give the rest of the forest a better chance to grow. This practice is called ***thinning.*** Sometimes crooked trees, diseased trees, and trees of less desirable species are removed, leaving only the more valuable trees to grow and produce lumber. This is ***improvement cutting.*** Thinning and improvement cutting can be combined in a single operation.

thinning

improvement cutting

Reforestation is an expensive process, but it is sometimes used to speed up the return of forests where there are no seed trees to do the job naturally. Young trees are started in nurseries and then set out to grow in the forest areas. This job used to be done by hand, but now there are tractor-drawn machines that can plant the seedlings much more rapidly.

reforestation

When the time comes to cut trees for lumber, there are three good ways to do it. One of these is ***selective cutting.*** Only the

Enrichment: Airplanes can be used to reseed burned-off forests.

selective cutting

Fig. 15-4
Block cutting. Seeds blow in from the surrounding areas and start new trees in these cut blocks. Is block cutting a good method to use in all forests? (American Forest Products Industries)

block cutting

clear cutting

Reference: See "America's Wilderness: How Much Can We Save?" in *National Geographic*, February, 1974.

wildlife

extinct

mature trees are cut. Other trees are left to grow larger. With this method there is always a forest.

When all of the trees in a forest are about the same age, a second method called *block cutting* may be used. One whole area, or block, is cut off completely. Trees surrounding the block shed seeds into it, from which new trees will grow. The block must be small enough so that the seeds can reach all of it. See Fig. 15-4.

Each block of land reforests itself if fire is kept out. By the time the last block is cut, the first one has regrown. This method provides a steady supply of lumber and jobs.

The third method used in harvesting trees is *clear cutting,* which is followed by replanting. All trees are cut, as in the block cutting method, but the areas from which the trees are cut are larger. In this case the foresters do not depend upon natural seeding to replace the forest. They plant young trees to replace the old ones. These young trees are carefully selected types that will grow rapidly and produce superior lumber.

Future forests. Actually, good forest management is simply a matter of treating trees as a crop. The only real difference between a forester's crop and a farmer's crop is that trees take much longer to grow. Like farmers, good foresters must think about the next harvest when they are gathering the present one.

There is always a temptation for the owner of forest lands to cut everything and make as much money as possible. In the past, most lumber companies have operated in this way. They left entire regions bare and burned. Then they moved on to other areas. Today, much of the forest land they destroyed is government owned. Expert foresters are hired to manage these lands, and to see to it that the forests grow again. There are also private companies that are managing their lands well. These companies must be fairly large. If they are using the block-cutting method, for instance, they must own enough land so that a new block can be cut each year for one hundred years or so. It will take that long for the first block to reach cutting size again. The companies must be managed by people who are interested in conservation as well as a steady income.

PRESERVING OUR WILDLIFE

The term *wildlife* means all of the wild animals in a region. Game, fur, or food species often can die or be killed faster than they can replace their numbers by reproducing. If nothing is done to prevent this from happening entire species could become *extinct*. That means that there would be no living members of a particular species left to reproduce. In order to prevent the extinction of certain species these animals must be protected by law. Animals such as insects, mice, and gophers, which reproduce quickly and in large numbers, do not need such protection.

Conservation and the environment. No living thing is independent of its environment. If a plant or animal is to survive, it must live in an environment that supplies *all* of the things it needs. If these conditions for survival are not present in a region, the plant or animal cannot live there.

Conservation is not just a matter of protecting animals from hunters. Even with no hunting, many animals die out when old environments are replaced by farms and cities. Their kind of environment has been destroyed. The conservationist tries to supply all of the conditions needed for the survival of such animals.

Rabbits are an interesting example of the effect of environment upon species. In many northern parts of the Central hardwood forest, the native type was the snowshoe rabbit. When the forests were cleared, this animal disappeared. It could not survive in open country, but the cottontail rabbit could. The cottontail moved up from the south and became the common rabbit of the farmlands. It did this in spite of the rabbit hunters.

Food is not a problem for rabbits because they eat almost any plant material. Cover is important. There must be hiding places where they can escape from enemies. Tall grass and brush along fence rows help them to survive.

Cottontail rabbits do well in areas where there are woodchucks. These rabbits do not dig holes, but they need holes as protection in very cold weather. So the rabbits often take refuge in woodchuck holes. Some states now protect woodchucks from heavy hunting. This benefits both the woodchucks and the cottontail rabbits.

Enrichment: *The Red Data Book* of the International Union for Conservation of Nature and Natural Resources lists more than 1,000 species threatened with extinction. More than 100 of these species are native to the U.S. Some of these creatures are the whooping crane, the Southern bald eagle, the California condor, the bighorn sheep, the Eastern timber wolf, and the blue whale.

Fig. 15-5
Left: The groundhog (woodchuck); Right: The cottontail rabbit. How does the presence of the groundhog help the rabbit to survive in northern environments? (Lynn M. Stone; Charles Palek/ Animals Animals)

Fig. 15-6
The bear (top) and the elk (bottom) are two of the large animals that cannot live in settled areas. (Stouffer Productions Ltd./Animals Animals; Lynn M. Stone)

Reference: See *And Then There Were None: America's Vanishing Wildlife* by Nina Leen, which describes endangered species and what is being done to save them.

There is one basic fact of wildlife conservation: **Animals must have a total environment that favors their survival.** In many places woodland species have had a hard time because the forest environment that favors their survival has been destroyed.

At the same time, other species have been favored by human activities. Deer are more plentiful than they ever were. Cut-over land produces more deer food than the great forests did. Meadow mice have become abundant, feeding on farm crops. Foxes have also increased in numbers, feeding on meadow mice. By controlling the environment, we control the species that can live on the land.

Big game. Most big game animals have a hard time surviving in a world dominated by people. Today, a type of succession is taking place; people and their tame animals are replacing the large wild animals in many communities of the world.

Bears, wolves, mountain lions, and other big game are not tolerated in settled country because they endanger farm animals. The large plant-eating animals compete with farm animals for food, so they also have been killed off. Deer are also big game animals, but, as we mentioned, they are an exception. Deer live quite successfully in many farm areas where there is at least some woodland cover nearby.

Pronghorn antelope are also a special case. Originally they lived on the plains, eating weeds while the bison ate grass. The two types did not compete with each other. When the bison were killed off, the grass grew so tall that it crowded out the weeds. Settlers also hunted the antelope very heavily at this time. The antelope nearly became extinct. Then cattle were brought in. Over-grazing killed off some of the grass, allowing more weeds than ever to spring up. The antelope then made a comeback. They are now present in large numbers in Wyoming and other states of the cattle country. As weed eaters, they do not compete with the grass-eating cattle.

This comeback of the antelope was aided by laws to control hunting. Conservationists trapped antelope alive and turned them loose on ranges from which they had disappeared. Cooperation by the local ranchers also was an important part of this program.

Other big game animals survive only in the wilder parts of the country, where farming is not possible. Mountains, forests, deserts, and tundra are the last homelands of the remaining big game types. Hunting is allowed in such areas, but it must be regulated so that only the excess animals are taken. Enough must be left at the end of the hunting season to maintain each type in balance with its food supply. Rangers and wardens protect wildlife from *poachers*, persons who kill wild animals illegally.

Bird conservation. Songbirds are not big enough to be important as food, but are useful as eaters of insects and weed seeds. In most of the United States and Canada songbirds and their eggs are protected from hunting. Only a few, such as the starling and the English sparrow are unprotected.

Conserving the fish population. Fish conservation is based on the same general rule that governs the conservation of other wildlife: The total environment determines how many and what kinds will survive. You may have heard it said that a certain lake has been "fished out." Over-fishing can cause the fish to disappear, but this can also happen as a result of changes in the environment.

Sometimes people who own houses on the shore cause changes in the environment. They clean the weeds, logs, and brush out of the water. Marshy shores are filled in. This makes a nice place for people to enjoy themselves, but the breeding and feeding areas of the fish are ruined.

Some lakes are improved by more fishing. Sunfish, bluegills, and similar panfish often breed too abundantly. There are so

many small fish competing for a limited food supply that few can grow to a large size. Catching some of these small fish gives the rest a chance to grow bigger. Introducing game fish like bass or pike into such a lake has much the same effect. They eat many small fish, allowing the others to get larger.

Dams are often used for power and for flood control. These dams prevent fish from migrating upstream in their normal way because the fish cannot swim over the dam. Often *fish ladders* of some sort are placed around these dams. One common type is a kind of stairway of low waterfalls. The fish jump each level in turn and bypass the dam.

Trout streams are sometimes improved by planting trees along their banks. These give shade that keeps the water cool enough for trout. Logs and brush are sunk in the edges of the water to give the fish hiding places.

WHAT YOU CAN DO FOR CONSERVATION

Conservation requires the cooperation of everyone. People in government must pass the laws needed to manage forests and to protect wildlife. Public support is needed to make these laws work. Be sure you understand them and that you encourage others to obey them.

People living in rural areas influence wildlife because they control land use. When using the land they should keep wildlife preservation in mind. For example, pheasants live best in certain northern, grain-growing areas when these areas have some ground left wild. This gives pheasants the cover they need to protect themselves from the winter storms. Weedy hedgerows along the fence lines and patches of grass and reeds in marshes provide cover. When all of these hedgerows are destroyed and the marshes are drained, there is no cover left for pheasants. In such "clean farming" country, the birds are sometimes found dead after the winter storms lying next to a tree trunk or fence post where they tried to find protection. Their nostrils are plugged with balls of ice formed from their frozen breath.

Highway departments sometimes use weed-killing sprays to keep brush from growing along the roadsides. This practice makes the roads look better, but it destroys valuable cover that wildlife might otherwise use. Bushes and grass along fence rows and roadways are almost the only good cover that is left in some areas.

You should always be ready to support conservation programs. If you hunt or fish, learn what the law is and obey it. These laws are set up to preserve our wildlife.

Hunting and fishing, of course, are not the only ways to enjoy wildlife. Many people get a great deal of pleasure from watching and studying wild animals and plants. Wildlife photography

Fig. 15-9

The pheasant (top) and the raccoon (bottom) are two of the species of smaller game that can live successfully in settled areas. People should keep these animals in mind when using the land. (Z. Leszczynski/Animals Animals; Lynn M. Stone)

is a sport that calls for the same sort of skill as hunting with a gun, but it has no closed season.

In general, the aim of wildlife conservation is to see that people in the future will have at least the same opportunity as we have to enjoy the out-of-doors and to see wild animals in their natural environment.

SUMMARY

Forests supply us with lumber, paper, and many other products. There are seven main forest regions in the United States and Canada.

Forest fires can completely destroy a forest, but sometimes fire is used purposely to control the type of forest growth. Forests are thinned to allow rapid growth of the remaining trees. Poor quality trees are removed. New trees are sometimes planted.

Selective cutting, block cutting, and clear cutting with replanting are three methods of tree harvesting that allow for future forest growth.

Each species of wild animal must have the right kind of environment in which to survive. We conserve them by supplying such environments. Many small species survive in farming country if some wild spots are left for them to hide in. Most big game animals survive only on lands not fit for farming.

Activity

Local Game Laws
In this activity you will make a study of the game laws in your state.

A. Write to the department in your state or province that is responsible for the conservation of wildlife, or go to a place where hunting licenses are sold and ask for information on the hunting laws.

1. What are the game animals of your region?

2. Which mammals and birds are completely protected from hunting?

Can you explain why?

3. Which ones are not protected at all? Why?

4. Which ones are protected part of the time and hunted part of the time?

5. How many of each species may a hunter take?

6. What time of year does most hunting take place?

7. Try to explain the reasons for each of these regulations.

Word Quiz

From the new words you learned in this chapter, write the word that correctly completes each statement. Do not write in this book.

_____is the practice of removing some trees so that the rest can grow better. When poor quality trees are removed from a forest, the practice is called _____. _____ is accomplished by planting young trees. In _____ only the mature trees are harvested and others are left to grow. In _____ small areas are cut and left to reseed naturally. When a large area is cut completely and then reforested, the practice can be called _____. All of the animals in a region make up its _____. If all of the members of a particular species died off, we would say that the species had become _____.

1   2   3   4   5   6   7   8

Check Your Facts

Science
Reading Skill
5. *Cause and Effect*

1. What are the main forest regions of North America?
2. Name several ways in which forests are useful to us.
3. What is the most important thing that must be done to proctect growing forests?
4. Under what conditions is each of the following useful in forest management: improvement cutting, reforestation, selective cutting, block cutting?
5. Is fire ever good for a forest? Explain.
6. What does the environment have to do with the conservation of wildlife?
7. Why is conservation of big game animals a difficult problem?
8. Why are hawks and owls often protected from hunters?
9. What can you do to help conserve wildlife?

Thought Questions

Science
Reading Skill
1. *Problem Solving*

1. Which of the forest cutting methods would produce the least danger of erosion, the best homes for wildlife, and the best recreation areas for people?
2. What would happen if there were no hunting of deer, elk, antelope, and other big game animals?

CHAPTER 16 Pollution and Energy

When harmful materials enter the environment we call the process **pollution**. Any such harmful material is called a *pollutant*. Pollutants can be solid materials, liquid chemicals, or gases that can harm people, plants, and wildlife. These pollutants can not only spoil the natural beauty of our land, but they can also seriously affect our health and the way we live.

Air, soil, water, and food can all become polluted. The same pollutants may enter more than one of these. Let us examine some different types of pollution and see what kinds of damage each causes.

TYPES OF POLLUTION

Pollution by sewage. Cities and towns often dump *sewage* into streams or into the ocean. This sewage is the water coming from sinks, bathtubs, toilets, factories, and drainpipes.

Sewage endangers humans and other plant and animal life because of the disease-causing germs, and harmful chemicals it can carry. It endangers fish also because of the human wastes and other materials it contains. These materials are very good sources of food for decay bacteria. Having this abundant source of food allows these decomposers to multiply rapidly and soon use up all of the available oxygen. With the oxygen gone, the fish suffocate. By the time the sewage has moved farther downstream, its decay is complete. Bacterial growth slows down, oxygen is not used up as rapidly, and fish can live downstream in the river from that point on.

Pollution by sewage can be avoided by installing sewage treatment plants. See Fig. 16-1. When sewage enters these plants it is put into large holding tanks where the solid particles are allowed to settle out. The water is then drawn off into other tanks where bacteria will break down the remaining wastes.

_____ objectives _____

After you read this chapter, you should be able to:

—**List** the common types of pollution

—**Describe** solutions to each kind of pollution

—**Explain** why there is an energy crisis

—**Describe** ways of solving the energy crisis

pollution

sewage

Science Reading Skill: *Cause and Effect* and *Problem Solving* are skills that appear often in this chapter.

Reference: See *An Introduction to Pollution* by Harold E. Schlichting, Jr. and Mary Southworth.

Before the water is sent out into a nearby river, lake, or bay it is treated with chlorine, which kills much of the bacteria. This type of sewage treatment should make the water clean, but there is still another problem.

The water that is released from sewage treatment plants usually contains a rich supply of nitrates and phosphates. These minerals act as fertilizers, which stimulate a very rapid growth of algae in the water. This type of heavy growth is called an algal *bloom*. These algae once again provide food for the decay bacteria that multiply and quickly use up the oxygen. Also, when some algae die they release poisonous substances into the water. These are harmful to fish and other water organisms.

In order to prevent this type of pollution an extra step to remove excess minerals from the water can be added to the sewage treatment process. This is costly and time consuming. For this reason many sewage treatment plants have not yet added this step.

Solid wastes. Garbage, old paper, tin cans, scrap metal, and other kinds of rubbish are known as *solid wastes*. These materials are often dumped into the ocean, where they cause pollution. In other areas they may be dumped on land. Rats breed in dumps and may carry disease. The dumps often catch fire, and foul-smelling smoke pollutes the air.

Dumps can be improved by making them into *sanitary land fills*. Bulldozers push fresh soil over the trash as soon as it is dumped. See Fig. 16-2. Bad odors do not reach the air, and rats do not reach the buried garbage. But there is still danger that poisonous chemicals in the rubbish will seep down into the groundwater. From there the chemicals enter wells, streams, and lakes. Also, land fills are often developed in areas that were originally swamps or marshes. This destroys the breeding grounds of many fish.

Enrichment: Another type of water pollution is thermal pollution. Nuclear power plants and industries use water to cool machinery. When this water is returned to a stream or lake, the temperature of the nearby water environment increases, killing plants and animals.

References: See *Recycling* by R. Fegely, R. Reemer, L.N. Rinehart, and E. Papp and *Recycling: Reusing Our World's Solid Wastes* by James and Lynn Hahn.

Fig. 16-1
(Left) The holding tanks in a sewage treatment plant. (Dorr-Oliver, Inc.)

Fig. 16-2
(Right) In a sanitary land fill, the solid wastes are covered quickly with soil. (Grant Heilman)

160

So many solid wastes are being produced that we are running out of places to dump them. The volume of these materials could be reduced by recycling many of the waste materials. Metals and glass can be melted and used over again. This is also an aid to the conservation of important natural resources.

Another way of disposing of garbage is to burn it. The heat from this burning process is sometimes used to generate some electric power, but this is not a major source of energy. Burning garbage can also pollute the air.

Chemical pollution. The modern world around us is filled with chemicals. They are used in making clothes, foods, the cars, homes, and nearly everything else we use every day. Many of these chemicals are harmless, but some are poisonous. When these poisonous chemicals are allowed to escape into the environment they can cause serious pollution. Here are a few examples of how chemicals have polluted our environment.

A factory in Virginia allowed a poisonous chemical called Kepone to enter the James River. All the way from Richmond, Virginia to the Chesapeake Bay the fish in this river were unsafe to eat, because they contained too much of this poison.

Also in Virginia, the South Fork of the Shenandoah and the North Fork of the Holston Rivers have been polluted by mercury. One plant used this poisonous metal in the manufacture of clothing fiber. The fish in these rivers were not safe to eat because they contained too much mercury. Unfortunately, this mercury will remain in the river for many years.

Chemicals called PCBs have been used in the manufacturing of a variety of products in the United States and other countries. These PCBs have entered several bodies of water. Scientists have found that PCBs can stop photosynthesis in some algae and kill certain other organisms such as shrimp. PCBs are also harmful to fish and the sea birds that eat these fish. It is not yet certain what effects PCBs will have on humans. But people have been advised not to eat the fish from waters such as Lake Michigan, which contains PCBs, more than once a week.

PCBs are no longer used except in controlled situations, so they will not enter the environment. This means that the situation should slowly improve.

Another example of chemical pollution occured at Love Canal, near Buffalo, New York. A company buried large amounts of left-over chemicals. This was done simply to dispose of useless chemical wastes. Years later the movement of groundwater carried these chemicals into the soil, well water, and basements of nearby houses. This made people sick and could have caused harmful mutations in their offspring. A whole neighborhood had to be abandoned. See Fig. 16-4.

Fig. 16-3
This photo shows drums of hazardous chemical wastes. How could these wastes be disposed of safely? (Tom Stack & Associates)

Fig. 16-4
Chemicals in the groundwater like this caused the neighborhood at Love Canal to have to be abandoned. (Kathleen Foster)

pesticide

herbicide

Word Study: "-cide" means to kill

Fig. 16-5
Water pollution. Left: Many fish are killed because of the large amounts of pesticides entering the water; Right: Sea birds are some of the victims of oil spills. When covered with oil they cannot fly to find food. (EPA Documerica)

In most of the examples just covered the pollution was accidental. No one understood or foresaw the danger at the time. Deliberate pollution does occur, but more often it is the result of a lack of knowledge. Many dangers are now better understood, but there is still more to be learned.

Safe disposal of dangerous waste chemicals is a very difficult problem. Some of them can be changed chemically to make them harmless. Some can be burned at very high temperatures and others can be placed in concrete containers and buried very deeply, where groundwater cannot move them into the environment. All of this adds to the cost of products manufactured by companies that must dispose of these wastes, but it is necessary to take these precautions.

Pesticides and herbicides. A chemical used to kill troublesome insects or other pests is called a *pesticide.* A chemical used to destroy weeds is called an *herbicide* (HUR-buh-side). Pollution of the environment by such chemicals has become a serious threat to fish, birds, and other wildlife. At times, it threatens human populations. Whenever people are faced with damage by some troublesome insect, they are tempted to spray the whole area with poisons. This may or may not control the pests. The result may be that the entire living community is injured.

The use of carefully selected insect sprays can be very helpful in controlling pests in limited areas. When, however, poison sprays are used over large areas, it is hard to predict what the final result may be.

Pesticides are not all alike. Some, like DDT, do not break down easily. They last for a very long time. Such chemicals are called *hard pesticides.* There are also *soft pesticides.* These do their job and then break down chemically so that they disappear after a short period of time. They are much less likely to poison the environment permanently.

When hard pesticides are used in large amounts they accumulate in the soil. From there they wash into streams. They kill fish and many of the small living things the fish feed on. Thousands of fish have been killed in the lower Mississippi River by poisons washed down from farmlands farther upstream. See Fig. 16-5.

Eagles, hawks, herons, and many other water birds have eaten these poisoned fish. Some have died. Others have passed the poison on into their eggs, which then fail to hatch. In the case of eagles, the poison causes the eggs to have very thin shells, so they break when the parent birds sit on them. Eagles have become rare birds all over eastern North America. The pelicans in California suffered in a similar way because pollution was entering the sea from a DDT factory on the shore. The pollution has stopped and the pelicans are returning.

Because of cases like these, DDT has been banned by law for most uses. But many other kinds of hard pesticides are still in use. Cases in which they do damage to wildlife and to people still keep turning up.

Herbicides have been a great help to farmers in controlling the weeds that compete with corn and other crops. They kill the weeds and allow the corn to grow. Other herbicides have been used to kill brush where it was not wanted.

Traces of herbicides and pesticides enter the food chains and get into human tissue. The use of certain ones has now been banned, because they were found to increase the cancer rate. This is part of the continuing effort to check the chemicals in our environment to be sure they are not harmful.

Oil spills. Petroleum is often carried across oceans in ships. There have been several cases where these ships were wrecked at sea, and the oil leaked from them. In such cases the floating oil covers a wide area of the water's surface. It sometimes washes up on the beaches, making them unfit for swimming and other recreational uses. The feathers of sea birds become covered with oil, so they cannot fly to find food. Thousands of birds starve to death as a result. See Fig. 16-5.

Oil can leak from storage tanks and factories to pollute rivers and harbors. Several years ago the Buffalo River in New York State was so polluted with oil that it caught fire. Since then there has been a program to clean up the river. Pollution of all kinds has been reduced, and now the fish are returning.

Air pollution. The pollutants that enter the air are either *gases* or they are tiny solid particles, called *particulates*. We can see smoke because of the particulates it contains.

Particulates may consist of tiny bits of ash or of carbon particles (soot) produced during the burning of fuels. They may be materials released during manufacturing processes. Steel mills, cement plants, and other industries often produce them. In the past, most people heated their houses with coal. The burning coal released more smoke into the air than the gas or oil burning furnaces that are common today.

Enrichment: In 1974, vinyl chloride pesticides were banned from use in homes, food-handling establishments, hospitals, and all enclosed areas. This substance has been associated with a rare form of liver cancer.

particulates

163

The main gases that pollute the air are sulphur dioxide (SO_2), oxides of nitrogen, hydrocarbons, carbon monoxide, and carbon dioxide. These gases are mostly produced by the burning of certain fuels in factories, homes, automobiles, and power plants.

Air pollution can have minor or serious effects on plants and animals. In humans air pollution can cause simple irritation of the nose, eyes, and throat. It can also cause more serious diseases such as bronchitis (brahn-kite-us) and emphysema (em-fa-SEE-ma), which cause shortness of breath. Pollution can even be a cause of lung cancer.

Let us take a look at some of the particular effects air pollution has on the environment.

smog

Smog. The word *smog* was first used as a slang word to describe a combination of smoke and fog. Smoky air often becomes foggy because moisture from the air condenses on particulates to make fog droplets. The famous fogs of London formed in this way because of heavy pollution by coal smoke. In 1952, about 4 thousand persons died in London due to a very heavy "killer smog" that covered the city for five days.

Dangerous *smog compounds* are also formed when sunlight reacts with unburned hydrocarbons. These are compounds of carbon and hydrogen found in fuels such as gasoline.

The smog in places like Denver and Los Angeles is due to these types of smog compounds. During the daytime the sunlight and the exhaust fumes from industry and heavy traffic create the conditions that form these compounds. These cities have mountains nearby that can often prevent wind from moving the air from over the city. So, a pool of smog-filled air can lie over the city for days. See Fig. 16-6.

Fig. 16-6
Breathing polluted air can be very harmful to human health. What can we do to prevent this pollution? (Tom McHugh/Photo Researchers, Inc.)

Depletion of the ozone layer. About 20 to 44 kilometers above the earth is an area called the *ozone layer*. This layer contains a chemical known as ozone (O_3), which has a very important property. It can screen out most of the ultraviolet rays from the sun. Too much exposure to ultraviolet rays can cause severe sunburn and even skin cancer. See Fig. 16-7.

Certain types of pollution can cause the ozone layer to break down and allow more ultraviolet radiation to reach the earth. Nitrogen oxides are one of these pollutants. Nitrogen oxides form when nitrogen and oxygen react at high temperatures. This happens in automobile engines, the engines of supersonic transports, and large electrical power and industrial boilers. The nitrogen oxides formed this way can then react with ozone to form nitrogen dioxide. Besides depleting the amount of ozone in the atmosphere, this is dangerous because nitrogen dioxide can be harmful if it is inhaled.

Other compounds that can harm the ozone layer are called fluorocarbons. These compounds were used in almost all aerosal cans. They are no longer being widely used.

Although we should try to protect the ozone layer in the upper atmosphere, we should also try to prevent ozone from being formed in the air that we breathe. If ozone is inhaled it can be harmful. When gasoline burns it gives off compounds of hydrogen and carbon called hydrocarbons. In the presence of sunlight these hydrocarbons can react with nitrogen oxides and form ozone. If we breathe this ozone, even in small amounts, it can irritate our noses and throats. In larger amounts it can have more serious effects. Ozone can also cause damage to crops and rubber products such as automobile tires.

Acid rain. Some pollutants eventually become acids in the atmosphere. Nitrogen oxides can become nitric acid. Sulphur dioxide, which comes from the burning of coal and petroleum, can become sulphuric acid. These are washed out of the air by rain. Rainwater is always a weak acid, but it becomes more strongly acid when these pollutants are added to it. The result is called *acid rain*. This acid rain corrodes metals and speeds up the weathering of stone and concrete. It can also kill fish in some lakes. The effect can be felt, for several hundred kilometers down wind from industrial areas. Acid rain from air pollution hundreds of miles away has eliminated the fish in some lakes in the Adirondack Mountains of New York.

The greenhouse effect. Carbon dioxide is always present in small amounts in the air. It is harmless to us, and plants use it to make food. It is produced whenever we burn fuels.

Plants are not able to use carbon dioxide as rapidly as it is being produced. Some of the carbon dioxide dissolves in the

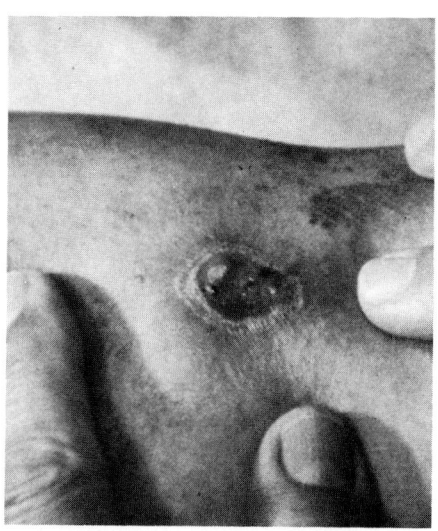

Fig. 16-7
Skin cancer can be caused by too much exposure to the sun. How does the ozone layer help to protect us from this? (American Cancer Society)

ozone layer

Fig. 16-8
If the temperature of the earth rises too much, what could happen to these polar ice caps? What harm would that cause? (Jen & Des Bartlett/Bruce Coleman, Inc.)

ocean, but the amount in the air is slowly increasing. There is fear that this may change the climate.

Sunlight heats the earth. This heat then radiates off into outer space. An increase in the amount of carbon dioxide in air would slow down this escape of heat. The climate could become hotter. This would cause many important changes. The Great Plains could become too dry to grow food crops. The western Antarctic ice sheet could melt. This would raise the sea level enough to flood the seacoasts. Cities such as New York and San Francisco would be underwater.

greenhouse effect

We do not know how much extra carbon dioxide would be needed to cause a serious degree of this *greenhouse effect*. (It is called that, because the sun would heat up the earth like a greenhouse.) If we continue to burn coal, oil, gas, and other fuels too rapidly, the amounts of carbon dioxide in the air will continue to increase.

Certain particulates in the air have the opposite effect. They cause some of the heat from the sun to be sent back into the outer atmosphere before it reaches the earth. We cannot count on having these two types of pollution problems "cancel each other out," however. We must try to solve both of these problems as soon as possible.

PREVENTING POLLUTION

You have already read about some of the things that have been done to help stop pollution. In areas where the government and the citizens together have made a conscious effort to clean up the environment positive results can be seen.

We have already mentioned the return of the pelicans off the coast of California and other similar examples. In New York,

Fig. 16-9
These teenagers are using a vacant lot in the city to plant a garden. How can you help to improve your local environment? (Ann Marie Rousseau)

certain species of fish have returned to the Hudson River because the levels of pollution have been lowered.

The steel mills in Pittsburgh, Pennsylvania used to give off a large amount of sulphur dioxide into the air. Special equipment was installed in the smoke stacks to remove the sulphur dioxide and particulates. Now most of the smoke that used to "hang" over the city of Pittsburgh is gone. These are only a few examples. There are many more like these.

Most pollution problems can be solved only by community effort. Each citizen must be concerned with these problems and be willing to do his or her share to eliminate them. We must all be willing to invest time, effort, and money to pay the cost of cleaning up our environment. This cost may be high, but the cost of not cleaning up our environment is high also. It includes a higher death rate and higher medical costs to treat illness. It also includes the destruction of wildlife and the loss of a pleasant environment to live in.

TYPES OF ENERGY

We use energy in many ways. It heats our buildings and cooks our food. We use it to drive our motors and to carry on our manufacturing processes. We use it to generate electricity. We get most of our energy by burning fuels. This leads to many forms of pollution. There is another problem, also. We are running out of fuel.

Fossil fuels. Coal, oil, and gas were formed from the remains of organisms that died millions of years ago. For this reason these are called *fossil fuels*. Today fossil fuels supply us with most of our energy. Since it takes so long for fossil fuels to form

Reference: See *The Energy Crisis* by Tad Szulc and *The Whole Earth Energy Crisis—Our Dwindling Source of Energy* by John H. Woodburn.

fossil fuels

beneath the earth, we are in danger of using up all the fossil fuels we have available. There are ways of obtaining more fossil fuels than we have available now, but these methods are costly and it will take time to put them into use.

Oil is present in rocks called *oil shale*. Ways may be found to get this oil from the rock, but fuel produced in this way will be expensive. Some very *thick forms of oil* are found in some places. This oil can be used, but getting it out of the ground is difficult. Fuel from this source will be expensive.

Natural gas is an excellent fuel that produces little pollution. Scientists believe that there is probably a great deal of natural gas yet to be discovered. More exploration for this gas is needed.

There is enough *coal* in the United States to supply our energy needs for several hundred years. Ways must be found to burn coal without causing serious air pollution. Pollution control is expensive, so energy from coal will be expensive. It may even be possible to produce *gasoline and gas from coal*.

Trying to increase our available fossil fuels is not the total answer to the energy crisis. We must also seek alternative sources of energy.

Energy from plant life. The use of wood to heat homes has increased slightly. Scientists are also investigating the possibility of using wood as fuel for energy-generating plants. In order to do this, large areas of fast-growing trees would have to be developed.

Alcohol can be used as fuel in automobile engines. It can be made from any plant materials. Grains are often used, but they are needed mostly for food. Sawdust, woodchips, grass, brush, or any other plant material could be used to manufacture alcohol.

Geothermal energy. Energy from underground heat is called *geothermal energy*.

In volcanic regions it is possible to drill wells into hot rock layers underground. Water is pumped into these wells and comes out as steam. The steam is used in steam turbines, which generate electricity. Geothermal power plants are used in Iceland, Italy, and several other places. They are being developed in California. We do not know yet just how much energy can be obtained in this way.

Wind energy. New types of windmills are being developed to generate electricity or heat. The smaller ones could light or heat a single building. A large one might supply power for a small town. Storing energy for use during windless days is a problem. It is too early to say how much energy may be produced in the future by windmills. There is much research yet to be done. See Fig. 16-10.

geothermal energy

Word Study: "geo-" means earth and "thermal" means heat

Assignment: Have students research recent events in the news that link nuclear test sites with the occurrence of cancer in humans.

Nuclear energy. Nuclear power plants are producing about thirteen percent of our electric power. Some people want this to be increased. Others want us to stop using ***nuclear energy*** completely.

In nuclear plants, uranium is allowed to break down into other elements under controlled conditions. Heat is produced, which makes steam to drive electric generators. The core of a nuclear power plant contains radioactive elements. These are elements that give off invisible radiation.

Some people worry about the possible danger to nearby populations if there should be a serious accident at a nuclear power plant. Most of this concern deals with the possibility of deaths caused by explosions and of cancer caused by radiation.

Biologists are also concerned with the danger of mutations, which could be caused by the type of accident called a melt down. A melt down could leave nearby rivers and land areas heavily polluted by radiation for thousands of years. This radiation could cause mutations. See Fig. 16-11. No such accident has occurred so far, but there have been some close calls, and some radioactive gases have escaped into the air.

Even if accidents are avoided there is still another problem. The nuclear wastes produced by power plants give off large amounts of radiation. No safe way to dispose of these wastes

nuclear energy

Fig. 16-11
The core of a nuclear reactor. This must be kept cool to avoid a "melt down". What problems could a "melt down" cause? (U.S. Department of Energy)

has been agreed upon. The wastes are continuing to accumulate in temporary storage areas; several have leaked into the environment. The whole problem of nuclear power is now being seriously discussed, both within the government and among private citizens. This could be a very valuable source of energy to us, but we must be able to use it safely.

Solar energy. If we make use of the energy in sunlight we say we are using *solar energy*. There is far more energy in sunlight than we could ever need. The problem is finding a way to use it.

Solar panels are placed on roofs. They allow sunlight to pass in through clear, window-like coverings to heat water or other liquid in dark-colored pipes. This water heats the house. *Solar cells* change light directly into electricity. See Fig. 16-12. In both cases there is the problem of how to store the energy so that it can be used when the sun is not shining. Solar cells are not efficient enough now to produce the amounts of electricity needed for general use.

CONSERVATION OF ENERGY

We can make our energy supply last much longer if we use it more carefully. Lighter cars with more efficient engines are being developed. We may soon have automobiles that go four or five times as far on a liter of fuel as they do now. That will have the same effect upon our energy supply as the discovery of several big new oil fields.

We can also save huge amounts of energy by thoroughly insulating our houses. It will then take much less fuel to heat them. Public transportation systems can help. It takes much more fuel to carry several people in separate cars than it does to move them all in a single bus or subway car. If we cannot solve our energy problem we will not just be forced to go back to a simpler standard of living, our whole economy will suffer.

Fig. 16-13
By using mass transportation instead of automobiles we can help to conserve fuel. (HRW photo by Brian Hamill)

SUMMARY

Sewage may pollute lakes, streams, and oceans. Disposal plants can prevent this. Garbage can be burned but this may also cause pollution. Scrap glass and metals can be reused. Chemical wastes can be destroyed or deeply buried. Pesticides and herbicides must be used with care. Air is polluted by particulates and gases. Air pollution is causing problems such as smog, depletion of the ozone layer, the greenhouse effect, and acid rain. These are all very serious problems that cannot be allowed to continue. Progress is being made in cleaning up the environment.

We are running out of clean burning fuels to produce energy. We can find ways to safely use more coal, discover more gas, get energy from plant materials, and develop geothermal energy, solar energy, and safer nuclear energy. We must also conserve the fuels we have.

ACTIVITY

Collecting Solid Wastes

A. At the end of each day collect all of the trash that is in your classroom. Place all of this trash in a large plastic bag and put it where it will not be thrown away. Do this each day for one week. At the end of the week weigh the trash.

1. How many kilograms of trash did you collect in one week?

B. Divide the number of students in your class into the weight of the trash.

2. What was the average amount of trash collected from each person?

C. Find out how many students there are in your entire school.

3. If each student created the average amount of trash you calculated in question 2, how much trash would collect in your school in one week?

4. How much trash would collect in your school during an entire school year?

D. Have a class discussion on ways you could cut down on the amount of trash produced in your classroom. Practice these methods for one week and collect the trash each day as you did before.

5. What does the trash weigh this time?

6. What is the new average amount of trash collected from each person in your classroom?

7. Using this new average, calculate the amount of trash that would be collected in your school during an entire school year if everyone used your methods of cutting down on trash.

E. Compare your answer to question 4 with your answer to question 7.

8. Do you think it would be worthwhile if everyone tried to cut down on the amount of trash they create? Why?

Word Quiz

Choose the number in **Column B** that best matches each letter in **Column A.** Do not write in this book.

Column A	Column B
a. geothermal energy	**1.** the entering of harmful materials into the environment
b. pollution	**2.** a poison used to kill harmful animals
c. herbicide	**3.** a poison used to kill weeds
d. particulates	**4.** small pieces of solid material in the air
e. pesticide	**5.** coal, oil, gas
f. fossil fuels	**6.** heat or electricity obtained from hot volcanic rocks
g. solar energy	**7.** heat or electricity from sunlight
h. nuclear energy	**8.** waste water that comes from sinks, bathtubs, drainpipes, etc.
i. smog	**9.** a combination of smoke and fog containing harmful chemicals
j. sewage	**10.** an increase in the temperature of the earth due to an increase in carbon dioxide in the atmosphere
k. ozone layer	**11.** the part of the atmosphere that contains a chemical that prevents ultraviolet rays from reaching the earth
l. greenhouse effect	**12.** electricity from radioactive substances

Check Your Facts

*Science Reading Skills
2. Problem Solving
3., 5., & 8
Cause and Effect*

1. How can solid wastes be disposed of?
2. What can be done to prevent chemical pollution?
3. How do pesticides endanger wildlife?
4. What is meant by the term *greenhouse effect?*
5. How does smog form? Why is it harmful?
6. What is the energy crisis?
7. List some ways we can help to solve the energy problem.
8. Why is the ozone layer important?

Thought Questions

*Science Reading Skills
1. & 2. Problem Solving*

1. Plants use sunlight to separate water into oxygen and hydrogen. If we could learn to do this, how would it help in the energy crisis?
2. All of the main air pollutants are produced naturally in larger amounts than they are produced by people. If this is true, why is there a problem?

Ecological concerns are both rural and urban and generate a fantastic range of jobs.

Forest and timber management, with job opportunities confined primarily to the western states, is an area of growing specialization. The need to protect our natural resources, to produce paper and pulp products, and to accommodate the growing public use of our forests and parks for recreation has created a demand for personnel skilled in forestry and conservation as well as in public relations. Careers in this field are highly competitive and openings are limited. Forest mana-

gers, rangers, and forest-fire managers increasingly have college degrees that combine course work in botany, agronomy, and forestry management with extensive practical experience. In fact, most conservation positions require a 4-year degree. If a college degree is not a possibility, you can start as a timber management technician, a position in which practical skills rather than theoretical knowledge are most important. Basic outdoor skills and good physical condition are re-

quired. Nonprofessional, but experienced, persons are still used as tower lookouts to spot fires not visible from the air. "Smoke jumpers" (firefighting parachutists) are also used. However, helicopters are beginning to replace both these functions. Summer work experience as a fire observer plus one year of firefighting experience is a necessary background.

For information on the Youth Conservation Corps, write:

The U.S. Department of
 Agriculture
Forest Service
Human Resource Program
P.O. Box 2417
Washington, DC 20013

Pest management is another important ecological task. Career opportunities in this area require different kinds of background and training. An entomologist is a life scientist who studies insects and usually has a master's degree or doctorate in this specialization. Finding ways to protect plants from insects, finding insects that prey on other plant-devouring insects to create "natural pesticides," and discovering better

Earl Roberge/Photo Researchers, Inc.

Nicholas Devore III/ Bruce Coleman, Inc.

ways to protect food in storage from damage are only a few important results of entomological research.

Agribusiness (a name for the food- and fiber-industry) is a major user of pesticides, a factor which generates many jobs. Chemists are employed in private industry to develop new pesticides. Pesticide salespeople, distributors, crop sprayers, and chemical technicians who test water quality for toxic levels are all necessary. These positions do not necessarily require a college degree: practical experience and technical education are helpful.

Integrated pest control managers supervise and direct the use of pesticides. They must have knowledge about plants, insects, bugs, soil, and weather conditions to organize this task competently. They usually have a college degree and/or a technical education in soil science and pest management.

In the cities, pest management problems are different. Controlling and eradicating populations of rodents, termites, silver fish, and roaches are the jobs of the exterminator. Working in heavily populated locations, the exterminator has to be aware of the toxic effect of pesticides upon humans as well as upon the pests people want to be controlled. Exterminators spend time training on the job as helpers; in some states they must then pass a state licensing exam. This industry shows an above-average employment growth.

The disposal of wastes is an immense societal problem which creates many work opportunities. The people who help get rid of household wastes (garbage, rubbish, and trash), are the most visible. Most work for the state or local government, although there are a growing number of private companies in this business. A few cities employ private firms to collect all household waste. Opportunities for employment depend on population size. Waste water treatment plants are a growing necessity, and people are needed to operate equipment and meters and to monitor the removal of waste materials from sewers or to make the material harmless to humans. There are over 100,000 people employed in this field. The skills are learned on the job, although a high school diploma is a minimum requirement in some states and a competitive civil service exam may also be required if the plants are run by the local or state government. The continuing construction of new plants makes the growth of this industry much faster than average.

For further information write to:

The U.S. Environmental
   Protection Agency
Office of Water Programs
   Operations
Manpower Development Staff
401 M Street SW
Washington, DC 20460

Peggy Kahana/Peter Arnold, Inc.

Arthur d'Arazien/ Image Bank

YOUR LEISURE TIME

UNIT THREE

Understanding basic ecology is essential for the informed voter.

Robert DeMicco

Courtesy/Clark Gardens

large numbers of people and complicated technologies, it can be easier to observe the cause and effect of human action upon plants and animal life and land and water formations.

People who study the relationships of nature and who are active in preserving its balances are called environmentalists and conservationists. While we often hear or read about the professionals in this field, many environmentalists are people who volunteer their time and energy to community, state, and federal efforts. They may staff the offices of wildlife organizations, answer phones, and respond to requests for information. They may help lobby for legislation or write on behalf of environmental issues.

Participating in community recycling efforts is another way of directing ecological interests. Churches, schools, and social organizations of different kinds organize to collect old newspapers or clean up areas of the community where debris has been thrown. Creating an environmental watch is also a way every community can monitor the dumping of industrial wastes and other pollutants.

Increasing your understanding of ecology—that is, the way the system of balances in nature works—is an important first step in protecting the environment. Many issues will present themselves to you as a voter, and knowledge of ecological relationships could be a factor in your decision-making.

Camping, wilderness survival courses, and such activities as white water canoeing, can bring people closer to the basic elements of nature. Away from

Hunters who abide by wildlife, game, and fisheries regulations can be an important part of maintaining the natural system of balances. While certain species are in danger of becoming extinct, others may thrive because their natural animal predators have been wiped out. Vegetation in the area may then be in danger. Hunters can act to cull the herd. Hunting is done for both food and sport and, for many people, it is a controversial activity.

THE KNOXVILLE
JOURNAL

Waste expert warns U.S. 'still creating Love Canal'

By TERRY McWILLIAMS

As much as 90 percent of America's toxic garbage is dumped into landfills without treatment.

Years from now, a hazardous waste expert says, that practice may haunt us the same way decades-old toxic chemicals did when they oozed from burial pits at Love Canal, N.Y., into a subdivision.

"We have far and away more hazardous waste than any (country) — 60 million tons," says Gary Davis, a former California hazardous waste specialist and now a Knoxville environment lawyer.

"We're still creating Love Canals, legally, because current regulations say you can dump anything into landfills less secure than Love Canal. And Love Canal was pretty secure as it goes."

Davis traveled through Europe last summer and talked to government officials, environmental groups and industry consultants to learn what system those countries use to dispose of hazardous waste. He found a discernible difference from the way America disposes of its waste.

"They (Europeans) don't even consider putting hazardous waste in the ground," said Davis. "They don't consider it a viable technology...They know that liquid waste can migrate into the groundwater and they depend on groundwater for drinking water."

Two of the leading countries, West Germany and Denmark, burn or detoxify more than 60 percent of their special wastes. He said the reasons are simple.

"Europeans recognized years ago that landfills leak. They recognize they do not have as much land as we do to waste. They decided the technology was there and they were going to use it. They pioneered (new) technology and put the weight of government behind it," he said.

> **W**aste disposal is an important global issue. As the technologies of industry become similar from country to country, so do the waste products of these processes. However, since social and political conditions differ, different alternatives for disposal are being chosen. Comparison through observation is a good method for learning how these different options work.
>
> What factors in our own society influence our methods of disposal for waste products? What waste products do we produce? Is there any method of production which does not produce waste material? How do you think methods might differ if we were a country the size of Switzerland?

Only the residues left behind from incineration or other treatment are buried in landfills, Davis said.

Davis, one of the authors of a California assessment on alternatives to burying hazardous waste, believes Denmark is 10 years ahead of the United States in the way it disposes of waste. He blames lax U.S. environmental regulation for the lag.

"There's no God-given right to pollute," he said. But business will take more regulation "as an affront."

Most European governments establish quasi-public entities to finance and build hazardous waste treatment facilities for industry. Then all industry generating toxic waste must pay to use the plants, he said.

In Denmark, municipalities are responsible for collecting and processing hazardous wastes — just like U.S. cities maintain sewage treatment plants, he added.

The government in Denmark subsidizes new treatment systems by 25 percent. Other countries, like Germany, discourage the production of new hazardous wastes by requiring more efficient processes. There, a panel reviews all new processes to determine if the industry can reduce pollution even more, Davis said.

Environmental regulation of industry is less in Europe, he said, but the attitudes of industry officials are generally such that factories "are better citizens." The relationship is "less adversarial" and more informal, he said.

One of the problems in the United States, he contends, is the financial advantage some companies gain by rejecting incineration for the less expensive dumping into the ground.

UNIT 4 Classification and Simple Forms of Life

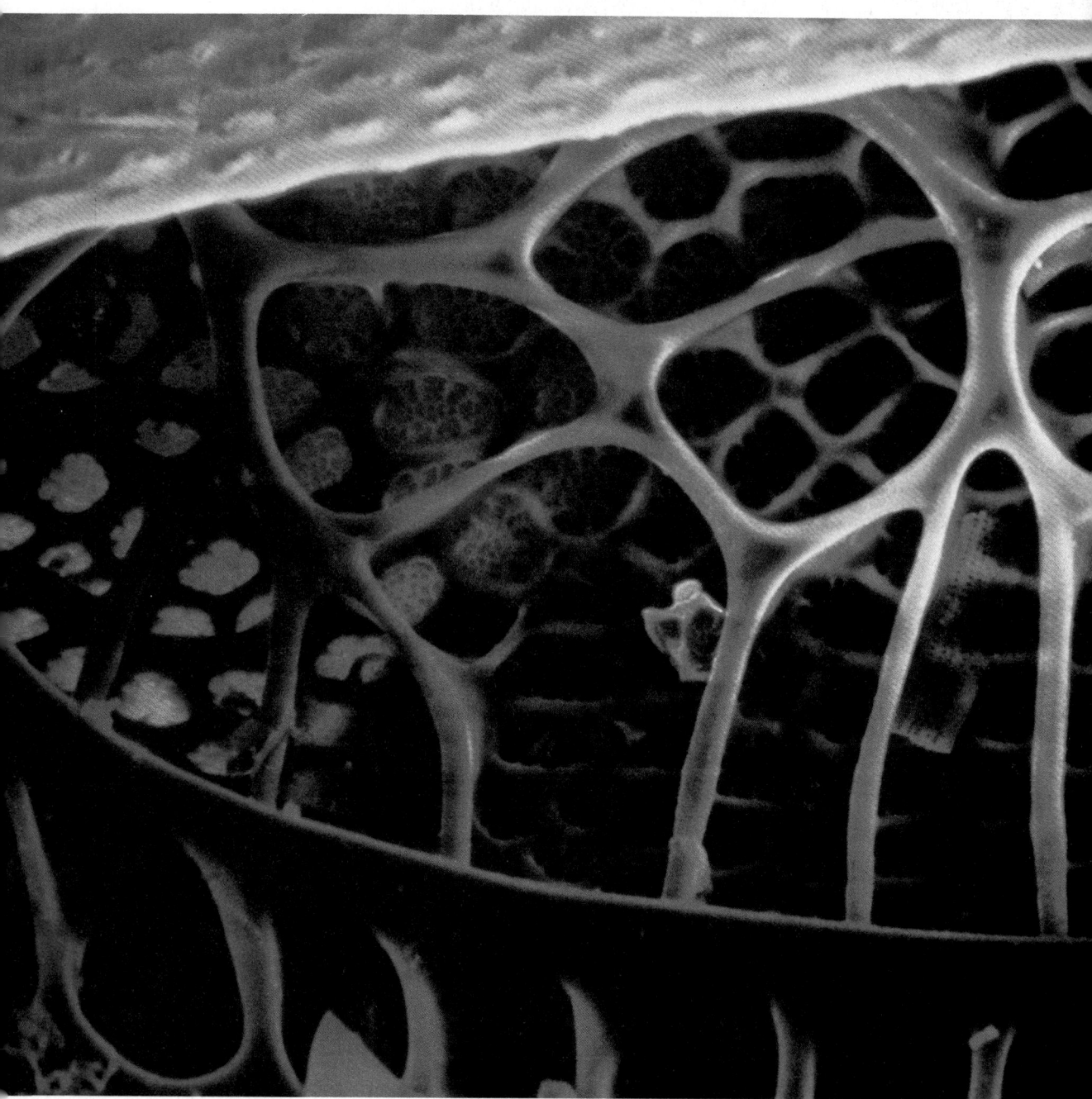

CHAPTER 17 Scientific Names

objectives

After you read this chapter, you should be able to:

__**List** the divisions into which living things are classified

__**Explain** how we decide what groups to place living things in

__**Name** the four kingdoms used in this book

Science Reading Skill: This chapter reinforces the skill of *classification* in which organisms are grouped together on the basis of likenesses and differences.

Enrichment: This can be expanded to include differences within a country, i.e., soda, tonic, "pop."

The scientist in the photograph above is closely studying a variety of insects. How do you suppose scientists decide how to name an insect or other organisms? How can they tell which organisms are related to each other? In this chapter you will find out.

Suppose your teacher asked you as a homework assignment to write a description of all the living things on the earth. How long do you think this would take you? If you worked very hard you might get the job done in four or five hundred years. After all, there are well over 2 million known kinds of living things. This would not be a fair homework assignment for you, but it is a job which must be done by biologists. Living things must be named and described so that each scientist can know exactly what plant or animal another scientist is talking about.

About 200 years ago, a man named Linnaeus (luh-NAY-us) set out to name and describe all living things. He did not finish in his lifetime. But what is important is that he established a system of naming and classifying that biologists still use.

NEED FOR SCIENTIFIC NAMES

You may wonder why the naming of living things is a problem. Why not use the common names people have always had for living things? There are several reasons. For one thing, many living things have no common names. There are over 200,000 kinds of beetles alone. Only a few of them have common names.

Another problem is that the same plant or animal may be called by different names in different places. The same bird is called a nighthawk in some regions and a bullbat in others. There is a woodpecker with four names—flicker, highholder, high-hole, and yellowhammer. See Fig. 17-1. The hog-nosed snake has thirteen common names in the southeastern United States. A scientist might study an animal and use one of its common names in his written report. Another scientist who knew it by a different name could read the report and not be sure what animal the first scientist was writing about.

Another problem is that the same common name is often used for different living things. "Mayflower" is used for several different plants. At least three different birds are called robins (Fig. 17-2 shows two of them.). This, again, leads to confusion.

Linnaeus wanted to work out a system of scientific names that would be in the same language throughout the world. He did not expect people to use these names in daily conversation. They were to be used by scientists to identify the types of living things that they worked with. He used Latin, because in those days all educated people knew Latin.

The naming system. In Linnaeus' system, each *species* of living thing is given a double name, in much the same way that you have a first and last name. The first word in the double name tells to which group of similar species this particular kind of living thing belongs. Such a group is called the **genus** (GEE-nus); plural, genera (GEN-uh-ruh). The second word is the name of the one particular species in the genus.

Let us take a few examples. The scientific name for a dog is *Canis familiaris*. The timber wolf is *Canis lupus*. The coyote is *Canis latrans*. You will notice that the word *Canis* is used in all of these names. This is the Latin word for dog. It is used here as the genus name. This genus includes all the dog-like animals. The second word in each case is the species name, which indicates the particular kind of dog-like animal. Notice that the genus name is alway capitalized. Also notice that the names are in *italic* type. In scientific writing, the scientific names are usually printed in italics.

Here is another set of examples: *Felis domesticus*, *Felis leo*, *Felis tigris*. Do you know what the common names of these

genus

Word Study: "genus" means kind or sort

Enrichment: This system is known as binomial nomenclature.

Enrichment: Some members of a species vary slightly but not enough to be considered a different species. We call such slightly different organisms varieties. The variety name is added as a third part of the scientific name.

Fig. 17-2
Left: The familiar American robin; Right: The South Island bush robin. What differences do you see in these two birds with the same common name? (U.S. Fish and Wildlife Service; C. B. & D. W. Frith/Bruce Coleman, Inc.)

Fig. 17-3
The lion, Panthera leo, *and the tiger,* Panthera tigris, *are so much alike that they belong to the same genus. Why do they have different species names? (David C. Fritts/Animals Animals; Z. Leszczynski/Animals Animals)*

Activity: Any method of classification is devised to aid the scientist and is not necessarily a natural method. Have students group items in a supermarket as groceries, dairy products, produce, meat, and poultry.

Enrichment: New species of wildlife are usually discovered about once every thirty years. A new species of marsupial was found in South Australia and a wild pig was found in Paraguay.

animals are? They are the cat, lion, and tiger. These animals are so much alike they were put in the same genus. Bobcats were not in this genus. They were considered different enough to be placed in a separate genus of their own, the genus *Lynx*.

These names for the cats seemed reasonable at the time. Later on, a more careful study of the animals caused scientists to change their minds. House cats are actually more like the bobcat than they are like the lion, so the names were changed. Bobcats and house cats are now placed together in the genus *Felis*. Lions and tigers are in a new genus, along with all other cats that roar. They are called *Panthera leo* and *Panthera tigris*. See Fig. 17-3. As you see, the naming system is flexible. It can be changed as we increase our knowledge. All such changes are on record, so it is always possible to identify a species correctly even if we run across the older name in print.

The scientific name for human beings is *Homo sapiens*. In Latin *Homo* means man and *sapiens* means wise. We were not very modest when we named ourselves, were we?

Who names the species? You may wonder who decides what the scientific names of the various living things shall be. The answer is that any person can name a species. Anyone who identifies a kind of living thing that has never been identified before can name it. The first step is to write a careful description of the new species and give it a name. If the organism also belongs to a new genus, the discoverer can "invent" that name also. Then the description is sent to a scientific journal to be published. The first scientific name in print is the one that is used. The name must be in the form of Latin, though it need not be made from real Latin words. There are such names as *michiganensis, canadensis,* and *jeffersonianum* in use. These, of course, are not Latin words, but they have all been given typical Latin endings.

THE CLASSIFICATION SYSTEM

If you were writing a book about different types of organisms, you probably would want to organize it in some way. You would not put an elephant on one page, a mouse on the next, a seaweed next, and then a beetle. The human mind remembers things best if it has them organized. In biology we *classify* living things. We put similar things in groups. A group is easier to remember than a lot of separate types.

This grouping begins with the scientific names. When we look at the name *Homo sapiens,* we know at once that we belong to the genus *Homo.* Next, several genera that are much alike are grouped to form a **family**. For instance, lions are in the cat family (*Felidae*), along with house cats and bobcats. Foxes are not in the dog genus, but are in the dog family *(Canidae).*

As Linnaeus looked at all of the living things, he decided that the first division he would make was to classify every living thing as either a plant or an animal. So he described two large groups—the Plant Kingdom and the Animal Kingdom.

Of course, a **kingdom** is a very large group, so a kingdom is divided into **phyla** (FY-luh); singular, *phylum* (FY-lum). Examples of the main plant and animal phyla are shown in Fig. 17-5. Each phylum is divided into **classes,** each class into **orders,** each order into families, and each family into genera, and each genus into species. The system is simply a method of organizing our knowledge of living things. See Table 17-1.

Assignment: Have students report on a new species that has been discovered in recent times.

family

kingdom

phyla

Word Study: "phylum" means tribe

classes

orders

TABLE 17-1 THE SCIENTIFIC CLASSIFICATION OF HUMANS	
KINGDOM	Animalia
PHYLUM	Chordata
CLASS	Mammalia
ORDER	Primates
FAMILY	Hominidae
GENUS	*Homo*
SPECIES	*sapiens*

Problems in classification. You may think it is an easy matter to decide what is an animal and what is a plant. Horses, crabs, and worms are obviously animals. Elms and ferns are obviously plants. So far, so good, but not all groups of organisms can be so easily classified.

There are some tiny single-celled organisms that live in the water. Some of them feed on particles of food the way animals do. Some have chloroplasts and make their own food, like plants. Some absorb dissolved food materials the way decomposers do. There are species of these organisms that feed in more than one of these ways. All this makes it hard to say that this group is either plant or animal. Many other types of living things are equally as difficult to classify as plant or animal.

Some scientists have tried to solve this problem by putting all of the more primitive types into a third kingdom, the protists. This has not worked very well, because some protists are less like each other than they are like certain plants or animals. In time we may agree on several new kingdoms, but we cannot say what they will be at this time. One new kingdom should contain the bacteria and the blue-green algae. Their cell structure is so different from that of all other living things that they cannot be closely related to any other group.

In this book we will consider the bacteria and blue-green algae to be in a kingdom by themselves, which we will call the *Bacteria and Blue-green Algae Kingdom.*

We will place all other living things with simple body organization in the *Protist Kingdom* (Kingdom Protista). Many of these protists are single-celled.

Each type with a more highly organized body will be placed in either the *Plant Kingdom* or the *Animal Kingdom;* plants being multi-cellular producers and animals multi-cellular consumers.

The basis of classification. The idea behind the classification system is to show actual relationships among groups. A dog and a wolf are close relatives. Their genes are nearly alike. They came from the same ancestors as the result of natural selection. We show this close relationship by placing them in the same genus. A fox is not as closely related to the dog as the wolf is, yet dogs and foxes are a great deal alike. We show this degree of relationship by putting them in the same family but not in the same genus. We use this same sort of reasoning when we say people are first, second, or third cousins. Such an arrangement shows how closely related they are.

In trying to decide how to classify living things we look at their basic structures. Two species may look alike and yet not be closely related. A whale certainly looks like a fish, yet a study of its structure shows that it is a mammal. It has the same

Bacteria and Blue-green Algae Kingdom

Protist Kingdom

Plant Kingdom

Animal Kingdom

warm blood, lungs, type of heart, type of reproduction, and many other features that are found in the land mammals. Whales came from ancestors that lived on land. Only its shape is fish-like. This shape is simply an adaptation to life in the water.

Look at Fig. 17-4. Would you say that the bat and moth are related because they both have wings? Before you answer this question notice the bones in the bat's wings. They are the same bones that are found in your hand. The moth's wings are formed by wide, thin folds of its outer body covering. The whole inside structure of the moth shows that it is an insect. The bat's structure shows that it is a mammal. This means that the bat and the moth are not closely related. In fact, the bat is more closely related to you than it is to the moth!

In the following chapters we shall study the main groups of living things. Most of the time we shall deal only with phyla. In only a few cases will we talk about smaller groups.

Fig. 17-4
A bat and a moth. Both have wings and a similar body shape. Does this mean that they are close relatives? (Sdevard C. Besserôt/Bruce Coleman Inc.)

SUMMARY

Each living thing is given a scientific name consisting of two Latin words. The first word is the genus name, which tells what closely related group the species is a member of. The second word is the name of that particular species within the genus.

Living things are grouped into larger and larger groups according to how much they resemble each other and how closely we think they are related by ancestry. This is the classification system. From largest to smallest its divisions are kingdom, phylum, class, order, family, genus, species.

This book uses four kingdoms: bacteria and blue-green algae, protists, plants, and animals.

Bacteria and Blue-Green Algae Kingdom

Bacteria

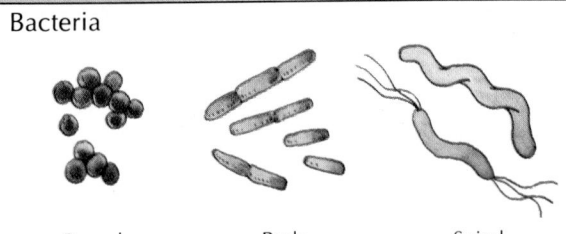

Round   Rods   Spiral

Blue-Green Algae

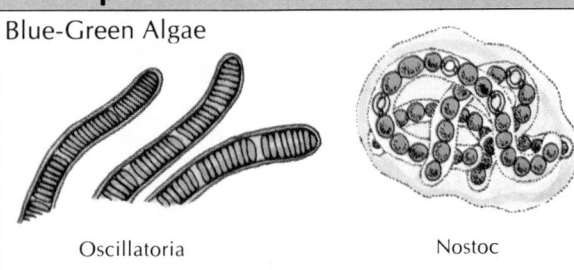

Oscillatoria   Nostoc

Protists Kingdom

Protozoans

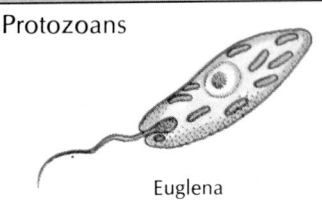

Euglena

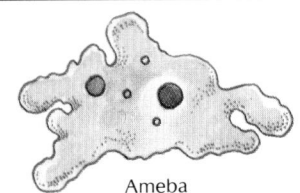

Ameba

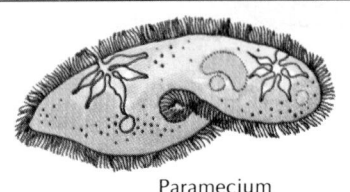

Paramecium

Algae

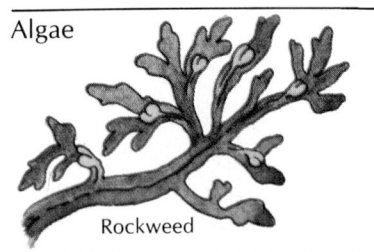

Rockweed

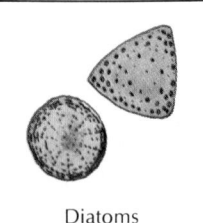

Diatoms

Fungi

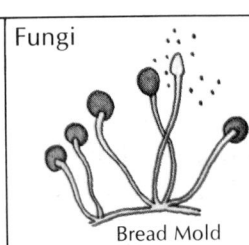

Bread Mold

Mushroom

Plant Kingdom

Non-Vascular Plants

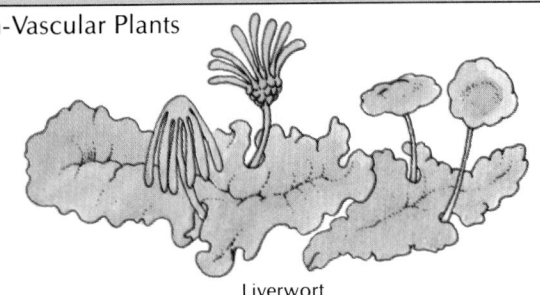

Liverwort

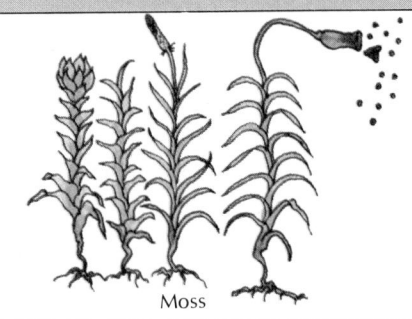

Moss

Vascular Plants

Fern

Spruce Tree

Maple Tree

Daisy

Animal Kingdom

Sponges

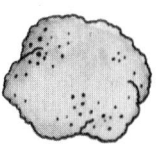

Bath Sponge

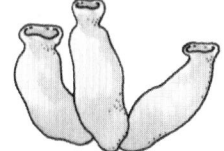

Vase Sponge

Coelenterates

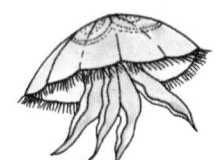

Jellyfish

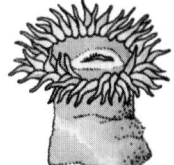

Sea Anemone

Flatworms

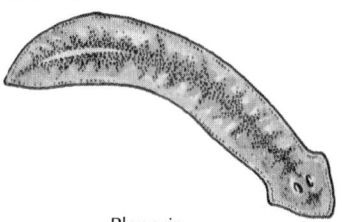

Planaria

Roundworms

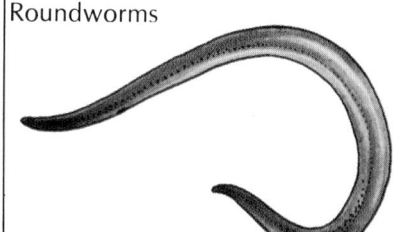

Ascaris

Segmented worms

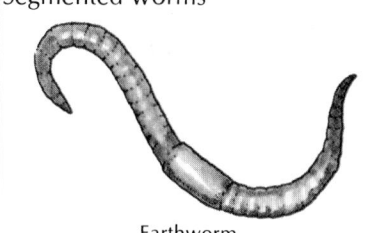

Earthworm

Mollusks

Snail

Octopus

Echinoderms

Starfish

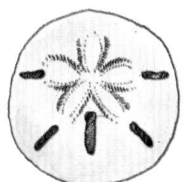

Sand Dollar

Arthropods

Crustacean

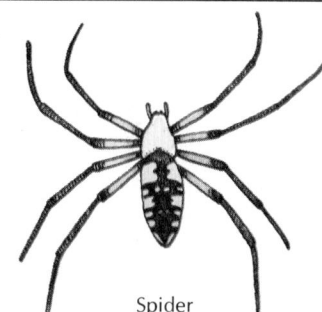

Spider

Insect

Chordates

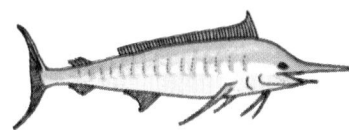

Fish

Amphibian

Reptile

Bird

Mammal

181

Activity

An Experience in Classification

This activity will give you a chance to understand the kinds of problems experienced by scientists when they classify living things.

A. Collect as many examples of plants and animals as possible.

B. Try to decide how these specimens should be placed in classification groups. Try to base your groupings upon differences and similarities in actual physical structure. Place the groups in different locations around the room. Do all of this as a class activity. Then try to answer the following questions:

1. What difficulties do you have in deciding how to group your examples?

2. Is it easy to get everyone to agree?

3. Do you suppose scientists had trouble in setting up their classification system?

4. Using reference books see how much your classification is like the one scientists use.

Word Quiz

The following new words appear in this chapter. Place them in order, starting with the one representing the largest group in the classification system and going to the one representing the smallest:

order, genus, kingdom, species, phylum, family, class

Check Your Facts

1. Why are scientific names needed in the study of living things?
2. Who started our present system of scientific names? Why was this done?
3. What makes up the scientific name of any one kind of living thing?
4. Why was Latin used for scientific names?
5. If you discovered a new organism, what would you have to do to decide how to classify it?

Thought Questions

1. Suppose the microscope had never been invented. Would we then have any trouble classifying all living things as plant or animal? Explain.
2. Does a plant or animal care what name we give it? Who is benefitted by a classification system?

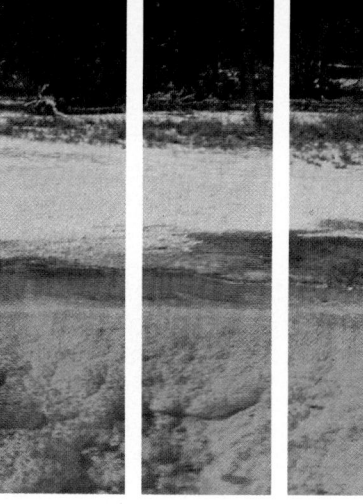

CHAPTER 18 Bacteria and Blue-Green Algae

The photograph above is a hot spring in Yellowstone National Park. What do you think is causing all of the different colors? Could someone have dumped some paint or other chemicals into this water? Actually, nothing has been dumped here. The colors are caused by the presence of bacteria and blue-green algae living in the water.

Cells of the **bacteria** (bak-TIR-ee-uh); singular, bacterium (bak-TIR-ee-um) and **blue-green algae** have a much simpler organization than the body cells of most other organisms. They have a cell wall and cell membrane. Inside, the protoplasm is *not* divided into cytoplasm and nucleus. Each cell has a strand of DNA, which acts as a chromosome, but it is not built like the chromosomes of other organisms. It has no protein around it, but similar proteins are present in the protoplasm. There is no complicated process of mitosis, but the DNA duplicates and the cells divide by pinching in two.

This primitive cell structure makes the group very different from all other living things. Its members have been on the earth much longer than those that have nuclei in their cells. Bacteria were already present over 3 billion years ago. Cells with nuclei did not appear until about one and a half billion years ago.

BACTERIA

The bacteria are a very large group of living things. There are more kinds of bacteria than of all other living species put together. They are smaller than most other single-celled forms, and they are found wherever living things exist. Many of these bacteria are of great importance. Life as we know it could not continue without them. The living world could get along without people, but not without bacteria. You probably think of them as disease germs. In fact, some bacteria do cause disease, but many others are very useful.

objectives

AFTER YOU READ THIS CHAPTER, YOU SHOULD BE ABLE TO:

___**Compare** cells of bacteria and blue-green algae with those of other organisms

___**List** ways in which bacteria are important

___**Explain** the nitrogen cycle

___**Describe** the structure and reproduction of a virus

bacteria

blue-green algae

Reference: See *Microbe Hunters* by Paul Dekruif.

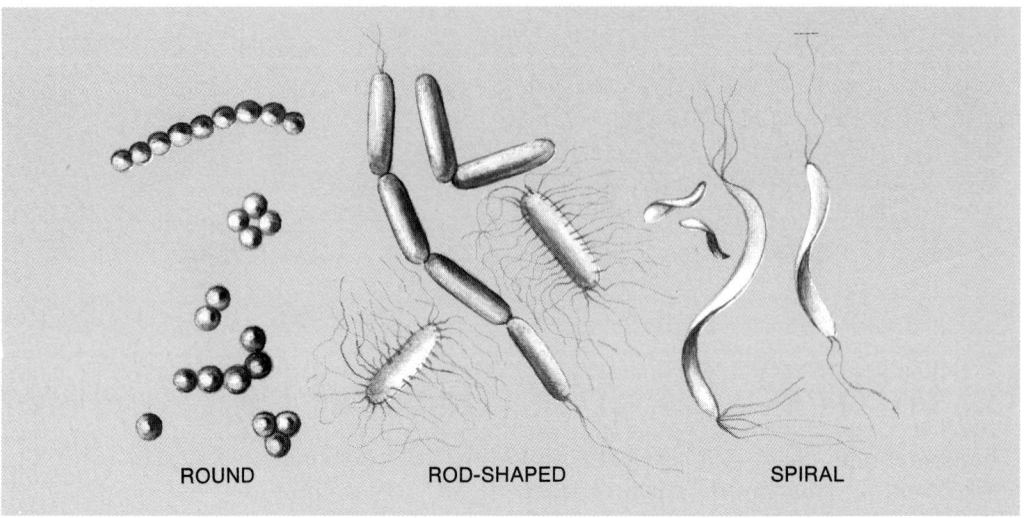

ROUND   ROD-SHAPED   SPIRAL

Fig. 18-1
The three common shapes of bacterial cells. Which one shows flagella?

Enrichment: More than 500 million of certain types of bacteria could fit on a postage stamp. Several thousand bacteria could fit on a small dot made by a pen.

colony

flagella

Word Study: The singular of flagella is "flagellum," which means whip.

Bacteria are very small. They are smaller than the nucleus of an average plant or animal cell. One trillion bacteria of average size weigh only one gram. That many cells of ordinary size could make up a 12-year-old child.

Bacteria are all around us. They are in the air, soil, lakes, ponds, streams, and oceans. There are bacteria on our floors, walls, ceilings, and furniture. They are on our skin, in our mouths, and intestines.

Characteristics of bacteria. Bacteria come in three common shapes—round, rod-shaped, and spiral. See Fig. 18-1. Bacteria are single-celled, but sometimes they are grouped together in pairs, chains, or clusters. Bacteria are named according to their shape and the way they are arranged. A large group of bacteria is called a *colony.* A colony can become large enough to be seen without a microscope. It looks like a spot or film on the surface of some decaying material.

In some species of bacteria, one or more very thin, hair-like structures stick out from the cell. These hair-like parts are called *flagella* (fluh-JELL-uh); singular, flagellum (fluh-JELL-um). They lash about in liquid like tiny whips. This causes the bacteria to move. They never go far in one direction. They just seem to move about in the same general area. This movement brings them in contact with food molecules.

Some bacteria are producers. Some species of bacteria contain a substance similar to chlorophyll. These bacteria carry on a form of photosynthesis in which no oxygen is released. They use the energy of sunlight to make food. They are not all green, however. Many are purple.

184

Another group of bacteria makes food in quite a different way. They do not get their energy from sunlight. They get it from chemicals. For example, one type combines oxygen with iron to get energy. Others get energy from compounds of sulphur. The energy is then used to combine water and carbon dioxide to make food.

You can see, then, that it is not quite correct to say that green plants make all of the food in the world. Bacteria also make food. Some use sunlight and some do not. When bacteria are eaten, the food these bacteria make is passed along in food chains just as other foods are. However, the amount of food made by bacteria is very small compared to that made by green plants.

Most bacteria are consumers. Most bacteria get their food ready-made. Among these are the bacteria of decay. When something decays, it is due to bacteria living in the material. These bacteria give off digestive juices that dissolve the material. Then they absorb this dissolved food through their cell membranes.

Each species of bacterium is adapted to live in some special type of environment and to use some particular food. Some bacteria need oxygen. Some do not. Some are actually killed by oxygen, so they must live deep inside their food supply—in such places as the mud of a lake bottom or inside the body of a rotting animal. Some use the same sorts of foods that we do. Others use all sorts of other foods, like humus, wood, cloth, petroleum, soap, manure, dead animals, and many others. In fact, every substance in nature that has possible food value is used by some kinds of bacteria. This is what makes bacteria so important. In their activity as decomposers bacteria break down these substances, and the elements they contain can be used over again.

When green plants need simple substances to build up into food materials, they could not get them if dead things did not decay. The needed materials would be tied up in dead bodies. Remember that elements in protoplasm are used over and over again by living things. Decay bacteria play an important part in breaking down dead substances into forms that green plants can use. Of course, they do not do this to be helpful. They are simply using dead things as food.

Bacteria and the nitrogen cycle. The *nitrogen cycle* is one example of how important bacteria can be in the day to day events of the living community. Decay bacteria in the soil break down proteins in dead material and produce nitrogen compounds called *nitrates*. Green plants in turn use these nitrates to make more proteins. When green plants die, bacteria begin

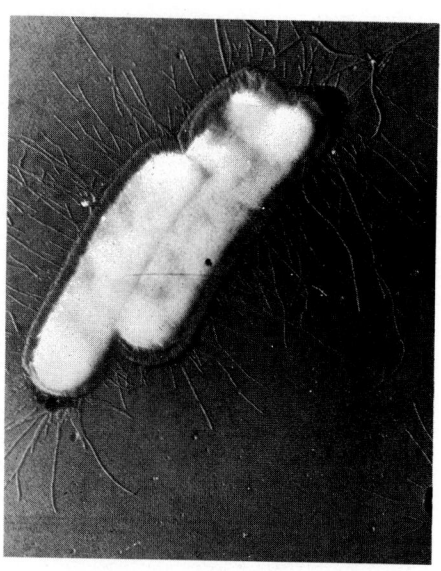

Fig. 18-2
Highly magnified bacteria showing flagella. (Walter Reed Army Institute of Research, Walter Reed Army Medical Center, Washington, DC)

nitrogen cycle

Enrichment: Louis Pasteur was a pioneer in the field of bacteriology. He concluded that heat would destroy microorganisms and in this way developed the process of pasteurization.

185

Fig. 18-3
The swellings on the root of this plant contain nitrogen-fixing bacteria. (William E. Ferguson)

their work of decay and release nitrates, and the cycle starts over again. If the plants are eaten by animals the result is not much different. Animal wastes and the bodies of dead animals are decayed by bacteria to release nitrates into the soil.

There is a larger nitrogen cycle also. The air is nearly four-fifths nitrogen gas, but most green plants are not able to use it. Certain bacteria, different from the bacteria of decay, take nitrogen directly from the air and use it to build their own protoplasm. Later this nitrogen is released into the soil in the form of nitrates. These bacteria are called *nitrogen-fixing bacteria.* Some of them live in the soil. Others live in the roots of plants. They obtain glucose from the plants, and the plants obtain nitrates from the bacteria. This is one more example of cooperation among living things.

Plants of the pea family are very likely to have nitrogen-fixing bacteria in their roots. If you pull up a clover or alfalfa plant you may see swellings on the roots where these bacteria live. See Fig. 18-3. Farmers often plant clover, soybeans, or alfalfa for this reason. Not only do the leaves and stems of these plants make good food for animals, but the bacteria in their roots add nitrates to the soil.

A third kind of bacteria break down nitrates in the soil and release free nitrogen into the air. In this way the nitrogen makes a complete cycle, starting in the air and coming back to the air in the end. Fig. 18-4 will help you to understand the nitrogen cycle.

The action of nitrogen-fixing bacteria is the most important way in which nitrogen from the air is combined to form compounds. It is not the only way, however. Some kinds of blue-green algae can also fix nitrogen, both in the soil and in the water. In addition, electric discharges such as lightning cause nitrogen to combine with oxygen to form compounds in the air. After some other chemical changes these compounds become nitrates. Small amounts form during thunderstorms and are washed into the soil by falling rain.

Growth of bacteria. To live actively, bacteria must have moisture, food, and suitable temperature. When conditions are not right for bacteria they become inactive. Some can remain alive on dry surfaces for a long time. When moisture and other conditions become right, these bacteria soon begin to grow and reproduce.

Activity for bacteria is a matter of feeding, growing, and reproducing. If the conditions of life are good, bacteria reproduce very rapidly. Each cell grows to full size and then divides. The two new cells in turn grow and divide. Divisions may take place as often as every twenty minutes. This means that a single bacterium can produce millions of offspring in a few hours. At

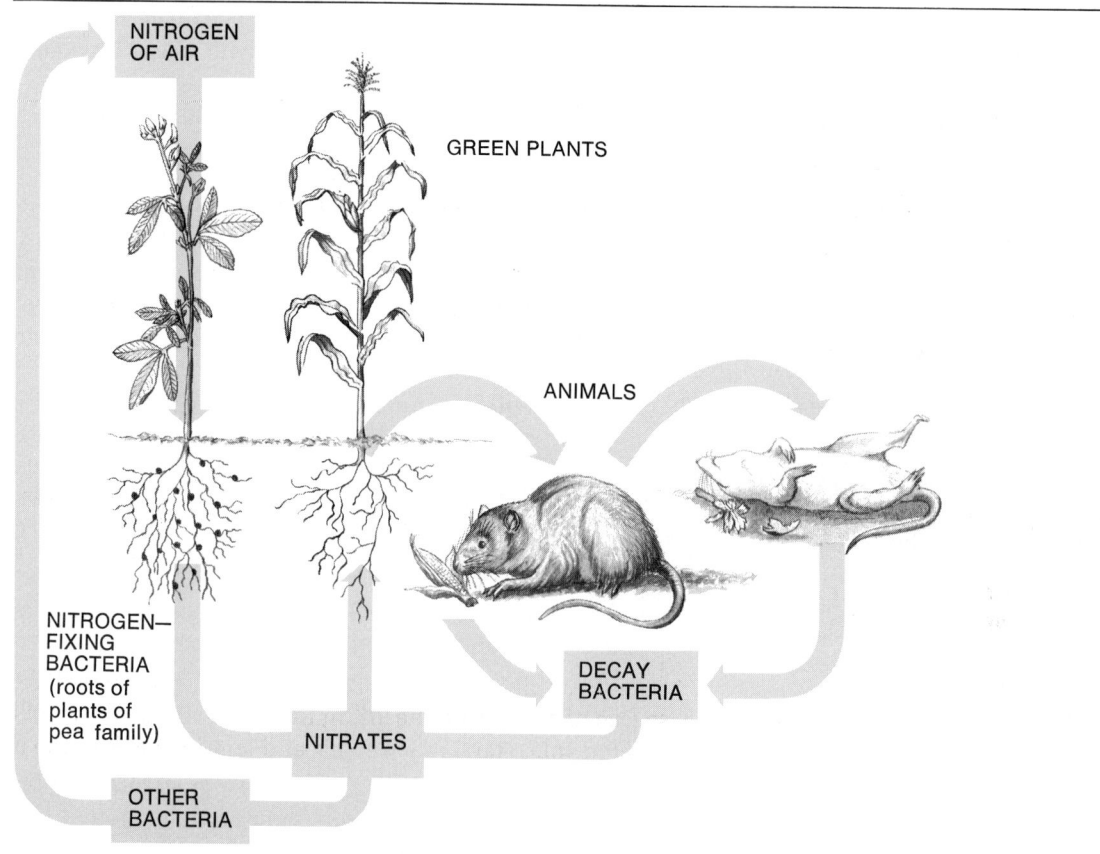

NITROGEN OF AIR

GREEN PLANTS

ANIMALS

NITROGEN—FIXING BACTERIA (roots of plants of pea family)

DECAY BACTERIA

NITRATES

OTHER BACTERIA

Fig. 18-4

The nitrogen cycle. The nitrogen supply is important to organisms. Nitrogen forms a part of proteins, nucleic acids, and ATP. Could we live without nitrogen?

this rate you might think the world would soon be full of bacteria. This does not happen because their environment never stays perfect for very long. They may use up their food supply, they may become too crowded, or they may dry up. When they are crowded their own waste materials slow down their growth. Also, bacteria have natural enemies that feed on them.

If the environment is unfavorable some bacteria form spores. A bacterial spore is a tiny cell developed within the parent cell. Often there is only one spore to a cell. A spore can continue to live when conditions would kill the parent cell. Some can even survive in boiling water for a while. When the environment becomes favorable again, a spore can develop into a full-sized bacterial cell.

Food spoilage. Since bacteria are present everywhere, any food material will soon decay. If we wish to protect our foods from bacteria, there are several methods we can use. *Refrigeration* does not stop decay, but it does slow it down. All chemical changes are slowed as the temperature is lowered. Food will keep about four times as long in the refrigerator as at room

Fig. 18-5
How did bacteria take part in preparing this lunch? (HRW photo by Ken Karp)

Enrichment: Botulism is a fatal disease that results from improperly canned foods. Students should be on the look out for "swollen" cans or canned foods that seem to have an unusual smell.

Enrichment: Bacteria are also used in the production of yogurt, cheese, cream, and butter and in the curing of tobacco.

temperature. *Freezing* does not kill bacteria, but it stops their activity almost entirely. Frozen food will keep for many months.

Salting preserves foods. Bacteria cannot live in a very salty environment, because salt causes water to be drawn out through their cell membranes. *Sugar* is sometimes used to preserve foods for the same reason. Often foods are *pickled* in spiced vinegar, salt, and water to keep bacteria from spoiling them.

Drying is another way of preserving foods. Without water, bacteria cannot grow. Crackers, flour, and breakfast cereals keep for this reason. Cloth and wood, as well as food, will rot if they stay wet for too long.

When we *can* foods we do two things. We cook the food at a high temperature to kill the bacteria already present. We use steam pressure cooking because ordinary boiling will not kill all of the bacterial spores that may be present. Finally we seal the can to keep any other bacteria from entering and to keep out oxygen. Properly canned food will keep for years.

Disease bacteria. Some bacteria do not wait until we are dead to start using us as food. They move right in while we are still alive. In other words, they are parasites. Many serious diseases are caused by bacteria, including tuberculosis, pneumonia, and diphtheria.

Practical uses for bacteria. Bacteria use a great many different materials for food. In doing this they produce many kinds of waste products. These waste products are often very useful to us. *Vinegar* is formed when bacteria change sugar into a weak acid. *Pickles* and *sauerkraut* are preserved by another acid that bacteria produce in the food material. *Silage*, used to feed cattle, is produced in the same way. Plant materials such as corn plants or grass are chopped up and stored in a deep bin called a silo. Bacteria change sugar from the plant juices into an acid that preserves the silage.

Bacteria are useful in producing several other foods by changing them in some way that will make them better to eat. These include black tea, coffee beans, vanilla pods, and cocoa beans, from which cocoa and chocolate are made. Some vitamins are produced by bacteria when they are grown in a soupy material containing ground-up grains. After the bacteria have grown for a while, the vitamins are removed from the liquid. It may soon be possible to place certain human genes into bacteria cells. Such bacteria could then be used to produce useful materials like insulin (for treating diabetics) or thyroxin (thie-ROCK-sun) (for people with thyroid trouble).

Decay bacteria are used to destroy sewage in sewage disposal plants. They are also used to weaken the structure of the plant stalks so that linen fibers can be separated from flax stems. Jute fibers (used for burlap) and hemp fibers (used for rope) are produced in the same way.

Industry is using bacteria more and more in the production of useful chemicals. Some of these chemicals include acetone, methyl alcohol, butyl alcohol, propionic acid, citric acid, vitamins, antibiotics, and several enzymes. In each case the chemical is formed by allowing bacteria to grow in some food material such as molasses, sawdust, or starch. The useful product is usually formed as a waste by the bacteria during their growth. The chemicals just mentioned are used in the making of many products, such as industrial solvents, soft drinks, dyes, plastics, leather, explosives, perfumes, drugs, and anti-mold materials.

Blue-green algae. The term *algae* (AL-jee); singular, alga (AL-guh), refers to any living thing that contains chlorophyll and that has a simple body structure. Trees and ferns are not called algae because they have complicated bodies containing many kinds of cells. *Chlorella* and *Spirogyra* are both algae. They have simple structure, and they contain chlorophyll. Their cells have nuclei and their chlorophyll is contained in chloroplasts. Therefore, they belong to the Protist Kingdom.

Blue-green algae have no nuclei, so they are not placed in the Protist Kingdom. They have a blue-green color, which gives this group its name. Chlorophyll is spread all through the protoplasm, and there are no chloroplasts. The blue-green algae are really just a special kind of bacteria.

Blue-green algae are not always blue-green. Some are various shades of green, blue, yellow, or red. Many are single-celled. Others form clusters of cells. Still others grow into long rows of cells. See Fig. 18-6.

Blue-green algae are common in water, on damp soil, and in the soil, where they usually live as decomposers.

algae

Fig. 18-6
Left: Oscillatoria, a blue-green alga that grows in a single filament. (Eric V. Grave/Phototake) Right: Nostoc, a blue-green alga that grows as a mass of filaments surrounded by a jellylike sphere. (L. S. Stepanowicz/Bruce Coleman, Inc.)

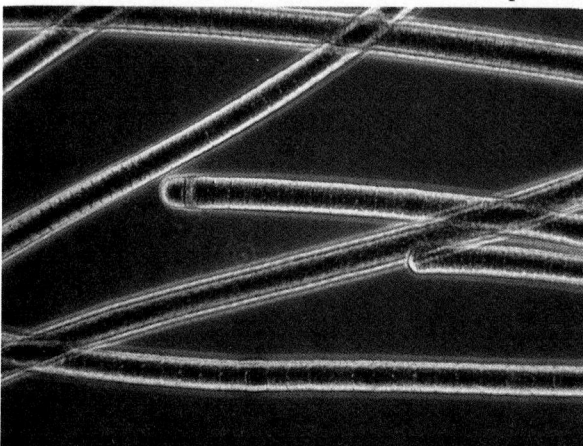
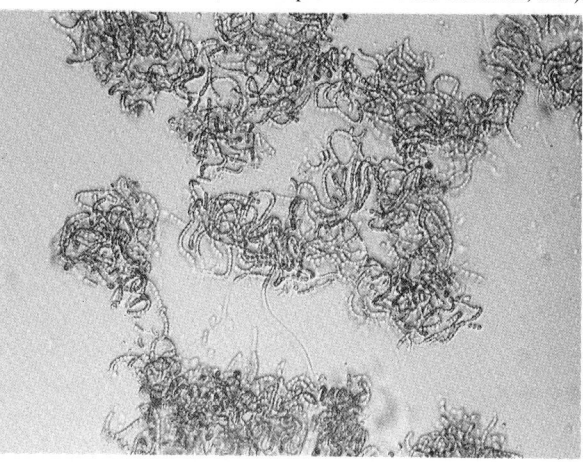

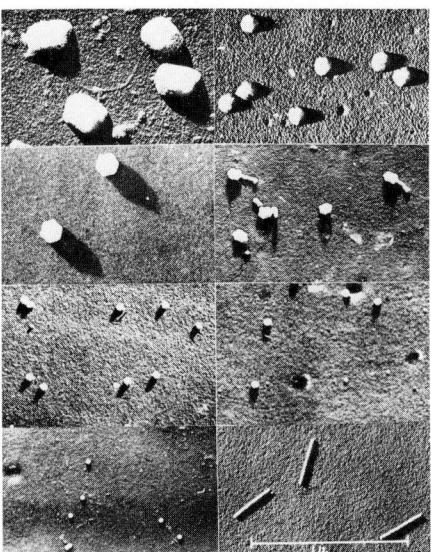

Fig. 18-7
These are several different types of viruses magnified 50,000 times. (R. C. Williams Virus Laboratory, University of California, Berkeley)

viruses

Enrichment: Point out to students that most of our knowledge about viruses has been learned through the use of the electron microscope.

Reference: See *The World of the Virus* by Steward M. Brooks.

Viruses, viroids, and plasmids. The **viruses** are not related to bacteria. In fact, they are not truly alive. They do not belong to any of the kingdoms. They are listed here because they are far more simple and much smaller than the bacteria. They can be seen only by using an electron microscope.

Viruses are not organized into cells, and they do not carry on most of the life activities. A virus is made of a small bit of DNA or RNA surrounded by a layer of protein. It is a small group of genes in a protein wrapper. Viruses are parasites in the cells of other living things. When not in the host cell, they are like dead chemical substances.

When a virus particle contacts a host cell, the gene material slips inside. There it acts like genes in the cell. It directs the cell's cytoplasm to produce the proteins and nucleic acids to make more virus particles. When the cell becomes loaded with these new viruses, it breaks open, and the new viruses are released. They can infect other cells.

You can see, then, that viruses really possess only one characteristic of life: they reproduce. But, *they cannot reproduce by themselves.* The host cell must do the work for them. A virus cannot carry on any other life activities.

Viruses cause serious diseases in everything from bacteria to people. Measles, mumps, yellow fever, and colds are examples.

Viroids are even smaller than visuses. They cause several diseases in plants. Each viroid is made of a tiny bit of RNA. There is no protein sheath. They reproduce in the host cell just as viruses do.

Plasmids are small units of DNA found in cells of many bacteria. A plasmid is not part of the bacterial chromosome. Plasmids reproduce in bacteria, and they can be passed on to other bacteria. They do not seem to harm their host, and they often carry genes that are useful to the host.

Scientists have learned to put genes into plasmids and then have these plasmids taken in by bacteria. For example, a human gene for producing insulin can be placed into the cells of bacteria in this way. The bacteria will then produce human insulin, which can be purified and sold by drug companies. We can expect many useful chemicals to be produced in this way.

Origin of viruses, viroids, and plasmids. It is possible that each of these parasites started as part of a living cell from which it escaped and became a separate unit. When it enters the same kind of cell, it can reproduce there. If this is true, then each parasite came from the type of cell it infects. Perhaps new kinds are being produced from time to time.

SUMMARY

The bacteria and blue-green algae form a kingdom of living things having no nucleus or chloroplasts in their cells. Bacterial cells are usually very small, living alone, or in pairs, chains, or clusters. Their form is generally round, rod-shaped, or spiral. They are the most important decomposers, though some are producers and some are parasites. In the nitrogen cycle, bacteria make nitrogen available to plants. Foods can be preserved if bacteria are prevented from growing in them.

Blue-green algae are generally larger than bacteria, and they contain chlorophyll. They are common producers in the water. Some live on and in damp soil.

Viruses are extremely tiny particles. Each is made of a few genes wrapped in protein. All are parasites, which direct the host cell to produce new viruses. They have no other life activities.

ACTIVITY

Growing Bacteria

This activity will show you how bacteria can be grown for study. It will also show how to test for the presence of bacteria.

A. Obtain or prepare some agar plates, which will be exposed to collect bacteria. Different members of the class can expose agar plates in different ways, to test for the presence of bacteria. In each case, open the Petri dish quickly, expose the agar to the possible source of bacteria, and close the dish. Keep it closed from then on and tape it shut. Place the dishes in an incubator or other warm place for a few days. Watch (through the lids) for spots or film to appear on the surface of the agar. These will be colonies containing millions of bacteria.

Be sure to leave a few dishes unopened. These are the controls.

1. Which cultures show the most bacteria? The fewest?

2. What happened in the controls?

3. Why do we use controls?

B. CAUTION: Most of these bacteria are harmless, but do not take chances. Return the dishes unopened to the teacher. They will be dropped into boiling water before they are opened and cleaned.

4. Why is this done? Aren't the same kind of bacteria present in the environment anyway?

Word Quiz

Choose the letter from **Column B** that best matches each number in **Column A.** Do not write in this book.

Column A	Column B
1. bacteria	**a.** any simple species having chlorophyll
2. blue-green algae	**b.** important mineral in soil
3. colony	**c.** a process in which an important element passes from air to soil and back again
4. flagella	**d.** algae without chloroplasts
5. nitrogen cycle	**e.** reproduces only in a host cell
6. nitrate	**f.** most common type of living thing
7. algae	**g.** a group of organisms living together
8. virus	**h.** whip-like structures used by some cells for movement

Check Your Facts

Science
Reading Skills
1. *Compare and Contrast*
4. & 10. *Cause and Effect*
5. *Sequencing*

1. How are the cells of bacteria and blue-green algae different from those of other living things?
2. Describe the shape and size of bacteria.
3. What are three ways in which bacteria obtain their food?
4. What must decay bacteria do to solid foods before they can absorb them?
5. Describe the nitrogen cycle.
6. List and explain some examples of how food can be preserved.
7. In what ways are bacteria useful?
8. What are algae?
9. Describe the appearance of blue-green algae.
10. What is a virus? How do viruses affect us?
11. How do viruses multiply?

Thought Questions

Science
Reading Skills
1. & 3. *Problem Solving*

1. Suppose a bacterium divides in twenty minutes. Then these two cells divide in another twenty minutes, and so on. How many bacteria would there be at the end of one hour? Two hours? Four hours? Eight hours?
2. Imagine that all of the bacteria died tonight. Write a science fiction story describing what changes would take place in the world during the next hundred years.
3. There are spore-forming bacteria that live in the soil. They also grow in sealed cans of meat, fish and vegetables when they are not properly sterilized. People have died from eating small amounts of such foods. What method of canning would prevent this kind of poisoning? Since heat destroys the poison the bacteria produce, how can we protect ourselves if we are doubtful about a food?

CHAPTER 19 PROTOZOA

Four phyla of single-celled protists are commonly called *protozoa* (prote-uh-ZOH-uh); singular, protozoan (prote-uh-ZOH-un). Protozoa used to be called single-celled animals, because they move about and find food the way animals do. They are some of the most interesting things that can be seen under the microscope. They are tiny, but their effects can be large. The cliffs in the photograph above are made of chalk. This chalk was formed under the sea long ago. It is composed mostly of the tiny shells of protozoa.

THE PHYLA OF PROTOZOA

The ameba phylum. Ameba and its relatives move by a flowing of their cytoplasm. They do not have flagella or other swimming structures. They take in food by wrapping around it.

Some kinds of ameba living in the ocean have shells around their bodies. Many of these shells are very beautiful when seen under the microscope. See Fig. 19-2. These little animals are very common in the sea; they drift, suspended in the water, and send strands of cytoplasm out through pores in the shell. These strands catch bits of food and draw them into the cell. When they die their shells settle to the bottom and form a limey mud that may finally harden into chalk or limestone. Your teacher writes on the board with these skeletons.

The flagellates. Another phylum of protozoa is made up of species that swim with flagella. Some types have one flagellum. Others have several. Because of this characteristic, they are commonly called the *flagellates* (FLAJ-uh-lates). Remember that a flagellum is a long thread reaching out from a cell. Its whipping motion can send the flagellate through the water. The flagella of these protists are more efficient than those found in

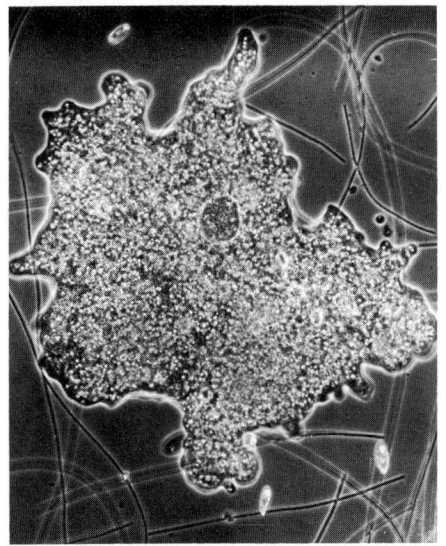

Fig. 19-1

An ameba. What parts can you iden-tify? (Walter Dawn)

Euglena

bacteria. The "tails" used by sperm cells for swimming are actually flagella just like those found in the flagellates.

A common species living in fresh water is ***Euglena*** (yoo-GLEE-nuh). Fig. 19-3 shows the structure of *Euglena*. It has a long body with a flagellum at the front end. *Euglena* swims by vibrating this flagellum. The cell contains a nucleus and several chloroplasts. There is a red *eyespot* near the frond end. This is not an eye that sees things. It is simply a structure that is sensitive to light. *Euglena* is able to respond to light because of this eyespot.

Euglena is able to make its own food with its chloroplasts. It can also take in food molecules from its environment by absorbing them through the cell membrane. Some other flagellates have another way of feeding. They take in solid particles of food and digest them inside the cell. This food is not taken in through just any part of the cell membrane, as in ameba. It enters through a definite mouth opening at the front end of the cell.

So you see that flagellates have three ways of feeding. Some of them make food, like the green plants. Some eat solid foods, as animals do. And some absorb dissolved foods, as bacteria and fungi do. Some species use more than one of these methods.

Some biologists believe that all of the true plants and animals developed from flagellates long ago. You can see how this could have happened. Through natural selection, some groups of flag-ellates would come to depend entirely on photosynthesis and finally develop into green plants. Others would specialize in absorbing dissolved foods. These would develop into fungi.

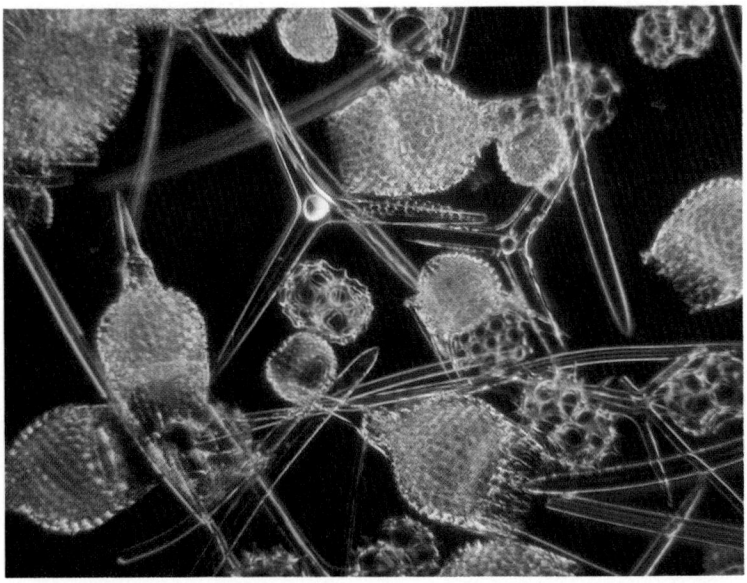

Fig. 19-2

These are the shells of single-celled or-ganisms that belong to the same group as ameba.

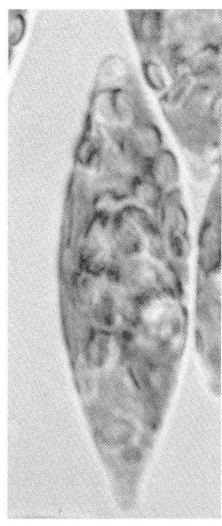
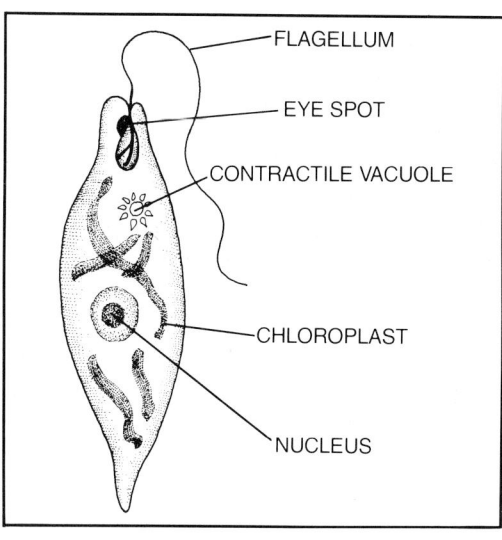

FLAGELLUM

EYE SPOT

CONTRACTILE VACUOLE

CHLOROPLAST

NUCLEUS

Fig. 19-3
Left: An actual photograph of a magnified Euglena; Right: A diagram showing the parts of Euglena. (Photo courtesy Carolina Biological Supply Company)

Still others would specialize in searching for food and eating it. The animal kingdom could have started from flagellates in this way. Even today, the sperm of most higher animals look very much like flagellates.

The ciliates. In almost any pond the most common protozoa of all are members of the *ciliate* group. *Cilia* are simply very short flagella, and a ciliate's body is usually covered with cilia on all sides. This gives the ciliates a better swimming ability than the other protozoa. A very common example of the ciliate group is *Paramecium* (par-uh-MEE-see-um); plural, *Paramecia* (par-uh-MEE-see-uh). In Fig. 19-4, you can see how cilia cover the entire outside of its body. They are not just around the edge as they may seem to be in Fig. 19-4. They beat the water like little oars sending *Paramecium* through the water.

If you look at a live *Paramecium*, you will see that there are definite front and back ends of the cell. Its body is flexible, but it keeps a more or less permanent shape. It does not change form as ameba does. Along one side of the cell is a *mouth opening*. Cilia send water that contains bacteria and other small bits of food through this mouth opening into the cytoplasm. In the cytoplasm, bits of food collect into a little ball with a membrane around it. This is called a *food vacuole.* When it becomes full of food the food vacuole breaks away from the base of the mouth opening, and it is carried slowly around the cell by the moving cytoplasm. Several food vacuoles may be present in the cell at the same time. The food in each vacuole is digested and absorbed by the cytoplasm and materials that cannot be digested are forced out of the cell through a weak spot in the cell covering, not far behind the mouth opening.

ciliate

cilia

Paramecium

food vacuole

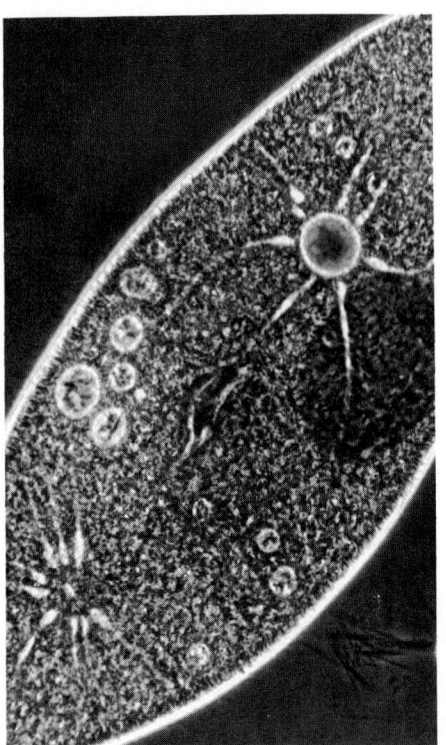

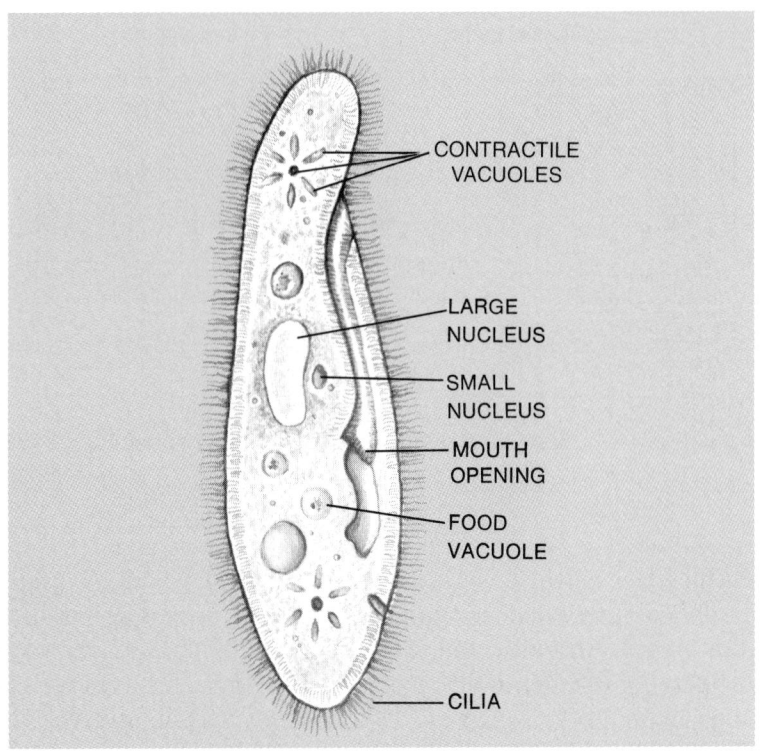

Fig. 19-4

Left: An actual photograph of a magnified Paramecium; Right: A diagram showing the parts of Paramecium. (Walter Dawn)

Something else to look for in *Paramecium* is a star-shaped set of vacuoles near each end of the cell. These vacuoles act like pumps to get rid of extra water that comes in through the mouth opening and the cell membrane. Many protozoa have vacuoles of this general type. As these vacuoles fill with water they get larger and larger. Then one of them contracts suddenly and the water is forced out through an opening in the cell covering. For this reason these are called *contractile vacuoles*. In *Paramecium*, one vacuole contracts while the other vacuole fills with water. Then the second vacuole contracts while the first takes on water again.

Paramecium is larger than many other single cells. If you hold up a drop of water containing several *Paramecia*, you can just barely see them moving around. They look like tiny white specks in the water. Most other cells cannot be seen without a microscope. When you examine *Paramecium* under a microscope you soon see that it is not as simple as ameba. Besides the things we have mentioned, *Paramecium* has more than one nucleus. One is a small nucleus, which contains chromosomes and seems to contol reproduction and heredity. The other is a very large nucleus containing many sets of the same chromosomes. These most likely control the growth and activities of the *Paramecium* cell.

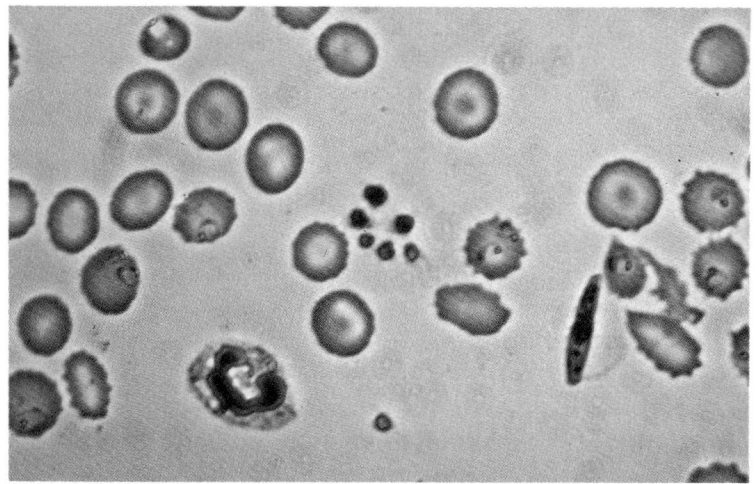

Fig. 19-5
*Malaria parasites in red blood cells.
What other diseases are caused by protozoa? (Eric V. Grave/Photo Researchers, Inc.)*

Paramecium and its relatives are just about as complicated as it is possible to get and still exist as one cell. The larger, more complex living things all have bodies that are made up of many cells.

Spore-forming protozoa. The fourth group of protozoa is made up of the *spore-formers*. Members of this group cannot move about, the way other protozoa do. They receive their name from the fact that they almost always form spores. A cell divides into many tiny cells, which can be scattered to spread the species into new environments. The spore-formers are all parasites that live in animals. The best known are the ones that cause malaria in people. See Fig. 19-5. As you probably know, they are carried by certain mosquitoes, so their lack of ability to swim does not matter. The mosquitoes carry these parasites from one host to another.

spore-formers

Enrichment: *Paramecia* also have a defense response. Just inside the outer membrane are structures called *trichocysts*, which are long thread-like strands of protoplasm. They are expelled when capturing prey and for defense. In some species the trichocysts contain poison.

Enrichment: The malaria protozoan attacks red blood cells. At regular intervals the blood cells burst as spores of the protozoan and waste products are released into the bloodstream. These substances cause chills, high fever, and sweating.

cyst

CYST FORMATION

Since most protozoa live in water they have a problem when the water dries up. How can they live until rain comes and fills the pond again? Many protozoa in various phyla solve this problem by forming a heavy wall around themselves. This is called a *cyst*. A protozoan inside a cyst cannot take food, nor can it be very active in any other way. The cyst keeps the cell from drying out. When the cyst has water around it again, the protozoan breaks out and resumes an active life.

IMPORTANCE OF PROTOZOA

Some protozoa form limestone and oil deposits in the bottom of the sea. Layers like this, which were once on the sea floor, have now become part of the dry land. We get oil and building stone from them. But when we speak of the importance of a group of living things we do not mean just their importance to humans. We are interested in how they fit into the whole living community. What is their effect on all the living things around them?

We do not really know how important the protozoa are. They probably are links in a great many food chains. They are common in all of the oceans, streams, ponds, and lakes. There are a great many of them living in damp soil. They feed on bacteria, algae, small bits of plant material, and smaller protozoa. They, in turn, are eaten by larger water animals. The fish that you eat may have eaten something that ate protozoa. This puts you in the same food chain with these tiny living things.

A few harmful human diseases are caused by parasitic protozoa including malaria, African sleeping sickness, and amebic dysentery. Besides these dangerous protozoa, there are many others that may live in humans without doing any harm at all. One type of protozoan lives in the intestine of termites. These tiny protists help to digest the woody material that termites eat. Termites would not be able to digest wood without the help of these protozoa. Certain other protozoa and bacteria live in the stomachs or intestines of plant-eating mammals. They help these animals digest plant foods.

Fig. 19-6

Vorticella (left) and Stentor (right). These are two of the many types of protozoa found in the water. (Courtesy Carolina Biological Supply Company)

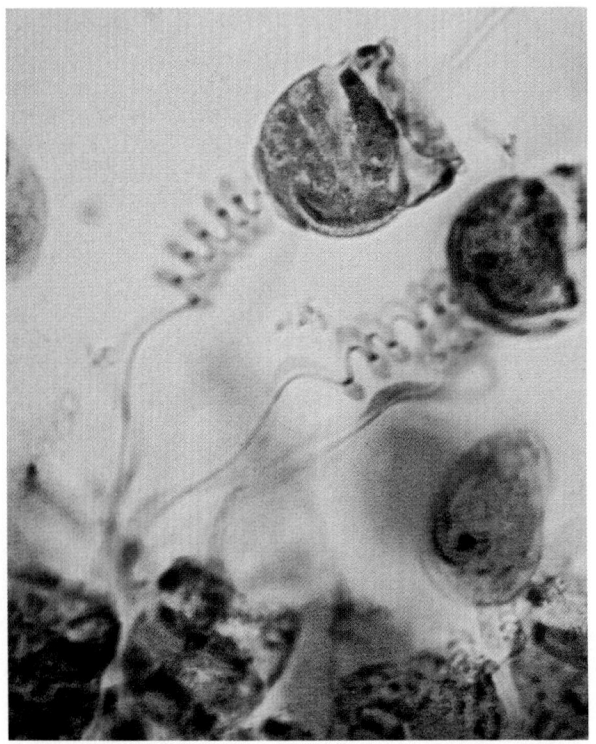
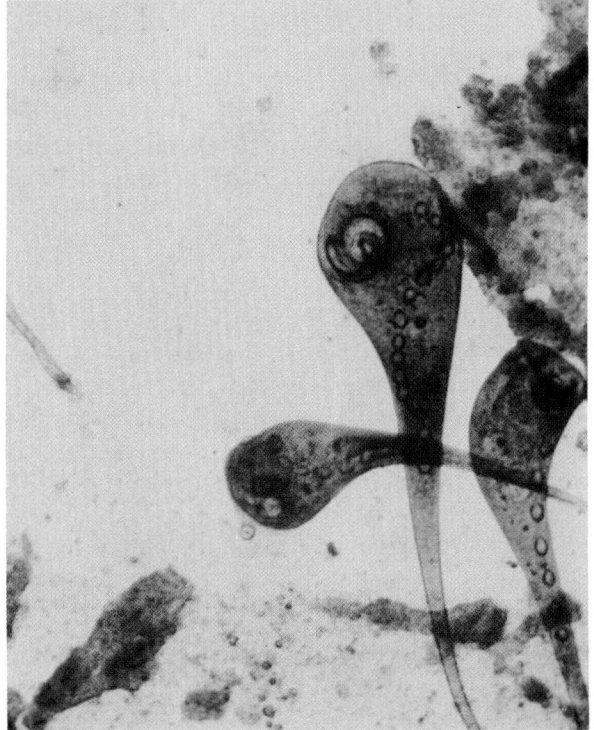

SUMMARY

Protozoa are single-celled, animal-like protists. They are common in the ocean, in fresh water, and in damp soil. Their principal food is bacteria. Food is taken into the cell and digested there.

The four phyla of protozoa are:

1. *Flagellates*. These have all of the cell structures found in the higher forms of life, such as nucleus, mitochondria, vacuoles, and chloroplasts. Example: *Euglena*.

2. *Ameba-like protozoa*. These use a flowing movement of the protoplasm to take in their food. Example: ameba.

3. *Ciliates*. These have a very complicated cell structure and a body covered with cilia. Example: *Paramecium*.

4. *Spore-forming protozoa*. These have no method of movement and live as parasites. Example: malaria parasite.

ACTIVITY

A Pond Culture

In this activity you will observe a population of living protozoa.

A. Take drops of water from the surface film of some pond culture and examine them under the microscope. You should see many different protozoa. You may also see *rotifers*. These are small multi-celled animals. There may also be some small *roundworms* and *segmented worms*. These are described in later chapters. Tiny *crustaceans* may also appear.

As you see, a pond culture is a whole community of living things. You may observe this community over a period of time and keep track of the succession that takes place in it.

1. What is the food supply in this pond culture community?

2. What are the first consumers that use this food?

3. What eats them?

4. What changes take place in the populations of living things during the next several weeks?

5. Does this community ever become balanced? If so, what are the producers in the balanced (climax) community?

Suggestion: Protozoans can be collected from ponds and quiet streams.

Suggestion: Certain protozoans are best seen under dark-field illumination. In order to achieve this without an expensive microscope, turn the mirror enough to get normal illumination and then turn the outer edge toward your face. The field will become dark and the protozoans will appear clearly in view.

Word Quiz

Choose the number from **Column B** that best matches each letter in **Column A.**
Do not write in this book.

Column A	Column B
a. flagellates	**1.** a common example of the flagellate group
b. ciliates	**2.** the place where digestion takes place in many protozoa
c. spore formers	**3.** a phylum of protozoa that swim with long whip-like threads sticking out of the cell
d. *Euglena*	
e. *Paramecium*	**4.** short, hair-like threads sticking out of a cell
f. cilia	**5.** organisms that swim by using short hair-like threads
g. food vacuole	**6.** a small round case that holds an inactive protozoan
h. cyst	**7.** protozoa that are always parasites
	8. a common protozoan belonging to the ciliate phylum

Check Your Facts

Science
Reading Skills
1. *Compare and Contrast*
2. *Reading Illustrations*

1. List the four general types of protozoa and describe what each is like.

2. Make drawings of ameba, *Euglena,* and *Paramecium.* What characteristics are different in each of these three cells?

3. We might call the ciliates and the amebas single-celled animals. Why not call *Euglena* an animal?

4. What are some diseases caused by protozoa?

5. Of what importance are protozoa in the environment?

6. Would *Euglena* swim toward or away from light? Why?

Thought Questions

1. Why is it generally believed that single-celled life was present on the earth before the larger, multi-celled types appeared?

2. How would it be possible to classify the flagellates as protozoa, as algae, and as fungi?

CHAPTER 20 Algae—The Plant-like Protists

Kelp is the largest seaweed. A single alga in an underwater bed of kelp can be over 30 meters long.

The term, *alga*, is not used in scientific classification. It is a common word used for any type of organism having chlorophyll and a simple body structure. Calling two things algae does not mean they are closely related. You have seen how the blue-green algae are related to the bacteria. They have no nucleus and no chloroplasts. Four phyla of algae do have nuclei and chloroplasts. You will learn about them in this chapter.

We are listing these algae as members of the Protist Kingdom. The simplest, single-celled types are easily thought of as protists, but the large, multi-cellular types could be called plants and are sometimes classified that way.

TYPES OF ALGAE

Green algae. The members of this phylum are the most common in freshwater and can be found in salt water also. Their cell structure and their type of chlorophyll are the same as those found in the higher land plants. It is thought that the first members of the Plant Kingdom developed from green algae long ago.

An individual green alga may be formed of a single swimming cell. Such algae look very much like flagellates. Some single-celled green algae cannot swim. *Chlorella* is an example. Some green algae form small clumps of cells. Still others form long rows of cells. *Spirogyra* is one of these thread-like types. Various green algae are shown in Fig. 20-1.

The thread-like types of green algae may be many centimeters long, but they are only one cell thick. You would need a microscope to study all the green algae described so far, but others are larger. *Sea lettuce* forms broad sheets of green tissue. It can often be seen growing on rocks in the edge of the sea. See Fig.

objectives

After you read this chapter, you should be able to:

__**Explain** why algae are sometimes classified as plants and sometimes as protists

__**Describe** the types of organisms found in each phylum of algae

__**Discuss** the worldwide importance of algae

Enrichment: Most of the approximately 30 thousand species of algae are marine. Algal growth can be found on trees and even on snow. One type of algae thrives in the 85°C (185°F) temperature of the pools and springs of Yellowstone National Park.

Science Reading Skill: *Compare and contrast* the blue-green algae with the algae of the Protist Kingdom.

Enrichment: The green algae are members of the phylum Chlorophyta.

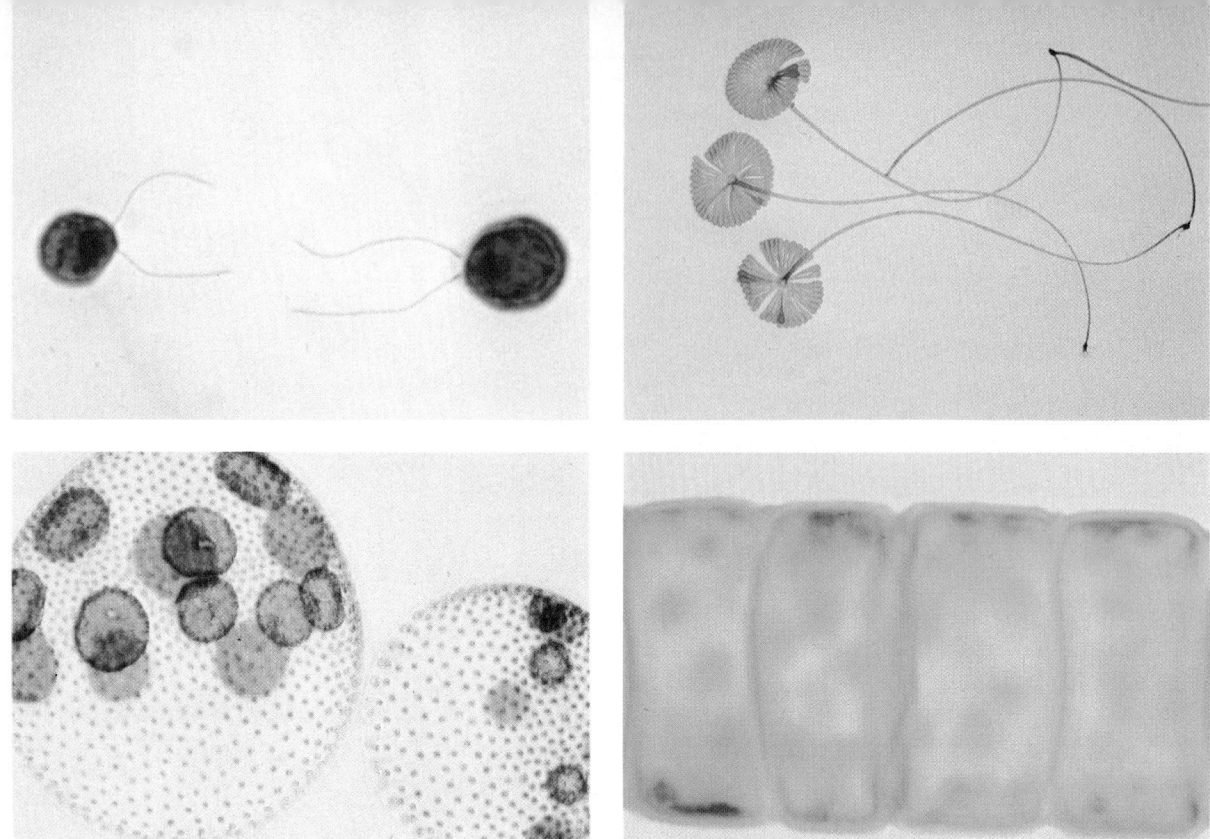

Fig. 20-1

Green algae. Top left: A single-celled swimming alga; Top right: A multi-cellular nonswimming alga; Bottom left: Algal cells growing in clumps; Bottom right: Thread-like algae. (All photos courtesy Carolina Biological Supply Company)

silica

Enrichment: The members of this group belong to the phylum Chrysophyta.

20-2. *Stoneworts* have the most complicated structure of any algae. They are usually about 10 centimeters high, and they grow in freshwater lakes. They have root-like, stem-like, and seed-like structures, but there is no vascular tissue. Lime collects on the surface of the stonewort. When the algae die this mineral is deposited on the lake bottom. There it forms deep layers of limey mud. See Fig. 20-3.

Yellow-green algae, golden-brown algae, and diatoms. These algae have large amounts of yellow-colored pigments mixed with the chlorophyll in their chloroplasts. Most of them have at least some *silica* in their cell walls. Silica is the mineral that makes up sand. It is also the main material in glass. Silica has a rough or *abrasive* texture.

The yellow-green and the golden-brown algae are found mainly in fresh water. A few live in the sea, and some grow in damp places on land. You can tell them from similar green algae by their color. See Fig. 20-4 on page 204.

Diatoms live in fresh and in salt water. They are the most common of all the algae that drift in the ocean. In some diatoms, the cells form into threads or clumps, but most are single-celled. They have a very curious type of cell wall formed mostly of silica and made of two separate halves. One half-wall fits over the other half-wall like the lid on a box. The protoplasm is inside this "glass-box."

The cell walls of diatoms come in many shapes. Pits and lines in their surfaces often form very beautiful designs. They have been compared to jewels, but you must use a microscope if you wish to see them. See Fig. 20-5 on page 204.

Brown algae. These algae have so much brown pigment that it often hides the chlorophyll. They look either brown or dark olive-green. Nearly all of them live in the ocean. Four types of brown algae are shown in Fig. 20-6 on page 205.

Brown algae are always multi-cellular. They grow anchored to the bottom of the ocean, and some are very large. The *giant kelp* may be over 50 meters long. It has a long tough stalk that is anchored to the bottom of the ocean. The upper end has flat, strap-like parts that carry on photosynthesis. One kind of kelp, called *gulf-weed*, floats in the warm waters of the Atlantic. It is not usually anchored in one place like other brown algae.

Since most of the brown algae are anchored to the ocean floor they cannot grow where the ocean bottom is too deep. Light for photosynthesis reaches only about 60 meters into the water. The kelps grow in the shallow waters near shore. *Rockweeds* are about 50 centimeters long and are very tough. They often cover the rocks at the edge of the ocean. At high tide they take the

Fig. 20-2
This multi-cellular green alga is known as sea lettuce. (Runk & Schoenberger/ Grant Heilman)

Enrichment: Brown algae belong to the phylum Phaeophyta.

Enrichment: Stress to students that holdfasts of marine algae are not functional roots, although they may resemble roots of land plants. Holdfasts are a means of attachment and do not provide nourishment to the algae.

Fig. 20-3
A green alga called a stonewart. Its structure is very complicated for an alga. (Runk & Schoenberger/Grant Heilman)

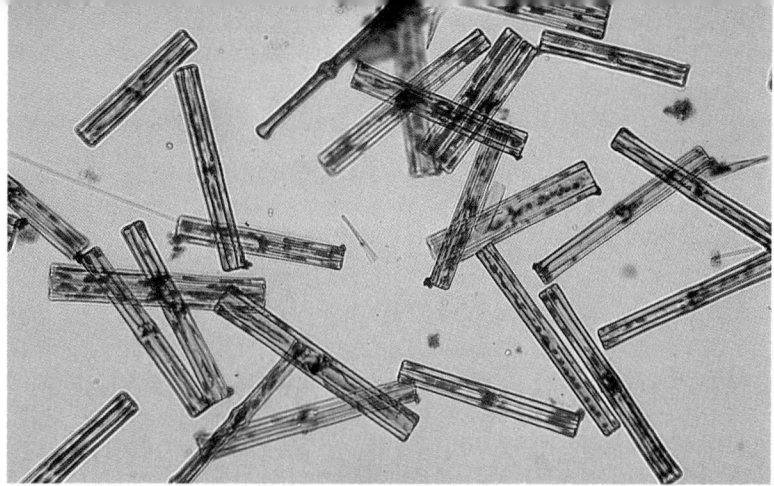

Fig. 20-4

A yellow-green alga called Tabellaria. This alga grows in filaments. (Dr. E. R. Degginger, FPSA/Bruce Coleman, Inc.)

air bladders

Enrichment: Red algae belong to the phylum Rhodophyta.

Science Reading Skill: *Classification—* Have students note the characteristics that algae of each phylum have in common.

Enrichment: Stress to students that color alone is not sufficient for purposes of classification. Structure and mode of reproduction are also involved in classification.

Fig. 20-5

Many different patterns are found in the cell walls of diatoms. Why are diatoms said to live in ''glass boxes''? (Manfred Kage/Peter Arnold, Inc.)

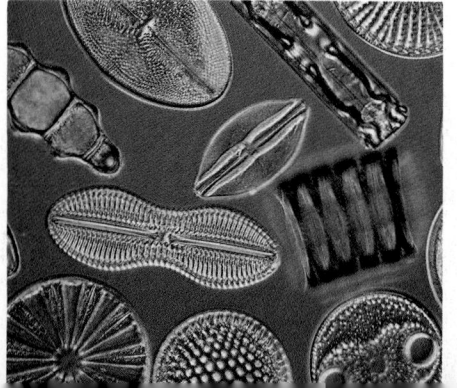

pounding of the waves. At low tide they lie exposed to the air. Tough outer layers keep them from drying out. The upper ends of rockweeds and many of the kelps have gas-filled hollow spaces, or **air bladders,** in them. These help them to float upright in the water, where they can catch the sunlight. Fig. 20-6 shows different types of brown algae.

Red algae. Because of special pigments, these algae are often red. However, some contain other pigments also that make them appear dark green, brown, or purplish-black. Most of them are multi-cellular and live in the ocean. A few small ones live in fresh water. The common ones are medium size (about 10 centimeters). They are often called *sea moss*. Red algae can survive in dimmer light than other algae. Some of them grow as much as 160 meters below the surface of the ocean.

The red algae grow anchored to the ocean bottom. You may find some of them on rocks among the rockweeds. Some red algae have a rock-like appearance. They form heavy deposits of lime among their cells. Together with corals and sponges, they help to build limestone reefs in the sea. See Fig. 20-7 and Fig. 20-8 on page 206.

REPRODUCTION OF ALGAE

Algae reproduce in many different ways. The single-celled types reproduce by *cell division*. The multi-cellular ones often use *vegetative reproduction*. Most groups produce *asexual spores*. Many of these spores swim by means of flagella. Some simply drift in the water.

You may recall how *Spirogyra* reproduces sexually by a union of ordinary alga cells. Some algae produce equal-sized *swimming* sex cells. Some produce sex cells that crawl like amebas to find each other and unite. Many algae produce *sperm and eggs* in much the same way as the hydra does. The golden-brown algae are not known to have any kind of sexual reproduction.

IMPORTANCE OF THE ALGAE

Algae are the producers in the sea communities. How important does this make them? The ocean covers nearly three-fourths of the earth's surface. Algae are the food makers of this vast area. Many small animals, such as shrimp and their relatives, eat simple algae. Fish eat the shrimp, and larger fish eat the smaller fish. In this way, the entire ocean community is fed directly or indirectly by algae. Photosynthesis carried on by algae also adds to the world's oxygen supply. Scientists think that algae may produce 50 percent of the oxygen found in the atmosphere.

Not all parts of the sea are equally productive. Shallower waters near shore contain a good supply of dissolved minerals. These minerals are needed for growth. They include nitrates, phosphates, and potassium salts. These are the same minerals that land plants get from soil and fertilizers. These minerals wash into the ocean from the nearby land, and they are used by the algae. Algae are the food-makers of the ocean food chains, so these off-shore communities are often very rich in marine life.

In the ocean, far from land, the tiny free-drifting diatoms and other algae often use up all available minerals. When they die

Enrichment: Some algae live inside other organisms. One single-celled species lives inside certain paramecia. Usually, algae do not harm the organisms in which they live and in fact help by supplying food via photosynthesis.

Reference: See "Where Would We Be Without Algae" *National Geographic*, March, 1974.

Fig. 20-6

Brown algae. Top left: Devil's apron; Top right: Giant kelp (Dr. E. R. Degginger); Bottom left: Gulf weed; Bottom right: Rockweed. (Courtesy Carolina Biological Supply Company)

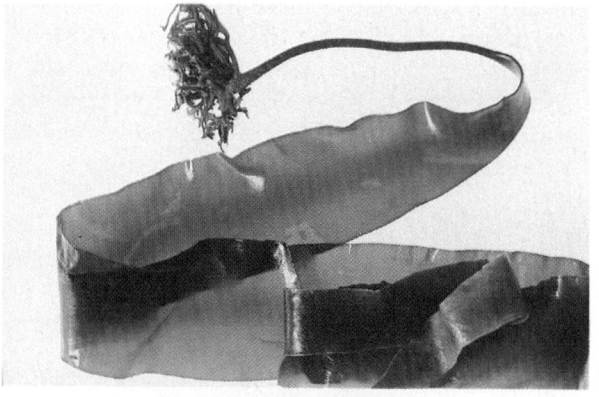
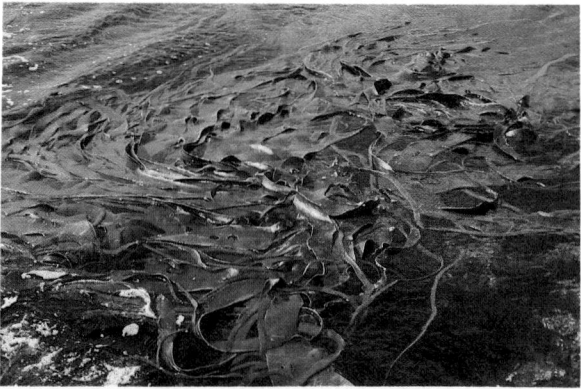

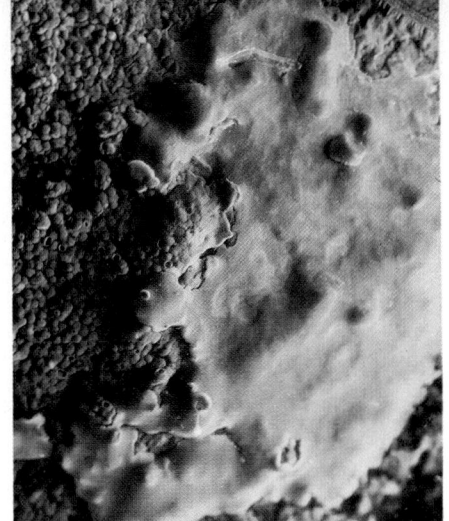

Fig. 20-7
A red alga that helps corals to build reefs. (William H. Amos/Bruce Coleman, Inc.)

diatomaceous earth

Fig. 20-8
A coral reef that has been built from red algae and minerals. (Dr. E. R. Degginger)

and sink to the bottom of the ocean they take the minerals in their bodies with them. They may be eaten by larger animals before they sink, but these larger animals will eventually die and sink also. Therefore, the bottom mud contains minerals, but the surface layers of the ocean are depleted, just as soil becomes depleted on land. Only the surface layers of the ocean have light for photosynthesis. For these reasons food production is no better in most of the ocean than it is in a desert on land.

In certain places ocean currents force deep water upward to the surface. This brings a rich mineral supply into the sunny top layers where algae can grow. These areas are very rich in sea life. Many fish are caught there and used for human food.

The uses of algae. The importance of algae is hard for us to realize because we live on land. Seafood is used by people in many places. The food material in the fish, lobsters, oysters, and shrimp, which are eaten by so many people, was first produced by photosynthesis in the algae of the sea. Countries such as Norway, Japan, and Greece depend very heavily on seafoods to feed their people. They have much sea coast but little farmland, so to them the sea is an important source of food.

Certain algae are eaten directly by people. In the past, a jelly-like material taken from a red alga, called *Irish moss*, was used as a dessert. Now it is more often dried and added to things like ice cream, cosmetics, and medicines to give them a smooth texture. You have almost surely eaten Irish moss in some food. See Fig. 20-9. *Agar* is another jelly gathered from red algae. It is used in many of the same ways as Irish moss. It is also very useful as a base for growing bacteria cultures in the laboratory. Iodine can be obtained from kelp. Kelp is also often spread on land as fertilizer.

Both red and brown algae are fed to cattle and are eaten by people in many countries of Asia. They add important amounts of vitamins and minerals to the diet. In some cases they are deliberately grown and harvested. This form of underwater farming may become much more common in the future.

Also, in the future, we may use algae like *Chlorella* for food, either for cattle or for humans. These single-celled algae can grow very rapidly in tanks or glass tubes under the right conditions. Then they can be filtered out of the water and packed into solid cakes. Scientists are studying such new sources of food because the human population is growing so rapidly.

Diatoms are very useful also. The silica found in the walls of diatoms does not decay easily. For this reason the walls of dead diatoms have collected on the floor of the ocean in large numbers. These make up a substance called *diatomaceous* (die-ah-toe-MAY-shus) *earth*. Since silica has a rough texture, this

diatomaceous earth is used in the manufacturing of silver polish, cleansers, toothpaste, air and water filters, and in lightweight concrete and brick.

Some troublesome algae. Some algae become a nuisance. They form heavy growths in reservoirs where cities store their drinking water. And certain golden-brown algae, especially, give the water an unpleasant taste and odor. Algae may also grow in swimming pools, forming a scum on the walls. When this happens, a chemical called copper sulfate can be added to the water to kill the algae. It is not good to use copper sulfate in lakes or ponds, however, because it poisons fish. Algae can also become a nuisance in aquariums. There can be so many algae cells that the water turns green. Other algae grow on the sides, so that you cannot see through the glass. This will not happen if the aquarium is kept in dim light.

Fig. 20-9
Left: A red alga known as Irish moss; Right: A dish of ice cream. What do these have in common? (Courtesy Carolina Biological Supply Company; HRW photo by Ken Karp)

Enrichment: Diatomaceous earth is also used as a filtering agent for gasoline and sugar and as insulation around steam pipes.

Enrichment: There is a type of algae that causes "red tides" that sometimes occur in warm ocean waters. Water conditions may favor a population explosion of this algae. Cells of the algae contain a pigment that is poisonous to fish and other sea animals.

SUMMARY

Algae with nuclei in their cells are placed in four phyla: 1. green algae, 2. yellow-green algae, golden-brown algae, and diatoms, 3. brown algae, 4. red algae.

The structure of algae varies from single-celled types to giant, multi-cellular ones. They are the food-makers of the sea. All other sea life depends on them. Algae also supply some of the food in freshwater communities.

People eat seafood (animals whose food come from algae). From algae they also get edible algae, Irish moss, agar, animal feed, fertilizer, iodine, and diatomaceous earth.

Activity

Studying Algae

You can get some firsthand experience with algae in the following activity.

A. Find some algae and bring samples to school. Look in ponds, lakes, streams, or ditches. Look also for green color on tree trunks. If you live near the ocean, collect some of the larger algae such as rockweeds and kelp. Examine all of these algae. Use a microscope for small ones.

1. Make a drawing of each kind to record what you see. Write names under all of those that you are able to identify from reference books.

B. Place a small amount of silver polish on a microscope slide. Add a drop of water to spread it out into a thin film. Cover with a coverslip. Observe it through, first, low power and then, high power. The silver cream contains diatomaceous earth. You should be able to see many pieces of diatom cell walls. Try to find some unbroken ones.

2. How many kind of diatoms are in the silver polish? Why are they there?

3. When and where did these diatoms live?

4. The silica in diatoms is very hard. Why does it not scratch the silver?

Word Quiz

Write the meaning of each of these terms. Do not write in this book.

silica   air bladders   diatomaceous earth

Check Your Facts

Science Reading Skill 1. Compare and Contrast

1. How are algae different from land plants?
2. Name each of the four phyla of algae. What do the algae in each phyla look like?
3. What is the general importance of algae?
4. In what ways may algae be used by human beings?
5. How do some algae become pests?

Thought Questions

Science Reading Skills 1. & 2. Problem Solving

1. Could people farm the sea bottom? What are some of the problems they would have to solve if they tried it?
2. Giant kelp resemble land plants in many ways. They even contain food-carrying vascular tissue. Why have they never developed water-carrying vascular tissue?

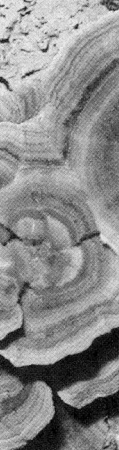

CHAPTER 21   Fungi—The Decomposer Protists

The fungi shown above are growing on dead wood. Fungi grow in many other places also. The sudden appearance and disappearance of mushrooms and other fungi has puzzled people. You will understand these organisms better when you have studied this chapter.

The *fungi* (FUN-ji); singular, fungus (FUN-gus), make up an interesting phylum of protists. They have no chlorophyll. Most are decomposers that cause decay. Some are parasites. Some fungi are single-celled and some are multi-cellular. Some grow in water, but most are found on land. They include yeasts, molds, rusts, smuts, mushrooms, and many others.

In getting food, fungi first give off digestive juices to break down the food substances. Then the dissolved food molecules diffuse through their cell membranes. Parasitic fungi cause many diseases in plants and some in animals.

THE YEASTS

Yeast is a single-celled fungus. The cells look like those in Fig. 21-1 on page 210, and they reproduce by *budding*. First, the nucleus divides by mitosis. Then one nucleus moves out against the cell membrane. The cell wall bulges outward, and the nucleus moves into the bulge. Slowly the bulge, or the bud, grows and becomes a new cell. When yeasts are growing rapidly there may be more than one bud on a cell at a time, and buds may grow on buds. This results in a multi-cellular appearance, but soon the cells separate and become single-celled yeast plants again. Yeasts also reproduce by means of spores.

The main energy food of yeasts is sugar. They break it down as we do, producing carbon dioxide and water as wastes. Oxygen is needed in this form of respiration. Yeasts can also break down sugar when oxygen is not present. The products of this

objectives

After you read this chapter, you should be able to:

__List the three uses for yeasts

__Describe the way in which molds grow

__Explain the unusual structure of a lichen

__Explain how fungi obtain their food

fungi

Enrichment: The classes of fungi discussed in this chapter belong to the phylum Eumycophyta. You may also wish to mention the phylum Myxomycophyta, which includes slime molds.

Enrichment: More than 250 thousand kinds of fungi have been identified.

Reference: See *The Wonders of Fungi* by Lucy Kavaler.

Enrichment: Yeasts belong to the class Ascomycetes.

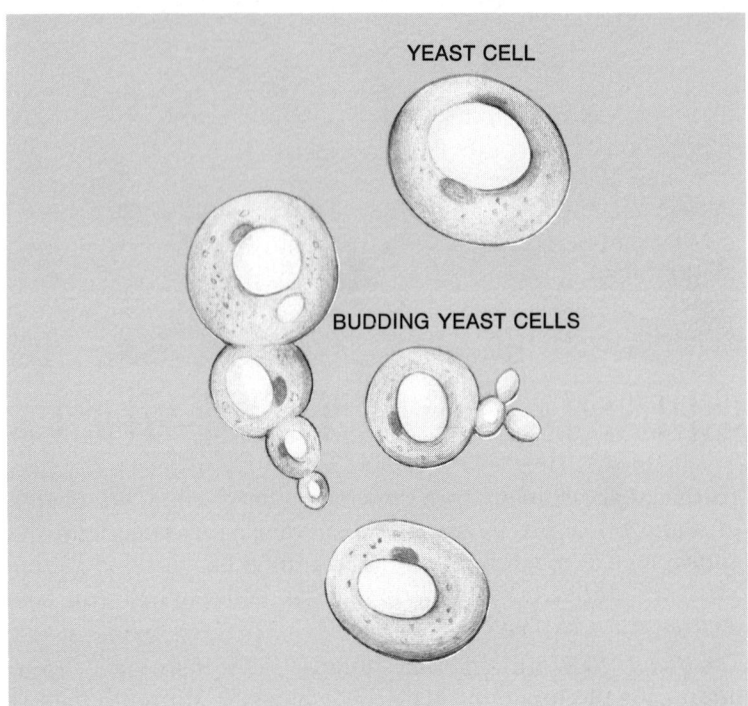

YEAST CELL

BUDDING YEAST CELLS

Fig. 21-1
*Enlarged cells of ordinary baker's yeast.
Why are some of the cells larger than
others?*

fermentation

Enrichment: Wine yeasts can form a
high alcohol content of up to 14
percent.

Enrichment: Molds belong to the class
Phycomycetes.

mold

process are carbon dioxide and *alcohol*. Molecules are changed
to release energy without using oxygen. This process is called
fermentation.

Yeast is useful both for the carbon dioxide and for the alcohol
it produces. When bread is made, yeast is added to the dough.
Because the bread dough is gummy, the carbon dioxide produced by the yeast cannot escape. It collects in bubbles, causing
the dough to swell up, or "rise." The alcohol in the yeast escapes into the air during the baking process.

Alcohol produced by yeast plants may be used either as industrial alcohol or as beverage alcohol. Alcohol is one of the
most important chemicals used in industry. Large amounts of
it can be produced when yeast is used to ferment molasses.
However, much industrial alcohol today is made synthetically
without yeast. Beverage alcohol is made by allowing yeasts to
ferment fruit juices to produce wine, or grain products to produce beer. The more concentrated drinks, like whiskey, are
produced from fermented grain mixtures by a process called
distillation. This process is used to concentrate the alcohol.

MOLDS

The word *mold* is used for any fungus that has a thread-like
form of growth. Molds have a fuzzy appearance. A common

type that grows on bread develops from spores that drift in the air. If a spore falls on damp bread, it grows by sending out branching threads. Each thread is surrounded by a cell membrane and cell wall. It contains cytoplasm and many nuclei. These threads grow down into the bread and produce digestive enzymes that change the bread into a liquid form. This liquid is then absorbed by the growing mold.

Meanwhile, other threads of the mold grow across the surface of the bread. At their ends, a new set of branching threads grows into the bread, and so the mold spreads. See Fig. 21-2.

Some threads grow up into the air. The ends of these threads swell up and form little balls. These are the spore cases. Cell division inside these cases produces spores. Large numbers of these spores are carried away by air currents.

The threads of bread mold are white. Its spore cases are black, so the moldy bread has a gray appearance. There are many other molds besides bread mold. Some are green, but not because they contain chlorophyll. Others are orange, yellow, pink, or black. Some produce their spores in long chains, like strings of beads. Some molds live in the water. See Fig. 21-3.

Importance of molds. Molds cause decay, like bacteria. They grow on bread, fruit, cheese, or other foods that we eat. They also grow on leather and cloth, and even on damp wood. Things will not mold if they are kept dry. Bakers put a chemical in bread to slow down mold growth, but in time mold appears anyway. As decay producers the molds are less important than the bacteria.

Certain cheeses, such as Roquefort and Camembert, get their special flavor from harmless molds that grow in them. You probably know what a useful drug *penicillin* is, but did you know that it is made from green mold called *penicillium* (pen-uh-SILL-ee-um)? This mold produces penicillin to kill the bacteria that compete with it for food. We use penicillin to kill certain parasitic bacteria that have entered our bodies. For this reason penicillin is called an ***antibiotic*** (an-ti-by-OT-ik). We now get a number of antibiotics from both molds and bacteria.

THE HIGHER FUNGI

The higher fungi are larger and more complex than the molds. They include the mushrooms, puffballs, and shelf fungi. Some are parasitic on living trees, but most of them use dead materials for food. They cause decay of dead leaves and wood in forests. Fungi are the major decomposers in the forests because forest soils are often too acid for bacteria to grow well.

In the rotting leaves on the forest floor the strands of fungus tissue digest and absorb their food. Fig. 21-4 shows such a

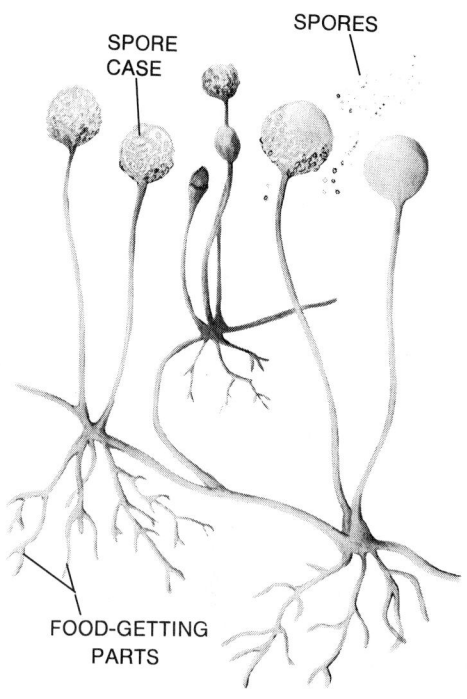

SPORE CASE

SPORES

FOOD-GETTING PARTS

Fig. 21-2
A view of what magnified bread mold looks like.

Activity: Have students grow fungi by placing food in covered jars. Use fresh fruit, cheese, crushed grapes, dried beans in water, etc.

antibiotic

Fig. 21-3
A typical water mold growing on a seed floating in the water. (Carolina Biological Supply Company)

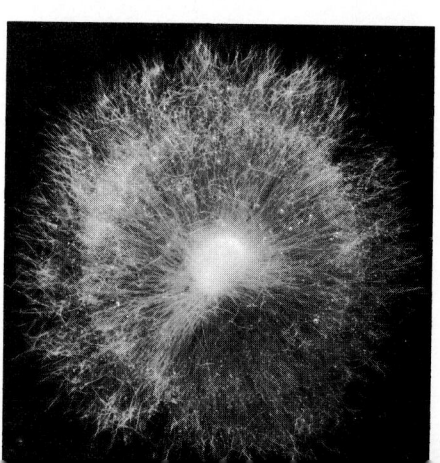

fungus mass. When this fungus plant has grown strong enough its cells multiply and form the *fruiting body*. This is the spore-producing organ of the fungus. One fruiting body you are familiar with is the common mushroom. Spores are formed on the underside of the mushroom cap. When the spores have been shed, the fruiting body dies, but the main part of the mushroom plant goes on living below the ground. It may send up more fruiting bodies another year. Often mushrooms appear in the same spot on a lawn each year where some old root or log lies buried. This is the food of the hidden fungus body.

The mushrooms that you can buy in the grocery store are the fruiting bodies of common field mushrooms. They have been grown in sheds or caves by experts. Some wild mushrooms are good to eat, but others are poisonous. Some of these poisonous mushrooms will merely make you sick, but others can kill. The death angel, see Fig. 21-5, is so deadly that one mushroom can kill a whole family. It is dangerous to eat wild mushrooms because some of the poisonous species look almost exactly like the safe species. Even scientists who were mushroom experts have made mistakes. One mistake is all you can make.

You may have seen fungi that looked like shelves growing on the side of trees or logs. These are the fruiting bodies of the *shelf fungi*. See Fig. 21-6. They cause wood decay but they are

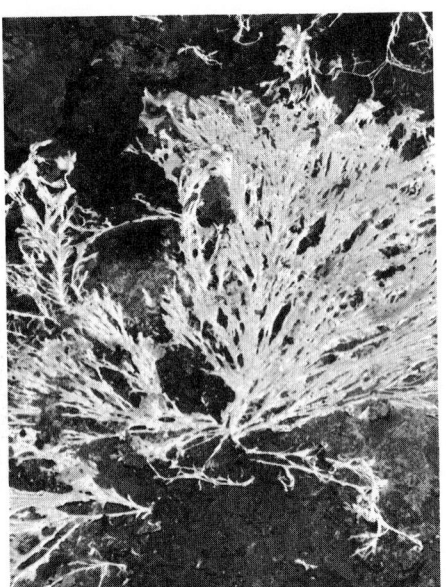

Fig. 21-4
The main body of a mushroom growing underground. Later this may produce mushrooms that grow up above the ground and give off spores. (Harold E. Teter)

fruiting body

shelf fungus

Enrichment: Toadstool is a word commonly used to describe a poisonous mushroom.

Enrichment: These are members of the class Basidiomycetes.

Fig. 21-5
The death angel mushroom. It looks pretty growing in the woods, but it can kill an entire family. (Dr. E. R. Degginger)

usually not parasitic. The wood in the inner part of a tree is dead. Fungi often rot this inner wood without damaging the living part of the tree around the outside.

Puffballs are the globe-shaped fruiting bodies of still other fungi. Some are small, but others may grow as big as your head. They look like very large white marshmallows lying on the grass. When these puffballs ripen, they turn darker and the entire inside becomes filled with spores. If you should happen to kick such a ripe puffball, a cloud of spores would rise into the air like dust or smoke.

Many other types of fruiting bodies are formed by fungi. Fig. 21-7 shows some of them.

DISEASES CAUSED BY FUNGI

A few kinds of fungi attack humans. There is an infection known as *ringworm* that is not caused by a worm at all; it is a fungus infection. One type of ringworm is athlete's foot. It is called this because people usually pick up the fungus by walking on damp locker room floors with bare feet. Athlete's foot causes itching and cracking of the skin and can leave it open to other types of infection. Fungi can also attack other parts of the body such as the lungs or the ears. Some of the diseases caused by fungi can be very serious.

Most parasitic fungi attack plants. Rusts, smuts, and blights are mold-like fungi that grow in or on plants. Their threads grow into the tissues of the host plant and use its protoplasm as food. The results can be very serious. The American chestnut used to be an important forest tree in the eastern part of our

Fig. 21-6
These are shelf fungi growing on a tree in a forest. Where is the feeding part of this fungus? (HRW photo by Russell Dian)

Enrichment: Disease-producing fungi belong to all classes of fungi.

Fig. 21-7
The fruiting bodies of four different types of spore-producing fungi. Have you ever seen any of these? (Harold E. Teter; Courtesy Carolina Biological Supply Company; HRW photo by Richard Weiss; Harold E. Teter)

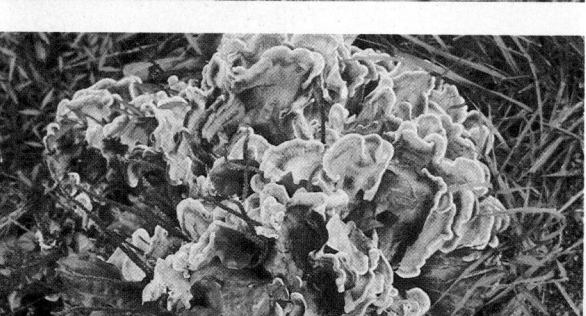

country. A fungus blight wiped out this valuable lumber species. Now most of our American elm trees are dying from damage done by Dutch elm disease, which is also caused by a fungus. This disease is carried from tree to tree by beetles.

The potato blight in Ireland in the 1840s completely destroyed the valuable potato crop. Many Irish people starved to death, and many others fled to America. The potato blight is still a problem, but we now know how to protect the plants by spraying them.

Another serious fungus disease is *wheat rust*. It is called a rust because some of the spores are a rust-red color. They break out through the surface of the wheat plants, giving them a rusty look. Plant breeders use artificial selection to produce new types of wheat that will not be damaged by the rust.

THE LICHENS

Out in the country you often see flat, silvery-green growths on tree trunks, posts, or rocks. These are one kind of **lichen** (LIE-ken). A lichen has such a simple structure you might think that it belongs to one of the algae groups. You would be only half right. A lichen appears to be a single plant, yet it is actually made up of a fungus and an alga growing together. The threads of the fungus give the lichen its shape. The cells of the alga are mixed in among these fungus threads, giving the lichen its green color. The fungus absorbs and holds water. This allows the alga to grow where it could not ordinarily survive. The fungus gets food from the alga. This is an example of very close cooperation between two living things.

One type of lichen is a very important food for the caribou herds of Canada. Since caribou are a kind of reindeer, this silvery lichen is often called "reindeer moss." See Fig. 21-8.

Lichens are often the first things to grow on bare rock surfaces. They can even be found growing on tombstones! Their fungus threads produce weak acids, in much the same way that plant roots do. These acids cause the rock surface to slowly crumble and to release minerals. The crumbling rock mixes

Science Reading Skill: *Classification*—compare the similarities and differences of the various groups of fungi.

lichen

Enrichment: You may want to introduce the term mutualism.

Fig. 21-8

Left: Lichens growing on a tree trunk. Lichens can be green, orange, red, or other colors; Right: Reindeer moss. This lichen is important food for caribou. (Harold E. Teter; Manuel Rodriguez)

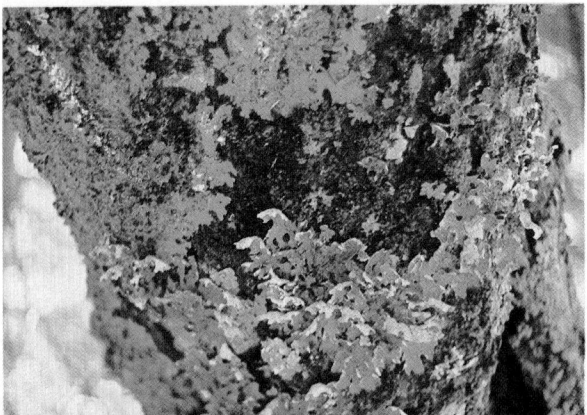

with humus from dead lichens to form the beginnings of soil. Small plants grow on this and produce more soil, so that still larger plants can grow. In this way lichens are often the pioneers that start the succession from bare rock to plant-covered land.

Science Reading Skill: *Sequencing*— trace the steps in the formation of soil, starting from the effect of lichens on rock.

SUMMARY

Fungi are nongreen protists having one, or many cells. Most are decomposers. Some are parasites. Higher fungi include mushrooms, shelf fungi, and puffballs.

Yeasts ferment sugar, forming alcohol and carbon dioxide. Molds are thread-like decomposers or parasites.

Lichens are composed of fungi and algae living in such close cooperation that they appear to be single organisms.

Some fungi may be eaten. Wild ones are dangerous, because poisonous ones can be mistaken for edible ones. Certain molds are used in producing some cheeses. Others produce useful antibiotics.

ACTIVITY

Studying Molds

A. Wet a slice of bread. Place it in a glass dish at least 6 centimeters deep. Leave it uncovered for half an hour. Then cover the dish with a glass plate (any handy closed container will do). Set the dish aside for several days in a warm, dark place and see if any mold grows on the food.

1. How many kinds of mold grow?

2. Can you see a center point from which each patch of mold grew outward? How does mold start at these points?

3. What does this tell us about the air we breathe?

B. With tweezers, pick tufts of each kind of mold and examine them under the microscope.

4. Can you see the spores? Are they produced the same way by each kind of mold?

5. Make a drawing to record what you see.

C. Take the mold plants from one patch and rub them lightly back and forth over the entire surface of a slice of wet bread in a dish. Cover the dish and look at it a week later.

6. This time, how many types of mold are growing on the food?

7. Explain what has happened.

Word Quiz

From the new words you learned in this chapter, write the word that correctly completes each statement. Do not write in this book.

Members of a large phylum of protists that live as decomposers are called _____. Any members of this group with a thread-like, fuzzy appearance are called _____. Some of these produce useful drugs that will kill bacteria in humans. Such drugs are called _____. The spore-producing part of a fungus is called a(n) _____. A fungus that grows like a shelf on the side of a tree or log is called a(n) _____. An organism composed of a fungus and alga growing together is a _____. The process of getting energy by changing sugar to alcohol is _____.

Check Your Facts

1. What are three industries that use yeast plants? How is the yeast used in each case?
2. How does a mold begin to grow in a new location?
3. Give an example of a case in which molds are a nuisance. Describe another case in which molds are useful.
4. What is the function of the part of a mushroom that you see above ground?
5. How do fungi obtain their food?
6. What are lichens? Where do they grow?

Thought Questions

1. Where is the best place to get mushrooms we can eat?
2. One boy heard about penicillin in green molds. He ate some mold and got sick. Can you explain this?

Microbes may provide the foundation of the next food revolution.

Even the smaller elements of life form the substance of people's careers. Bacteria, algae, and protozoa are important in the food, cosmetic, linen, and leather industries. Bacteria, for example, which we often think of in relation to disease, are also a fundamental ingredient in cheesemaking. Food technicians in that industry must know the type of reaction the bacteria produce. They must also be aware of how that food interacts with other nutrients and chemicals. Many people have experienced side effects from the mixture of prescribed mood-altering drugs and the enzymes in certain cheeses. Consumer specialists in both the food and drug industry have a responsibility to alert the public to this interaction.

Algae and bacteria, a problem for both public and private swimming pool owners, create the need for the services of pool testers. These are chemical technicians who test water quality and determine the appropriate balance of chemicals. They usually work for a private business and take courses with the chemical company that sells the testing equipment,

which is generally computerized. Pool owners bring in samples of water to be tested for chlorine–pH balance and the presence of a variety of other substances. In this profession, too, the seller must be aware of chemicals in the environment that can cause negative side effects. A common complaint is that people's hair turns green while swimming. This usually happens because minerals, especially iron or copper found in the pipes carrying the water source, mix with hair dye. Knowing how to remedy this situation is important to the pool sellers' business. While pool problems are hardly the major concern of most chemical technicians, they give you an idea of how private enterprise uses the services of science-oriented specialties.

Bakers, too, are involved with chemistry and small organisms, notably yeast and other leavening agents. Both on-the-job experience and technical training in culinary institutes or cooking schools will provide the appropriate background. Some culinary institutes have been established by the large food processing

HRW: Ken Karp

companies. Retail bakeries, restaurants, and institutions all employ bakers, chefs, and cooks. There are, of course, different levels of expertise. Some people in the food preparation industry simply follow supervisory directions. Others outline the methods of preparation and storage, where taste, color, combinations of ingredients and, most important, nutrition must be considered. More and more people in the industry obtain some technical training, and food management is likely to be part of the curriculum.

Lawrence Migdale/ Photo Researchers, Inc.

phers, who hold either a master's degree or doctorate in sociology and demography, classify and follow the movement of human populations. They are employed by the government, and by universities and colleges. Actuaries, two-thirds of whom work for insurance companies, analyze statistical data on death, injury, and property and contract losses. They have a college degree in math or statistics. The insurance field is a growth industry and prospects for entry are excellent.

For more information, contact:

The American Statistical Assn.
806 15th Street NW
Washington, DC 20005

or write to:

The Society of Actuaries
208 South LaSalle Street
Chicago, IL 60604

The brewery industry (see photo) is also dependent upon the action of yeast. Here again, technicians need and develop understanding of the chemical interaction of ingredients.

Classification is an activity directed not only toward the plant and animal kingdom. Human beings undergo classification in many ways. We are counted each time the government takes a census. Census takers are usually seasonally employed by federal, state, or local governments to count individuals and households. It is primarily a part-time job requiring that you be able to read and write, have good interpersonal skills, and be willing to persevere. Census information is important in determining po-

litical representation as well as taxation. Accuracy and integrity are very important in this job. False or incorrect information provides an unstable foundation for future political decision-making.

Good interviewing skills are important to pollsters and market researchers of all kinds. These companies may advertise for part-time or seasonal help, but doing telephone interviews can be a way of getting to know a company and finding out what other jobs are available. It is also a way to develop telephone skills which are both personally and professionally useful.

On the professional level, there are occupations centering on counting. Demogra-

U.S. Bureau of the Census

Collecting and classifying sharpen our eye through observation of similarities and differences.

Taxonomy is the science of classification and a form of recreation. Collecting and classifying things such as butterflies, bugs and insects, and specimens of plant life sharpen our eye for the variations in the species and teach lessons of classification through observation of similarities and differences.

One of the most famous collections of samples of plant life is on display at the Botanical Museum of Harvard University in Massachusetts. The precision and beauty of these handblown glass flowers, which are exact replicas of the actual plants, draw thousands of visitors each year. What often starts out as one person's interest becomes a source of information for many.

A home aquarium offers an opportunity to collect living varieties of marine life. It is unique in that a number of species can be enclosed within a relatively small area. An aquarium is a small ecosystem that can be studied easily and with interesting results.

At first glance, it may not seem that people focus any of their leisure activities on bac-

HRW: Ken Karp

teria and yeast. A second look revises that impression. Bread-making is a source of enjoyment and pride for many people. Even today, when commercial bakery products supply the bulk of the country's bread, making a loaf or two, at some time or other, is not an unusual effort for both men and women. Home cheesemaking was not unusual either. However, commercial production of milk has reduced the kinds of bacteria and enzymes present in milk culture, making it difficult or impossible to produce cheese at home. Making yogurt at home has become popular, however, and special equipment has been marketed toward this end.

Yeast is used in home brewing as well. Beer and spirits are sometimes made for family consumption from traditional recipes. Knowledge of the processes of fermentation are then handed down from generation to generation.

Home canning and preserving are, of course, important activities, especially in our nation's agricultural areas. There are over 2.3 million farms in the United States and while the bulk of this country's produce comes from the larger farms, many are moderate-sized, family-run operations where a portion of the land will provide some of the family food supply. Done well, the summer's fruits and vegetables make winter easier to withstand.

Mimi Cotter/International Stock Photo

Dallas Times Herald

Migration to non-urban areas reshaping rural U.S., report says

Associated Press

COLLEGE STATION—A Texas A&M University study shows that a substantial migration from cities to rural America could lead to fundamental changes in society, officials said.

The study concludes that Americans should prepare to shape these changes or be ready to face social stress that could lead to serious disintegration of society.

"It is time we took a new, fresh look at rural America and what it is becoming," said Dr. William Kuvlesky, a rural sociologist and professor of sociology. "Rural America, the vast array of diverse non-metropolitan areas which cement the large metroplexes of our society together geographically and socially, is being reshaped rapidly."

He said statistics show that, between 1970 and 1978, almost three million more people moved out of metropolitan areas than moved in, and three-fourths of all non-metropolitan counties in the United States gained population.

"This complex transformation of truly great historical significance is taking place with little public notice and unbelievable lack of concern," Kuvlesky said of research done for the National Institute for Work and Learning.

The Washington Post

Victories few as species fight for survival

By DALE RUSSAKOFF

WASHINGTON—Sometime over the weekend an unsung species on Earth probably became extinct. It happened in 1914 to the passenger pigeon, in 1977 to the tacopa pupfish.

Some scientists estimate there could be two million such losses by the year 2000, mostly because man has intruded in nature's environment, claiming vertebrates as well as plants and fish.

Ten years ago Congress committed the federal government to making an unusual stand against this worldwide trend. On Dec. 28, 1973, President Nixon signed into law the Endangered Species Act, creating a program to rescue the nation's plants and animals from extinction.

The bald eagle, almost felled by poisoning from the pesticide DDT, now soars in several states because of research and projects authorized by the act. A rare primrose, with seeds believed to harbor cures for serious disease, is holding on in the Antioch Dunes. An obscure green parrot whose habitat was almost wiped out by developers is rallying in the forests of Puerto Rico. The manatee is swimming despite continuing threats in Florida estuaries.

But these victories are tiny in a global context. In the tropical rain forests of South America, Africa and southeast Asia, development projects have driven the extinction rate to a species a day.

Scientists estimate that there are 10 million species of plants, animals and microorganisms, and that thousands will die out before they can be identified.

This loss of species and genetic diversity was recently cited by Harvard University biologist Edward O. Wilson as "the one process ongoing in the 1980s that will take millions of years to correct...the folly our descendants are least likely to forgive us."

UNIT 5   The Plant Kingdom

CHAPTER 22   Types of Plants

___nonvascular plants___

Reference: See *The Plants* by Frits Wendt and the Editors of *Life* magazine.

Plants have many uses. We decorate the insides of our houses or apartments with them. We plant them around the outside of our homes and on our streets. We use them as a source of food. We need the oxygen that plants make during photosynthesis. Just as plants have many uses, they come in many different types. In this chapter, you will find out about the structure and function of various types of plants.

NONVASCULAR PLANTS

Some plants have structures that resemble true roots, stems, and leaves, but they have no true vascular tissue. These are called ___nonvascular plants___. Two examples of nonvascular plants are the mosses and the liverworts.

__Mosses.__ If you have ever walked in the woods, you have probably seen moss growing on fallen logs and at the bases of trees. You might even find some in a yard on the shady side of the house. This moss probably looked to you like a green carpet. If you were to examine moss closely you would discover that it is made up of many little green plants

Fig. 22-1
A "green carpet" of moss growing on the ground in the forest. (Manuel Rodriguez)

Fig. 22-1 shows moss growing in the forest. Fig. 22-2 shows a close-up of a moss plant. As you see, it looks like a small model of the higher land plants. It has structures that look like a stem, roots, and leaves. Actually, these structures do not have the highly developed parts of real stems, roots, and leaves. The leaf-like parts of the moss are just thin sheets of cells—often only a single layer. They do not have the complicated structure of true leaves.

The root-like parts of the moss are single rows of cells growing out into the soil. No part of the plant contains any vascular tissue. Water is absorbed by the root-like cells, and often by the rest of the plant, also. It passes from cell to cell through the stalk to the "leaves," which carry on photosynthesis. Mosses contain chlorophyll and make their own food.

The life cycle of the moss. Many organisms go through a series of different stages in their development. This is known as a *life cycle.* In one stage of the life cycle of the moss, sperm and egg cells form at the tops of these plants. Some species of moss can produce both sperm and eggs on the same plant and other species produce them on separate plants. A sperm swims through the dew to reach the egg cell. Fertilization takes place inside the female structure of the plant. After fertilization, a thin bristle-like stalk begins to grow on the top of the moss plant. At the end of this stalk is a spore case. Spores fall out of this case and are carried by the wind. If these spores fall on moist soil they may grow into new moss plants. See Fig. 22-2.

When a moss spore begins to divide and grow into a multi-cellular structure it does not form a regular moss plant at first. The cells line up to form branching green threads that spread over the surface of the ground. From these, the typical moss plants grow upward. In this way one spore produces many moss plants. New plants continue to form by vegetative repro-duction from the bases of older ones. In time, a moss "carpet" is formed.

Mosses as land plants. You may have noticed that a moss is no more complicated than some of the algae. If mosses always grew in the water, we might think of them as one phylum of algae. But mosses grow on land, and only a few are found in fresh water. Mosses are an example of primitive land plant development. Their root-like structures absorb water from the soil and their stalks are strong enough to hold their simple "leaves" up in the light. These are adaptations to land condi-tions.

Their structure has enabled mosses to live on land, but they are not nearly as well adapted to that environment as the higher land plants. The water-absorbing cells of mosses cannot reach

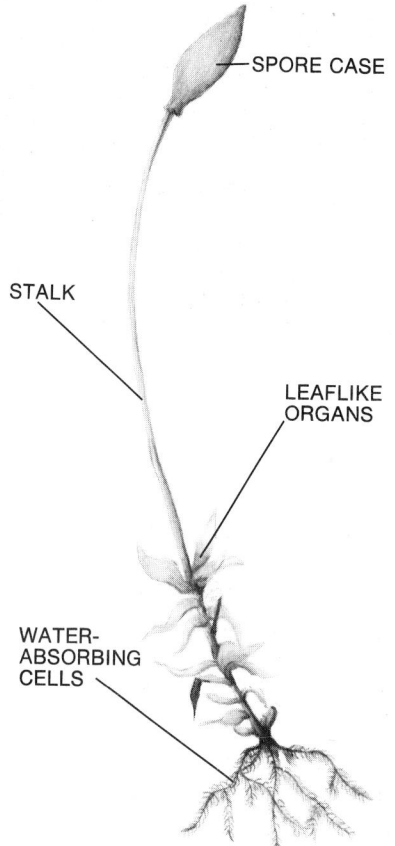

SPORE CASE

STALK

LEAFLIKE ORGANS

WATER-ABSORBING CELLS

Fig. 22-2
A single moss plant.

life cycle

Enrichment: An organism that has two stages in its life cycle is said to show alternation of generations. This type of life cycle involves a sporophyte generation alternating with a gametophyte generation. You may want to introduce these terms.

Enrichment: These root-like structures are called rhizoids.

219

peat

liverworts

more than about one centimeter into the ground. If this top layer of soil dries up, the mosses either die or stop growing until it rains again. Their lack of vascular tissue is also a handicap. If a moss should grow one meter high, the top would dry out in the wind. Without vascular tissue to carry it, water could not rise rapidly enough through the stalk to keep the plant alive. The thin, delicate structure of the "leaves" is also a handicap because they lose water too easily.

As you see, one large problem for land plants is getting and holding water. Mosses are poorly adapted to do this, so most of them live only in shady, damp locations. Shaded soil is less likely to dry out at the surface than soil in sunny locations. Some mosses have the ability to become inactive when they are dry. They wait until it rains, then they soak up water and begin to grow actively again.

Importance of mosses. There are a few ways that mosses affect the communities in which they live. They supply food and shelter for insects and other small animals. They sometimes cover steep slopes in the forest. This protects soil and keeps it from washing away during rainstorms. Mosses also aid in forming soil. They are often the first plants to grow on the small amounts of soil formed by lichens. Acids produced by the root-like parts of the moss continue the breakdown of rock. Dead moss parts turn into humus, forming soil for larger plants to grow in.

In northern bogs, a large type called *peat moss* grows so thickly that it is often a dominant plant in the community. As the moss grows upward, its lower parts die and pack down into the bog water. This mass of dead moss does not decay much because the water is cold and acid. It becomes a *peat* deposit. Peat is simply a mass of partly rotten plant remains. Any plants growing in wet places may produce peat.

Peat is dug up, dried, and used for fuel in northern Europe. We have many peat bogs in the United States and Canada, but we do not use peat for fuel. As a result of the energy crisis do you think we may give some consideration to the use of peat as fuel in the future?

Gardeners mix peat moss into their soils to make them looser and more absorbent. Nurseries use it to pack around the roots of plants when they ship them. The moss holds water and keeps the roots moist.

Liverworts. Another example of nonvascular plants are the *liverworts*. Liverworts are relatives of the mosses. You can find them in damp, low woodlands and along shady stream banks. Some types grow floating in the water. They are not nearly as common as mosses, and you might live your whole life without ever noticing them, even when they were under your feet!

The plant bodies of most liverworts are flat, ribbon-shaped structures. See Fig. 22-3. The ribbon branches every so often as it grows across the surface of the soil. On its underside the liverwort has a large number of root-like cells growing into the soil to absorb water. But, like the mosses, liverworts have no true roots, stems, or leaves. They are not very well adapted to life on the land. That is why they live only in damp places.

Fig. 22-3
These are liverworts. They have a ribbon-like structure and live in damp places. (Courtesy Carolina Biological Supply Company)

VASCULAR PLANTS

The higher plants have true stems, roots, and leaves. They contain vascular tissue, so they are called the *vascular plants*.

vascular plants

General structure of vascular plants. You already know that getting and holding water is necessary for the survival of land plants. You have seen how mosses are limited in this respect. The vascular plants have become much better adapted to land conditions. They have roots that can go deep into the soil to absorb water and minerals. They have leaves that make food. These leaves have an outer covering that holds in the moisture, so they do not dry out too easily. The stems are strong enough to hold the leaves up where they can receive sunlight.

Water and dissolved food reach all parts of the plant by way of the vascular tissues. The long, hollow cells of these tissues carry liquids much more rapidly than they could travel through ordinary cells. The vascular tissues also give strength to stems. Wood is simply a solid mass of vascular tissue. Its cells have especially thick and tough cell walls.

Science Reading Skill: Compare and contrast the nonvascular and vascular plants.

221

As you can see, nearly all the things that make the higher plants more complicated than the algae are adaptations to land conditions. Any mutations that have allowed them to survive better on land have been preserved through natural selection.

club mosses

The club mosses. Fig. 22-4 shows one example of the *club mosses.* They look something like big moss plants, and they have club-like structures that give them their name. Some people call them ground pines, but they are neither mosses nor pines. Common club mosses are about 8 to 15 centimeters high. Several types grow in the Northern forests. Sometimes large numbers of these club mosses are gathered for Christmas decorations. Some larger club mosses are found in the tropics.

Fig. 22-4
Club moss. Notice the simple, small leaves and the "clubs" that produce spores at the top. (Harold E. Teter)

Fig. 22-5
Horsetail plants. The side stems carry on photosynthesis. The brown scales are modified leaves. (Harold E. Teter)

A club moss has simple, scale-like leaves that stay green all year. The "clubs" at the top are cones that produce spores. The spores are carried by the wind and may produce more club mosses if they fall in a good location.

horsetails

Horsetails. Fig. 22-5 shows another type of vascular plant. It is a member of the group known as the *horsetails.* You can tell that they are horsetails by their rough, bushy stems. The stems are in sections that can be pulled apart easily. There are little ridges running lengthwise along the surface of the stem. The leaves are small scales and have no chlorophyll. The finely branching stems are green and carry on photosynthesis.

There are also unbranched, reed-like horsetails known as *scouring rushes*. Mineral deposits in their stems make them quite rough to the touch. In frontier times, they were used for scrubbing kitchen pans.

Ferns. You probably have seen potted ferns many times. *Ferns* also grow wild in the woods. They reproduce by means of spores, just as the club mosses and horsetails do, but in most other ways they are more highly developed. Their leaves, for instance, are large and flat, with well developed vascular tissue. Ferns are closely related to seed plants. They often have a creeping, underground stem. From this stem many roots grow down into the soil and large, finely divided leaves grow up into the air. These leaves are often called *fronds*. You already know how spores are produced on the undersides of fern leaves. Water absorbed by the roots passes through vascular tissues to all parts of the fern plant. Fig. 22-6 shows ferns growing on the side of a tree.

The underground stem and roots live through the winter. In most northern ferns new leaves develop each year. A fern that you see growing in the woods may be as old as any tree around it. In warm climates the leaves stay on all year. Even in the North a few species have evergreen leaves.

Importance of ferns. One kind of fern is very common on sandy soils of the United States and Canada. It is called the *bracken fern*. Wherever Northern pine forests have been cut or destroyed by fire, the bracken fern is likely to appear. In these cut-over areas, the new trees that come in first are often a small species of poplar, called aspen. Under the poplars, the bracken grows waist high, covering the ground completely. See Fig. 22-7. The yearly rotting of bracken leaves helps to build up the soil for the return of the pines.

ferns

Fig. 22-6
In a wet climate, some ferns are able to grow on the sides of trees. (Harold E. Teter)

Enrichment: The spore case of ferns is called a sorus (plural: sori).

Enrichment: The young developing fern is commonly called a "fiddlehead" because of its resemblance to the musical instrument. These fiddleheads can be gathered and eaten.

Fig. 22-7
Ferns are some of the first plants to appear in a succession that takes place after a forest fire. (EPA-Documerica)

223

In certain parts of the tropics, ferns are more common than they are in North America and some grow to be as large as trees. A tree fern looks something like a palm tree. It has an upright stem, and a crown of very large leaves at its top.

The seed plants. Ferns, club mosses, and horsetails have one main disadvantage. They reproduce by means of spores. It takes a long time for one spore to produce a plant as much as 2 or 3 centimeters high. All during that time the young plant is in danger of drying up.

Seeds are the answer to this problem. The seed contains an embryo and a food supply. By the time this food is used up, the young plant is well established. It needs some moist weather to start growing, but after that, ordinary dry spells will not destroy it. Plants that produce seeds are called *seed plants*.

The presence of a pollen tube that brings the sperm to the egg cell is one of the seed plant's adaptations to land life. This prevents the sperm from drying. It makes fertilization possible where no water is present.

There are two main groups of seed plants. These are the *flowering seed plants* and the *nonflowering seed plants*. See Fig. 22-8. The most common examples of nonflowering seed plants are trees of the pine family. These include pine, spruce, fir, hemlock, juniper, cedar, cypress, and redwood. These trees have cones instead of flowers. The seeds develop in one kind of cone. The pollen is produced in smaller cones that fall off the trees after the pollen is shed. Pollen is carried by the wind.

seed plants

Fig. 22-8
Seed plants. Left: Nonflowering fir trees; Right: Flowering plants; pear tree, grass, and dandelions. (Harold E. Teter)

The leaves of plants in the pine family are needle-like and look as if they had been varnished. These leaves are able to hold water well. This is one reason why pines so often grow on dry, sandy soils where water must be conserved. In Fig. 22-9 you see leaves of a ginkgo tree. This nonflowering plant is unusual because it has flat leaves instead of needles. It is frequently planted in public parks and sometimes in backyards.

Flowering plants are the most abundant land plants on earth. They include our common broad-leaved trees, such as oaks, maples, beeches, hickories, ashes, elms, and fruit trees. They also include the common small plants like grasses, weeds, flowers, fruit trees, and farm crops. They are the main plants in the tropical rain forests, the broad-leaved temperate forests, the grasslands, and the deserts.

Many useful products besides food come from seed plants. These include lumber, rubber, turpentine, clothing fiber, paper pulp, oils, drugs, and dyes.

Fig. 22-9
Nonflowering seed plants. Left: The flat leaves of the ginkgo tree; Right: The cycad tree. This is not a relative of the palms. (U.S. Forest Service; Julia Morton)

HISTORY OF THE VASCULAR PLANTS

The first land plants to become really abundant were the spore-producing vascular plants. These formed rich forests on the land over 300 million years ago. Club mosses grew into large trees over 30 meters high and 2 meters thick. Tree ferns were common, and even the horsetails were tree-sized.

The climate was mild, and there was much swampy ground. In these swamps great peat deposits were formed. Later the peat hardened and became coal, so this period is called the *Coal Age*. See Fig. 22-10.

There were also some early seed plants, called seed ferns, which looked like ferns but produced seeds. The seed plants of today probably developed from these.

Science Reading Skill: *Sequencing*— Note the changes in the dominant plant types over the last 300 million years.

The Coal Age came to an end about 225 million years ago, and the climate gradually became very cold. There was a long *Ice Age*. The cold climate and competition from seed plants wiped out the Coal Age forests.

The nonflowering seed plants became the most common in the warmer age that followed. Redwoods were one of these common plants. There were many ginkgoes and many species of a group called *cycads*. See Fig. 22-9. Only one ginkgo and a few cycad species are left today. This period could be called the Age of the Nonflowering Seed Plants, but it is more often called the *Age of Reptiles*, for this was the time of the dinosaurs.

Then, about 135 million years ago there came another change. The flowering plants appeared on the earth. The flowering plants have more efficient vascular tissues. Their reproduction is also more efficient. They have been able to crowd out the nonflowering seed plants over most of the earth, but the pine family is still abundant in some places.

From this brief history, you will see that our club mosses, horsetails, and ferns are the last remnants of what were once important groups. The nonflowering seed plants are less important than they once were. Today, the flowering plants have taken their place as the most important forms on the land. Meanwhile, the old, primitive algae continue to be the main "plants" of the sea.

Fig. 22-10
This is what a Coal Age swamp may have looked like 300 million years ago. The peat that formed in the swamps became the coal we use today. (Buffalo Museum of Science)

226

SUMMARY

Mosses and liverworts are no more complicated than some algae, but they are adapted to living in damp places on land. Mosses have upright stalks with simple leaf-like structures for catching light and carrying on photosynthesis. Special cells, or lines of cells, grow into the ground and absorb water. There is no vascular tissue.

Liverworts grow as flat ribbons of green tissue across the surface of the ground. Water-absorbing cells grow into the soil from the under side.

Both mosses and liverworts reproduce by means of spores.

Vascular plants have vascular tissue, stems, roots, and leaves. All of these structures adapt them to life on land.

The principal groups of vascular plants are the club mosses, horsetails, ferns, and seed plants. The first three reproduce by means of spores. Seed plants have seeds, which give them an advantage in land reproduction.

Seed plants are divided into the nonflowering seed plants and the flowering plants.

Activity

Observing Plants
This activity will give you personal experience with different groups of plants.

A. Carefully pick single plants from a moss "carpet" and examine them under a good hand lens or a binocular microscope.

1. Are all types of mosses alike? If not, how do they differ?

B. If some of your mosses have spore cases on them, crush one on a microscope slide. Make a temporary mount of it and examine it under the microscope.

2. Can you see the spores? Describe them.

C. Examine a single moss "leaf" under the microscope. Focus up and down. See how many layers of cells come into focus, one after the other. This is a way of seeing how many layers of cells make up the leaf.

3. How many layers of cells are there?

D. Carefully examine specimens of club mosses, horsetails, ferns, nonflowering seed plants (pine, spruce, etc.), and flowering seed plants.

4. Write a paragraph describing, in your own words, the plants of each group.

Suggestion: Horsetails are commonly found in early spring and appear in clumps.

Word Quiz

A. In your own words, give a brief definition of each of the following:
1. nonvascular plants   2. vascular plants   3. life cycle   4. peat

B. From the following plants make two groups; one of vascular plants and the other of nonvascular plants. Then give a brief description of each plant.
mosses, ferns, seed plants, liverworts, horsetails

Check Your Facts

Science
Reading Skills
3. & 5.
Generalization

1. Describe the structures of a moss plant. Of what importance are mosses?

2. What is a liverwort? Where might you find it growing?

3. Why is vascular tissue very important to land plants? What other adaptations of vascular plants make them more successful than the mosses?

4. How would you recognize a club moss? A horsetail?

5. How do mosses, club mosses, horsetails, and ferns reproduce? How does this handicap them in competition with seed plants?

6. What fern is fairly important in some forest areas today? Why is it important?

7. Which group of land plants is most important today? How is this group useful to us?

8. What were the three main periods in the history of land plants on the earth? In what order did they take place?

Thought Questions

Science
Reading Skills
1. & 3. *Problem
Solving*

1. Mosses are common in forests. If we were to cut the trees, would this help or hurt the moss?

2. People often speak of "moss covered rocks" under the water along a shoreline. Are these really mosses? If not, what are they?

3. Why are there no ferns in the desert?

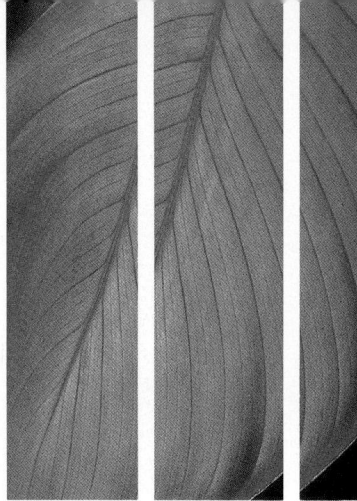

CHAPTER 23 Leaves of Flowering Plants

Since flowering plants are the most common and useful plants, they are worth some careful study. The photograph above shows the pattern formed by the veins of a leaf. In this chapter you will learn about these veins and other leaf structures.

Leaves are the food-making organs of flowering plants. Remember that light is the source of energy for photosynthesis. Leaves are thin and flat, so most of their cells are exposed to the light. The thin shape also provides good contact with air. Leaves must get their carbon dioxide from the air and give off oxygen into the air. Their veins are made of vascular tissue. Veins bring water in from the stem, and carry food that the leaf has made to other parts of the plant. Veins also give strength to the leaf.

THE STRUCTURE OF A LEAF

Fig. 23-1 shows how the cut edge, or cross section, of a leaf might look under a microscope. A *surface layer* of cells covers both the top and bottom of the leaf. Such a surface layer of cells is called an *epidermis.* The epidermis of the leaf is tough and gives strength to the leaf. A waxy material on the outside prevents evaporation of water. Cells of the surface layer are clear, so light shines through them to reach the green cells inside.

Between the upper and lower layers are the *food-making cells,* containing chloroplasts. The upper cells are long and narrow. The cells lower in the leaf have more irregular shapes. These inner leaf cells are not packed tightly together. All through the leaf there are many *air spaces* between the cells.

The air spaces in the leaf are connected with the outside air through small openings, or *pores,* in the lower leaf surface. A single leaf has thousands of these tiny pores.

objectives

After you read this chapter, you should be able to:

___List the different parts of a leaf and give their functions

___Explain how leaves prevent the loss of water

___Explain what causes coloring in leaves

epidermis

Word Study: epi- means upon, dermis means skin. Dermis is considered the true skin and the epidermis is the covering "upon" this skin.

food-making cells

Enrichment: These pores are also called stomates. Pores can appear in both upper and lower leaf layers but in most woody plants they are found in the lower layer.

pores

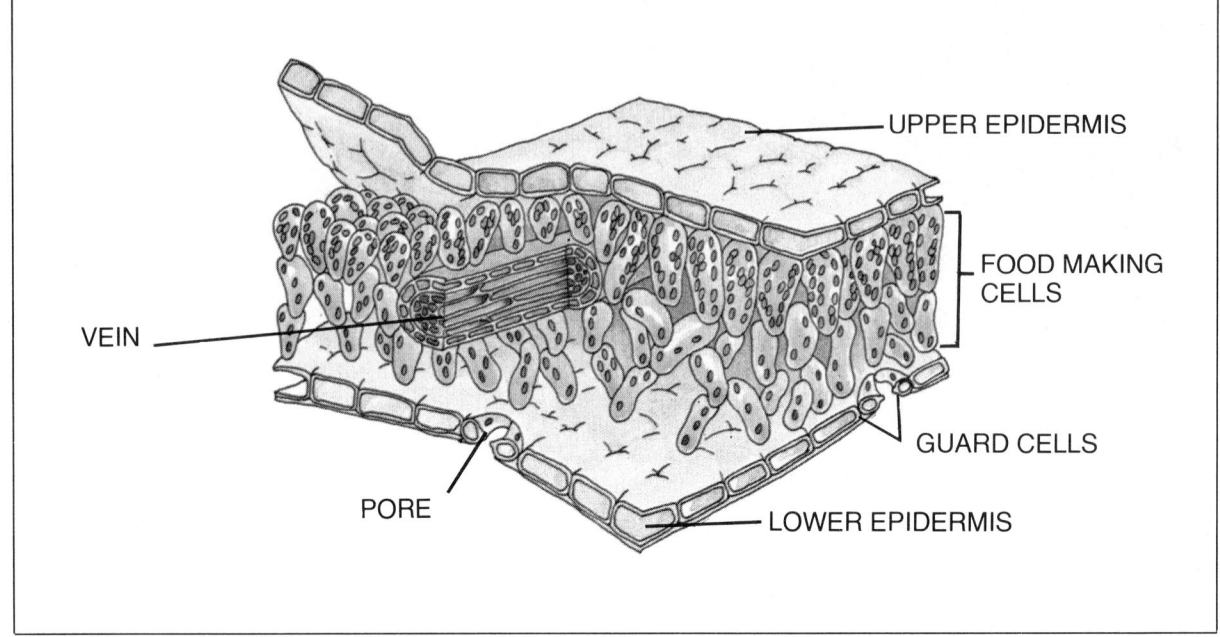

UPPER EPIDERMIS

FOOD MAKING CELLS

VEIN

GUARD CELLS

PORE

LOWER EPIDERMIS

Fig. 23-1

The inner structure of a leaf. Where does the water enter? Where does the food leave the leaf? Where does the carbon dioxide enter?

Question: Ask students why the pores of water plants such as pond lilies are found on the upper leaf layer and not on the lower leaf layer.

Question: Ask students how the structure of the leaf enables it to effectively carry on photosynthesis.

Water enters the leaf through the *veins*. In fact, the vein is like a bundle of small water pipes. Its vascular tissue cells are long and hollow. The part of the vein located toward the upper surface of the leaf carries water. The part of the vein located toward the lower surface of the leaf carries dissolved food from the leaf to the other parts of the plant. Water molecules pass from vein cells into the food-making tissue. As you saw in the photograph at the beginning of this chapter there are so many small veins in the leaf that these molecules never have very far to travel.

THE LEAF IN ACTION

Photosynthesis. Now try to imagine the leaf in action. Sunlight shines on the cells. Some of its energy is absorbed by the chloroplasts. The light energy is changed into chemical energy, which can be used to carry on photosynthesis. The chloroplasts in the food-making cells take water from the veins and carbon dioxide that entered the air spaces through the pores. The atoms from the water and carbon dioxide are rearranged to form glucose. Oxygen is given off as a waste into the air spaces and then out through the pores. The glucose, dissolved in water, moves into the veins and is carried to other parts of the plant for use by the cells, or it is stored as starch.

Fig. 23-2
What would happen if these leaves lost too much water? (HRW photo by Russell Dian)

The water problem. The food-making cells are moist, and water evaporates from them into the air spaces. This water vapor then passes out through the pores into the air and is lost. Such loss of water is not serious as long as the roots can replace it. But sometimes the ground is too dry, and the roots cannot absorb water rapidly enough. The plant loses water to the air faster than it can get it from the soil. If this goes on very long the plant begins to wilt. It may dry up and die.

The leaf can prevent this water loss by closing its pores. This keeps the water vapor inside the leaf. Carbon dioxide cannot enter through closed pores, so food-making slows down. But this is better than dying for lack of water.

The closing of a pore is brought about by the action of the two *guard cells* that surround it. These guard cells are curved, and when they come together the pore in the leaf is closed. Fig. 23-3 gives a surface view of a pair of guard cells. Notice that they are the only cells in the epidermis that have chloroplasts.

guard cells

Science Reading Skill: *Cause and Effect*—The cause of wilting and of the opening and closing of pores that is regulated by the guard cells.

EPIDERMAL CELL   GUARD CELL

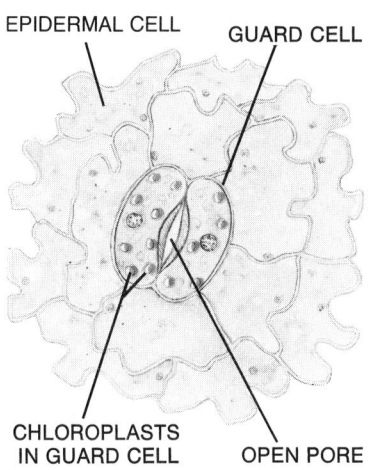

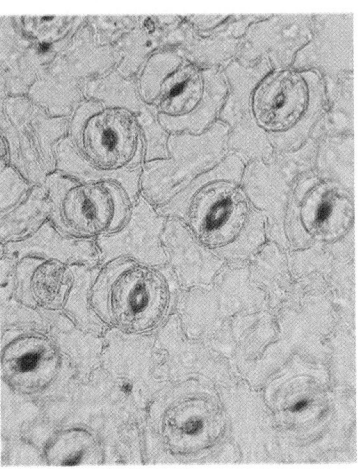

CHLOROPLASTS IN GUARD CELL   OPEN PORE

Fig. 23-3
Left: A diagram showing the guard cells surrounding the pore in the epidermis of a leaf; Right: An actual photograph of guard cells. (William E. Ferguson)

When a leaf is drying out, the guard cells become flabby from loss of water. Ridges in their walls act like springs, causing them to straighten out. This closes the pore. When enough water once more reaches the leaf, pressure builds up inside the guard cells, and they are forced back into the curved shape, opening the pore again.

Now you can see what the guard cells are guarding against. They prevent the plant from losing too much water. They regulate water loss by opening and closing the pores.

The pores also close at night. This is a good adaptation for land plants, because it helps prevent the drying out of the soil. If the pores did remain open, water would evaporate through them all night long. This water would be removed from the soil without doing the plant any good. It would not be there later when it was needed.

The action of the pores is automatic. In darkness the chloroplasts in the guard cells no longer get sunlight. Food-making has stopped. When there is less sugar in these cells, they hold less water in their protoplasm. They get flabby, and the pores close.

modified

Some desert plants, such as the cactuses, do not have leaves like the ones we have been discussing. Such leaves would lose too much water. The only leaves left on a cactus are modified to form spines. *Modified* means changed from the original form. The stems of cactuses are the food-making structures, and they also serve as storage organs for water.

FALLING LEAVES

pigment

Fig. 23-4
Cactus spines. These are the modified leaves of a cactus plant. (W. H. Hodge/ Peter Arnold, Inc.)

Some plants grow in warm areas where there is a long, dry season. They may shed their leaves during this dry weather. Winter is a dry season for northern plants. The ground freezes, and the roots cannot absorb ice through their cell membranes. Even cold, unfrozen water is not absorbed easily. With so little water available, leaves would lose too much water even with the pores closed. Guard cells cannot prevent all loss of water. Therefore, the shedding of leaves in the fall is needed to help prevent the drying up of the plant during the winter months.

LEAF COLORING

Most leaves are green during the growing season because they contain chlorophyll. But there are other pigments in leaves also. The word *pigment* means coloring material. We also speak of pigments in paint. Yellow pigments are always present in leaves. They help in the process of photosynthesis. They do not show in the growing season because they are hidden by the chlorophyll. In the fall, the temperature becomes too low for chlorophyll formation and the old chlorophyll begins to break down. This allows the yellow pigments to show. These yellows account for much of the autumn coloring. The red pigments form in the presence of sunlight from some of the breakdown products of chlorophyll. We get our brightest autumn colors in the years with the sunniest fall weather. See Fig. 23-5.

KINDS OF LEAVES

Leaves come in many sizes and shapes. Some of these are shown in the photographs in the margins of this chapter. In the pines they are needle-like. In the flowering plants they are usually flat and thin. They may be large or small, narrow or wide, long or short. They may have smooth, saw-toothed, or wavy edges. Some are thickened for water storage. Some leaves are divided into several leaflets, looking like many separate small leaves. All of these differences can be used to identify plants. The easiest way to learn the names of trees, for instance, is to look at their leaves and then compare them with the drawings in a tree guide.

Looking down on top of a plant you will be impressed with how well its leaves are arranged. Hardly any bare ground shows through the spread of leaves. They catch all the light shining on that particular area without much shading of one leaf by another.

Science Reading Skill: Cause and Effect—The cause of leaves turning color in autumn.

Fig. 23-5
What causes these leaves to ''turn'' color in autumn? (Manuel Rodriguez)

Fig. 23-6
Various types of leaves. Top left, Japanese maple; Top right, strawberry; Bottom left, palm. (HRW photos by Russell Dian) Bottom right, sugar maple. (HRW photo by Ken Lax)

SUMMARY

Leaves are the organs of a flowering plant that carry on photosynthesis. Cells of the epidermis are clear and covered with wax. The inside of the leaf contains food-making cells that contain chloroplasts. Leaf veins are made of vascular tissue, which brings water to the leaf and carries food away.

Air spaces in the leaf connect with the outside air by way of pores in the leaf surface. Carbon dioxide enters and oxygen leaves through the pores during photosynthesis. Guard cells close the pores during drought conditions. This prevents too much water vapor from escaping from the leaf.

Leaves are discarded in the autumn so they will not lose water when it cannot be replaced from frozen ground. Yellow colors already in the leaf show up when the chlorophyll breaks down. Red colors develop from products of this breakdown of chlorophyll.

Activity

Observing Leaves

You have just read about leaf structure. Now try to find these leaf parts yourself.

A. Peel a bit of the lower surface from a geranium leaf. You know when you have just the surface layer because it is colorless. If any green shows, you have taken too much. Put this bit of leaf surface on a slide and add a drop of water. Cover it with a cover glass. Now study it under the microscope.

Under the microscope you can see the pores. They look like dark slits. On each side of the slits are the guard cells that contain chloroplasts. You can also see the other cells of the surface layer. Notice how their wavy edges lock together.

1. Make a drawing to record what you see.

B. Observe a prepared slide of a leaf section under the microscope. It will show the same type of view as Fig. 23-1.

2. Can you find the veins? Are they all sectioned straight across like the one in the drawing?

3. How many layers of long food-making cells can you see? Do you think this leaf grew in sunlight or shade?

4. Can you see the air spaces in the leaf? Where are they located?

5. Are there any guard cells?

Word Quiz

From the new words you learned in this chapter, write the word that correctly completes each statement. Do not write in this book.

Part of a living thing that is different from the way it was in earlier generations is said to be _____. A coloring material is called a(n) _____. Cells containing chloroplasts are the _____ of the leaf. Air enters the leaf through the _____. Each of these has a(n) _____ on each side of it. The surface layer of cells covering the leaf is called the _____.

Check Your Facts

Science
Reading Skills
1. Reading
illustrations 4.,
5., & 6. Cause
and Effect

1. Make a drawing that shows the internal structure of a leaf. Label all of the parts.

2. Give the function of each of the parts you labeled in question 1.

3. What advantage is it for a tree to shed its leaves in the fall?

4. Explain how the guard cells function.

5. Why do leaves turn yellow in the fall?

6. Why do leaves turn red in the fall?

Thought Questions

Science
Reading Skills
1. & 2. Problem
Solving

1. Suppose you buy a potted plant with thick, fleshy leaves and a glossy surface. What kind of area are these plants adapted to? How should this influence the kind of care you give it?

2. Potatoes are found growing under the ground. Will beetles eating the leaves have an effect on the crop? How? Why?

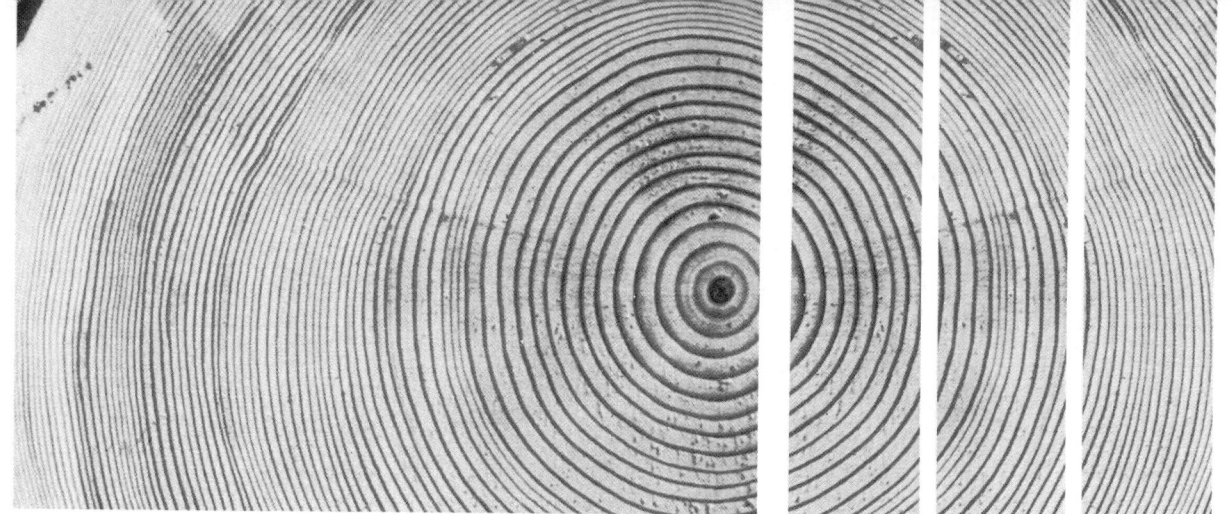

CHAPTER 24 STEMS AND ROOTS

The photograph above shows a cut section of the stem of a huge tree. You may know that if you count the "rings" in this stem you can tell how old the tree is. Have you ever wondered what causes these rings to form each year? After you read this chapter, you will know the answer to this and many other questions about the roots and stems of trees and other plants.

STEMS

The functions of stems. In most plants the stem has two main functions. It **supports the leaves,** holding them up to the light; and it **carries water up to the leaves** from the roots and **carries food from the leaves to the rest of the plant.**

Some stems have special uses. As we said in the last chapter, cactus stems *store water*. Stems may also be modified to *store food*. A common white potato is an underground stem that stores food and reproduces the plant. In fact, nearly all stems store some food. Sugar cane stores sugar in its stem. This is unusual, for most plants store their food in the form of starch. Some stems are important as *food-making* structures. Some are modified to carry on *vegetative reproduction*. Certain vines have small branch stems that are adapted for *climbing*. We shall study only the three most common types of stems that grow upright and support leaves.

Structure of stems. Stems must have strength, and they must be able to carry liquids. The carrying of liquids is, of course, the job of the stem's vascular tissue. Also, most of the strength is supplied by the thick walls of vascular tissue cells. Other strong types of cells may help.

There are two main types of vascular tissue. One kind carries water and dissolved minerals. The other carries food. Both types are often found together in long strands of tissue passing

objectives

After you read this chapter, you should be able to:

—**Explain** the functions of stems and roots

—**Diagram** three kinds of stem structure

—**Compare** the growth of stems and roots

—**Describe** how roots get water

Enrichment: Other examples of modified stems include onions and flower bulbs.

Enrichment: You may want to introduce the terms xylem and phloem.

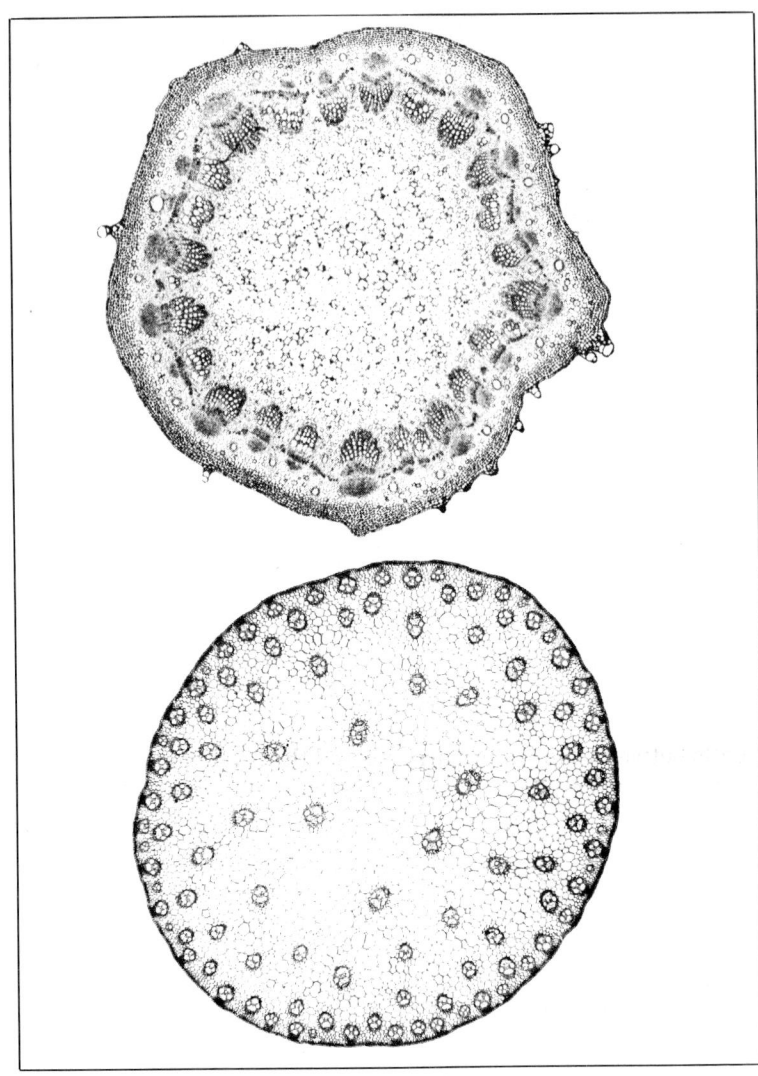

Fig. 24-1
Arrangement of the vascular bundles in different stems. Top: A tomato stem; Bottom: A corn stem.

vascular bundles

up through a stem. Such strands are called *vascular bundles.* Fig. 24-1 and Fig. 24-2 show two common ways vascular bundles may be arranged in stems. Now look more closely at Fig. 24-1.

The stem on the top has its bundles arranged in a single circle. Tomatoes, cucumbers, buttercups, geraniums, and most common weeds are a few of the plants having this type of arrangement. The stem on the bottom has its vascular bundles throughout the stem. This arrangement is found in grasses, reeds, lilies, and many other plants. In both types of stems the water-carrying vascular tissue is on the inner side of the bundle, and the food-carrying type is toward the outside. In northern climates these two types of stems commonly live for only one season.

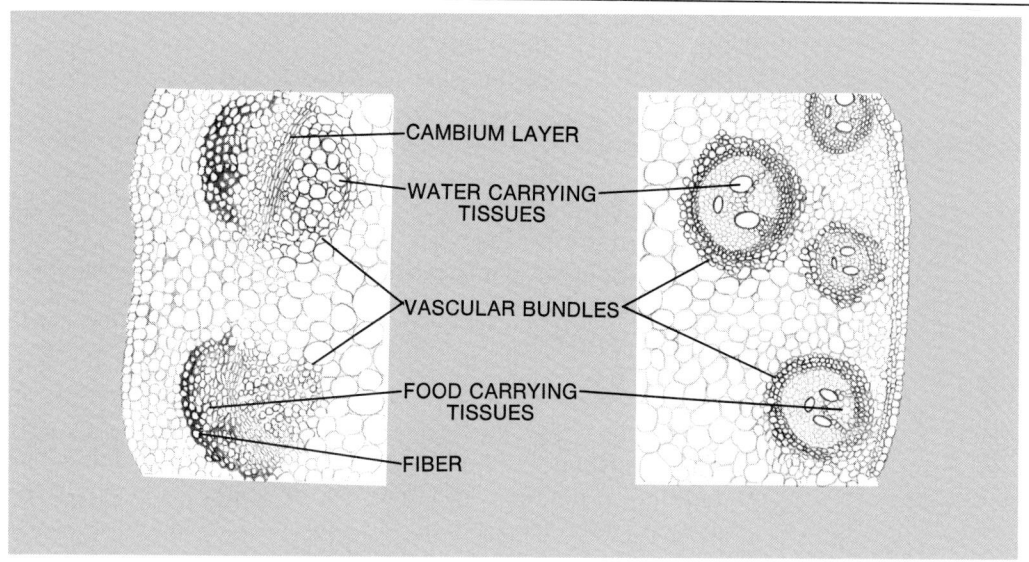

CAMBIUM LAYER

WATER CARRYING
TISSUES

VASCULAR BUNDLES

FOOD CARRYING
TISSUES

FIBER

A third type of stem may live for many years. This is the **woody stem,** which is the type found in trees and shrubs. Woody stems start out as the form of stem with a single circle of vascular tissue. Such stems have a *cambium,* or growing, layer of cells between the water-carrying and the food-carrying regions.

Cambium cells are the growing tissue in stems. Growth in the cambium makes the stem become thicker. All the other cells of the stem perform special functions. When these specialized cells are fully developed, they can no longer divide.

Only the cambium cells continue to divide. As these cells multiply, some become specialized. Others remain unchanged. These unchanged ones are still cambium cells that can divide again. The ones that specialize may develop into water-carrying vascular tissue or into food-carrying vascular tissue. The cambium lies between these two kinds of tissue. New water-carrying tissue forms toward the inside of the cambium. Layer after layer is added, and the stem becomes thicker and thicker. Many stems thicken a little bit in this way, and then die at the end of the season. Woody stems do not die. Their cambium adds a new layer of water-carrying vascular tissue each year. This is the wood of the tree.

If you look at the cut end of a log you can see the layers of wood. In this view they look like rings, and they are often called **growth rings.** You can count them in a log or stump and tell how old the tree was when it was cut. Each ring usually represents one year of growth.

During the growth process in a woody stem, the cambium itself is pushed outward. The food-carrying cells and the outer

Fig. 24-2
An enlargement of the vascular bundles found in a tomato stem (left) and a corn stem (right).

woody stem

Science Reading Skill: *Cause and Effect*—The cause of increase in the thickness of a tree that produces growth rings.

growth rings

layers of the stem are also pushed outward. Many of these outer cells are crushed and die. New cells form and take their place. This outer area of the stem is the *bark*. Old bark keeps dying and splitting off, but new bark forms underneath.

The innermost part of the bark, next to the cambium, contains the food-carrying vascular tissue. Food is carried up or down the stem through the inner bark. In summer the food moves downward toward the roots. In spring food stored in roots and stem moves upward through the inner bark to nourish the growing stem tips. Water goes up the stem through the long, hollow cells of the wood.

Suppose you were to peel a strip of bark off all the way around a tree. What would happen? Water would still rise through the wood. The leaves would remain green and go right on making food. But food moving down through the inner bark could not reach the roots. In a year or two the roots would use up their stored food, and then would die. So would the rest of the tree. You see, then, that damage to the bark of a tree can be serious.

You cannot tell the age of a tree by its size alone. Some trees grow very rapidly. A cottonwood grows about one meter thick in twenty years if growing conditions are unusually good. Some other trees may take 500 years to become that thick. Oaks or hemlocks can live a few hundred years. Douglas fir may live over a thousand years and some redwoods over 3 thousand. So far as we know, the oldest trees on earth are bristle-cone pines that live in the Sierra Nevada mountain range. They are not very large but they live over 4 thousand years.

In looking at some logs, you may notice that the wood in the center is darker than the wood around the edge. This darker wood is dead. All of the living wood is in the outer, light colored part of the stem. Even there many cells are dead. Their protoplasm disappears, leaving open channels through which water can rise. Only this outer wood carries water. The dark, center wood is plugged up with gummy waste materials. So you see that although a tree may be very old, its living parts are all fairly young. They may not be as old as you are.

Different kinds of trees form wood cells of different sizes and shapes. This is why lumber differs in appearance. You can learn to identify wood by its *grain*, which is simply the pattern formed by the layers of wood in the growth rings.

Growth in length of stems. Look at Fig. 24-3. It shows a twig, which is the end of a tree stem. The trunk and all its branches form the stem system of a tree. The drawing shows the twig as it would appear in winter, when no leaves are present.

Activity: Collect stems from several kinds of trees. Identify the parts shown in Fig. 24-3 and determine how much the stem has grown during the year.

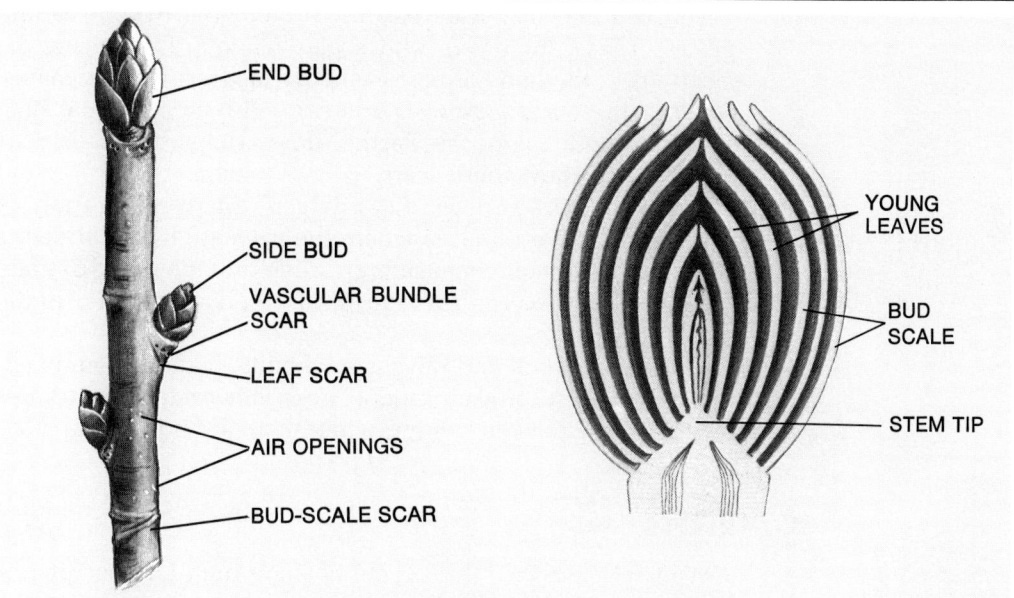

Labels on left diagram:
- END BUD
- SIDE BUD
- VASCULAR BUNDLE SCAR
- LEAF SCAR
- AIR OPENINGS
- BUD-SCALE SCAR

Labels on right diagram:
- YOUNG LEAVES
- BUD SCALE
- STEM TIP

Fig. 24-3

Left: Outside view of a tree stem with a bud; Right: Inside view of a tree bud. How does this structure make it possible for a tree to produce leaves very quickly in the spring?

On the end and along the sides are the *buds*. The brown *scales* covering each bud are really modified leaves. See Fig. 24-3. If you should split one of these buds open, you would find that the stem ends in a little point. This *stem tip* contains thin-walled, simple cells, very much like those in the cambium. They can grow to produce new stem tissue.

Arranged around the stem tip are small, well formed *leaves*. They already have stalks and veins and many of their other cells. When spring arrives, cells in the stem tip begin to multiply. The stem begins to lengthen, the bud scales fall off, and the small new leaves grow rapidly. In only a week they reach nearly full size. Most of this rapid leaf growth is a matter of absorbing a great deal of water into cell vacuoles. The stem continues to grow in length for one season. Then all growth stops for that year, and new buds begin to form.

Notice that all growth in stem length is at the tip. It is from the bud outward. Older parts, below the bud, never grow in length again. Of course they do grow thicker each year.

Buds on the side of the stem form branches if they grow, but many of them never do. They are the stem's reserves. If the end of the stem dies, then the side buds grow out and take its place.

Buds form at the bases of the leaves on a stem. You may not notice them during the summer because the leaf stalks hide them. The bark of the twig shows scars just below each bud. These scars are the places where the leaves grew. You can even see small dots in these *leaf scars* where vascular bundles came through from the stem into the leaf stalk.

You can also see scars entirely around the twig at certain places. These are where bud scales fell off. They mark the locations of the buds of past years. By measuring the distance between these *bud-scale scars* you can tell how much the stem grew in length each year. Farther down the stem, the bark is too rough to show these scars.

Over the entire surface of the twig are small rough spots in the bark. These are places where the cells are loose, allowing air to enter the stem. Stem cells are alive, and they need oxygen. They get their oxygen through these small *air openings* in the bark.

Enrichment: These air openings are called lenticels.

The arrangement and the shape of buds, scars, and air openings is different for each kind of tree. This arrangement helps us identify species in winter, when there are no leaves.

ROOTS

We have described three types of stems found in seed plants. There are several other kinds of stem structures among the vascular plants. Roots do not vary in this way. All roots have the same structure whether they grow on club mosses, horsetails, ferns, or seed plants. **The main job of roots is to absorb water and dissolved minerals from the soil and to anchor the plant.** A large tree with a thick trunk will bend in the wind. Imagine what strength its roots must have to hold it against such force.

Structure of roots. The big, old roots on a tree look very much like old stems. They are covered with bark. They have food-carrying vascular tissue in the inner bark. Wood is laid down in growth rings by the cambium, which lies between bark and wood. Of course there are no buds on a root.

Old, bark-covered roots do not absorb liquids. The young, delicate branch roots do this job. In this respect roots are something like stems; they must grow each year. Leaves are found only on the new growth of the stem, so the plant must grow a new stem every season. Only young roots can absorb water, so there must be new growth of roots each year.

Growth of roots. All growth in the length of a root takes place at the very tip. About one centimeter back from the tip there is no further growth in length. In trees the roots also grow in thickness, but in small plants, which do not live for a long time, the roots do not become very thick.

The cells of a root tip grow and divide and repeat this division over and over again. This increase in size develops enough

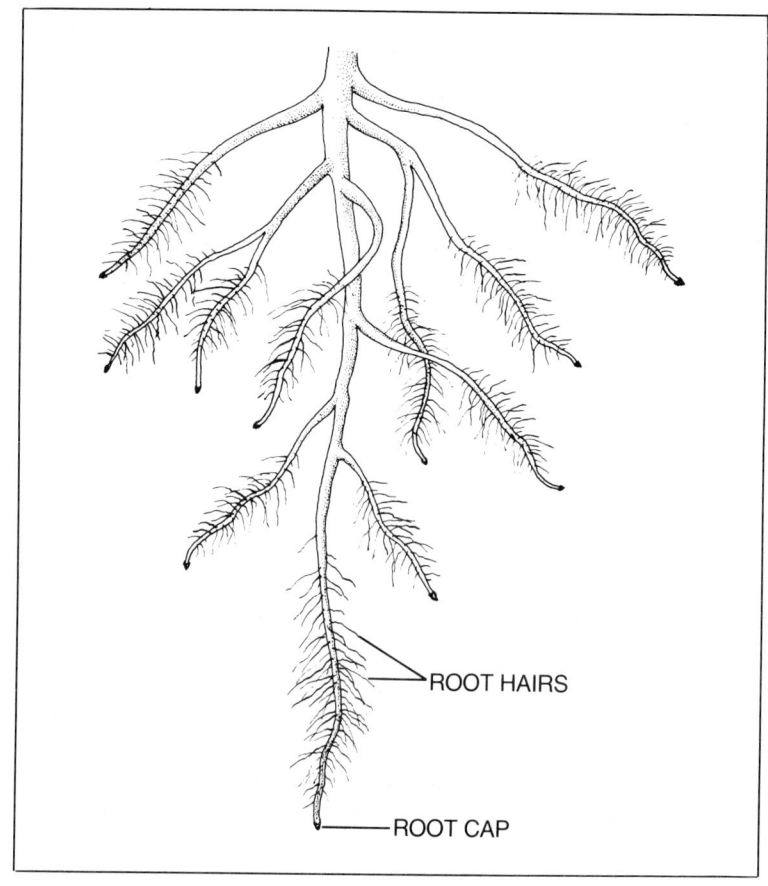

ROOT HAIRS

ROOT CAP

Fig. 24-4
Root tips. What is the function of the root cap? What do the root hairs do?

pressure to drive the root tip forward through the soil. You may wonder why these delicate, thin-walled growing cells are not killed as they scrape against the soil particles. They are protected from this danger by the ***root cap.*** The root cap really *looks* like a cap. It is a layer of thick-walled cells covering the end of the root. See Fig. 24-4. As the root pushes through the ground, the outer cells of this cap are worn off. They are replaced by new cells that form underneath.

root cap

Water absorption by roots. The cells that make up the outer surface of a young root are thin walled and are able to absorb water molecules from the soil. The amount of water that can be absorbed depends upon the amount of root surface that is exposed to the water. The surface area of the root is greatly increased by the growth of structures called ***root hairs.***

A root hair is a long, slender bulge extending outward from a cell at the surface of a root. See Fig. 24-6. It is very well suited

root hairs

for absorbing water. Water diffuses inward through the thin cell membrane that covers the root hair. Then it diffuses inward from cell to cell until it enters the vascular tissue.

The center of a root contains vascular tissue. See Fig. 24-5. Branch roots start from this central region and grow outward through the outer layers of the root. Water and minerals pass up through the vascular tissue of the root and stem and into the leaves. Food from the leaves moves downward through the vascular tissue to feed cells in the stem and roots. Some of the food may be stored in root or stem cells to be used at a later time.

Roots are made up of living cells. These cells need oxygen. If the roots of an ordinary land plant were to grow into the ground water below the water table, they would not get enough oxygen.

Study Fig. 24-6 carefully. It shows the soil particles, which are mostly tiny bits of rock. The water film that clings to the surface of each soil particle is in color. Air is shown in white. See how the root hairs grow in among the soil particles and

Fig. 24-5

A cross section of a young root. Notice that there is a root cap on the branch root even before it breaks out through the surface of the root.

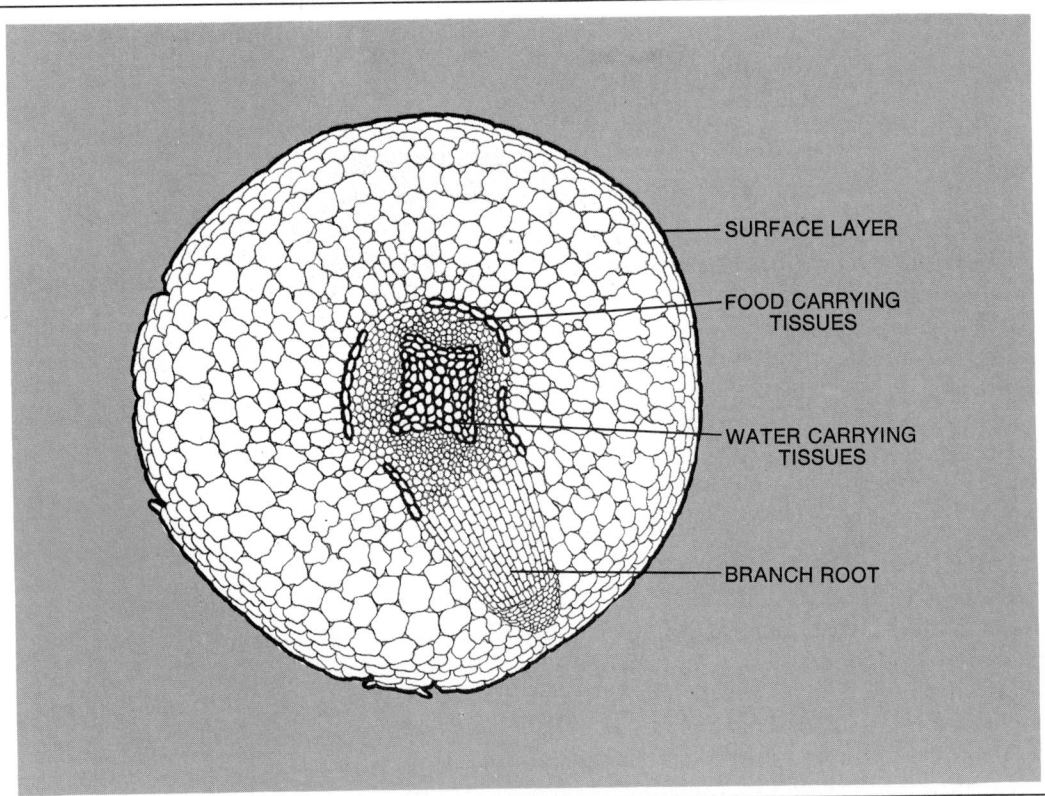

SURFACE LAYER

FOOD CARRYING TISSUES

WATER CARRYING TISSUES

BRANCH ROOT

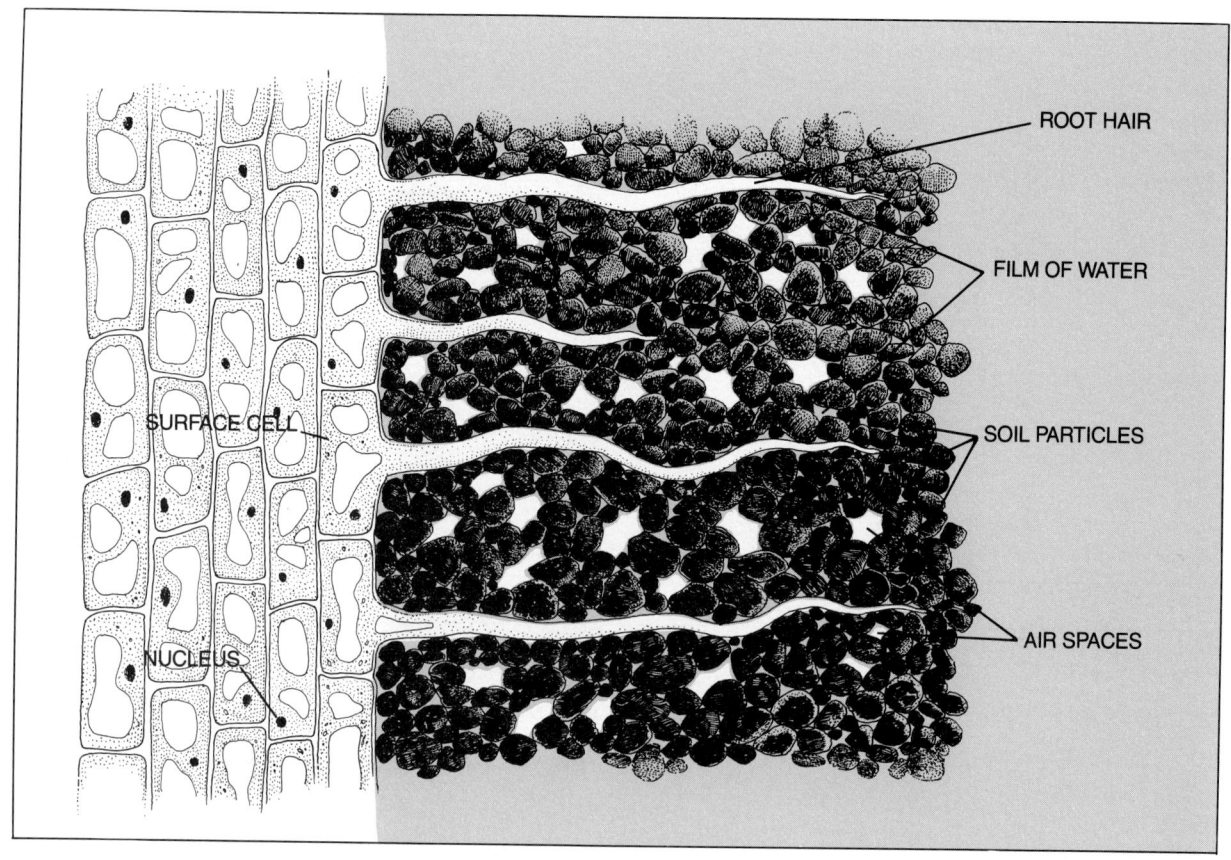

Labels on figure: ROOT HAIR, FILM OF WATER, SOIL PARTICLES, AIR SPACES, SURFACE CELL, NUCLEUS

make close contact with the water film. Water diffuses from these thin water films into the root hairs. Root hairs are surprisingly efficient. They continue to get water from soil that feels quite dry to touch.

Since root hairs are so small and delicate, they do not last very long. The plant must grow some new roots each year to produce new root hairs.

Types of root systems. The root systems of plants have a variety of forms. Some plants have many roots that are almost the same size. Grasses are an example of this type. Others have one main root growing straight down into the soil. All roots branching from it are much smaller. Burdock and dandelion have roots of this type. Some roots are thick and fleshy. These are modified for food storage. The carrot, radish, and turnip are roots of this type.

In general, plants adapted to high ground with dry soil have a root system that goes down deep. In these locations the water table is deep also. If roots go far down, they reach damp soil near this water table.

Fig. 24-6
Root hairs in the soil. Notice how much more area they add to the surface of the root. Why is this important?

Science Reading Skill: *Compare and contrast* the different types of root systems.

Enrichment: Numerous plant roots have an economic importance as food. Taro plants are grown in the Pacific areas and the cassava is popular in tropical countries. Farina and tapioca are produced from the roots of the cassava plant.

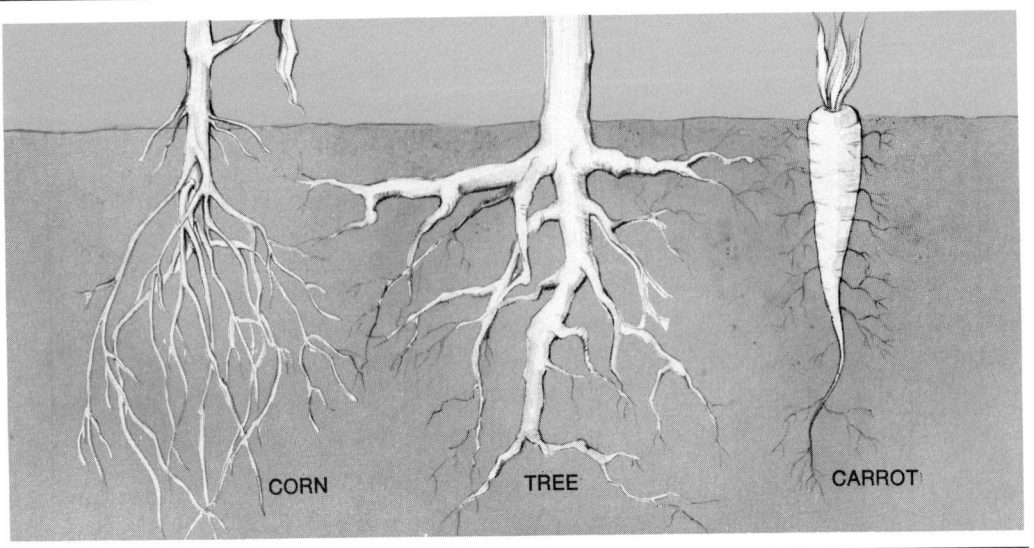

WATER TABLE

Fig. 24-7

Top: Three different types of root systems. To what type of location is each adapted? Bottom: Why do roots stay above the water table?

On low ground the water table is often very near the surface. During a wet spring it may even be above the surface. In other words, the ground becomes flooded. Plants adapted to these locations have shallow, spreading root systems. If you have ever walked through a lowland forest you have probably stumbled over tree roots lying at the surface of the ground, where they can obtain oxygen. See Fig. 24-7.

In many deserts there is no water table anywhere near the surface. Rain falls, and the water evaporates without soaking very far into the soil. Cactus plants are adapted to this situation by having a shallow, widely spreading root system. They soak up water when it is available and store it in their stems. Then they live on this stored water for weeks or months until there is more rainfall.

Absorption of minerals. Many elements a plant needs must come from the soil. These elements come in the form of compounds that we call *mineral salts*. Dissolved mineral salts diffuse into the root hairs in the same way that water does. Topsoil contains the best supply of minerals, so most roots are in the upper 70 centimeters of soil. Roots that go deeper are used more for water absorption than mineral absorption.

SUMMARY

The stems of land plants hold the leaves up to the sunlight. Vascular tissue in stems transports food and water between the leaves and roots. Some plants have vascular bundles throughout the stem. In others they form a ring inside the stem. Stems with bundles in rings may become woody by adding new layers of wood each year. Water travels through the wood, and food travels through the inner bark.

Roots anchor plants to the soil. They absorb water and dissolved minerals through their root hairs. The vascular tissue of roots connects with the vascular tissue of stems.

Activity

Observing Stems and Roots

A. Study prepared microscope slides showing cross section views of the three types of stems discussed in this chapter. Locate the water-carrying and the food-carrying vascular tissues.

1. Make drawings of each type and label the parts.

B. Study slides showing long sections of onion root tips. Look at the root cap at the end of the root.

2. Draw and label a diagram of what you see.

C. Obtain some radish seeds that have been soaked in water and allowed to sprout. Study the roots under a hand lens or a binocular microscope. Notice the fuzzy appearance of these roots.

3. What is this fuzz?

4. What does it do for the plant?

Word Quiz

Choose the number from **Column B** that best matches each letter in **Column A.** Do not write in this book.

Column A	**Column B**
a. vascular bundle	**1.** protects root tip
b. woody stem	**2.** contains tissue that carries food and water throughout the plant
c. growth ring	**3.** absorbs water
d. root cap	**4.** contains a solid mass of water-carrying tissue
e. root hair	**5.** layer of wood formed in a single year

Check Your Facts

1. What are the main functions of stems?
2. What are the two types of vascular tissue?
3. Describe the arrangement of the vascular tissue in the three main types of stems. You may use drawings as part of your explanation.
4. What part of a tree trunk carries water and what part carries food?
5. How can you tell the age of a tree?
6. What do tree buds contain?
7. What part of a stem grows in length? In thickness?
8. What are the main functions of roots?
9. What part of a root grows in length?
10. What is a root cap? What is its function?
11. What are root hairs? What is their function?

Thought Questions

Science Reading Skills 1. & 2. *Problem Solving*

1. A walnut tree has been growing 20 centimeters taller each year. If you drive a nail into the trunk one meter above the ground, about how high will it be in 20 years?
2. Explain why it is easier for the wind to blow down trees in a swamp than trees growing on higher ground.

Plants are not just important, they are essential.

The number of jobs connected with plants is immense. We have already touched on those dealing with forest management. The paper and pulp industry supports thousands of people as well, while several million more work directly on farms. Available jobs vary widely depending on the level of technical skill involved. For details write:

The Department of
    Agriculture
Washington, DC 20250

The number of people involved in agribusiness is even larger. For an overview see:

Jobs in Agribusiness
by Robert J. Houlehen
New York : Lothrop, Lee and
    Shepard Company   1974

Jobs exist that involve direct contact with plants and require knowledge of plant varieties and their growing conditions. On a small to moderate business level a florist is a retailer who is the final distributor of nursery grown flowers and plants. Florists must have good color vision,

HRW: Ken Karp

manual dexterity, and the ability to arrange flowers. Estimating the customers' demand on a weekly basis is essential, too. Bad business sense as well as poor horticultural knowledge will result in damaged plants. Basic arithmetic and accounting ability will be helpful. Summer work in a florist's shop, or in a nursery where plants are developed, or volunteer work in a public botanical garden will provide good background. Experience in businesses with perishable items, e.g., food

products, would also be useful since learning to assess the demand and to market an item as quickly as possible is essential to the selling of cut flowers particularly.

The title of gardener and groundskeeper may be applied in several ways. Positions of this nature may be found in public parks, as well as in landscaping companies. Many large industries have relocated away from cities to sites where there is a great deal of land. They sometimes hire seasonal help to assist in basic maintenance tasks. This may be a way to acquire early experience. Usually, however, major landscaping is "contracted out" to a professional landscaping company. Some of these deal with the outdoor settings and others are concerned with interior design.

Interior landscape design is more concerned with plants, smaller trees and flowers. This particular industry is growing because building designers have started to include spaces

sign are required. Training in flower arranging would be useful.

Plants and plant by-products are the raw material of the textile industry. Some synthetic fibers are entirely chemically produced, but we still produce many fabrics using natural fibers. Plant and animal fibers are the sources of cotton and linen, while cellulose from cotton linters and wood pulp are used for certain synthetic fabrics. Corn produces plant protein which can be spun into fibers. Most of the weaving of fabric, whether natural or synthetic, is done by various kinds of machinery. Jobs involve operating and tending these machines, e.g., quiller operator, precise winder, reeling tender, cloth printer. The technical knowledge necessary to this work is learned on the job.

New fabrics are, of course, created and new designs developed. A chemistry background is important in the research end of the synthetic fiber industry, while training and education in fashion and textiles is esssential to a career as a designer.

appropriate to vegetation. "Atriums," which are patios with greenery and an outdoor appearance, perhaps even a waterfall, are being built into immense urban skyscrapers. Designers and plant specialists work together to produce green spaces within the framework of cement, glass, and steel.

Another industry that has grown up around interior landscaping is that of artificial plants and flowers. While many, though not all, of these items are manufactured overseas, there are retail businesses that specialize in creating "permanent" floral designs for private homes and businesses. Artistic intuition and a sense of interior de-

Acquire skill and expertise with plants through your leisure activities.

HRW: Ken Karp

There are many leisure activities through which a person can acquire skill and expertise with plants. Gardening is, of course, the most obvious. You may not have land available to you. Perhaps a neighbor does or perhaps a vacant urban lot could be used. Often public gardens have volunteer programs where people both learn about and care for the plants on the premises. Scouting groups and 4-H clubs often organize projects in cooperation with public gardens, where group members can play a special role in the construction of new sections.

Using plants for decorative art is both a recreational pastime and, for some, a business as we noted in the previous section. Pine cones and wreaths spell winter's seasonal greetings, while leaves and boughs make handsome displays at anytime of the year. The skill of making decorative arrangements produces welcome gifts as well as giving you a craft you can teach at summer camp or in community programs.

Few people, except those interested in traditional crafts, know how to spin yarn, but many weave, and many more knit, crochet, and sew. Needlepoint has also increased in popularity, especially since it has received publicity as an activity which includes male enthusiasts.

Skill in designing, making, or repairing clothes can be valuable, whether used on your behalf, or on behalf of others. Social organizations receive donations of clothing for the needy that must be repaired. This work is usually done by volunteers.

Art work done with fabric and fiber can be both beautiful and practical. Macrame, the knotting of rope and yarn into elaborate designs, is derived from sailors' necessary skill with knots. Collages of cloth and batik are other examples of how fabric can be used creatively. Each of these arts requires a commitment to learning about the materials with which you are working. Besides that, artists usually look for an audience who will appreciate their work and learn from it. The skill and art can be shared in this way.

Your desire to learn can become a means of serving your community, first by volunteering in order to acquire a skill or expertise, and then through the teaching or expression of your craft.

Bruce Roberts/Photo Researchers, Inc.

CONSUMER SCIENCE IN...

The Miami Herald
Educators eye way to humanize ag education

GAINESVILLE—(UPI)—Down on the farm, Shakespeare doesn't mix too well with sloppin' the hogs...and vice versa.

Educators from 60 of the nation's 100 land-grant colleges and 60 of the 100 top-ranked liberal arts colleges will gather Jan. 3-7 in Gainesville to discuss how to humanize agricultural education and include more about agriculture in other courses.

Among the featured speakers will be U.S. Agriculture Secretary John R. Block.

"Two years of testing at the University of Florida has proved that including agricultural information in history, English, anthropology and philosophy classes makes UF students more aware of the ethical, cultural and philosophical aspects of agriculture, food and nutrition," Dr. Richard Haynes, who directs UF's Humanities and Agriculture program, said Thursdsay.

"With issues like pesticides Temik and EDB and water purity facing the American public, that awareness is increasingly important," Haynes said.

On the other hand, Dr. Ray Lanier, a UF agriculture professor, said that there is much ignorance about farming among the nonfarming population.

"Just 20 years ago, agricultural issues were decided by legislators who knew something about agriculture," Lanier said. "Now, only 3 per cent of the American population is in agriculture. And most people, including legislators, are estranged from it."

Meanwhile, voters' market baskets are feeling the effects of political, social and moral issues related to agriculture—such as the environment, water quality, water quantity, pesticides, herbicides, land use and tax laws.

"City and country folk have a common set of problems," said Lanier. "They both need water, clean water, and they both have environmental effects that affect each other."

While only 3% of the population is "technically" involved in agriculture, a much higher percentage is involved in the food and fiber industry and 100% of the population eats. New food technologies, herbicides, and changing lifestyles are only a few factors that affect eating habits.

If you were designing a course of study for an agricultural school, what subjects would you include? If you were creating a course for urban students to help them understand agriculture and related businesses, what would you include?

Austin American-Statesman
State finds suspected carcinogen in grain products

By JOHN C. HENRY

State agriculture officials last month found ethylene dibromide, a suspected carcinogen known as EDB, in grain products sold in Texas and 10 days ago asked health officials to help in deciding if the items need to be removed from grocery stores.

However, information about the chemical and its potential dangers was not issued by Agriculture Commissioner Jim Hightower's office until most of the Capitol press corps returned Tuesday from the Christmas-New Year's holidays.

AFTER FINDING food products with levels of the chemical up to 90 times higher than allowed in Florida, Hightower and his staff wrote state and federal officials 10 days ago urging them to set standards "so our consumers can know what processed grain products are safe to eat."

"Rather than use Texas consumers as test mice, let's get this stuff off the shelves of Texas grocery stores," Hightower said in a statement delivered Tuesday. Assistant Agriculture Commissioner Ron White stressed that "there is no need for any consumer panic since one box of grits won't kill anyone."

Given the contamination described in this article, some states have taken the affected cereals off the shelves. Others have not established safe levels for this particular carcinogen, ethylene dibromide, and have decided not to take action until more information is available. What dilemmas does this present to the consumer? How do consumers decide which authority is correct? What information would you need to make a balanced decision?

CHAPTER 25   Sponges and Coelenterates

objectives

After you read this chapter, you should be able to:

__**Explain** the importance of sponges and coelenterates

__**Compare** the structure of sponges and coelenterates

__**Explain** what is meant by division of labor

__**Describe** how filter feeders get their food

Science Reading Skill: *Compare and contrast the different methods of obtaining food.*

The photograph above shows corals and sponges growing in the sea. They may look like plants to you, but they are actually members of the two animal phyla that you will study in this chapter.

WHAT IS AN ANIMAL?

There are three ways in which living things can "make a living." That is, there are three ways in which they get food.

There is the *food-maker* method. The green plants and the algae do this when they produce food by photosynthesis.

Fungi and most bacteria use the *decomposer* method of getting food. They give off digestive enzymes that dissolve food. Then they absorb this liquid food by diffusion through their cell membranes. This results in decay of the food material.

Animals and protozoa use a third method of food-getting. They take food into their bodies and digest it there. Nearly all of the characteristics by which we recognize animals are simply adaptations for finding and eating food.

Most animals can move around to search for food. Most have good sense organs, so they are able to detect the presence of food. Animals usually have mouths for eating. Most have a nervous system to control their complicated actions during food-getting.

Plants do not need and do not have any of this equipment. Animals, on the other hand, do not need roots or leaves, therefore they do not have them.

Notice that there are protists using each of these methods. Some use more than one of them. It is only the more complicated, multi-cellular food-makers that are called *plants*. The complicated, multi-cellular types that digest food inside their bodies are called *animals*.

Some protozoa live in colonies. A colony is a group of cells living together in a ball or cluster. Could we call this a multi-cellular animal? We do not do so because each cell in a colony does everything for itself. In a multi-cellular plant or animal, different cells do different things. The simplest animals that can be called multi-cellular are the sponges.

THE SPONGE PHYLUM

You may be surprised to learn that *sponges* are animals. For many years, they were thought to be plants. They grow attached to some object in the water and cannot move from place to place. They look like plants more than they look like animals. You will understand sponges better if you study a simple one first. The simple sponge in Fig. 25-1 is about one centimeter high and is hollow with an open top. The base is attached to a rock. A closer look shows that there are many pores opening through the wall of the sponge.

If you dropped a little ink in the water next to a simple sponge, you would find that the sponge was pumping water. The inky water would go in through the pores and out through the top opening. The water the sponge takes in through its pores brings oxygen and food into its body.

Enrichment: Sponges are members of the phylum Porifera.

Enrichment: A single sponge only one centimeter wide and 10 centimeters high can pump more than 20 liters of water through its system in one day.

Fig. 25-1

Left: A simple sponge; Right: A cutaway view of the simple sponge. Notice that the feeding cells have flagella that force water through the body cavity. The water passes out through the top opening (see arrows).

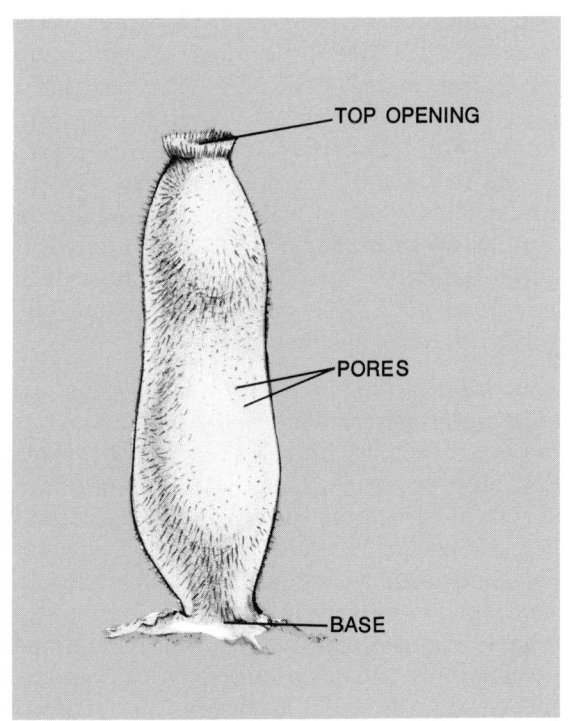

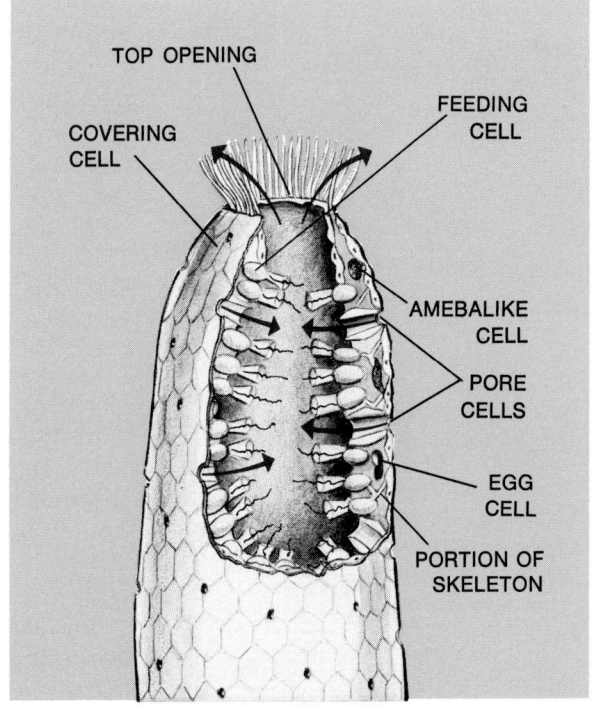

Cells of the sponge. Fig. 25-1 shows a simple sponge with the upper part cut away. Notice the outside layer of *covering cells*. The pores are formed by *pore cells*. The pore is a passage through the middle of the pore cell, like a hole in a doughnut. The cells that line the hollow space are *feeding cells*. Feeding cells have flagella that whip water up through the hollow cavity and out through the top opening.

The feeding cells take in food the same way protozoa do. Bits of food are taken directly into the individual cells and digested. The food used by the sponge is any living or dead material that may be drifting in the water. This includes tiny single-celled algae, protozoa, or bits of dead plants and animals. There are animals in many different phyla that also obtain their food by straining the water. These animals are called *filter feeders*. Filter feeders are especially common in the sea. Sponges are the first of several that we will mention in this book.

Notice that there is another kind of cell, between the covering layer and the feeding cells. It looks something like an ameba. This type of cell moves around in the same way that an ameba does. It takes digested food from the feeding cells and carries it around to other cells of the sponge. Sponges have still another type of cell that forms little spiny structures that make up their skeletons. The skeleton is in the body wall, and it is strong enough to hold the sponge upright in the water.

How sponges reproduce. One way sponges reproduce is by means of sperm and egg cells. The fertilized egg develops into a swimming *larva*; plural, larvae (LAR-vee). The word larva refers to any young animal that is very different from the adult. A caterpillar is the larva of a butterfly. A tadpole is the larva of a frog. The sponge larva is a little saucer-shaped mass of cells covered with flagella on one side. It swims around for a while and then settles down on some solid object, where it grows into an adult sponge. A sponge may also reproduce asexually by developing buds near its base. This explains why you often find whole masses of sponges growing together.

Types of sponges. Besides the simple little sponge that we have described, there are many others. Some of them are as large as your head, and some are bigger. They may be black, red, orange, yellow, blue, or purple. These large sponges have the same kinds of cells as the simple sponge, but they are built up into large masses of tissue. There are passages throughout this mass of cells. These passages connect with many small, hollow spaces, each lined with feeding cells.

Both large and small sponges can live in warm parts of the ocean. In colder waters only smaller sponges can be found. A few sponges even live in fresh water. The freshwater sponges

filter feeders

Enrichment: The ameba-like cell is called an amebocyte. The spiny structures that make up the sponge skeleton are called spicules.

larva

look like a rough coating all over the outside of a sunken stick or rock. Some are green because algae grow among them.

The skeletons of sponges may be made of a large number of simple, three-pointed spines, or they may be more complicated. Some sponge skeletons are made of lime. Some are of a stiff glass-like mineral. Still other sponges have a framework of a softer, flexible material. The sponges that are useful to people all have soft skeletons. The skeleton is the part that we use.

Commercial sponges. The kinds of sponges sold in stores may grow in shallow water just beyond the low tide mark. These can be gathered by wading, or by using hooks on the ends of long poles. Others, in deeper water, are collected by divers. Some are found at depths of over 40 meters. Many sponges are found off the coast of Florida. The West Indies and the Mediterranean also produce commercial sponges.

After live sponges are gathered, they are left in the air for several days while the cells die. Then they are cleaned and trimmed. People buy sponges for cleaning purposes because sponges hold water so well.

Division of labor. In any single-celled organism the cell does everything for itself. It gets its own food and oxygen, and gets rid of its own wastes. The multi-cellular plants and animals produce several different kinds of cells, each with its own job to do. The work of the body is divided up among the different cells. We call this *division of labor*.

The sponge is a simple example of division of labor. It has seven kinds of cells. They are epidermal cells, feeding cells, pore cells, skeleton-forming cells, ameba-like cells, sperm-forming cells, and egg-forming cells. Covering cells give protection, feeding cells obtain food, skeleton-formers provide support, and so on. No single sponge cell can do all of these things.

Division of labor is much more complicated in higher animals than it is in the sponges. Higher animals not only have more kinds of cells, but the cells are organized in more complicated ways to form organs and systems. There are no organs or systems in the sponge.

Because a sponge is so simple, it can sometimes survive when higher organisms can not. If you were to cut a sponge into little pieces and throw the pieces back into the water, many of them would live. Each piece has all the cell types in it. Each can live and grow into a new sponge if it is in a suitable environment.

Fig. 25-2
Sponge fishermen. Cleaned sponges are hanging. Newly caught sponges lie on the deck awaiting cleaning. (Higgins/ U.S. Fish and Wildlife Service)

division of labor

coelenterates

THE COELENTERATE PHYLUM

The *coelenterates* (suh-LEN-ter-ates) include the *Hydra, jellyfish, sea anemones* (uh-NEM-oh-nees), and *corals*. Most of them

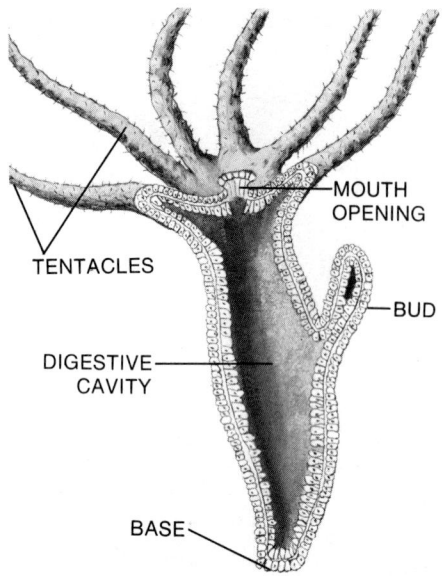

Fig. 25-3
The cell structure of Hydra.

Enrichment: The *Hydra* exhibits a feeding response when sugar is placed into the water. It is this awareness of chemicals that makes it aware of the presence of food.

Science Reading Skill: *Compare and contrast* feeding methods in sponges and *Hydra.*

live in the sea. They often have delicate, filmy-looking bodies. *Hydra* is the only member of this phylum that is really common in fresh water. We shall study it first, because of its simple structure. It will help you to understand the basic body organization of all the animals in the phylum.

The structure of the *Hydra*. In Chapter 7 you studied the types of reproduction carried on by *Hydra*. You may recall that *Hydra* has a tube-shaped body with a mouth opening at the top. A ring of tentacles surrounds the mouth. This small animal is common, but you have probably never noticed it. The *Hydra's* body is only about 2 millimeters long.

Fig. 25-3 is a diagram of *Hydra*. As you see, its body wall has two layers of cells. The tentacles are hollow and have the same two layers. Tentacles are really long, slender bulges in the body wall that extend upward around the mouth.

The hollow space inside *Hydra* is its digestive cavity. The mouth is the only opening into this digestive cavity. Food must enter through the mouth, and undigested materials leave through it. Cells lining the cavity produce digestive juice, which dissolves the food. Dissolved food molecules can then enter the inner layer of cells through their cell membranes. These cells also have the ameba-like ability to take in small solid particles. Some of the digested food diffuses from the inner cells to the outer cell layer.

Hydra can move around, but most of the time it stays in one place with its base attached to some support, with its tentacles trailing out into the water. It can slide along slowly on its base. It can also turn "handsprings." The body bends over, and the tentacles touch the ground. Then the body swings up and over. In this way *Hydra* moves end over end.

How *Hydra* gets food. *Hydra* is a hunter. It lives on other small animals. *Hydra* kills its victims by stinging them to death. Stinging cells are found all over the body of the *Hydra* but are especially numerous on the tentacles. One type of stinging cell shoots out a tiny barbed thread that sticks into the prey and injects poison. Another type of cell shoots out a thread that becomes tangled in the legs or little hairs of the victim.

If you put a live *Hydra* in a small dish of water with a water flea, you will see the hunter in action. You need a magnifying glass or a low-power microscope to watch the water flea swim around and around until it accidentally touches a tentacle. Instantly, it stops swimming and never moves again. The stinging cells have done their work. The water flea is dead, and it is stuck to the tentacle. See Fig. 25-4. Then *Hydra* slowly takes the dead animal in through the mouth opening to the cavity, where it is digested.

Jellyfish. If you could turn *Hydra* upside down and pull the sides of its body out into an umbrella shape it would resemble a jellyfish. Jellyfish are adapted to drifting in the open sea. They are called jellyfish, because there is often a jelly-like material between their two cell layers. The mouth of a jellyfish opens downward, with tentacles hanging around it. Fig. 25-4 shows one kind of jellyfish.

Jellyfish sting and eat small fish and other sea animals that swim into their tentacles. Many jellyfish are just a few centimeters wide, but there are some giant ones with bodies 2 meters across and tentacles up to 15 meters long.

Jellyfish are often so transparent that people fail to see them in the water. If the filmy, cellophane-like tentacles touch a swimmer's skin, they cause a burning sensation. The thousands of stinging cells that puncture the skin produce painful blisters. Usually this is not very serious, but a few people have been killed when they got too many stings all at once.

Enrichment: This jelly-like material is called mesoglea.

Enrichment: The Portuguese man-of-war is a common poisonous jellyfish that is actually made up of a colony of smaller jellyfish. Each member of the colony has a special function.

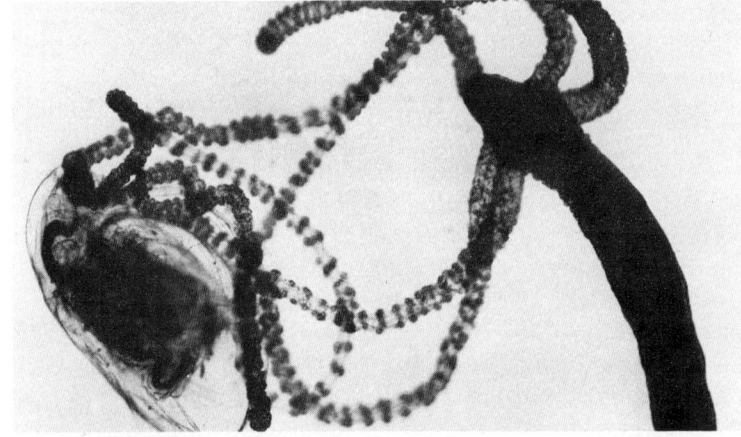

Fig. 25-4
Top: Hydra *catching food; Bottom: A jellyfish. What do these organisms have in common? (Eric V. Grave; Douglas Faulkner)*

255

Sea anemones. The *sea anemones* look like giant *Hydras*. They may be several centimeters across with hundreds of tentacles around the mouth. Many are brightly colored, and look like beautiful, big flowers. This is why they are named after a flower—the anemone. Like *Hydra* they feed on small animals coming near the tentacles. See Fig. 25-5.

You can find sea anemones along some seacoasts at low tide. A few people have kept them alive in large tanks of sea water. They are very attractive and a group of them looks like a flower garden, but this garden must be fed a bucket of live minnows now and then!

The corals. *Corals* are important because they build coral reefs. A *reef* is a shallow, rocky ridge under water. The individual coral animal is like a very small anemone or a large *Hydra*. Each animal forms a limestone crater around its body. The coral animals use these craters for protection. They reach out from the craters to catch food with their tentacles. The little coral animals multiply by budding until they form very large groups. Their craters are connected, forming one big mass of limestone. This is the reef. The thousands of coral animals continue to build the reef upward until it may even stick out above water at low tide. Sponges and algae also help to build up these reefs. See Fig. 25-6.

Coral reefs appear only in shallow, warm seas. Bermuda, Florida, and the West Indies all have coral reefs, but the best examples are found in the Pacific. Most of the islands of the South Seas have coral reefs around them. These reefs act like *breakwaters*, which break the force of the waves and protect the shore from erosion. The biggest coral reef in the world is the Great Barrier Reef of Australia. It is 1,700 kilometers long. Coral reefs have many caves and hollows in them. They are good hiding places for all kinds of sea animals—fish, crabs, octopus, and many others.

reef

Science Reading Skill: *Compare and contrast* the structure of sponges and coelenterates.

Reference: See *Animals Without Backbones* by R. Buchsbaum and *The Lower Animals: Living Invertebrates of the World* by R. Buchsbaum and L. Milne.

Fig. 25-5
A sea anemone. It looks very much like a giant tough-bodied Hydra. *The tentacles can sting small fish to death.* (Dr. E. R. Degginger)

Fig. 25-6
This photograph shows several coral animals. Notice how each is similar to Hydra. *Each animal forms a limestone crater around itself. (Douglas Faulkner)*

A coral reef in the right place can make a protected harbor. In the wrong place, it can wreck a ship. Some land masses have also been formed from coral reefs. The long line of islands known as the Florida Keys is really a line of old coral reefs that were formed many years ago when the sea level was higher than it is today.

SUMMARY

Sponges are simple, multi-cellular animals. In a sponge different cells perform different jobs. Sponges are filter feeders, taking bits of food from the water that they pump through their bodies. Feeding cells, which line the hollow body, take in the food and digest it. The sponge skeleton is a simple framework that stiffens the body.

Hydra, jellyfish, sea anemones, and corals are members of the coelenterate phylum. Coelenterates have tentacles armed with stinging cells that are used to capture food. Corals form large colonies that are supported by lime skeletons. Coral skeletons, together with skeletons of algae and sponges often form reefs, sometimes of great size.

Activity

Observing Hydra

A. Place a live *Hydra* in a small glass dish with water, and examine it under a hand lens or a microscope at low power.

1. Can you see it move?

2. How long is it?

3. What does it do when you poke it with a probe? Notice how it stretches out its tentacles when it is left undisturbed.

B. Place a water flea or other small animal in the dish and observe it. The animal swims around the dish. Sooner or later it will touch the tentacles of the *Hydra.*

4. Describe what happens then.

Word Quiz

From the new words you learned in this chapter, write the word that correctly completes each statement. Do not write in this book.

When different cells do different jobs in the body the arrangement is called _____. A young animal that looks very different from the adult is called a(n) _____. A(n) _____ takes tiny bits of drifting food from the water. A(n) _____ is a place where the sea bottom lies near the surface. The *Hydra* and the jellyfish belong to a group of organisms called the _____.

Check Your Facts

Science Reading Skill

6. Compare and Contrast

1. What do we mean by division of labor? How does a sponge illustrate division of labor?
2. How do sponges get food? How do they digest it?
3. How are sponges gathered for commercial use?
4. How does *Hydra* get its food?
5. How many layers of cells are there in the body wall of a coelenterate?
6. Briefly describe the appearance of jellyfish, sea anemones, and corals.
7. What makes corals important to other sea life and to people?

Thought Questions

Science Reading Skill

1. & 3. Problem Solving

1. How could a sponge fisher make sure that there would be plenty of sponges to gather in the future?
2. What is it about the structure of sponges and coelenterates that makes them unlikely to be useful as food for people?
3. Why can't sponges or coelenterates live on land?

CHAPTER 26 WORMS

The photograph above shows the food-gathering organs of filter-feeding worms, which live in the sea. Do they look to you like relatives of the earthworm?

When we call something a worm, we are not classifying it. We are describing it. The word worm simply means that the animal we are talking about has a soft, narrow body. We will study three phyla of worms. Calling them worms does not mean that they are closely related.

FREE-LIVING FLATWORMS

Some members of the *flatworm* phylum are parasites and some are free living. A *free-living* animal gathers food for itself. It does not live on, or in a host organism.

Planaria. A good example of a free-living flatworm is *Planaria*. Fig. 26-1 shows what a *Planaria* looks like. This little worm is very flat and it is about one centimeter long. It lives under rocks and leaves in fresh water.

Planaria moves by means of two layers of muscles in its body wall and by thousands of cilia which cover its lower surface.

The mouth is on the underside, a little behind the middle. The throat is a flexible tube which can be pushed out through the mouth to search for food. *Planaria* eats decaying bits of plant or animal material. It also eats live insect larvae or other water animals that are small enough to swallow and slow enough to catch.

The throat opens into an *intestine*. The intestine of an animal is its main digestive organ. In *Planaria*, the intestine has three main branches, as shown in Fig. 26-2 on page 261. The mouth is its only connection with the outside, so undigested materials must pass out through the mouth opening.

In its head, *Planaria* has a mass of nerve cells, which form the brain. Two large nerves go back through the body from the

objectives

AFTER YOU READ THIS CHAPTER, YOU SHOULD BE ABLE TO:

—**Compare** the structures of flatworms, roundworms, and segmented worms

—**Trace** the life cycles of several parasitic worms

—**Suggest** ways to protect yourself from each parasite

—**Describe** the effect of earthworms upon the soil

—**Compare** earthworms, sandworms, leeches, and filter-feeding worms

flatworm

Enrichment: The flatworms belong to the phylum Platyhelminthes.

intestine

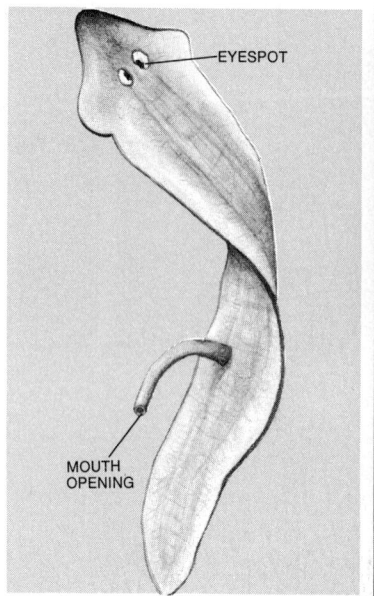
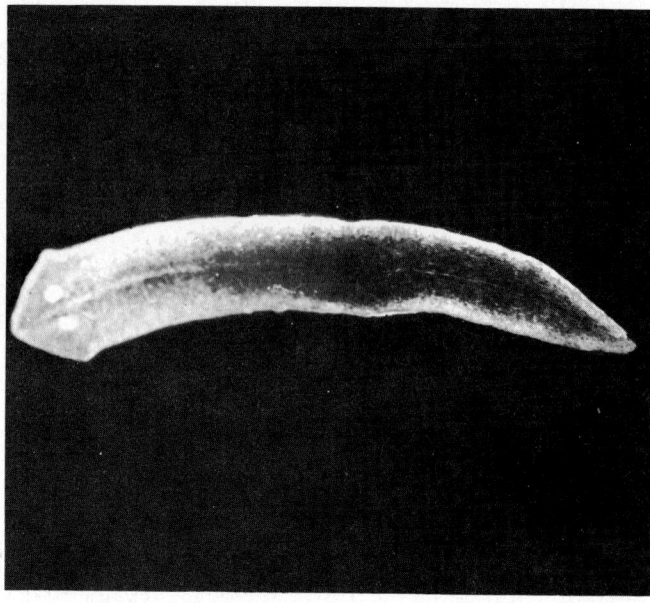

Fig. 26-1

Left: A drawing of Planaria. *Note how the throat sticks out to take in food; Right: A photograph of a living Planaria. (Hugh Spencer)*

brain, as shown in Fig. 26-2. These large nerves give off branch nerves to all parts of the body. This simple nervous system controls the various things that this animal can do.

The eyespots of *Planaria* allow the animal to tell light from dark and the direction the light is coming from. *Planaria* moves away from bright light. The eyespots are set at an angle, so this worm is sometimes called the "cross-eyed" worm.

As you see, *Planaria* is much better organized than *Hydra*. *Hydra* has some tissues, and its tentacles are simple organs. *Planaria* has several tissues, organs, and systems. We have mentioned the nervous and digestive systems. It also has covering, reproductive, and excretory systems.

PARASITIC FLATWORMS

Planaria and similar flatworms find their own food. They are variety eaters. There are other members of the flatworm phylum

Fig. 26-2

The digestive and nervous systems of Planaria. *Notice how the brain connects with the entire body by way of nerves. Notice also how the intestine branches all over the body.*

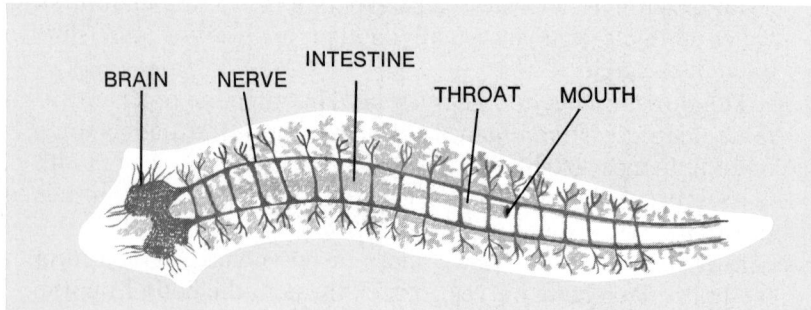

that live as parasites in larger animals. These include the flukes and the tapeworms. They get their food entirely from the animal host.

Dogs, cats, chickens, cows, and other domestic animals usually contain a variety of parasitic worms. So do all wild animals. These parasites may be tapeworms or flukes, or they may be members of the worm phylum that we will study next. Most of these worms are well-adapted to the host and do not cause great harm. After all, if the host dies, so does the parasite. Natural selection tends to get rid of parasites that handicap their hosts too much.

Tapeworms. Adult *tapeworms* live in the intestines of higher animals including people. They are adapted in many ways to a parasitic life. They have no mouth or intestine. They live in the intestines of the host and absorb molecules of food that the host has digested. They have no eyes and do not move around much. Their ability to reproduce is highly developed.

The life of a parasite is very easy, once it has found a host. The host gets food for both of them and the host avoids enemies. All of the ordinary dangers and problems that an animal must meet are taken care of for the parasite by its host. The one big problem a parasite has is finding a host. Most young parasites fail to do this. A high rate of reproduction is needed if the species is to survive. Tapeworms produce millions of eggs.

Fig. 26-3 shows part of an adult tapeworm. The head is really just a special organ for hanging on to the host. It has four suction cups and a ring of hooks. With these, the head clamps on the

Fig. 26-3
Left: A diagram of the species of tapeworm found in humans; Right: A photograph of an adult tapeworm. (Walter Dawn)

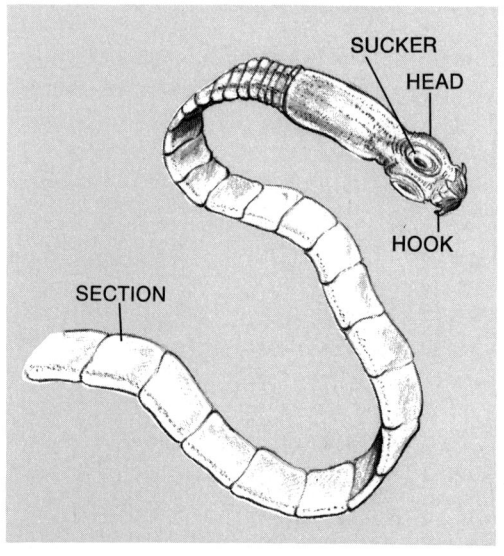

SUCKER
HEAD
HOOK
SECTION

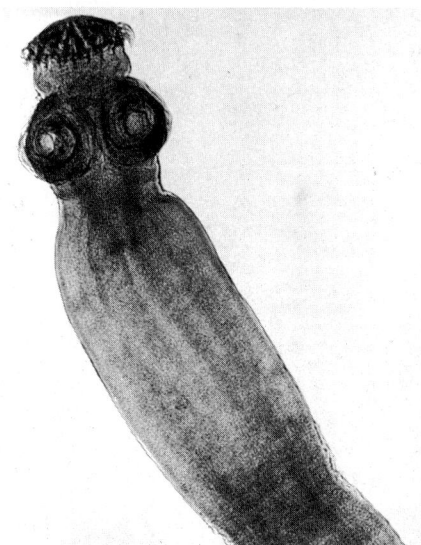

lining of the host's intestine. Food keeps moving through the intestine of the host. The worm needs to be anchored, so it will not be carried away with the food. New sections keep budding off just behind the head of the tapeworm. The older sections, which grow larger, are pushed farther and farther back as new ones form. The result of this is a long, flat worm that really *looks* like a piece of tape. There are many species of tapeworms. Some are so small you would need a microscope to study them. Others are much larger. One tapeworm that lives in the human body is less than 2 centimeters long, but another type can grow to 15 meters in length.

The life cycle of a tapeworm. Each section of a tapeworm contains a complete set of male and female reproductive organs. Old sections near the end of a tapeworm's body are much larger than new sections near the head. The old sections are large and swollen because they are filled with hundreds or thousands of tiny eggs. These egg-filled sections break loose when they reach the end of the worm. They are carried through the intestine with the digested food and leave the host in the *feces* (FEE-seez). Feces are the undigested wastes that come from the intestines of an animal. The feces fall to the ground and decay. So do the tapeworm sections, but the eggs remain alive. They get scattered all over the ground, and they can live a long time.

The eggs may happen to be swallowed by some animal with its food. If it is the right kind of animal for that kind of tapeworm, the eggs will hatch and grow inside the animal's body. Of course, most tapeworm eggs are not swallowed, or they are swallowed by the wrong animal. Then they do not survive. This is why a high rate of reproduction is one of the adaptations of a parasite.

Tapeworms usually have two hosts. The adult worm lives in one kind of animal, but the larva lives in another kind of animal. The host that has the adult worm is the *main host* (or primary host). The host with the larval worm is the *secondary host*.

People are main hosts for several tapeworm species; secondary hosts are such animals as pigs, cows, and fish. The most common tapeworm found in dogs has rabbits as its secondary hosts.

As an example of the life cycle of a tapeworm, let us study the *beef tapeworm*. It has people and cows as its two hosts. In the cow, the tapeworm egg hatches into a little larva that travels in the bloodstream for a few days and then settles in a muscle. There it forms a cyst, or capsule around itself. Inside the cyst the little tapeworm forms a head and a few sections. The cyst is small and not easily seen.

People may eat meat from cows infected with tapeworm. If such beef is not cooked enough, the worm may still be alive

Science Reading Skill: *Sequencing—* Trace the stages in the life cycle of the parasites discussed in this chapter.

feces

when it is eaten. The worm will come out of its cyst, clamp on to the lining of the intestine, and begin to grow. When the tapeworm is fully grown, segments containing eggs pass out of the person's body mixed in the feces. If any of these feces are left on the surface of the ground, some of the eggs may get on grass blades. Cows swallow the grass, and the cycle starts over again.

Tapeworm eggs of this species have little chance of reaching the secondary host if sewage is disposed of in sanitary ways. For this reason few people get beef tapeworms in countries with good sanitation. But sometimes human wastes are allowed to lie on the ground. Also, many people like to eat rare beef. So some people become hosts to this worm. Doctors can cure people of tapeworm by giving medicines that will either kill the worm or, at least, make it relax and let go of the lining of the intestine. It then passes out of the body with solid waste materials.

Some people do not even know when they have a tapeworm. It does not upset them enough to make them realize that anything is wrong. Others feel sick. It depends partly on which kind of tapeworm they have. Some types are much more dangerous than the beef tapeworm.

Flukes. Another group of parasitic flatworms is known as *flukes*. *Liver flukes* look somewhat like *Planaria*. They may be 2 centimeters long, and the adults are found in the livers of higher animals. One species is a serious parasite of human beings in Asia. Snails and fish are the secondary hosts. In America a liver fluke causes illness in sheep. See Fig. 26-4.

flukes

Enrichment: Liver flukes are more common in countries where raw fish is consumed. In Japan, vasts amounts of raw fish fillets called Sashimi are eaten.

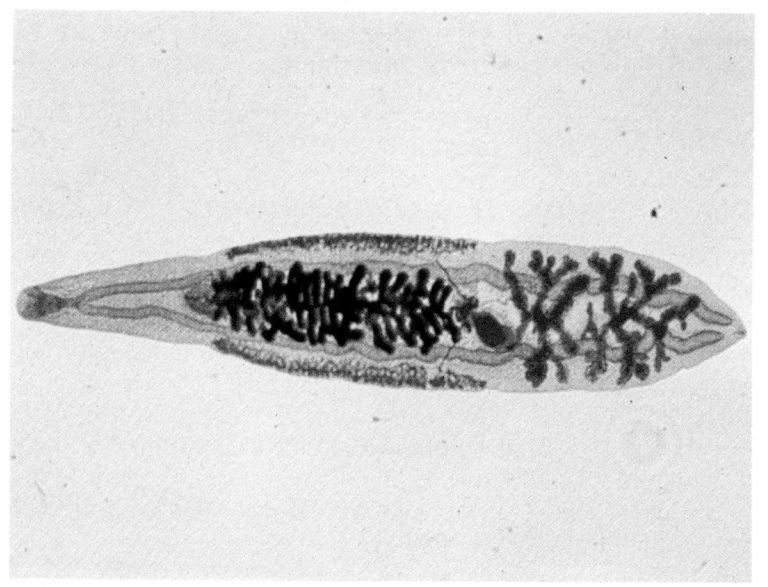

Fig. 26-4
A photograph of a liver fluke. This is a flatworm. (Courtesy Carolina Biological Supply Company)

Blood flukes are very tiny. Several species are really animal parasites. But they can attack humans by mistake and cause discomfort.

The main hosts for these tiny blood flukes are various species of birds (such as ducks or grackles) or some species of mammals (such as a mouse or a muskrat). Alternate stages in the worms' development take place in snails.

Worm eggs are present in the feces of the main host. If this material falls into water, the eggs hatch into tiny larvae. Each larva swims through the water and, if it finds a snail, it bores in. There it multiplies. For each larva entering a snail, many worms finally leave it and swim away. If a bird or mammal of the host species is wading or swimming in the water, some of these worms may burrow through its skin and enter its blood stream. They settle near the intestine and lay eggs, which enter the intestine and mix with the feces of the host.

If you wade in water where these worms are leaving their snail host, hundreds of them could burrow into your skin by mistake. You are the wrong host, so they could not live in you. They would die in your skin, causing a red rash that is commonly called *swimmer's itch*. Some resort owners have had to poison the snails along their lake shores to protect their guests from swimmer's itch.

Human beings *are* the main hosts for three species of blood flukes found in South America, Africa, and Asia. These cause serious disease and death. About 200 million people in 70 countries are infected.

This health problem is caused by water pollution. Snails are infected by worms from human feces entering the water. Later, worms leave the snails and infect people. These people get the flukes in drinking water or from wading in irrigation canals and in other places where the snails live.

Enrichment: This disease is known as schistosomiasis. Several new drugs have been developed in the treatment of this disease which it is hoped will lead to its control.

IMPORTANCE OF THE FLATWORMS

The free-living flatworms like *Planaria* probably do not have any great effect upon the communities in which they live. Of course, they do fit into various food chains. The parasitic flatworms are of more importance because of their effect upon the hosts they attack. They seldom kill their hosts, but they do slow them down. Once there were four or five tapeworm cysts counted in each cubic centimeter of meat taken from a caribou. This would mean there were about 150 thousand worms in the entire deer. This did not seem to handicap the caribou very much, but it surely would have felt better without all those worms!

ROUNDWORMS

It is likely that you have never seen a *roundworm.* Yet round-worms are a very common and successful phylum of animals. They live almost everywhere—in water, in soil, and as parasites in higher animals and plants. Most free-living ones are so small that we do not notice them. The parasites include some fairly large types, but they live out of sight inside their hosts.

The roundworm phylum does not include the common earth-worms, even though they are round. The roundworms are slen-der, round, and not divided into segments like earthworms. Many books and gardening magazines call the roundworms *nematodes* (NEM-uh-toads). This comes from the name of the phylum to which they belong.

The structure of roundworms. The roundworms are only a little more complicated than the flatworms, but their body or-ganization is more like that of the higher animals. There is a body wall made up of a layer of epidermis and two layers of muscle. Inside this, the worm is hollow. The intestine passes through this hollow space. It is a tube with openings at each end—the mouth and the *anus* (AE-nus). The worm takes in food through the mouth, digests it, and absorbs it as it moves along the intestine. Wastes pass out through the anus. There is no mixing of old and new food, as in *Hydra* and *Planaria.* This arrangement, called a "tube-within-a-tube," is found in all of the higher animals.

roundworm

anus

Fig. 26-5
This roundworm is called a vinegar eel. You can see its intestine through the body wall. (Hugh Spencer)

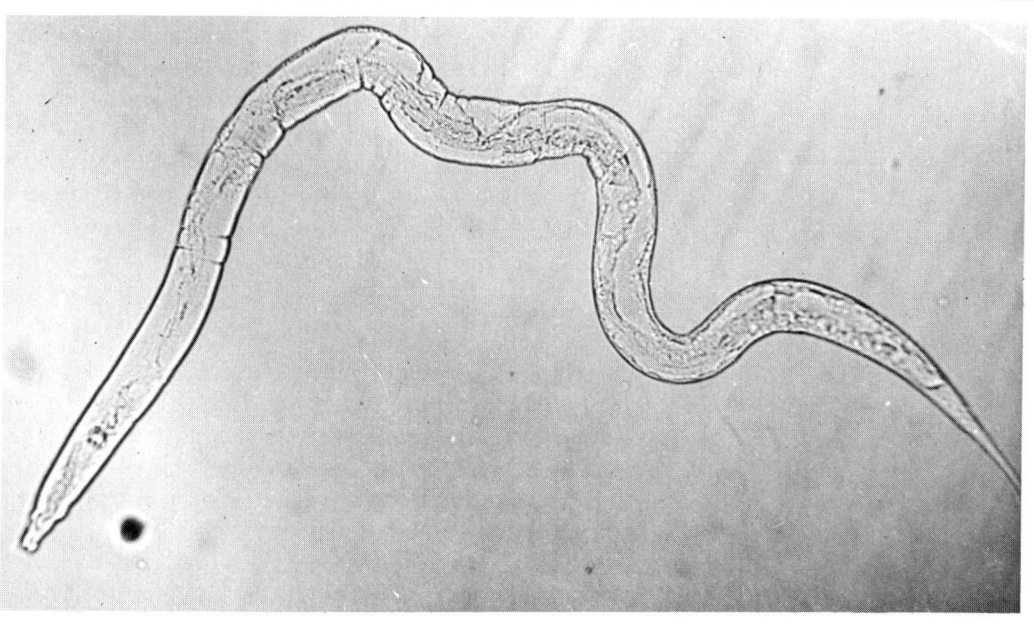

FREE-LIVING ROUNDWORMS

Many tiny roundworms live in the mud on the bottoms of lakes, ponds, and streams. Others live in soil. Some of those in soil attack the roots of plants and cause some damage. They would be far more destructive if it were not for certain molds living in the soil. These molds form sticky loops of fungal tissue that act as traps. They catch and digest the unfortunate roundworms that crawl into them. Fig. 26-5 shows a type of roundworm that may live in homemade cider vinegar. These roundworms are perfectly harmless. They are often called "vinegar eels."

PARASITIC ROUNDWORMS

The roundworms are by far the most numerous of all the parasitic worms.

Ascaris. The most common parasitic roundworm is *Ascaris* (as-CARE-iss). One species, about eight inches long, can live in the human intestine. *Ascaris* feeds on partly digested food.

The female worms lay millions of microscopic eggs, which leave the host in the feces. In regions without sanitary sewage disposal, the eggs are often present in the soil. People may swallow *Ascaris* eggs when they eat vegetables grown in such soil. Or they may get them from their own hands if they did not wash before eating. In some countries of Asia, human sewage is used as fertilizer for growing farm crops. This is good farming, for it returns minerals to the soil; but it is poor sanitation. Nearly everyone there has *Ascaris*. In America, children are the most likely to get this worm. They like to play in the dirt, and they are careless about washing their hands.

Ascaris usually is not dangerous, but it is sometimes quite harmful. Masses of worms can block a human intestine. The adult worms sometimes bore out through the intestine wall and into the main body cavity, causing infection. They may wander about in the body and be found in the liver or some other organ. In these cases death can result.

As in the case of tapeworms, doctors can use medicines to remove *Ascaris* worms from the intestine.

Hookworm. Adult *hookworms* are less than one centimeter long, but thousands may live in the intestine at once. They use blood from the lining of the intestine as food. This steady loss of blood weakens the host. Persons with hookworm may not feel well enough to do a full day's work. These people give the impression of being lazy when actually they are sick.

Eggs laid by these worms leave the host in the feces and hatch on the ground. The larvae rest on the soil or on blades

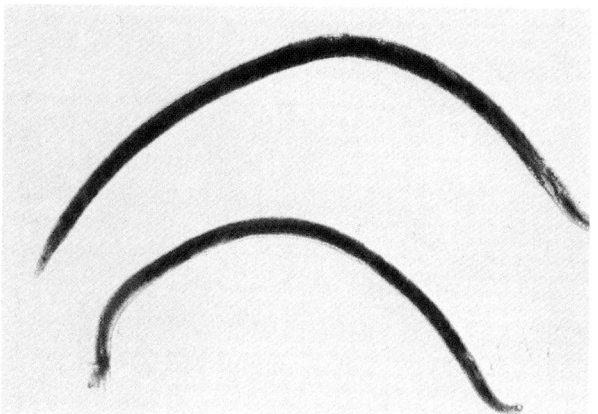 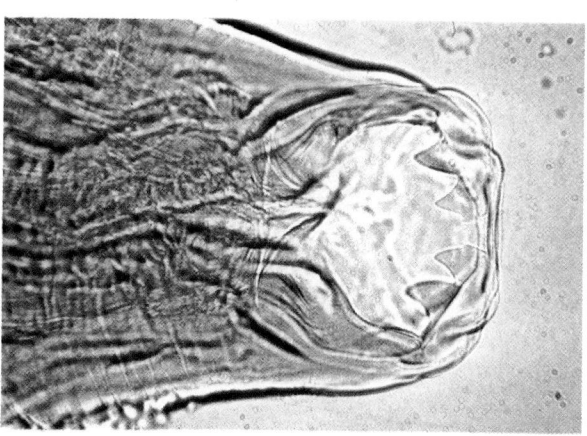

of grass. If someone brushes against them they stick to the skin. Then they bore through the skin and into the bloodstream. They are so small that they are not noticed entering the skin. After traveling through the blood and lungs of the host, they arrive in the intestine, where they feed on blood. See Fig. 26-6.

Hookworms are found in warm climates, including the southern United States. In colder regions, the worms are killed in the soil by freezing. In some warm countries, so many people have hookworm that the whole standard of living is lowered. There are not enough healthy people to do the work. A community can get rid of hookworm by establishing safe disposal of human waste thus preventing the worm eggs from settling on the ground. Hookworms usually enter through the skin of the feet, so people can protect themselves by wearing shoes.

Trichina. We said a well-adapted parasite does not destroy its host, but the *trichina* (trick-EE-na) is always dangerous to people. It is much less harmful to other species. The adult trichina worm is about 6 millimeters long and lives in the intestine of the host. The eggs hatch in the oviduct of the female worm. She injects the larvae into the blood vessels of the intestine. From there they travel in the bloodstream to all parts of the host's body. Finally they bore into muscles and form cysts. When another animal eats this host, the young worms come out of their cysts, grow up, and start producing young of their own. Notice that these worms never leave the body of a host. They are passed on as host eats host. See Fig. 26-7.

People attacked by trichina worms suffer from aching muscles and fever. This happens while the young worms are boring into the muscles to form cysts. If too many worms are present the person may die, or a part of the body may be paralyzed. Once the worms have settled down in the muscles, they usually cause little discomfort.

Fig. 26-6
A hookworm (left) and a close-up of its head (right). (Courtesy Carolina Biological Supply Company)

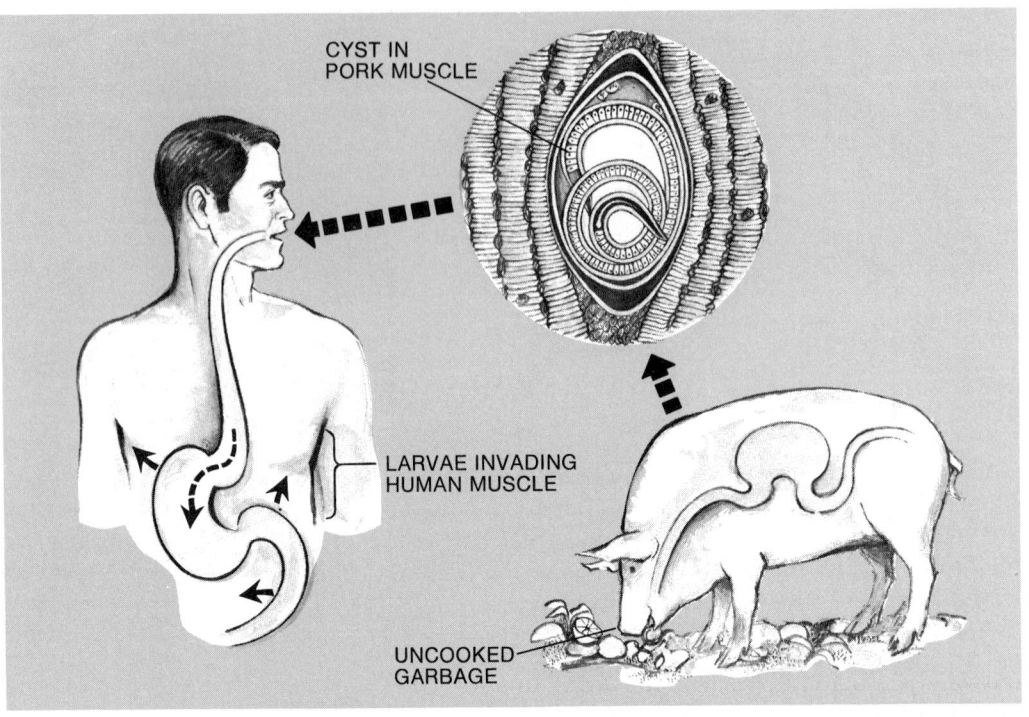

CYST IN
PORK MUSCLE

LARVAE INVADING
HUMAN MUSCLE

UNCOOKED
GARBAGE

Fig. 26-7
A young trichina worm in the muscle of a pig can be eaten by a human. For this reason pork should always be well cooked.

People get trichina worms from eating undercooked meat. This meat must come from an animal that has live trichina cysts in its muscles. Usually this is a pig. This is why pork must be thoroughly cooked before it is eaten.

Pigs, rats, bears, and a number of other animals are hosts of the trichina worm. Pigs may eat dead rats that contain the worm cysts. Pigs may also eat raw meat scraps in garbage. Most states now require that all garbage be steam cooked before it is fed to pigs. This has greatly reduced the number of people infected by trichina worms. More effective rat control has also helped. But even so, to be completely safe from these parasites meat should be cooked long enough to get it thoroughly hot all the way through. Smoked sausage and ham cause the most trouble, because they look and taste good even when they are undercooked.

SEGMENTED WORMS

segmented worms

The *segmented worms* include all worms having their bodies divided into sections (segments) except the tapeworms. *Earthworms* belong in this phylum. When you look at an earthworm, you can see the crosslines marking the segments. See Fig. 26-8. Members of this phylum are more complicated than animals you have studied so far.

Enrichment: Segmented worms belong to the phylum Annelida.

Structure of the earthworm. The earthworm has a *body wall* consisting of an outer *skin* and two layers of *muscles*. The muscles are used for movement. Fig. 26-9 shows the internal structure of the earthworm.

Inside the body wall is a *body cavity* through which the *intestine* passes. The front part of the intestine is modified to form special structures. There is a muscular *throat* that sucks food into the mouth, a **crop** that holds foods, and a **gizzard** that grinds food. Then the tube-shaped intestine runs the entire length of the worm. Above the throat there is a very tiny *brain* connected to a *nerve cord,* which lies under the intestine. This nervous system controls the actions of the worm.

An earthworm has red *blood* that moves around the body in *blood vessels*. It also has five *hearts* that help to keep the blood moving. Actually these hearts are just short sections of blood vessel that contract and force the blood along. The blood carries dissolved food from the intestine to all parts of the body. It carries oxygen from the outer skin to other cells of the body, and carbon dioxide from these cells to the skin.

The skin is thin and moist, so oxygen molecules diffuse through it from the air into the blood. Carbon dioxide diffuses outward through the skin into the air. In this way, the skin acts as the earthworm's respiratory organ.

Two little tubes are present in most of the worm's segments. Liquid wastes from the body cavity pass into these tubes. The tubes lead to pores in the body wall and the wastes pass out through them. These tubes are the excretory system of the worm. See Fig. 26-9, page 270.

crop

gizzard

Question: Ask students why a circulatory system in an organism as complex as an earthworm is so important.

Fig. 26-8
An earthworm. Notice how the body is divided into segments. (Courtesy Carolina Biological Supply Company)

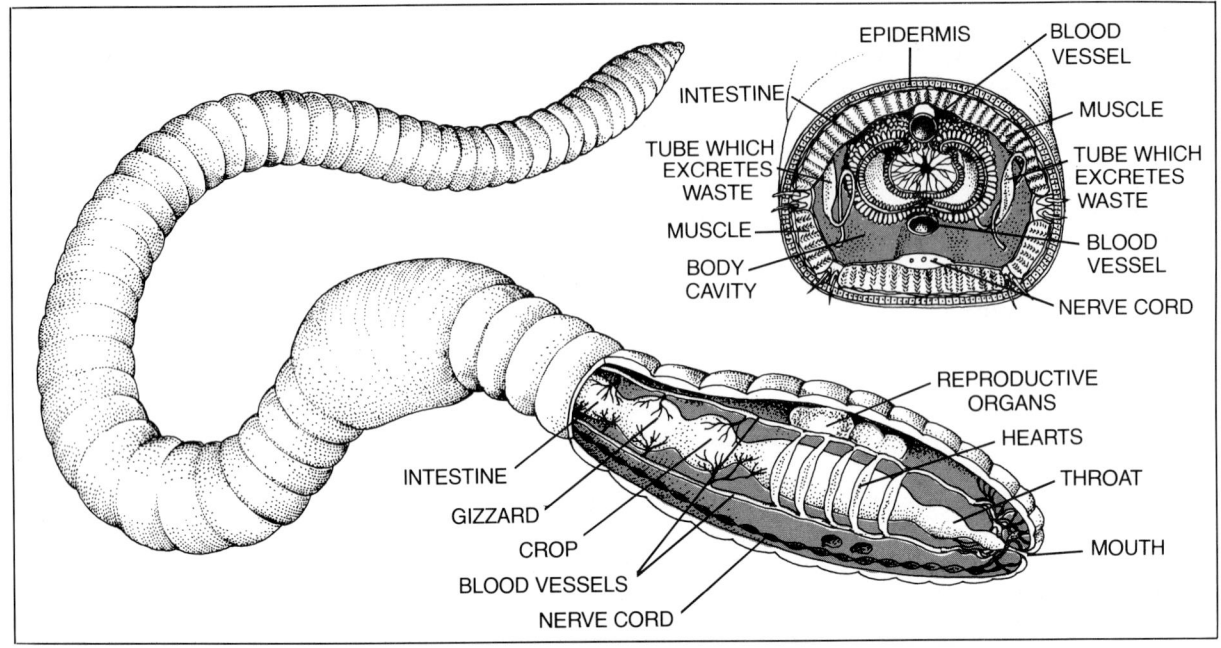

Fig. 26-9
A cut-away view of the internal structures of an earthworm and a cross section through its body.

Enrichment: An organism that has both male and female reproductive organs is called an hermaphrodite.

Activity: Have students dissect an earthworm and try to identify the parts of its body systems. Students can also test live earthworms for sensitivity to light and chemicals. Shine a flashlight on the worm and hold cotton swabs dipped in chemicals such as ammonia and vinegar near the worm. Make sure the earthworm is kept moist at all times.

Reproduction in the earthworm. Each worm has both male and female reproductive organs. Pairs of earthworms come together and exchange sperm. Eggs and sperm are placed in little "egg cases" that are left in the soil. Inside these cases the sperm fertilize the eggs. The eggs hatch, and the young worms leave the egg cases. The egg case is produced by a smooth band of thickened skin that can be seen about one-third of the way back from the front end of the worm.

As you can see, earthworms are quite complicated little animals. They have well-developed organs and most body systems. The one most obviously missing is the skeletal system. If they had skeletons they would not be worms, would they?

The importance of earthworms. An earthworm digs through the soil by eating dirt. The soil passes completely through the worm and out its anus. Bits of humus in the soil are digested. This digging and stirring loosens the soil and allows rainwater and air to enter. This improves soil for the growth of plant roots.

Wastes excreted by the earthworm are mixed with the soil passing through its body. This "worm manure" is rich in minerals needed by growing plants, so earthworms enrich the soil. Minerals in this worm manure come partly from the digested humus that the worm used for food. The rest is from the rock particles in the soil. They are ground to smaller sizes as they pass through the worm's gizzard. This speeds up weathering of these rock particles, and its causes them to release dissolved minerals.

Other segmented worms. Earthworms are about the only segmented worms living on land. Earthworms must live under ground to keep from drying out.

Some species of segmented worms live in fresh water. Most of these look like small earthworms. They eat dead plant and animal materials in the mud. Many lakes and streams have hundreds of small segmented worms living on every square meter of bottom. These worms often have their front ends in the mud and their back ends waving in the water to get oxygen.

Leeches are segmented worms that sometimes get on the skin of swimmers and suck blood. Perhaps you call them "blood suckers." They also attack turtles and fish. Snapping turtles often have several leeches attached to their undersides all of the time.

A leech has a suction cup around its mouth and another at its back end. These are used for clinging to any surface it crawls on. Inside the mouth are three small, hard jaws. These are used to cut the skin of the host so that blood will flow. You feel no pain when a leech bites you, because its saliva contains a pain killer. Most leeches are found in fresh water, but a few live in the sea, or in wet places on land. Not all leeches suck the blood of larger animals. Many of them feed upon water insects. When leeches are hunting their food they may crawl along like other worms. They also use a waving motion of their bodies to swim through the water. Fig. 26-10 shows what a leech looks like.

By far the greatest number of segmented worms live in the sea. Some live in burrows in the sea bottom and eat dead materials. Some live in the bottom but have long feather-like organs that they wave in the water to catch small bits of food.

Science Reading Skill: *Compare and contrast* the body structure of flatworms, roundworms, and segmented worms.

Enrichment: The salivary glands also secrete a substance that prevents blood from clotting while the leech feeds. Leeches were used medicinally in the Middle Ages and later because it was believed that blood loss healed the sick.

Fig. 26-10
Left: A photograph of a leech. This is one type of segmented worm. (Hugh Spencer from the National Audubon Society); Right: A sandworm. This segmented worm lives in shallow sea water along the shoreline.

Some of these *filter-feeding worms* are brightly colored and very beautiful like the ones in the photograph at the beginning of this chapter.

Still other segmented worms are active hunters. The *sandworm* is an example of this. It swims actively through the water and captures small animals with its pincer-like jaws. In Fig. 26-10 you can see that a sandworm has many little flaps on each side of its body. These are used by the sandworm for swimming and the absorption of oxygen from the water. These little flaps are simple respiratory organs.

A sandworm has four eyespots and a group of tentacles on its head. These are sense organs. Because the sandworm leads a more active life than an earthworm, it needs sense organs that are better developed. It must be able to detect food and see where it is going.

Science Reading Skill: *Compare and contrast the feeding habits of the different types of segmented worms.*

SUMMARY

Flatworms have several body systems and a single opening to the intestine. *Planaria* are free living in ponds. Tapeworms and flukes are flatworm parasites in larger animals and people.

Roundworms have an intestine, with a mouth at one end, and an anus at the other. Roundworms include both parasitic and free living species. *Ascaris*, hookworm, and trichina are roundworm parasites of humans.

Segmented worms include earthworms and similar small types that live in water. This phylum also includes sandworms, filter-feeding worms, and leeches. They have segmented bodies with many internal organs and several body systems.

ACTIVITY

Activity: Use a sharp knife to cut off the head or tail of *Planaria*. Observe how they regenerate over the next several days.

Observing Worms

A. Observe *Planaria* under a hand lens or with a low-power microscope. Observe the vinegar eels under the low power of a regular microscope.

1. Write a paragraph describing your observations of each of these worms.

2. In what ways do these worms differ from each other?

B. See if you can find roundworms or flatworms in pond cultures that have been set up in your classroom. Examine them under the microscope.

3. Can you see the intestines at work inside the transparent bodies of the roundworms?

4. Describe what you see.

Word Quiz

The following new words appear in this chapter. (1) Tell which of them are types of worms, and give the structure of each. (2) Which term is a waste material? (3) Which of them are body parts, and what is the function of each?

flatworm, intestine, feces, fluke, roundworm
anus, segmented worm, crop, gizzard

Check Your Facts

1. Compare the body structure of flatworms, roundworms, and segmented worms.

2. What systems does *Planaria* have that *Hydra* does not have?

3. What are flukes?

4. For each of the following worms, give its life cycle, tell how people can protect themselves from it, and how a community can try to eliminate it: tapeworm, *Ascaris,* hookworm, trichina worm, blood flukes.

5. How do earthworms dig? How do they improve the soil?

6. What structures does the sandworm have that help it to lead a more active life than an earthworm?

7. What do leeches eat? How do they get their food?

Science Reading Skill
1. *Compare and Contrast*
4. *Sequencing*

Thought Questions

1. Of what advantage is a flat shape to the *Planaria*?

2. What conditions could explain why the human blood fluke is a very serious problem in Egypt but not here?

3. We may see advertisements offering to sell us earthworms to put in our gardens to improve the soil. Would it be a good idea to buy such worms? Why?

Science Reading Skills
1. & 3. *Problem Solving*

CHAPTER 27   The Mollusks

In the photograph above you see a collection of shells. These shells were produced by *mollusks* (MOL-usks). Snails, clams, mussels, scallops, oysters, squid, and octopuses are all members of the mollusk phylum.

The mollusks make up a large and important phylum of animals. They have complicated soft bodies, with all of the body systems. Most mollusks live in the water, but some snails and slugs live on land.

THE CLAM: AN EXAMPLE OF A MOLLUSK

There are several classes of mollusks. Before we study these different classes we will describe one mollusk in detail. This example will help you to understand mollusks, in general. The *freshwater clam* is the example we will use.

Structure of a clam. The freshwater clam is protected by a double shell. This is its skeleton. The two parts of the shell hinge together and open like the covers of a book. A piece of muscular flesh can be pushed out through the shell opening. This is called the *foot*. The clam sticks this foot into the mud to pull itself along.

Loose flaps of flesh close off the gap between the outer edges of the open clam shell. At one end, two tubes lead to the space inside the shell. They are called *siphons* (SY-funs). The word siphon means tube. A current of water passes into the animal through one siphon and out through the other. Fig. 27-1 shows the outside view of a clam.

The clam does not move around much. Most of the time it stays still, half buried in the mud or sand on the bottom of the water. Like the sponge, the clam is a filter feeder, and it gets many small bits of food from the water that is always passing through its siphons.

objectives

AfTER you REAd This ChAPTER, you should be Able To:

__**Compare** the body structure in the clam, snail, and octopus classes

__**Describe** how the differences between mollusk classes can be explained

__**List** ways in which mollusks are used by people

---

foot

Word Study: The word mollusk comes from "mollusca" meaning soft, which refers to the soft bodies of these animals.

siphons

Enrichment: Some mollusks are called shellfish, but they are not fish at all. Many do not even have a shell.

Science Reading Skill: *Reading Illustrations*—Have students identify the functions of each of the clam's organs. See Fig. 27-2.

Fig. 27-2 shows the inside of a clam. See how the foot comes down from the main body of the clam. See, also, how two gills hang down on each side of the foot. These gills have many little holes through them. This gives them a great deal of surface area for absorbing oxygen from the water. The gills are also food-catching organs. Food particles in the water stick to the gills. Then the bits of food are pushed along by cilia from the gills to the mouth. Notice that the shell is lined with a thin layer of flesh. This layer is called the **mantle**. The word mantle means coat. The cells of the mantle produce the shell.

The space containing the gills and foot is called the **mantle cavity**, because it is surrounded by the mantle. The current of water entering the mantle cavity from one siphon passes through the gills and out through the other siphon. The mantle lining is covered with cilia. They beat the water to produce the current. The edges of the mantle form the flaps that cover the space between the edges of the shell when it is open.

A clam is a complicated animal. It has muscles and a skeleton, in the form of its shell. It has an intestine (digestive system); gills (respiratory system); and heart that pumps blood (circulatory system). The blood carries oxygen and digested food all through the body. The clam has an excretory system to remove wastes from the blood. It has a nervous system to control its movements. We cannot say that it has a brain, because there are three main nerve centers, all about the same size. It has a reproductive system that produces sperm or eggs. All ten of the possible body systems are present in the clam.

KINDS OF MOLLUSKS

There are six classes of mollusks. We will study the three most common ones: *clam class*, *snail class*, and *octopus class*. All

Fig. 27-1
An external view of a clam. (Bruce Coleman Inc.)

mantle

mantle cavity

Enrichment: Clams belong to the class Pelecypoda meaning hatchet-footed. Snails belong to class Gastropoda meaning stomach-footed. Class Cephalopoda includes octopus and squid and means head-footed.

Fig. 27-2
The inside structure of a clam. The shell has been removed to expose the various organs.

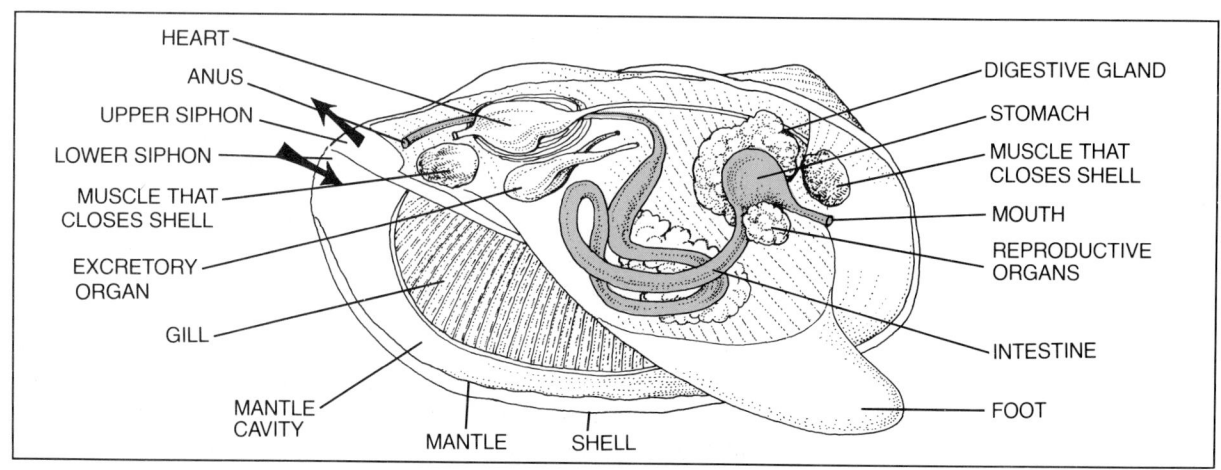

HEART
ANUS
UPPER SIPHON
LOWER SIPHON
MUSCLE THAT CLOSES SHELL
EXCRETORY ORGAN
GILL
MANTLE CAVITY
MANTLE   SHELL

DIGESTIVE GLAND
STOMACH
MUSCLE THAT CLOSES SHELL
MOUTH
REPRODUCTIVE ORGANS
INTESTINE
FOOT

of these mollusks have the same basic body organization as the freshwater clam, but each class has the parts arranged in a somewhat different way. This adapts them to different ways of living.

The clam class. The clam class includes all mollusks with a pair of shells that hinge together. The freshwater clam is a member of this group. This class also includes many other clams, oysters, mussels, and scallops. In the ocean, adult oysters and many of the clams and mussels do not move around at all. They live fastened in one place, to a rock or other object. An oyster has its hinge on one end of its long shell. This end is fastened to some support. Some types of clams have long siphons. They can hide 40 centimeters deep in the mud and still get clean water through these siphons. The giant clam of the South Pacific gets to be as large as 2 meters across and weighs about 100 kilograms. If you should put your foot into this clam you would frighten it into closing its shell. You would not be able to pull your foot free.

The eggs of a freshwater clam hatch in the mantle cavity. They develop into tiny larvae smaller than pinheads. When the shadow of a passing fish falls upon the female clam, she clamps her shell shut. This action shoots water out through her siphon. This water contains a cloud of larvae. They clamp onto the gills or fins of the fish and ride for a brief period. At this stage they are parasites. Then they drop to the bottom and grow up. This "hitchhiking" on fish spreads clams through our freshwater streams and lakes. In the ocean young clams and oysters do not do this. Instead, they swim about for some time under their own power, using cilia, before they settle down to life on the bottom of the ocean.

The snail class. Snails have a mantle that forms a single, twisted shell. The vital organs, such as the heart and intestines, are inside this shell. So is the mantle cavity, with its gills. Not all snails have gills. Some have a lung instead. The part that looks to us like the snail's body is mostly its foot. At the front end is the head, with tentacles on it. The snail can feel and taste with these tentacles. Some snails have simple eyes on the ends of them. See Fig. 27-3.

Snails have a tongue-like structure covered with tiny sharp teeth. They use this for scraping off bits of food from the surface they are crawling on. Some eat plants. Garden snails and slugs can damage land plants with their scraping tongues. Some water snails eat the film of slime off surfaces under water. This film may contain bacteria or algae. Snails like this help keep an aquarium clean. Some snails, called *drills*, cut holes through oyster shells and eat the flesh inside. When a snail is attacked, it pulls its foot and head into its shell.

Enrichment: Clams, oysters, scallops, and mussels are called bivalves because their shells have two halves called valves.

Enrichment: This class also includes conches and abalones. Since most of these animals have only one shell they are called univalves.

Fig. 27-3
The snail crawls along slowly and scrapes food into its mouth with its file-like tongue.

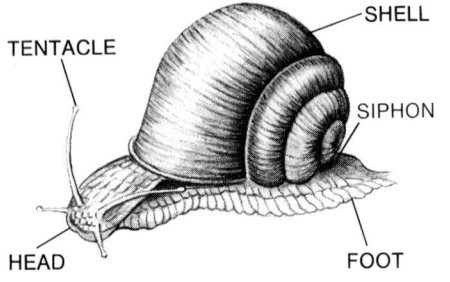

TENTACLE

SHELL

SIPHON

HEAD

FOOT

Fig. 27-4

A squid. Note that it has the same general parts as other mollusks, but they are arranged differently. (Runk & Schoenberger/Grant Heilman)

The octopus class. Octopuses and squid belong to another important class of mollusks. The members of this group have the same parts as other mollusks but are arranged in still another way. Look at the *squid* in Fig. 27-4. The squid has no shell on the outside. The mantle covers most of the body. A tough, muscular fin is located on each side of the mantle near the back end. Inside is the mantle cavity containing gills. A brain is located inside the head. The foot is closely attached to the head. It is divided into eight short and two long tentacles. All ten tentacles have suction cups to give them a firm grip. They are used to catch fish and crabs for food. The mouth is in the center of the ring of tentacles. It has a sharp beak with which it can bite off chunks small enough to swallow. On each side of the head is a large eye. These eyes are much like ours. Each eye has a lens and many of the other parts that our eyes have. This is the first animal we have studied with eyes that can see the shapes of objects.

An *octopus* has the same general structure as a squid, but its body is short and rounded instead of having the torpedo shape of a squid. See Fig. 27-5. An octopus does not have the two long tentacles. The shape of the octopus adapts it to life on the ocean bottom. A squid swims swiftly in the open sea. When either of these animals is in a hurry, it shoots water out through its siphon. This jet of water sends the octopus or squid streaking away like a rocket. These animals had jet propulsion long before our rocket scientists ever thought of it! A squid can also swim by rippling its fins.

Members of this class have a curious means of defense. They shoot out an inky substance that clouds the water. Hidden in

Fig. 27-5

An octopus. How does this animal resemble a squid? How is it different? What environment is each animal adapted to? (Jane Burton/Bruce Coleman, Inc.)

Enrichment: Giant squid grow to a length of 18 meters and weigh up to 1,800 kilograms making them one of the largest invertebrates in the world.

Science Reading Skill: *Compare and contrast the three classes of mollusks.*

this cloud, they can either swim away or they can attack their prey. What is better suited for wrestling in the dark than an animal with eight arms?

In spite of stories you may have heard, an octopus hardly ever attacks a human being. Octopuses in Atlantic waters are seldom more than 1¼ meters wide. This is measured from the tip of one tentacle to the tip of an opposite tentacle. Some octopuses in the Pacific grow to nearly 10 meters across. These could, of course, be dangerous, but they live deeper in the water than most swimmers are likely to go. Most squid are less than 30 centimeters in length. But off the west coast of South America are some species that are over 2 meters long.

Octopuses and squid are eaten by people in various parts of the world. They are popular foods in Europe and Japan. Squid are also cut up and used for fish bait.

Sailors used to return from their voyages with stories of sea serpents and monsters. One story told of a monster squid called the "kraken." It was supposed to be able to sink large ships and eat their crews.

No one took these squid stories very seriously. But finally giant squid were found and measured by scientists. Some were 20 meters long. Of course, this is not large enough to sink ships, nor is there any record that a squid ever ate a person. But those old sailors were not quite the liars that everyone thought they were. The kraken is real. Sailors today claim they have seen an enormous type of octopus in deep waters of the Caribbean. Do you think these stories will also prove to be true?

Giant squid live deep in the ocean and are rarely seen. There must actually be a good many of them, for they are the principal food of the sperm whale. The fact that these big squid went undiscovered for so long shows how little we know about the deep sea.

Divers have learned much about life in the shallower waters near shore and along coral reefs, but they cannot go much deeper than 100 meters, so they cannot explore the open sea. Depths there average from 3 to 5 kilometers, and a few places are over 11 kilometers deep.

We have better maps of the moon than we have of the sea bottom, but this situation is improving. Ships are using sonar to map the bottom. The sea covers nearly three-fourths of the earth's surface, so complete mapping will take a long time.

Scientists study life in the sea by dredging up samples from the bottom and by collecting them in nets dragged through the water. Cameras are lowered to take pictures of the bottom. Special, small submarines are able to take two or three people all the way to the bottom. There, they look out through small portholes at the life around them. In these ways our knowledge of the sea is slowly increasing, but there is still very much to be learned.

IMPORTANCE OF MOLLUSKS

The mollusks are a very numerous and successful group of animals. They affect the communities in which they live in many ways, and they enter into many food chains.

A successful phylum. Can you see why the mollusks are a successful phylum? Long ago the first mollusks must have been much alike—something like a flat-shelled snail. Then different groups of them began to live under different conditions. Natural selection brought change in these groups. Over the ages each group of mollusks has become better and better adapted to its way of life.

Animals that hunt for a living must be swift and quick to react. They must have good sense organs. A shell would weigh them down. The octopus class of mollusks fits this description of a hunter. Animals that wait for food to come to them can use a heavy shell. A body wrapped in a shell is well-protected. The clam class fits this description. The clams have even lost their heads. A head, with its collection of sense organs, is very useful to an animal that moves around. The original mollusks had heads. So do the snails and squid, but the filter-feeding clam does not.

The snails are in between the clam class and the octopus class in their way of life. They must move about to find food, but their rate of 3 or 4 meters an hour is fast enough. Their kind of food does not run away. They carry a medium-weight shell

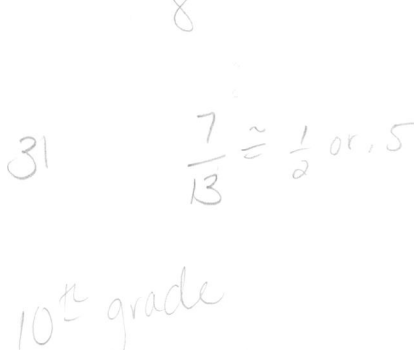

Assignment: Have students research the economic and ecological importance of mollusks.

279

with them to hide in when danger comes. So, you see, the different classes of mollusks are the result of adaptation to different ways of life. The mollusks have become numerous because they can live in so many different environments.

Useful mollusks. Clams, oysters, scallops, snails, octopuses, and squid are all eaten by people. If you live far from the sea, oysters are probably the only mollusks you may have ever eaten. They are commonly shipped inland and are sometimes canned. Many oysters are grown in shallow bays by a kind of underwater farming. Oyster gatherers prepare a bed of broken tile, scrap metal, or old shells for the young oysters to grow on. Then, when the oysters are large enough, they're sent to market.

We have always used the shells of mollusks. Primitive people used them as dishes and spoons. Sharp-edged shells were used to cut hair. The two halves of a clam shell were used as tweezers for pulling out whiskers. Beads and other decorations were made of whole shells and pieces of shell. In parts of Africa and in the South Pacific, certain shells were used as money. Mother-of-pearl is still used in some decorations. It is made of polished shell. You may have seen it on knife handles or as inlay on musical instruments.

Shells are also used to make buttons. The white pearl buttons used on some shirts are made of shells. Both freshwater clams and the pearl oyster shells are used. After button blanks have been drilled out, the rest of the shell is broken up and sold to farmers. Chickens eat this "oyster shell." The hen's body uses the lime in shells to make eggshells. Of course, plastic buttons are often used now instead of the pearl ones.

Pearls are made by mollusks. Because of their beauty they are used as jewelry. Any mollusk that has a shell can make a

Fig. 27-6
Mollusk fishermen bringing in their catch. (General Foods)

pearl. A pearl forms around some foreign object, usually a parasitic worm, which gets inside the flesh of the mollusk. On its way into the mollusk, the worm may pass through the mantle, and some of its shell-forming cells are dragged deep into the flesh of the mollusk. There these cells multiply and go right on producing shell. They coat the parasite with this shell material. The result is a pearl. Year by year the pearl becomes bigger as more layers are added to it. If the worm or other object has a regular shape, the pearl has a regular shape too, and is valuable for use as jewelry. Parasitic worms may form cysts inside mollusks. Since cysts are usually round, the pearls around them become spheres. Pearl formation is actually an unusual event. Only one oyster in a thousand is likely to contain one.

Pearls are made of the same material as shell, but we cannot make a pearl by polishing a piece of shell. It must have the natural surface produced by the mollusk. This surface looks like the lining of the shell. If the shell lining is beautiful, then that mollusk can make a beautiful pearl. The shell of the oyster we eat has a dull, gray lining. So you will not find riches in your oyster stew!

The Japanese use a way of forcing pearl oysters to make pearls. They take a bead made out of shell and cover it with live mantle cells. It is then thrust into the flesh of a pearl oyster with a pair of long, pointed tweezers. Oysters treated in this way are placed in wire cages and returned to the sea for several years. Pearls form around the beads. They are called *cultured pearls*. They look like natural pearls. Their outer layers *are* natural pearl, but they sell at much lower prices. People will pay more for a pearl with a worm inside it than for one with a bead inside!

Fig. 27-7
Pearls are used in making jewelry because of their great beauty. How are they formed? (Phillip Gendreau)

SUMMARY

The mollusk phylum includes the clam class, the snail class, and the octopus class. All of these animals have the same body parts, including foot, mantle, and mantle cavity. They have all ten of the body systems.

Clams, oysters, mussels, and scallops have shells that hinge together. They are filter feeders.

Snails have a single, spiral shell.

Squid and octopuses have no outside shell. They are swift, powerful hunters that eat crabs and fish.

Many mollusks are eaten by people. Buttons and mother-of-pearl are made from shells. Pearls form in mollusks.

Activity

Observing Snails

A. Place water snails in aquariums. Watch them crawl along the glass to feed on the algae and bacteria that have collected on it. This cleans the glass. Some water snails have gills in their mantle cavities, but others have lungs. See how these snails come to the surface once in a while and open a pore to take in air.

1. Can you guess where their ancestors lived long ago?

B. Place land snails and some lettuce leaves in jars. See how these snails use their file-like tongues to eat holes in the lettuce. Place a land snail on the table and watch it crawl. Touch one of its tentacles lightly.

2. What happens?

3. What does the snail do if you gently poke it several times with your finger?

C. Measure the time it takes a snail to crawl 10 centimeters.

4. At this rate how far would that snail go in an hour?

Word Quiz

Each of the new words listed below refers to a body part of a mollusk. Make a table listing the location of each part and its function in the clam, the snail, and the squid.

foot

siphon

mantle

mantle cavity

Check Your Facts

Science Reading Skill
3. Compare and Contrast

1. How do the young clams find new places to live?
2. How do clams, snails, and squid defend themselves?
3. Compare the body structures of a clam, a snail, and a squid.
4. What sort of food is eaten by snails, squid, and clams?
5. How does natural selection explain the differences between the mollusk groups?
6. Name some ways in which mollusks are useful to us.
7. How do pearls form?

Thought Questions

Science Reading Skill
1. Problem Solving

1. What structures would mollusks need to be more successful on land?
2. If you walk on the beach at low tide, you may see little jets of water spout from the sand around you. What causes this to happen?

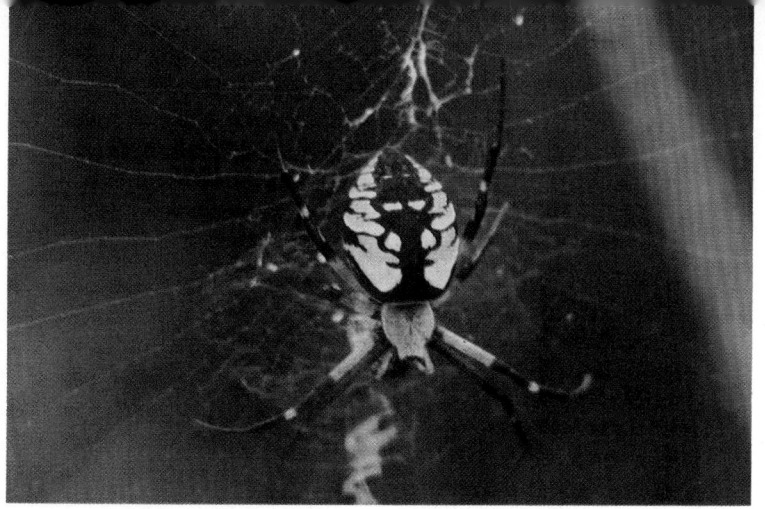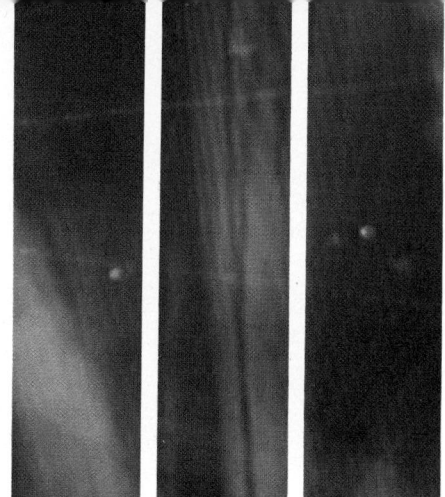

CHAPTER 28 Arthropods

Spiders are members of the arthropod phylum. In many ways, the **arthropods** are the most successful animals on earth. There are more kinds of them than of all other animals, and they give us serious competition. There are arthropod species adapted to nearly all possible environments. They live on land, in the air, in the soil, and in the water. The arthropod phylum includes insects, spiders, ticks, crabs, shrimp, centipedes, and many others.

CHARACTERISTICS OF ARTHROPODS

Arthropods are segmented, like earthworms. Their internal organs are also like those of earthworms in some ways. Many people believe that the arthropods developed from segmented worm ancestors long ago. Members of the arthropod phylum have bodies that are much more highly organized than those of the segmented worms. They have all of the body systems.

Appendages. Each arthropod has many jointed appendages attached to its body. An **appendage** is a special part that extends out from the body of an animal. Legs, feelers, wings, and mouth parts are some of the appendages found on arthropods.

These appendages are different in the different species. They adapt each one to its environment. Some species can swim, some can walk, and some can fly. Some arthropods can do all three. Mouth parts are modified in many ways. This allows different arthropods to eat many different kinds of food.

The arthropod skeleton. The arthropod skeleton is on the outside of the body, so it is called an **external skeleton**. It is made of a hard material with flexible joints. It is, in fact, a suit of armor. This suit of armor is moved by muscles attached to the inside of it. The human skeleton is arranged in exactly the opposite way. Our bones come together to form joints. The

objectives

After you read this chapter, you should be able to:

___**List** the main physical characteristics of the arthropods

___**Compare** the five classes of arthropods

___**Explain** two kinds of life cycles found in insects

___**Describe** several ways of controlling insect pests

arthropods

Word Study: "arthro-" means jointed and "-pods" means feet

appendage

Science Reading Skill:
Generalization—Describe the characteristics common to all arthropods.

external skeleton

Reference: See *Man and Insects* by L. H. Newman, *The Arachnids* by Keith Snow, *Insect Behavior* by Phillip Callahan, and *The Insects* by Peter Farb and the editors of *Life Magazine*.

muscles that move these bones are attached on the outside. These are the only two ways to build a movable skeleton. Arthropods have an external skeleton. We have an *internal skeleton*. Which of them do you think is more efficient?

Have you ever seen an ant dragging a caterpillar many times its weight? If you were as strong for your size as the ant is, you could grab a truck by the bumper and drag it up a steep hill. If you could jump as well as a flea, compared with your size, you could leap over a 100-story building.

As you see, the arthropod skeleton and its arrangement of muscles gives these animals great strength. It works well in small animals but not in large ones. You may have seen horror movies in which huge insects or spiders terrified the country-side. Actually, such animals could not exist. There is a limit to how large an external skeleton can be and still work. There is also a limit to the possible size of animals with internal skeletons, but the limit is much larger. Very large animals all have internal skeletons. For an animal the size of a bug, the external type of skeleton is more efficient.

Activity: Collect several different kinds of insects and observe the arthropod characteristics.

The success of the arthropods. Small size is not a disadvantage to the arthropods. It is an advantage. We have wiped out the bears, wolves, elk, mountain lions, and other large animals in all of the more settled parts of our country. You may ask why we do not wipe out insect pests. One reason is that their small size makes them difficult to control. They can hide easily, and they do not require much food. They also reproduce rapidly. If 99 percent of a species were wiped out, the remaining one percent could replace the loss in a short time!

Suppose that all of the descendants of just one pair of house-flies were allowed to reproduce at their greatest rate. At the end of one summer the total number of flies produced would completely fill all of a large high school building! But luckily insects do not actually increase in number so rapidly. A good many of them are eaten by their natural enemies. Some die because they do not find the right kind of food. Others are killed by cold, heat, or other unfavorable conditions.

Question: Ask students why arthropods have been so successful.

The many kinds of appendages adapt different arthropods to water and land environments. The external skeleton also helps adapt some species to land conditions. It keeps their bodies from drying out in the air.

We shall describe five classes of arthropods and see how each class is adapted to life on land or in the water.

THE CENTIPEDE AND MILLIPEDE CLASSES

Enrichment: Centipedes belong to the class Chilopoda and millipedes belong to the class Diplopoda. Both classes are often referred to as Myriapoda, which means many-footed.

The *centipedes* and *millipedes* make up two arthropod classes. Both of these types have long, segmented bodies. Both have

large numbers of legs. The centipedes have one pair of legs on most body segments. Millipedes have two pairs. The number of legs varies a great deal, but most centipedes have about forty, and most millipedes about one hundred. A centipede and a millipede are shown in Fig. 28-1.

Both centipedes and millipedes are commonly found in the ground or in almost any dark, damp place. Millipedes eat decaying plant materials. Even with all of their legs, they generally move slowly. Centipedes use pincer-like poison jaws to catch insects or other small animals. The poison from the jaws helps to overcome the victims. Common small centipedes are harmless to people. But some tropical centipedes are as much as 20 centimeters long. These big ones can give people painful bites.

THE CRUSTACEAN CLASS

The *crustaceans* (kruh-STAY-shuns) include crabs, lobsters, shrimps, prawns, and crayfish. Crustaceans are well known animals because many of them are good to eat. Freshwater *crayfish* have always been favorite animals for study in biology classes. A crayfish does not have the worm-like shape of a centipede. Its segments and appendages have become specialized to do different jobs. Look at Fig. 28-2. You can see that at the front end of the body the segments are all grown together, so that the head-chest region does not even look segmented. In the rear part of the body the segments show clearly. They form the *abdomen*. This is not a tail, for the intestine runs through its whole length. An abdomen is the part of an animal's body containing the digestive organs. See Fig. 28-2.

A crayfish has nineteen pairs of appendages. There are two pairs of feelers, six pairs of mouth parts, five pairs of legs, five pairs of appendages under the abdomen, and a pair of tail fins.

Fig. 28-1
A centipede (left) and a millipede (right). How can you tell them apart? (American Museum of Natural History; Walter Dawn)

crustaceans

Word Study: "crustacean" comes from the word "crustaceus" and means having a crust or shell

abdomen

Fig. 28-2

A crayfish. This arthropod has segments and appendages that have become specialized to do different jobs. (Courtesy Carolina Biological Supply Company)

The mouth parts are handy tools for holding food, cutting it up, grinding it, and pushing it into the mouth. The first pair of legs bear large powerful claws, or pincers. The appendages under the abdomen are used in slow, forward swimming, for sending water through the gills, and, in females, for carrying the eggs until they hatch. There are gills hidden in cavities on each side of the body.

Crayfish live in lakes, ponds, and streams. By day, they hide under rocks and logs. At night they come out and hunt. A crayfish eats tender plants and any live or dead animal material it can find. The big claws on the front legs are used to catch food and also for defense. If a crayfish is frightened, it swims backward by flipping the abdomen underneath the body.

If a pond where crayfish live dries up, they do not die. The water table is still near the surface, and they dig down to it. Each crayfish makes a neat, round hole going nearly straight down. During the day it stays in the water at the bottom of this hole, keeping its gills wet. At night it comes out and hunts for food, but it must return to the hole again before its gills dry out.

Lobsters and sea crawfish live in the ocean. They are very similar in structure to crayfish. Crabs are also somewhat similar. They have the same sort of legs and pincers, but their bodies look different. A crab's abdomen is very small and is carried under the chest where it does not show. The crab's body is often flattened so that it looks like a plate with legs. See Fig. 28-3.

The most important crustaceans are not big ones like lobsters and crayfish, but the small ones, which live in very large numbers in fresh water and in the oceans. These are the bug-like animals of the sea. Small crustaceans come in many sizes and shapes. See Fig. 28-4. Many are too small to see without a

Fig. 28-3

A crab. Its body structure looks like "a plate with legs." (U.S. Fish and Wildlife Service)

magnifying lens, and others are the size of small insects. Some are several centimeters long. They drift in the open sea, they crawl on seaweed, or they live on the ocean bottom. Some eat algae, and some eat other crustaceans. Some are parasites. The ones that eat the tiny, free-drifting forms of algae are the most important. They are the first step in many food chains in the sea. Even the largest whales eat them. These huge mammals swim through the sea with their mouths open, taking in millions of small crustaceans. Crustaceans are also eaten by many fish.

THE SPIDER CLASS

The spider class of arthropods includes spiders, mites, ticks, and scorpions. All of these have eight legs. This helps you to tell them from the insects. *Ticks* are round-bodied, spider-like types, a few millimeters to one centimeter long. They suck the blood of larger animals. They can carry several serious diseases. Texas fever in cattle and spotted fever in humans are examples of diseases carried by ticks.

Mites are similar to ticks, but very small. Some are only a few tenths of a millimeter long. Many eat humus, but others are parasites. Some mites cause *mange*, a serious skin disease in animals. The itch mite burrows into human skin.

Enrichment: Spiders belong to the class Arachnida.

Fig. 28-4
A few of the many kinds of small crustaceans. Some are so small you can hardly see them without a microscope. Do these tiny animals have any importance?

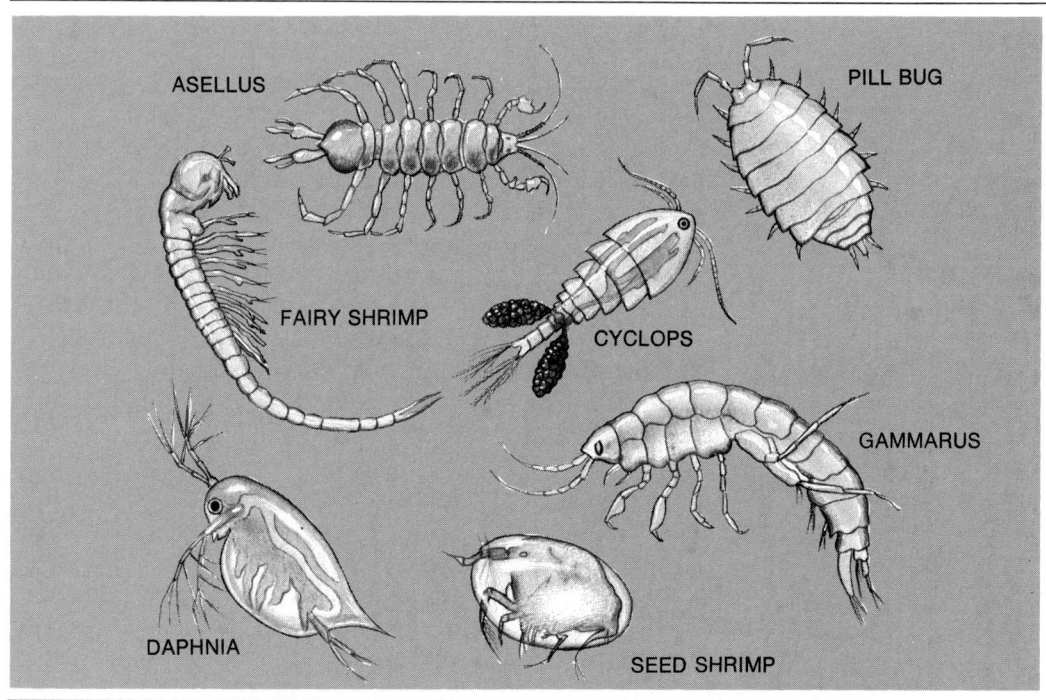

ASELLUS

PILL BUG

FAIRY SHRIMP

CYCLOPS

GAMMARUS

DAPHNIA

SEED SHRIMP

Fig. 28-5

A scorpion can give a painful sting. Notice the stinger on the end of the abdomen. (P. A. Knipping)

Scorpions live in warm regions, including the southern United States. They look a little like crayfish, but the two are not closely related. They have a long, segmented body with a slender abdomen. There is a stinger on the end of this abdomen. See Fig. 28-5. One of the two pairs of mouth parts is large, with pincers on the ends. These are used to take hold of insects. The abdomen whips over and the stinger injects poison into the victim, which the scorpion then eats.

Spiders have two body parts—a combined head and chest region in front and an abdomen behind. The eight legs are on the chest. The jaws are armed with a pair of poison fangs. These are used to kill insects. A spider sucks liquids from the body of the insect. It does not eat the hard parts.

Spiders catch insects in many ways. Some spiders run after insects. Some creep close to the insects and then jump on them. Some hide in holes and run out to catch passing insects. Many spin webs to use as traps. The web is made of silk that comes from glands at the back of the abdomen.

Many people think of spiders as dangerous animals, but spiders seldom harm people. They are important in the balance of nature. They keep insects in check, and insects are our chief competitors.

People do not get bitten by spiders very often, and when they do the bites are not usually serious. They just cause a little itching and swelling. But two spiders found in America are really dangerous. One is the famous *black widow*. She is a "widow" because she kills and eats the male as soon as they have mated. People have been known to die from black widow spider bites, but most victims recover. The other dangerous spider is the *brown recluse*. It is called a "recluse" because it hides in dark caves. This spider has a poisonous bite, which is painful and takes a long time to heal.

THE INSECT CLASS

Insects are the most common arthropods on land, just as crustaceans are the most common ones in the sea. They are very numerous, and they affect our lives in many ways. You can recognize an insect by its six legs and its three body parts—*head*, *chest*, and *abdomen*. Most insects also have two pairs of wings. Some have only one pair, and some have no wings.

Structure of insects. We shall use the grasshopper as an example of an insect. If possible, you may wish to examine a real grasshopper while you read this description.

The head has two long feelers. Some insects feel with these. Others detect sounds with them. Still others smell or taste with their feelers.

A grasshopper has two large eyes on its head. They are made up of many small eyes packed together. Eyes of this type are called **compound eyes.** Fig. 28-6 shows a close-up view of a grasshopper's eye. A grasshopper also has three small, simple eyes, one in front of each compound eye and one in the center of the head.

The mouth of the grasshopper has upper and lower lips, which move up and down. Two pairs of jaws have a sideways movement. With these mouth parts the grasshopper is able to bite off and chew bits of plant leaves. Many other insects have chewing mouth parts, also. Others have the same mouth parts modified to form tubes for drinking liquids.

The grasshopper's legs are jointed for efficient movement. There are hooks near the ends of the legs, used for hanging on to things. Most insect legs are of this general structure, but of course all insects do not jump like the grasshopper. The large back legs of grasshoppers are modified for jumping. Other insects may have legs modified for running, digging, swimming, or catching other insects.

A grasshopper folds its wings over the top of the abdomen. The front wings are narrow and leathery. The wide back wings are thin, like cellophane. They fold up under the front wings. When the grasshopper flies, the wings unfold and reach out sideways. Some grasshoppers fly well. Others use their wings just to help them jump a little farther.

A grasshopper gets oxygen through pores in its sides. There are pairs of these pores in ten of the grasshopper's segments. They connect with branching tubes that carry air to all parts of the body.

insects

Enrichment: Class Insecta contains 75 percent of all of the known species of animals.

Enrichment: Grasshoppers and other insects produce sound by scraping their legs against their abdominal sound organs. Some insects can detect frequencies that are 100,000 cycles per second as opposed to human ears that can detect 20,000 cycles per second.

compound eyes

Fig. 28-6
A close up of the compound eye of a grasshopper. (Hugh Spencer from the National Audubon Society)

Enrichment: Insects have no red blood cells. Their blood is clear, green, or yellow in color.

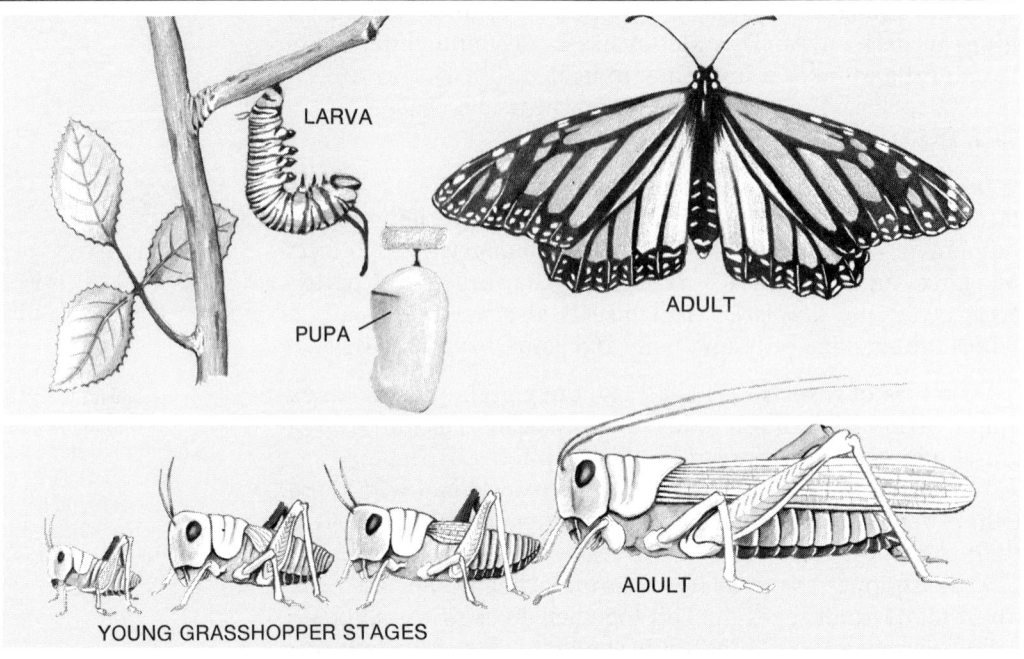

LARVA

PUPA

ADULT

YOUNG GRASSHOPPER STAGES

ADULT

Fig. 28-7

Top: The life cycle of a butterfly, which shows its development from larva to adult; Bottom: The life cycle of a grasshopper. How are these two types of development different?

nymph

Science Reading Skill: *Sequencing*— Describe the life cycles of insects.

pupa

Enrichment: The series of stages as insects develop from egg to adult is called metamorphosis. If metamorphosis is completed in three stages as in the grasshopper, this is known as incomplete metamorphosis. The four stages in the development of the butterfly is called complete metamorphosis.

Life cycles of insects. Female grasshoppers lay their *eggs* in the surface of the soil. They do this in the fall of the year. When an egg hatches in the spring, the young grasshopper is quite small. It has no wings or sex organs, but otherwise it looks and acts like an adult grasshopper. See Fig. 28-7. This young grasshopper is called a *nymph* (nimf). The nymph feeds and grows until its external skeleton stops further growth. Then it sheds its outer skeleton and increases in size a little before its new outer layers harden. This growth and shedding of the external skeleton continues over a period of time until an adult is produced.

The butterfly illustrates a different type of life cycle. The butterfly *egg* hatches into a *larva* called a *caterpillar*. This worm-like larva eats steadily and grows rapidly. It sheds its tough outer skin (external skeleton) several times. When it sheds its skin for the last time it does not look like either a caterpillar or an adult. It has become a *pupa*. This is the *resting stage* in the development of a butterfly. The pupa may be covered by a case called a cocoon.

During the resting stage the body of the animal is completely reorganized. Then the outer covering splits, and the *adult* insect flies out.

Kinds of insects. There are about twenty-five orders of insects. The following list gives seven of the most common ones. Insects in the first three of these orders have life cycles like that of the grasshopper, with egg, nymph, and adult stages. Those in the last four orders have egg, larva, pupa, and adult *stages*, like the butterfly.

1. *The grasshopper order*. Besides the grasshoppers this order includes crickets, walking sticks, cockroaches, and katydids. Grasshoppers and crickets are very destructive insects. They eat about 10 percent of our grain and pasture crops before they are harvested. The locust plagues of Africa and Asia were simply swarms of flying grasshoppers.

2. *The dragonfly order*. These insects are often called "darning needles." They are beautiful, long-bodied insects, with marvelous flying ability. They have two pairs of wings. They feed upon other insects, including mosquitoes. Dragonfly nymphs live underwater, where they eat mosquito larvae and other insects.

3. *The bug order*. Most people call any small arthropod a bug. There is nothing wrong with this. Common names are often used to refer to different organisms. There are many biologists who will use the word only for this one order of insects. Its members have sharp, tube-like sucking beaks, which they use for eating liquid foods. Stinkbugs, squash bugs, chinch bugs, and many others stick their beaks into plants and suck the plant juices. Assassin bugs attack other insects. Bedbugs suck human blood. You can recognize these insects by their peculiar wings (except the bedbug, which has no wings). The upper wings of insects in the bug order are thick in front but thin at the tips. They fold to form a crisscross pattern on the back. See Fig. 28-8.

Science Reading Skill: *Compare and contrast* the five classes of arthropods.

Fig. 28-8

From left to right; A grasshopper (grasshopper order) (Manuel Rodriguez), a dragonfly (dragonfly order), an assassin bug (bug order) (William E. Ferguson)

4. *The butterfly and moth order*. These large-winged insects are often very beautiful. Some adult moths have no mouths. Other moths and all butterflies have mouth parts that are modified to form long tubes through which they suck nectar from flowers. They do some pollinating of flowers. During the caterpillar stage most members of this order are plant eaters, and some species do serious damage to fruit trees, grains, vegetables, and forest trees.

5. *The beetle order*. These are the most heavily armored of the insects. Their external skeleton is very strong. The front wings are heavy shields that cover the abdomen. When a beetle flies, these front wings are raised, and the thin back wings unfold to do the flying. Beetles have chewing mouth parts, much like the grasshopper. See Fig. 28-9.

   This is a very large order of insects. It includes useful types, such as the ladybird beetles, the tiger beetles, and the ground beetles. These feed upon other insects. Many other beetles are serious pests. They feed on all kinds of plants, including some that are useful to us. The larvae of beetles are sometimes called *grubs*.

6. *The fly and mosquito order*. This order includes a few useful types, such as the robber flies, which eat other insects. But the order also includes some of our worst pests. They have sucking mouth parts, and many of them feed on blood. Some of them carry disease germs from one host to another. Malaria, yellow fever, sleeping sickness, and other deadly human diseases are carried by flies or mosquitoes.

   Houseflies are well known members of this group. They do not bite. A fly will eat sugar by first dissolving it with saliva and then drinking the liquid. Their young live in piles of animal manure. The adult flies walk about on our foods. Their bodies are covered with short hairs and bristles that collect dirt and germs wherever the flies go. Naturally, the flies leave dirt and germs on food.

   You can recognize members of the fly order by the fact that they have only two wings. Their larvae are usually called *maggots*.

7. *The ant and bee order*. This order includes bees, wasps, and ants. Bees, as you know, are the most useful of all insects in pollinating flowers. A great many plants could not reproduce if there were no bees.

   One of the most interesting things about this order is that many of its species live in well-organized groups, or societies. They are known as *social insects*. A beehive, a hornets' nest, or an anthill is like a city with many citizens

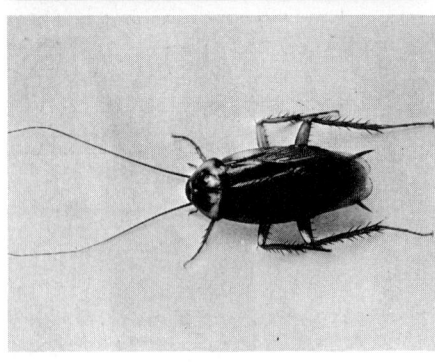

Fig. 28-9
Top: A monarch butterfly (butterfly-moth order) (Lynwood M. Chase); Bottom: A cockroach (grasshopper order) (USDA).

Fig. 28-10
Why are these called "social insects"?

living in it. See Fig. 28-10. A honeybee colony usually has just one egg-laying female, the *queen*. Most of the others are *workers*. These are females that never reproduce. A few males, called *drones*, may be present. The workers gather food, care for the queen bee, raise the young, and defend the group against enemies. Ants have the most complicated organization of all. They may have several kinds of workers that do different kinds of jobs.

Controlling harmful insects. As you have seen, some insects are useful to us, but many of them give us serious competition. They eat our food supply. Termites and clothes moths eat our building materials and clothing. Mosquitoes and lice feed directly on us and carry diseases.

Insect numbers are kept down mainly by their natural enemies. These include disease germs, centipedes, scorpions, spiders, frogs, lizards, shrews, birds, and bats. The insects that eat other insects are also important. In spite of all these enemies, insects still manage to be a problem to us.

If we wish to get rid of flies or mosquitoes in a neighborhood, we may attack them where they breed. Manure and garbage are food for fly maggots. Regular removal of manure piles and burying or burning garbage gets rid of most flies. Mosquito larvae live in water. Pools and swamps can be drained or stocked with small, insect eating fish.

We can encourage the natural enemies of insects. On farmlands that have some trees and shrubs along the fences, there is less insect damage to crops. Birds nest in the trees and feed upon grasshoppers and other insects in the fields nearby. Not

Assignment: Have students research the use of alternative forms of insect control and the long term effects of the use of chemical pesticides.

many birds can live in country where nothing but crop plants are allowed to grow, with no trees or shrubs nearby.

We have imported enemies of some insect pests. New species of ladybird beetles and of ground beetles have been brought into the country, because they eat certain kinds of undesirable insects.

In the southern United States there was a very serious insect pest called the screw worm fly. This fly laid its eggs in any wounds it could find on large mammals. The larvae grew in the flesh of the mammal, often causing its death. A great many cattle were lost because the flies laid eggs on the navels of new-born calves.

The fly has been eliminated in the southern states. This was done by raising and releasing many thousands of male screw worm flies. Before they were released these males were sterilized by exposing them to X-rays. They produced semen that contained no sperm. These males mated with normal, wild females which then laid unfertilized eggs. These eggs, of course, never hatched. The sterile males were produced in very great numbers. There were more of them than there were of the wild males, so most females mated with sterile males. Sterile males were released each year until no new larvae were being produced.

Biologists are now experimenting to see if this method can be used to kill off other pests. They are also studying the use of odors in controlling insects. Many male insects are attracted to the females by their odor. A male moth can smell a female moth 2 kilometers away. The idea is to use such odors to attract the males into traps. A different odor would be necessary for each species, so only one particular pest species would be destroyed.

biological controls

Such methods as the use of natural enemies, sterile males, or attractive odors are called *biological controls*. When they can be used, these biological controls are far better than poisons. Pesticides are useful in limited areas, but they can cause serious pollution if their use is not carefully controlled. See Fig. 28-11.

Biological controls affect one particular pest species without destroying other members of the community. The development of biological controls is a rather new field of science. Perhaps the time will come when we can always destroy just the pests without hurting other forms of life.

Insects as a hobby. Many people enjoy collecting and identifying insects. Butterfly and beetle collections are especially popular. These insects are often brightly colored, and the collections make beautiful displays. Your library has books that would help you to get started in an insect study hobby.

Fig. 28-11
Fruit trees being sprayed to prevent damage from insects and fungi. Why is it necessary to choose the chemicals for such spraying very carefully? (Dow Chemical Co.)

SUMMARY

The arthropods form a large and successful phylum of animals, both in the water and on the land. They include the centipedes, millipedes, crustaceans, spiders, and insects.

Arthropods have segmented bodies, external skeletons, and numerous, jointed appendages.

Crustaceans include crabs, lobsters, crayfish, shrimp, and many tiny forms that eat algae. They are very important in the sea.

Spiders, ticks, and scorpions make up another class of arthropods, living mainly on land. Spiders and scorpions are important insect eaters. Ticks suck the blood of larger animals.

Insects have three body parts and six legs. Most have wings. They are very numerous and important animals on land. There are about twenty-five orders of insects.

Insects may help us by pollinating flowers and by eating harmful insects. Harmful insects are controlled mainly by many natural enemies. We destroy insect pests with poisons and with biological controls.

ACTIVITY

Observing Crayfish

A. In this activity you will observe live crayfish. Always keep live crayfish in shallow water (about 5 centimeters deep). They will die in deeper water.

1. What do you think is the reason for this?

B. Place a live crayfish on the table top. Watch it walk. You can hold a crayfish by gripping its chest between your thumb and forefinger. It cannot reach far enough back to pinch you.

2. Does it try to pinch when you hold it?

3. Poke it gently with a pencil eraser as it rests on the table top. Does it defend itself?

4. Place it in a tank of shallow water and gently poke at it there. Does it fight or try to get away? How?

C. When a crayfish is resting quietly in the tank drop a worm or a piece of meat in front of it. Watch how it uses its pincers and mouth parts to place the food into its mouth.

5. Describe the action of the mouth parts.

D. Examine the outside structure of the crayfish. Feel the armor-like outer skeleton. Count the feelers, mouth parts, and legs. Examine the small appendages under the abdomen.

6. What kind of eyes does the crayfish have?

7. How many pairs of appendages can you find?

Word Quiz

Choose the number from **Column B** that best matches each letter in **Column A**. Do not write in this book.

Column A	**Column B**
a. arthropods	**1.** arms, legs, wings, feelers
b. appendages	**2.** class to which crab, lobster, and shrimp belong
c. crustaceans	**3.** type of skeleton found in arthropods
d. abdomen	**4.** type of skeleton found in people
e. external skeleton	**5.** releasing sterile males to mate with female insects
f. internal skeleton	**6.** a young insect that looks very much like the adult
g. insect	**7.** the resting form of an insect that is changing into an adult
h. compound eye	**8.** a phylum of animals with segmented bodies, outside skeletons, and numerous jointed appendages
i. nymph	**9.** one of the three body parts in an insect and other animals; it contains the intestines
j. pupa	**10.** an animal with three body parts and six legs
k. biological control	**11.** seeing organ made of many small units

Check Your Facts

Science Reading Skills

1. Generalization

4. Compare and Contrast

8. Sequencing

10. Problem Solving

1. Describe the general characteristics of arthropods.
2. How is small size an advantage to arthropods?
3. What are some advantages and disadvantages of an external skeleton?
4. Compare the structure of centipedes, millipedes, crustaceans, spiders, scorpions, and insects.
5. Discuss the importance of each arthropod class.
6. How do scorpions get their food? Are they dangerous to us?
7. What is the main importance of spiders? How do they get their food?
8. Explain the two kinds of life histories found in insects. Give examples of each.
9. Name some examples of each of the orders of insects listed in this chapter.
10. How are most insects prevented from becoming too numerous? How do we try to control insect pests?

Thought Questions

Science Reading Skills

1. & 2. Problem Solving

1. Small size is an advantage to the *species*. Is it an advantage to the *individual?* Explain.
2. Suppose you found that ladybird beetles were eating large numbers of aphids (plant lice) in your garden, but that the aphids were still doing some damage. Would spraying with poisons help the situation? Explain.

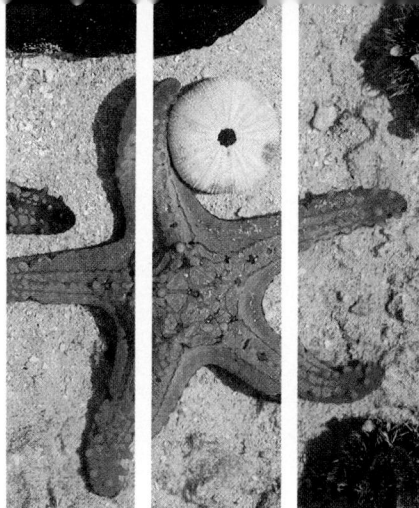

CHAPTER 29 Echinoderms

These starfish have been left exposed to the air when the tide went out. Soon the water will rise and cover them again.

Starfish are members of the ***echinoderm*** (ee-KINE-o-derm) phylum. Echinoderms seem very strange to us. Their bodies are formed by several, similar parts arranged in a circle, their method of movement is different from that found in any other phylum, and nearly all of them have some sort of spines on their bodies. In fact, the word echinoderm means "spiny skinned."

TYPES OF ECHINODERMS

The starfish. A starfish has no head. It has a circle of arms that come together at the center of the body. At this center there is a stomach that opens through a small mouth on the underside of the animal. The skeleton of a starfish is made of many small plates of hard lime. These plates are connected by softer material so that the whole animal is flexible. There are many short spines all over the body. All parts of the skeleton, even the spines, are covered by a thin layer of living material.

Inside the animal is a body cavity that extends out into the arms. The liquid in this cavity is mostly seawater. This liquid carries dissolved foods and oxygen around the body. It is kept in motion by cilia that line the body cavity. This is the circulatory system of the starfish.

On the upper surface, between the spines, are many little bulges. They are made of thin membranes. Oxygen molecules diffuse through the membranes into the body cavity of the starfish. These bulges are a simple form of gill. All together they make up the respiratory system of the starfish.

---

objectives

AFTER YOU READ THIS CHAPTER, YOU SHOULD BE ABLE TO:

__**Describe** the body structure of the common starfish

__**Describe** the unusual way in which this starfish eats

__**List** several other echinoderms

__**Discuss** the general importance of the echinoderms

---

echinoderm

Word Study: "echino-" means spiny or prickly and "-derm" means skin

Fig. 29-1
This starfish is about to open a clam. Notice the tube feet. (Courtesy Carolina Biological Supply Company)

tube feet

A starfish moves on **tube feet**. There are hundreds of these little flexible tubes lined up in the grooves on the underside of each arm. The outer end of each tube foot acts like a suction cup. The tube feet are connected by a system of water canals that are in the body cavity. Water can be pumped from these canals into the hollow tube feet. This makes them lengthen. Withdrawing the water pulls them in. Hundreds of tube feet can reach out at one time, clamp on to any solid surface, and shorten. This action pulls the starfish along.

The starfish has an amazing ability to replace lost parts. You can cut off its arm, and the starfish will soon grow a new one. You can even cut the animal in two, and the halves will grow into two complete starfish.

The common starfish eats clams and oysters. It crawls over the top of a clam, clamps on to both sides of the shell with its tube feet, and begins to pull. Of course the clam tries to hold its shell shut, but finally the steady pull of the tube feet wins out. The shell opens.

The mouth of a starfish is much too small to swallow a clam. It cannot bite off pieces either. A starfish meets this problem in a most unusual way. It turns its stomach inside out and pushes it out through its mouth. See. Fig. 29-1. The thin-walled stomach slides in through the open shell of the clam and clamps against the soft flesh. Digestive juices produced by the stomach lining kill and digest the clam. The starfish absorbs this dissolved food by diffusion through its stomach lining. Then it pulls in its stomach and moves away on its tube feet. An empty shell is left behind.

Starfish eat other food, such as dead fish, which they may find lying on the bottom of the ocean. There are many kinds of starfish, and the different species eat a wide variety of foods.

Animals related to the starfish. Other members of the echinoderm phylum include basket stars, sea urchins, sea lilies, and sea cucumbers. See Fig. 29-2. *Sea lilies* are filter feeders. They grow attached to the bottom by long stalks and reach their feathery arms into the water to catch food. *Basket stars* are similar, but they move freely through the water. They have no stalks.

Sea urchins are globe-shaped and covered with spines. They reach downward with a five-pointed set of jaws to pull in bits of food. *Sand dollars* are similar in structure, but not so spiny, and their shape is *very* flat. *Sea cucumbers* are long and soft and actually look something like cucumbers. In some countries they are gathered to use in soups and other dishes. The fleshy body wall is eaten.

Fig. 29-2

Members of the echinoderm phylum. Left: Sea urchin; Middle: Sea lily (EPA—Documerica); Right: Sea cucumber. (Courtesy Carolina Biological Supply Company)

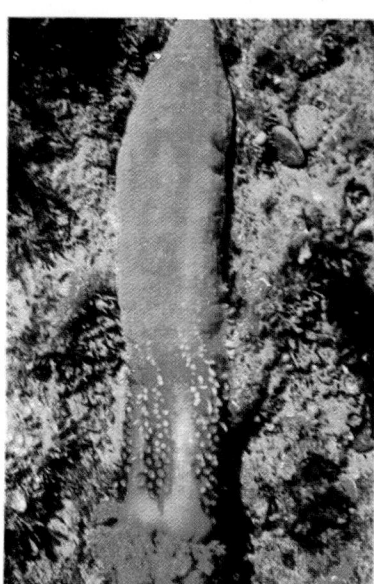

If a fish or crab attacks a sea cucumber, it gets a surprise. The sea cucumber contracts its body wall very forcefully. This breaks loose its intestines and blows them out into the face of the attacker. These intestines stick to the enemy. By the time the sticky mass has been rubbed off, the sea cucumber has moved away. In time it will grow a new set of intestines.

IMPORTANCE OF THE ECHINODERMS

Reference: See "Starfish Threaten Pacific Reefs," *National Geographic,* March, 1970.

Echinoderms are extremely numerous in the sea. They nearly cover the bottom in some places. Some of them live in shallow water. Other species are found at great depths. Any group so large in numbers must play a very important part in the food chains of the ocean communities.

Some people eat sea cucumbers, but these are not a major source of food. Starfish are a great annoyance to oyster gatherers, because one starfish can destroy as many as twelve oysters in a day. The spines of sea urchins can be painful to the feet of swimmers. Some sea urchins do damage to kelp beds, and certain starfish destroy living corals.

Strange as it may seem, the echinoderm phylum is thought to be more closely related to humans than any of the other animals we have studied so far. The larvae of the starfish are very similar to the larvae of certain primitive members of our own phylum.

SUMMARY

The echinoderms include starfish, basket stars, sea lilies, sea urchins, sand dollars, and sea cucumbers. Their bodies have arms or other parts arranged in a circle around the mouth. Most of them have spines. Some are filter feeders. Some eat clams. Others eat a wide variety of other foods. All of them live in salt water. They walk on tube feet.

The common starfish eats clams and oysters. It pulls them open with its tube feet and pushes its stomach inside the shell to kill and digest them. This makes starfish pests to oyster gatherers. Starfish can also harm corals and kelp.

Echinoderms are important to the ocean community as members of the food chain.

Activity

Observing Echinoderms

A. Examine a preserved starfish.

1. How many arms does it have?

2. Describe the body surface. What does it feel like?

3. Where is the mouth?

4. Describe the tube feet. Where are they located?

B. Examine all of the specimens of echinoderms that your school may have in its collection.

5. In what ways are they alike?

6. In what ways are they different?

C. Place live starfish or sea urchins in an aquarium of seawater. Watch how the tube feet carry the animals across the bottom and even up the side of the glass.

7. Describe the action of the tube feet.

Word Quiz

In your own words, write a definition for each of the following terms:

        echinoderms   tube feet

Check Your Facts

1. Describe the structure of the starfish.
2. How does a starfish move? How does it eat?
3. Name some other members of the echinoderm phylum.
4. Of what importance are the echinoderms?

Thought Questions

1. When starfish invaded a local shellfish bed, many people gathered the starfish and chopped them in two and threw them back in the sea. Do you think this solved the problem? Why?
2. Sea otters were wiped out along the California coast. Later, the kelp beds that are so important to fish and other sea life began to die out. Why?

Science Reading Skills:
1. & 2. *Problem Solving*

301

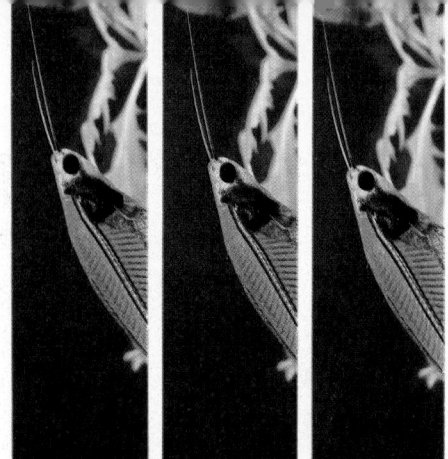

CHAPTER 30 Fish and Amphibians

vertebrates

chordate

Question: Before beginning the chapter, ask students if they can name some vertebrate characteristics.

Activity: Using a fish, demonstrate the characteristics of a vertebrate.

Enrichment: Jawless fish are members of the class Cyclostomata. Sharks belong to the class Chondrichthyes. Bony fish belong to the class Osteichthyes.

lamprey

What do you see in the photograph above? Is it a skeleton, or a fossil? It is neither of these. This is a photograph of a living fish called a glass catfish. The body of this fish is transparent, so you can see right through to its internal structures. Notice that this fish has a backbone. This means it is a member of an important group of animals called *vertebrates*. Other examples of vertebrates are frogs, lizards, birds, dogs, humans, and many more. Vertebrates not only have a backbone, they also have highly organized bodies with all of the body systems.

Vertebrates belong to the *Chordate* (CORE-date) phylum. This phylum is divided into four divisions called *subphyla*. Vertebrates form the only subphylum that we will study. The other three subphyla contain small numbers of primitive animals that live in the sea. They do not have backbones, but their body structure shows that they are related to the vertebrates.

Vertebrates have an internal skeleton moved by a complicated set of muscles. They have a head, neck, body, and usually a tail. A body cavity contains the intestine and other special organs like the kidneys, liver, and heart. The head has eyes, nostrils, and mouth. There are usually two pairs of movable appendages.

The members of three classes of vertebrates are called *fish*. There are four other classes: *amphibians* (am-FIB-ee-uns), *reptiles*, *birds*, and *mammals*. This chapter deals with fishes and amphibians.

THE FISHES

Jawless fishes. This primitive class includes the *lamprey*. See Fig. 30-1. There are several different species of these organisms.

Lampreys have no jaws and no paired fins. Their skeletons are made of cartilage instead of bone. Lampreys live by attacking other fish. Their mouths are big round funnels that they clamp against the sides of their victims. They use their spiny tongues to tear an opening in the skin. Then they drink the blood. Such an attack may kill the fish. A fish that escapes will have a round scar on its body.

The shark class. *Sharks* belong to an old and successful class of fishes. They also have skeletons made of cartilage. Most sharks are hunters, preying upon other sea animals. Their jaws are equipped with several rows of flat, sharp-edged teeth. Teeth in the front row do the biting. If one falls out, a tooth from the row behind it rises up to take its place. See Fig. 30-2. Most sharks never attack people, but individuals belonging to some species are sometimes very dangerous. Many people have been killed by them.

Some sharks are fairly small, but others are very large. The whale shark and the basking shark can grow to over 12 meters in length. The basking shark is the largest of all fish. You might think that these sharks eat large fish, or even people. But, this huge fish feeds in much the same way as whales do; by swimming with its mouth open. It strains tiny forms of sea life out of the water that flows into its mouth and out through its gills.

Sharks breathe by means of gills. There are several gill units on each side of the body. Each of these gill units has a separate opening to the outside. Sharks must keep swimming all the time, or they will "drown." This is because a current of water is kept moving over the shark's gills only by the motion of its swimming. Other fish can stay in one place because they have muscles in their heads that keep the water current moving. Sharks do not have these.

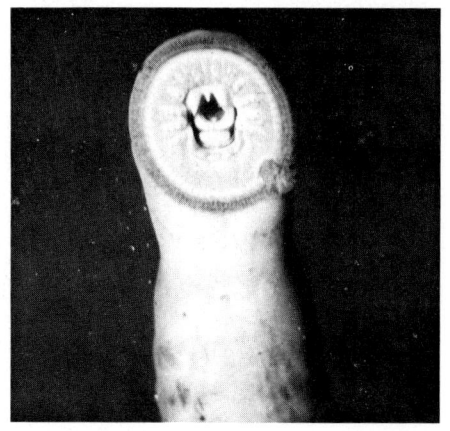

Fig. 30-1

A lamprey. This organism has a sucking mouth with no jaw. There are no paired fins. (American Museum of Natural History)

Fig. 30-2

Left: The jawbone of a shark. Notice the several rows of teeth; Right: A shark. Can you tell what species this is? (Museum of Natural History)

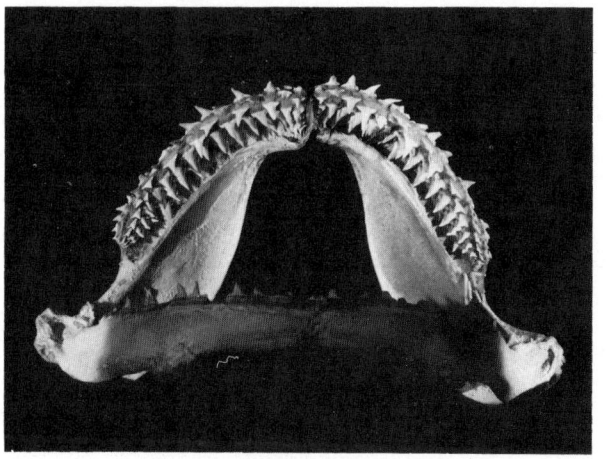

Fig. 30-3

A sting ray. This name comes from the poison spine at the base of the tail. These fish seem to "fly" through the water. Where is the mouth? (American Museum of Natural History)

Science Reading Skill: *Compare and contrast the three classes of fish.*

Fig. 30-4

The internal structure of a fish. How many of these same parts are found in man?

The *skates* and *rays* are also members of the shark class. They have very flat bodies, with the mouth on the underside. See Fig. 30-3. They are adapted to feed and hide on the bottom of the water. They swim by waving the big triangular fins on the sides of their bodies. This makes them look as if they are flying through the water. Their tail is a long whip. Often there is a poison spine near the base of this tail. This can give a painful wound to anyone who steps on the fish. People who are wounded in this manner should get an anti-tetanus shot, for these stingers often carry dangerous germs.

The bony fishes. This is the class of fishes that we are most familiar with. It includes perch, bass, pike, cod, mackerel, salmon, and many others. These fishes have skeletons of bone and most of them have a covering of scales.

Like all fish, the bony fishes get their oxygen by means of *gills*. A gill is an organ for absorbing oxygen from the water. The gills are fastened to bony arches along each side of the throat. A gill is divided into many small parts, which gives it a feathery look and provides a great deal of surface area for absorbing oxygen. The outside of a gill is covered with a thin membrane, and the inside contains blood. Dissolved oxygen molecules diffuse into the blood through this membrane, and carbon dioxide molecules diffuse out through it. Water passes in through the mouth of the fish and out through openings in the sides of the throat. This brings a constant supply of water to the gills. In the bony fishes there is a single *gill cover* on each side of the neck over all of the gills. Each gill cover is open along its back edge, where the water goes out.

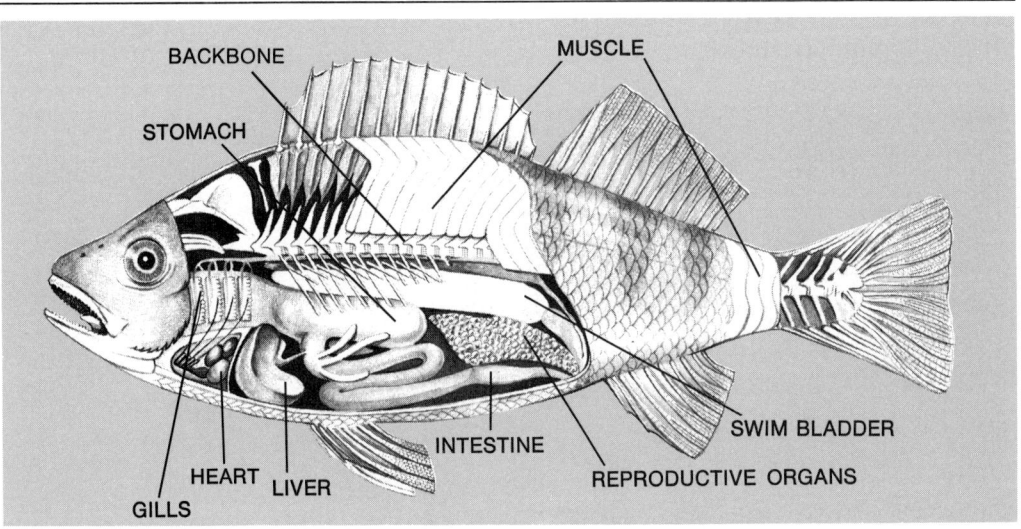

Fig. 30-5
Two types of bony fishes. Left: A damselfish; Right: A guppy. (Grant Heilman)

Fish swim mainly with their tails. The movable pairs of fins, which correspond to our arms and legs, also help in swimming. Most bony fishes have a *swim bladder* inside the body. This is a gas-filled structure that lightens the fish so that it can drift in the water without sinking. The other inside organs of fish are shown in Fig. 30-4. They are much the same in all vertebrates. We shall study our own body systems later on, so we will not go into detail on other vertebrates at this time.

swim bladder

Importance of fish. Fish are adapted to swim actively against water currents. This means that they can go wherever necessary to get food and to avoid enemies. Smaller animals like the free-drifting crustaceans cannot do this. Squid and whales are about the only water animals that can compete with the fish in this respect, and the fish far outnumber them.

The swimming ability of the fish has made it possible for them to dominate most water environments. They live in the oceans, right down to the deepest parts. They also live in rivers, lakes, and ponds. They eat both large and small plant and animal foods.

Fish are important to us as food. Large numbers are taken from the oceans and from inland lakes and streams. Fishing is also a very popular sport. It is one of the main attractions in many vacation areas.

Assignment: Fish rank next to agricultural products as the most important source of food for humans. Have students research the effects of pollution on the spawning grounds of fish.

THE AMPHIBIANS

The salamanders, frogs, and toads belong to the *amphibian* class of vertebrates. As adults most of them have legs and lungs, so they are able to live on land and breathe air. Yet their thin,

amphibian

moist skin loses water easily by evaporation. Because amphibians are always in danger of drying out, they usually live in moist places. Adults may be found on the land, but they return to the water to breed. This means that they are not likely to be far from a pond, lake, or stream. In other words, amphibians are partly adapted to land life, but they are not efficient land animals. They cannot compete with fish in the open water, or with higher animals on the dry land. They are animals of the shores, ponds, marshes, swamps, and low meadows. In these places, amphibians may be better adapted than other vertebrates. Frogs, especially, are often very numerous.

Salamanders are lizard-shaped amphibians. But they are not lizards. Lizards belong to the reptile class. Most of the salamanders are only a few centimeters long, though in the eastern mountain area there is one kind that may be as much as 45 centimeters long. Another type, found in Japan, can grow to 1½ meters.

salamanders

Fig. 30-6
Top: A toad; Bottom left: A frog; Bottom right: A salamander. These are all amphibians. How are the frog and the toad different from each other? (William E. Ferguson)

You can tell a salamander from a lizard in several ways. The salamander has the smooth, moist skin of an amphibian. A lizard's skin is dry and scaly. Salamanders have no claws. Lizards do. Salamanders have blunt noses and bulging eyes. They are generally slow in their movements. Lizards are usually swift. Salamanders hide in damp places; turning over old logs in the woods is one way to find them. They are perfectly harmless and can be handled. See Fig. 30-6.

Frogs are a more highly developed type of amphibian. Adult frogs lack the primitive, swimming tail of the salamanders, and their hind legs are specialized for jumping. This gives them a way of escaping their enemies, so you may find them in fairly open places. They often live along a shore and dive into the water when they are frightened. They hibernate in the mud, under water. There are more frogs than any other kind of amphibian. Some of them live in such unusual places as treetops.

Toads are related to frogs and look very much like them. Their legs are shorter, and their skin looks warty. They can absorb water through this rough skin, so that rain and dew keep them moist. They spend most of their adult life on land. They even hibernate on land in holes that they dig with their hind legs. Toads spend only a short time in the tadpole stage. They are already toad-shaped by the time they are one centimeter long. At this time they come hopping out of the water. A frog and a toad are shown in Fig. 30-6.

You may wonder how these harmless little animals that hop so slowly can escape their enemies on land. The two large warts on each side of the toad's neck are poison glands. If a dog or fox or bear swallows a toad, the animal becomes very sick from the poison and coughs up the toad. This teaches the animal to leave toads alone.

Importance of the amphibians. Amphibians are insect eaters. This makes them useful in controlling insect pests. Frogs may eat a great many grasshoppers and leafhoppers in low meadows. Toads are the most useful to us, because they live on higher ground. They come into our crop lands and eat harmful insects.

Enrichment: Certain frogs hibernate in winter in the mud at the bottom of a pond. The mud keeps the frogs from freezing during the winter months.

Enrichment: Surinam toads carry their young in separate pouches on their backs. Some male toads carry strings of eggs wound around their hind legs.

Enrichment: Many poisonous amphibians, including certain species of salamanders and frogs, warn other animals to stay away through their vivid colors and markings.

EARLY VERTEBRATES

The study of fossils suggests that the first vertebrates were jawless fishes of the lamprey class. Unlike modern lampreys, they had bony skeletons. Through natural selection, the descendants of some of them may have developed jaws and two pairs of movable fins. Some scientists believe that the sharks

Science Reading Skill: *Sequencing*—Trace the events in the development of the early vertebrates.

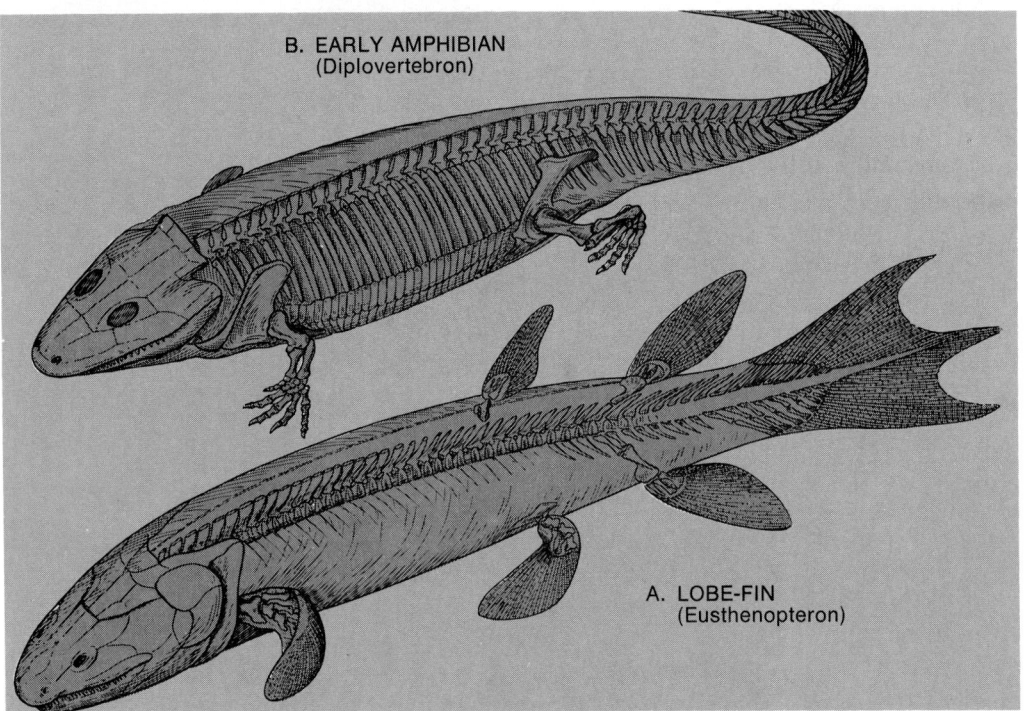

B. EARLY AMPHIBIAN
(Diplovertebron)

A. LOBE-FIN
(Eusthenopteron)

Fig. 30-7
Top left: An artist's idea of how early lobe-finned fish may have looked coming out of water onto land; Top right: Early Amphibians; Bottom: Drawings of the skeletons of these animals, which have been found as fossils. (American Museum of Natural History)

and bony fishes are both descended from these early jawed fishes. All of this would have happened hundreds of millions of years ago.

The first members of the bony fish class were specially adapted to life in swamps. They had gills like other fish, but they also had thin-walled bladders opening into their throats. These were used as lungs to absorb oxygen from the air. Such fish are called *lungfish*. Some scientists have proposed a theory that says that when the swamps dried up or the water became foul, these fish did not die. They breathed air with their primitive lungs.

One group of these swamp fish had *lobed fins*. Their movable, paired fins had solid fleshy bases next to the body. With fins like these, the lobe-finned lungfish could walk across land from one pool to another.

You may wonder why land animals did not eat these clumsy, walking fish. The theory states that at that time there were no land animals other than insects and spiders. The lobe-finned fish did not need to be very well adapted to land life, because they had no competition on land. On land they found new food supplies. Gradually, the theory continues, through mutation and natural selection, some of their descendants became better and better adapted to their new home. The lobed fins were modified to become legs. Adults lost their gills, and the lungs became more fully developed. The lobe-finned lungfish had become amphibians. See Fig. 30-7.

Other lungfish developed in the opposite direction. They moved into deep water. In their descendants the lungs became swim bladders, and they are the most common and successful of all the fishes living today.

There are still a few kinds of lungfish left. One species survives in the swamps of Africa. Another lives in Australia, and one lives in South America.

Amphibians first appeared on the land about 425 million years ago. They continued to be the main land animals until the end of the Coal Age, about 100 million years later. Huge salamanders as big as alligators crawled about the swamps among the giant horsetails and tree ferns.

During this time one group became much better adapted to land conditions than the rest of the amphibians. These modified amphibians gave rise to the first reptiles.

These reptiles spread out into many of the drier environments, where the amphibians had never been able to live. Even in the swamps the reptiles replaced many of the amphibians. At the end of the Coal Age most of the amphibians were gone. The great age of reptiles was about to begin. We shall study reptiles in the next chapter.

SUMMARY

The vertebrate subphylum is made up of animals with backbones. They have internal skeletons and all of the body systems.

Three classes of vertebrates contain animals that use gills for respiration throughout their lives. These are all called fish. They are the jawless fishes, the shark class, and the bony fishes. The lamprey has no jaw and no paired fins. The other fishes do have these structures. Sharks and bony fishes are the dominant, large animals of the water communities.

The amphibian class is made up of animals that have gills when young and lungs when grown. This class includes salamanders, frogs, and toads. Amphibians are only partly adapted to the land, so they are found in swamp or shoreline environments.

Amphibians developed from lungfish long ago. Reptiles developed from certain amphibians.

ACTIVITY

The Effects of Temperature on an Amphibian

A. Place a piece of masking tape about 6 centimeters long on the floor. Obtain a live frog and place it on the floor with its front limbs touching the tape. Touch the frog gently on its back so it will begin to jump. Allow the frog to jump about for exactly 15 seconds. While the frog is jumping, trace its path on the floor with a piece of chalk.

B. After 15 seconds place the frog in a container. Next measure all of the chalk lines on the floor to obtain the total distance the frog jumped.

C. Repeat steps A and B three more times and take an average of the four distances the frog jumped.

1. What is the average distance the frog jumped?

D. Place the frog in a container that is half full with ice water. After 8 seconds take the frog out of the water and place it on the floor with its front limbs on the tape.

E. Touch the frog on its back to make it move and begin timing for 15 seconds. Mark the path of the frog on the floor and measure the distance traveled as before.

F. Repeat steps D and E three more times and take an average of the four distances the frog jumped.

2. What is the average distance the cold frog jumped?

3. Did the frog travel farther when it was warm or cold?

4. What was the difference in the average distances traveled?

5. Would trying to live in a cold environment be an advantage or disadvantage to the frog? Why?

Word Quiz

From the new words you learned in this chapter, write the word that correctly completes each statement. Do not write in this book.

Any animals with backbones can be called _____. The _____ seems strange to us because it has no jaw. Lizard-shaped animals in the same class as the frog are called _____. All members of this class are called _____. Vertebrates belong to the _____ phylum. A(n) _____ helps fishes to drift in the water without sinking.

Check Your Facts

1. What are the vertebrates? How can you recognize a vertebrate? What are chordates?

2. What are the three classes of fishes? How are they different from each other?

3. How do fish get their oxygen?

4. What is the swim bladder used for?

5. What are amphibians? How do they differ from other vertebrates?

6. How can you tell a salamander from a lizard?

7. In what way are frogs and toads useful to us?

8. When did amphibians first appear on the earth? Where did they come from?

Thought Questions

1. Why do we say that the amphibians are only partly adapted to living on land?

2. We can think of the amphibians as being more highly developed than the fish. Why are the fish far more numerous and important than the amphibians?

Science Reading Skills:
1. & 2. *Problem Solving*

CHAPTER 31 Reptiles and Birds

objectives

After you read this chapter, you should be able to:

__**Compare** the reptiles and amphibians

__**Compare** the reptiles and birds

__**Describe** the kinds of reptiles living today

__**Explain** how birds are adapted to flying

Science Reading Skill: *Compare and contrast* the classes of amphibians and reptiles.

You have seen various birds and reptiles many times. You may even have some of these animals as pets. How much do you really know about these organisms? Are you aware of all the different ways they get their food? Do you know how they care for their young? In this chapter, we will answer these and many other questions about reptiles and birds.

REPTILES

Characteristics of reptiles. The reptiles were the first vertebrates that were well adapted to living on land. The reptile skin is dry and covered with scales, so that moisture is held in the body. The lungs and circulatory systems of reptiles are more efficient than they are in amphibians. Reptiles reproduce on land.

The reptiles have some disadvantages. They do not care for their young. Therefore these young do not have as good a chance of survival as birds and mammals.

Reptiles are cold-blooded. They cannot produce enough heat from the oxidation of food to keep their bodies warm on cool days. At such times they expose themselves to sunlight to absorb heat. On hot days they move into cool, shady places to prevent overheating. In very cold weather a reptile is unable to move or remain conscious, so northern reptiles must hibernate during the winter months.

The modern groups of reptiles. The main groups of reptiles living today are the lizards, snakes, turtles, and the crocodile group.

The *lizards* are mostly fast-moving insect eaters. They live mainly in warm countries. Some can break off their own tails when in danger of being captured. While an enemy holds on to the wriggling tail, the lizard escapes. Later, it grows a new tail. Most lizards are small, but a few tropical species are rather large. The biggest is the dragon lizard of Indonesia, which may reach 4 meters in length. See Fig. 31-1. The dragon lizard hunts animals as large as deer.

Snakes, with their long legless bodies, seem to survive well because of their ability to hide easily. They all eat small animals. A snake's small teeth cannot bite off pieces, so their victims must be swallowed whole. The jaws of snakes are very flexible and can stretch far out of joint. This is why snakes are able to swallow food thicker than themselves.

When a snake attacks another animal, it has no feet or claws to use. Its long body may be injured as the victim fights back. Different kinds of snakes have different ways of meeting this danger. Some eat only harmless animals. The garter snake, for instance, feeds on worms, insects, frogs, and toads. It simply grabs its victim and slowly swallows it.

Other snakes use a wrestling method. They wrap their bodies around the animal's chest and squeeze. This does not crush the victim, but it prevents it from breathing. The victim is soon dead, and the snake can go ahead with the slow business of swallowing it. The bull snake and fox snake take mice in this way. The biggest snakes of all, the pythons and boas, also use the squeezing method. The largest of these snakes are about 10 meters long. See Fig. 31-2 on page 314.

The third method by which snakes kill their victims is through poison. Poison quiets the victim quickly. The poison comes from glands in the snake's neck and enters the victim through hollow, needle-like *fangs*. There are three groups of poisonous snakes—the *rear-fanged serpents*, the *cobras*, and the *vipers*. The rear-fanged serpents have two poison fangs in the back of the mouth. Because of the location of the fangs, they cannot easily use them on large animals. These snakes are seldom dangerous to people. There are a few rear-fanged snakes in the southern United States.

Cobra-type fangs are short and hollow and they are located in the front of the mouth. These snakes grab their victims and chew the poison into their flesh. Two species of *coral snakes* are the only members of this group in North America, except for a few tropical sea snakes that may wander as far north as the southern California coast.

The vipers of America all belong to a group known as the **pit vipers**. They have a pit-like sense organ on each side of their heads. This looks almost like an extra nostril. Many vipers in other parts of the world do not have this pit. The pit is very sensitive to heat. The snakes use it to detect mice or other warm-blooded animals hidden in the grass. They can feel the heat radiating from the mammals' bodies.

Fig. 31-1
The dragon lizard of Indonesia. It is nearly as big as an alligator and it hunts food on land. (W. Suschitzky)

Enrichment: Snake poison, also called venom, has many medical uses, such as an anticoagulant and in the preparation of antivenom.

fangs

Enrichment: Discuss how to treat poisonous snake bites.

pit vipers

Fig. 31-2
Some snakes strangle their prey. (C. B. Frith/Bruce Coleman, Inc.)

The most common poisonous snakes in America are all pit vipers. They are the *rattlesnakes*, the *water moccasin*, and the *copperhead*. Their fangs are so long that they must fold back when the snake closes its mouth. See Figs. 31-3 and 31-4. The fangs are driven into the mouse, rat, or other victim with a very swift stabbing movement of the head and neck. Then the snake lies back and waits while the victim dies.

People fear the bites of poisonous snakes, and snakes fear being stepped on by people. They bite in self-defense when someone comes too close. Poisonous snakes are not common in most parts of the country. There are many areas that have none of them. Find out which dangerous ones live in your area. Look out for them, but do not fear the others.

More people are bitten by the copperhead than by any other North American snake. It matches the color of dead leaves on the forest floor perfectly, and people step on it by accident. Very few people die from such bites. The copperhead is not a large snake, and it does not have a great deal of poison. Any fairly good first aid or medical treatment will save the person's life, but there will be a great deal of pain.

The snake most likely to kill a person is the species of diamondback rattlesnake, which is found all through the southwestern states. It is large, and it has a great deal of poison. Most snakes fear people and try to keep out of their way, but the southwestern diamondback is different. Sometimes it will actually edge closer, to get in position to strike. At other times it will only try to get away.

Many snakes have a nervous habit of vibrating their tails when they are excited. The rattle on the tail of a rattlesnake produces a buzzing sound at such times. This helps you to avoid the snake, but do not depend upon this warning signal. Sometimes a snake will strike without rattling first.

The names turtle, terrapin, and tortoise all refer to members of the group we call *turtles*. Turtles have bodies specialized for defense. The shell is made of the ribs, backbone, and breastbone, all flattened out and grown together. The outside is covered by thin plates. Most of our turtles escape from land animals by sliding into the water. Few water animals are able to hurt them because of their strong shells. Some turtles such as the box turtle, the gopher turtle, and the giant land turtoise of the tropics spend all of their time on land.

The largest freshwater turtles in North America are snapping turtles. The northern species may weigh 20 kilograms and the southern 75. These snapping turtles are dangerous to handle. They have no teeth, but the sharp beak can cut off fingers. See Fig. 31-5 on page 316.

The biggest turtles of all are sea turtles. They may weigh as much as 700 kilograms. The toes of these reptiles have grown together to form flippers. Sea turtles spend their entire time at sea, coming ashore only to lay eggs. See Fig. 31-5.

The *crocodiles and alligators* are shaped like lizards, but are much bigger and heavier than the lizards. They are more highly developed than other reptiles and are related to the dinosaurs. Crocodiles and alligators live in swamps and along river banks. They eat fish and any land animals that come within their reach. The African crocodile sometimes eats people. The American alligator seldom bothers people, but a large one can be dangerous. The American crocodile is often vicious. See Fig. 31-6 on page 316.

Some members of the crocodile group are quite large. Alligators can grow to be 5 meters long. In other parts of the world some crocodiles may reach a length of 10 meters.

The American alligator is common in southern swamps. It was becoming scarce, because too many were killed by hunters.

Fig. 31-3

(Left) A pit viper. Note the pit-like sense organs on either side of the head. (C. B. Frith/Bruce Coleman, Inc.)

Fig. 31-4

(Right) The fangs of a rattlesnake (a pit viper) are like needles for injecting poison. These fangs fold back when the mouth closes. The other teeth are the common needle-pointed type found in all snakes.

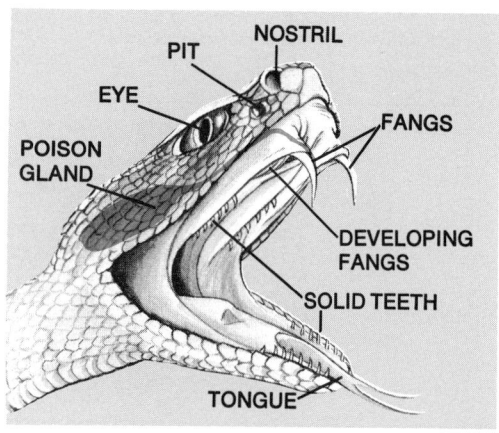

315

Fig. 31-5
The snapping turtle (left) is the largest freshwater turtle in North America. The sea turtle (right) is the largest of all turtles. (Leonard Lee Rue III/Animals Animals; Australian News and Information Service)

Fig. 31-6
Left: The American alligator; Right: The American crocodile. Both are members of the same family and are much alike. What differences can you notice? (Grant Heilman; Hal Harrison/ Grant Heilman)

The skins were sold for making leather. Conservation laws have protected alligators for several years now, and their numbers are now increasing.

Importance of the reptiles. Reptiles are not numerous in most places today. The effects they have on the environment are mostly good from our point of view. The smaller snakes and lizards eat many insects. Some other snakes eat large numbers of mice and rats that would otherwise destroy our crops.

The skins of alligators, crocodiles, and the larger snakes and lizards are used to make leather, but the numbers taken are carefully controlled. A number of turtles are eaten. People in some parts of the world eat snakes and lizards. In Florida the alligators dig ponds in the marshy areas. These become important water holes for all kinds of wildlife during dry seasons. This is one reason why the alligator is protected from hunting in that state.

History of the reptiles. The early reptiles seem to have developed from certain amphibian ancestors over 250 million years ago. Their ability to live on dry land made them a very successful group. There were many kinds of reptiles, both large

and small. Some were plant eaters, and some were hunters. They lived in many different environments. Some even lived in the sea.

The lizards and turtles of today are rather primitive reptiles. Their ancestors date far back to the early days of the reptile period. Snakes developed later from the lizards.

In the past there were more highly developed reptiles than any alive today. Some of them were probably warm-blooded. One group of advanced reptiles is thought to have become the early mammals. The most famous reptiles of all were the dinosaurs. Most of them were large, and some were *very* large. If they were alive today, the big ones would have to bend down to look into the upstairs window of a house.

This great age of reptiles is thought to have lasted about 160 million years and then suddenly ended. By suddenly we mean that most reptile groups seem to have become extinct within a few hundred thousand years. This is a very short time in the history of the earth. It appears that no land animal weighing more than 25 kilograms survived.

The numbers of dinosaurs were already much lower than they had been by the time these last dinosaurs died. There is evidence that a huge meteor, called an *asteroid*, hit the earth at this time. It would have been a piece of rock ten or twenty kilometers across and traveling thousands of kilometers per minute. This may have had something to do with the extinction of dinosaurs and many other species of living things.

The age of reptiles ended 65 million years ago. Only the lizards, snakes, turtles, and crocodiles remain.

Actually, one group of dinosaurs did not become extinct. It is a very common and successful group today. We do not call these animals dinosaurs. We do not even call them reptiles. We call them *birds*.

BIRDS

The birds are so different from modern reptiles that we place them in a separate class. When we compare birds to some of the small, two-legged dinosaurs we realize that the two are close relatives. A bird is simply a dinosaur adapted to flying.

Birds had already appeared on the earth before the age of reptiles ended. When most of the reptiles disappeared the mammals and birds increased in numbers. These are the important vertebrates on the land today. Insects continue to be the main *small* land animals.

Characteristics of birds. The wings of birds are modified front legs. Most of the wing surface is made up of long, stiff feathers. These form a large surface area that has little weight. Feathers

Assignment: The ancestors of the reptiles of today may have been the dinosaurs of the past. Have students report on one or more dinosaurs.

Assignment: Have students trace the history of birds and the possible transitional stages between reptiles and birds. Discuss the significance of the fossil bird *Archaeopteryx*.

Enrichment: Birds belong to the class Aves.

Science Reading Skill: *Compare and contrast* amphibians, reptiles, and birds. Discuss how they are adapted to life on land.

Fig. 31-7
Feathers are lightweight, which makes them very useful for flying. (Peter Turner/The Image Bank)

Activity: Ask students to bring in bird feathers that they may have at home. Examine them under a hand lens.

Enrichment: Birds have a poor sense of smell compared to fishes. Their eyesight is exceptionally good, however. An insect-eating falcon can spot a dragonfly from a distance of about 700 meters.

are probably modified scales. See Fig. 31-7. Reptile-type scales still cover the feet of birds. Small feathers cover the bird's body. This holds in heat. A high body temperature is needed to maintain the high energy output needed for flying.

Powerful heart, lungs, and flying muscles make up the "engine" of the "bird flying machine." All other body parts are reduced in weight. Many bones are hollow and contain air sacs. A lightweight beak replaces the heavy jaws and teeth of the reptile. The tail is reduced to a short stub. Long tail feathers do the steering.

The main foods of land birds are insects and seeds. Both of these foods provide the concentrated energy needed for flight. The ability of birds to fly makes them better insect catchers than either the reptiles or the mammals. Seeds often grow on the ends of branches and stalks, so flight also helps in gathering them. Other foods eaten by birds include small mammals, which are eaten by hawks and owls. Vultures eat dead meat.

Large numbers of birds get food from the sea. Some pick it up on the beaches. Some find it near the shore. Others fly many miles out to sea to find food. Crustaceans, molluscs, worms, and fish all become food for sea birds.

No flying bird eats grass or leaves. These are low-energy foods. Too much of them would be needed to power a flying animal. Also, flight would not help a grass-eating animal. The grass will not run away. Mammals are much better adapted to eating plants, so birds and mammals share the land environment. Each type lives where it can find food and escape destruction by its enemies. A few birds, like the ostrich, have lost the power of flight. They feed on the ground like mammals.

Bird study. Studying birds is a favorite hobby of many people. Birds are the only wild animals that are easily seen in the daytime in settled parts of the country. You can even watch them in a backyard or in a city park.

The beak and feet of a bird can tell you a great deal. Look at Fig. 31-8. Short, heavy beaks are used for cracking seeds. Slender, lightweight beaks are good for catching insects. Flat beaks can shovel in mud to catch worms and snails. Long, spear-like beaks can catch fish and frogs. Hooked beaks can tear small mammals into pieces that a bird can swallow. Can you name birds with each of these types of beaks?

Science Reading Skill: *Reading Illustrations*—Refer to Fig. 31-8. Discuss bird beaks, feet shape, and egg color as adaptations to special environments.

Fig. 31-8
Several kinds of beaks and feet. What is each adapted for?

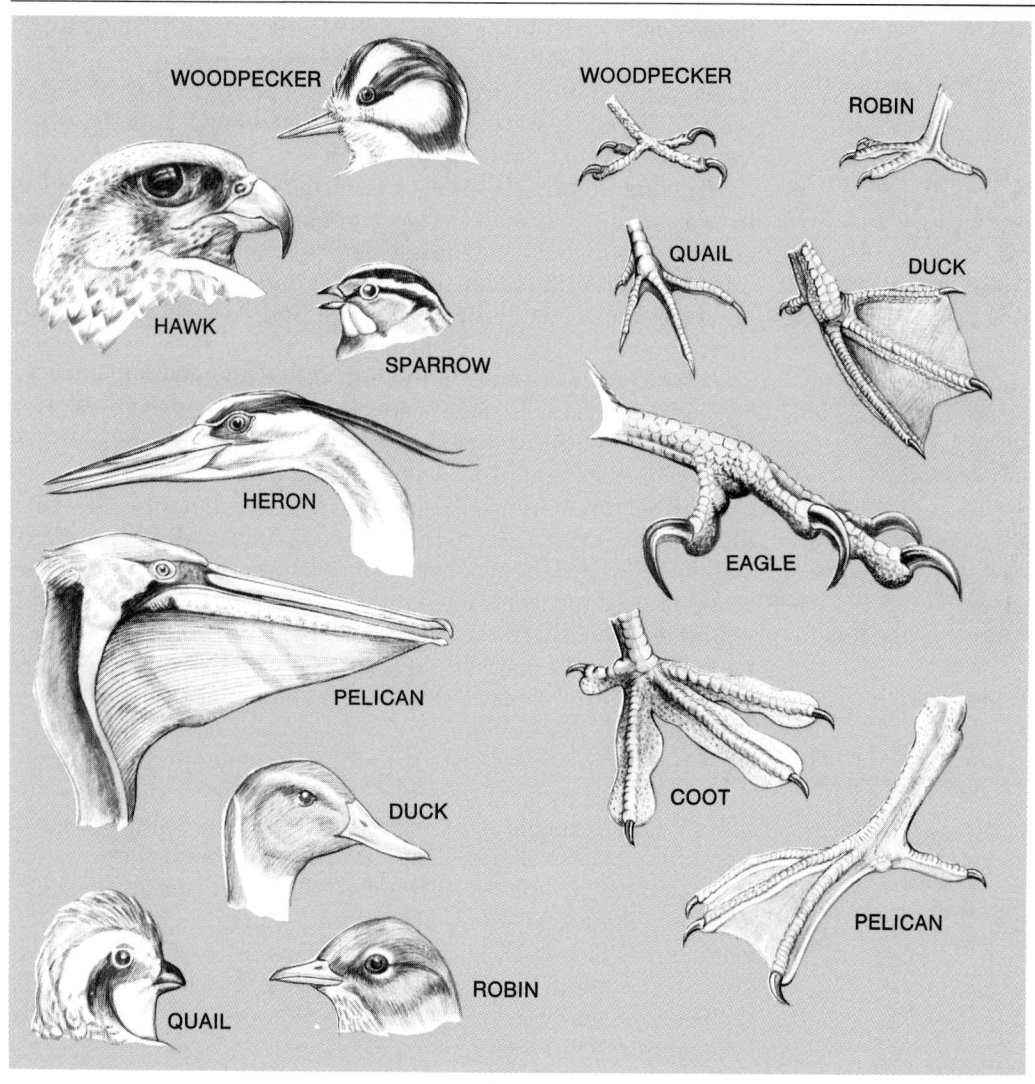

The feet of birds may also tell a story. They may be adapted for perching on twigs, for walking, for swimming, for wading, for scratching the ground to uncover food, for killing small mammals, or for clinging to tree trunks. See Fig. 31-9. Learning to observe such things as beaks and feet will help you to know the different birds.

migration flight

Migration of birds. One of the most interesting things about many birds is their *migration flight*. Most birds that live in the north fly south for the winter. They do not migrate to escape the cold. With their warm bodies and feather covering, birds can withstand cold weather very well. Birds fly south because their food supply is gone. Insects are not active in the winter. Some seed-eating birds, like the cardinal, can still find food. They stay in the north all winter. So do woodpeckers, owls, chickadees, field sparrows, jays, and several others. Each of these birds eats foods that are available all winter. Some species stay in a warm climate all year long.

Why do migrating birds return north again? When birds nest in the spring they need large amounts of food to feed their growing young. By spreading out over more territory, more birds can find this needed food. This is one important advantage of migration. It enables more birds to find enough food to raise their young.

Of course birds do not know why they migrate. They do so by instinct. When the days become a certain length each spring, the birds respond by flying north. In the fall, as days become shorter, birds respond by migrating back to the south.

The distances traveled by birds during migration are surprising. The house wren goes from Canada and the northern states to Florida, the wood thrush to southern Mexico, and the bobolink to Argentina. The arctic tern nests in the arctic, then migrates to southern Africa and South America, 18,000 kilometers away. The golden plover, a land bird, may fly nonstop from Labrador to Brazil—4,000 kilometers over water.

Flights like this call for great flying ability. They also call for good navigation. Biologists have always wondered how birds find their way on long flights. Experiments indicate that some birds watch the position of the sun and the stars. They recognize particular groups of stars. They judge their distance north and south by seeing how high these stars seem to be in the sky. They use their keen sense of time and watch the movement of stars across the sky. This tells them how far east or west they are. Ship captains navigate in the same way, but they need instruments and mathematical training to do so.

Some birds are able to sense the earth's magnetism. It is as if they have a built-in compass. This tells them in which direction they are going.

Importance of birds. Since there are so many birds, they affect the communities in which they live in many ways. As eaters of insects and of weed seeds, they are very useful to us. Hawks and owls are important in keeping meadow mice and other pests in check. Some birds, such as chickens and ducks, are raised as domestic animals. They furnish us with meat, eggs, and feathers.

Enrichment: Discuss oil spills and the disastrous effects they can have on bird populations.

SUMMARY

Reptiles include the lizards, snakes, turtles, and crocodiles. They are cold-blooded, lay eggs on land, and have dry, scaly skins.

Lizards are usually insect eaters. Snakes eat insects, amphibians, and small mammals. Poison is used by some snakes to kill their prey.

Crocodiles and alligators live along shore lines and in swamps. They eat fish and land animals.

Turtles are protected by shells. Some live on land, and some live in water.

Many more reptiles lived in the past. The great age of reptiles ended 65 million years ago.

Birds can be described as dinosaurs adapted to flying. They eat insects, seeds, and several other foods. Birds are a very successful group of animals.

ACTIVITY

Observing Reptiles and Birds
A. Keep some living turtles in the classroom. Prepare a cage that gives them a choice of being in or out of water. Feed them pieces of meat dropped in the water. Remove any food they do not eat before it spoils. Land turtles will eat on land. They need no water to swim in, but supply them with a dish of drinking water. You can feed them vegetables.

1. Are your turtles able to pull their heads and feet into the shell? When do they do this?

2. How has this ability helped turtles to survive?

3. What do the feet of your turtles look like?

4. Does this tell you anything about where these turtles live?

5. Place the turtle on its back. Describe how it turns itself over.

6. What adaptations does the turtle have that allow it to turn itself upright?

Suggestion: Be sure that the aquarium has at least 3 centimeters of water in it. Also, place some smooth stones in the aquarium for the turtles to climb on. The water should be kept at about 24°C.

Word Quiz

In your own words, give a definition for each of the following:

fangs   pit vipers   migration

Check Your Facts

1. In what ways are reptiles better adapted to land than the amphibians?

2. How are birds different from reptiles?

3. What are the four main groups of reptiles living today?

4. What are three ways in which snakes capture their food?

5. Name the three main types of poisonous snakes.

6. Are most snakes useful or harmful? Explain.

7. What are the main foods eaten by birds?

8. How do beaks, feet, and wings of birds adapt them to different ways of living?

9. Why do birds need to migrate?

10. Of what importance are birds?

Thought Questions

Science Reading
Skill
1. & 2. *Problem
Solving*

1. Some people enjoy putting food out for birds in the winter. Sometimes this encourages birds to stay north when otherwise they would have migrated south. How can careless bird feeding be dangerous to these birds?

2. Several kinds of snakes have heat sensing organs on the sides of their heads. They are very useful for detecting warm-blooded animals in the dark. Why are heat sensing organs not found in birds or mammals?

CHAPTER 32 The Mammals

Above is a photograph of organisms that belong to the phylum known as mammals. What makes mammals such a successful phylum? In this chapter you will find out.

CHARACTERISTICS OF MAMMALS

Mammals differ from reptiles in several ways. Like birds, mammals are warm-blooded. This enables them to be active when the weather is cold.

Just as all birds have feathers, all mammals have hair. There are mammals, such as humans, that do not have very much noticeable hair, but they do always have hair. The fur of the mammal helps to hold in body heat. Another characteristic of the mammals is their learning ability. A mammal's brain is more highly developed than that of the reptile, making them more intelligent.

You already know how mammals reproduce. They care for their young. This means that more of their young can survive. Female mammals have structures called **mammary** (MAM-uh-ree) **glands**. These produce milk to feed the young. No other animals do this.

All of these things give mammals an advantage over reptiles. The first mammals appeared during the Age of Reptiles, but their big chance did not come until most of the reptiles disappeared. For the last 65 million years mammals have been the most successful of all vertebrates on the land. There are mammal species adapted to almost every environment.

ORDERS OF MAMMALS

There are many orders of mammals. We will study several of the more important ones.

___ objectives ___

After you have read this chapter, you should be able to:

___**List** the general characteristics of the mammals

___**Name** several members belonging to each of the orders of mammals

___**List** specific characteristics of various orders of mammals

mammary glands

Science Reading Skill: *Compare and contrast mammals and reptiles.*

Reference: See *Mammals of the World* by Michael Boorer and *The Mammals* by Carrington and the Editors of *Life* Magazine.

Fig. 32-1
A spiny anteater. This is a primitive mammal species that still survives. It lays eggs like a reptile. (Australian News and Information Bureau)

Enrichment: You may want to introduce the term marsupial.

Assignment: Have students report on the unusual wildlife of Australia.

Enrichment: This order is called Insectivora referring to the diet of these mammals.

Egg-laying mammals. The spiny anteater and the duckbill are the only animals in this group. See Fig. 32-1 and Fig. 9-7, page 80. You studied their reproduction in Chapter 9. They are interesting because they are the most primitive mammals.

Pouched mammals. These are the main mammals of Australia and nearby islands. A variety of these pouched mammals have developed there, and they are adapted to many different ways of living. Female mammals of this order have pouches in which they carry their young. See Fig. 32-2. Kangaroos feed on grass; koalas climb trees and eat the leaves; wombats look and act like woodchucks; the Tasmanian wolf hunts other animals. There are squirrel-like, mouse-like, and mole-like forms of pouched mammals in Australia. In general this group is not able to compete well with the placental mammals. Their low intelligence is probably one reason. Pouched mammals have survived in Australia because the ocean has separated them from the placental mammals since very early times.

One group of pouched mammals does survive in competition with other mammals. This is the opossum group. The common opossum of the United States and southern Canada is a "living fossil." Opossums lived in the Age of Reptiles. Opossums are slightly larger today, but otherwise they have not changed much in the last 100 million years.

The shrews and moles. These mammals are insect eaters. You probably know that moles live underground. Shrews are not so well known, yet they are the most common mammals in many areas. An ordinary shrew is the size of a mouse, but its nose is much more pointed, and its fur is thicker. See Fig. 32-3. Many shrews are almost blind. They hunt insects on the ground and under dead leaves. They have many sharp little

Fig. 32-2
A koala is a mammal that carries its young in a pouch. The koala is also a tree-climber. (Australian News and Information Service)

teeth. The pigmy shrew is only half as long as a mouse. It is the smallest of all the mammals.

The bats. Bats are much like the members of the mole and shrew order, except that they have wings. The front toe bones are long, with skin stretched between them. The wing skin continues back to the ankles of the hind legs and on to the tail. See Fig. 32-4. Since front and hind legs are both part of the wing, bats cannot walk very well. They either fly or they hang head down by the claws of their hind legs. Bats fly at night, taking insects out of the air. By day they hang in dark places, such as caves, hollow trees, or attics. North American bats are considered very useful to us, because they eat so many insect pests.

Some tropical bats eat fruit. They are large, with wing spreads of one meter. A large bat in South America hunts and eats small mammals. The vampire bat, which is not very large, also lives in South America. It lands on the backs of large animals and cuts the skin with its sharp teeth. Then it laps up the blood that flows out. Sometimes vampire bats attack sleeping people.

Bats can fly in complete darkness. They find their way by listening to the echoes of their own high-pitched voices bounding back from the objects around them. This is the same principle as sonar. Sonar is used to detect submarines and to tell how deep the ocean is. The inventors of sonar got the idea from a study of bats.

Rodents. Mice, rats, squirrels, woodchucks, porcupines, muskrats, and beavers are a few kinds of *rodents*. Rabbits are very much like them, but they are placed in a separate order. Rodents have heavy gnawing teeth in the front of their mouths and strong grinding teeth toward the back. This gives them the

Fig. 32-3
A common type of shrew. There are a great many of these little insect eaters in North America, but they stay out of sight so you seldom see them. (R. H. Noailles)

rodents

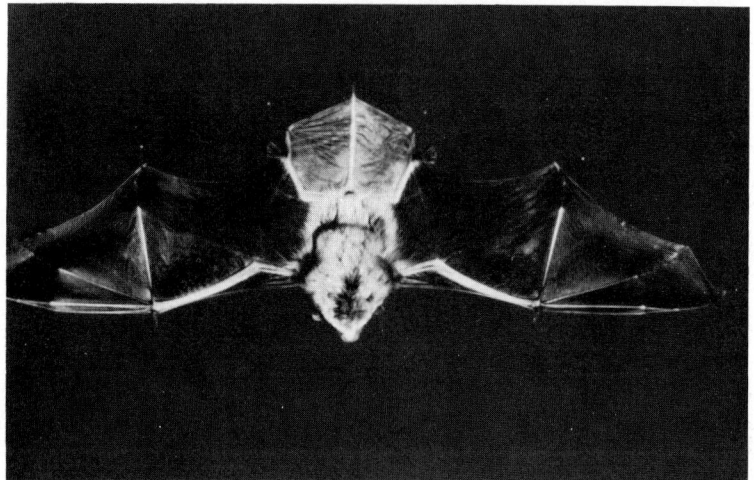

Fig. 32-4
The bat's wing seems to wrap around its body. The bones in the wing are similar to those in the human hand. (Courtesy Carolina Biological Supply Company)

325

ability to eat rough foods. Mice are mainly seed eaters. Squirrels eat nuts and twigs. Woodchucks eat grass. Most rodents also eat some insects.

The rodents are very successful animals. They can find food almost everywhere. They can hide easily, because of their small size. Many larger animals eat them, but this does not reduce the size of rodent populations, because they reproduce rapidly.

Some rodents are useful to us. The squirrel and guinea pig are eaten. The muskrat, beaver, and several others produce valuable fur. Many other rodents are harmful to us, because they compete with us for food. They eat our crops in the fields and in storage. They do damage to our buildings, and sometimes they carry diseases. The house mouse and the common rat are often serious pests in cities and around farm buildings.

The flesh eaters. This order includes the dog family, cat family, weasel family, bears, raccoons, and seals. These animals have teeth for catching and killing other animals, which they use for food. The flesh-eating mammals are more intelligent than most other animals. They serve a useful function in the wild communities, where they keep the plant eaters in balance with their food supply.

We have killed off the larger flesh-eating mammals in the settled areas of our country. The smaller ones, like foxes, raccoons, weasels, and minks continue to live in large numbers, even close to cities. The Alaskan brown bear and the polar bear are the largest members of the flesh-eating order.

Whales. Porpoises, dolphins, and whales are mammals shaped like fish. Although they breathe air, they live in the sea. They are believed to be descended from land mammals that became adapted to life in the water. The blue whale is the largest anmimal that has ever lived. It may be over 31 meters long and weigh over 140 tons.

The elephants. There are two species of elephants alive today, the Indian elephant and the African elephant. In the past there have been several others, including six species that lived in America. Mammoths and mastodons were elephants that died out in America within the last 9 thousand years. Early Indians hunted these species, and this may be what caused their extinction.

Reference: See "Wildlife in Danger" in *Newsweek*, January 6, 1975.

Assignment: Have students report on the intelligence of dolphins and porpoises.

Fig. 32-7
(Left) A blue whale is a mammal that lives in the sea. How is it adapted to breathe air? (Russ Kinne/Photo Researchers, Inc.)

Fig. 32-8
A drawing of a group of wooly mammoths. These organisms no longer exist on earth. (American Museum of Natural History)

Hoofed mammals. There are two orders of these mammals. The members of both have hoofed feet, and are plant eaters. They depend upon running away to escape their enemies.

The animals in one of these orders have an odd number of toes. These include the horse, rhinoceros, zebra, donkey, and tapir. Animals of the other order have an even number of toes—the so-called *"cloven-hoofed"* animals. Deer, sheep, goats, pigs, and cows are a few of the members of this group.

The flesh of hoofed mammals has been important food for the human race for hundreds of thousands of years. These mammals were hunted and later some of them were tamed. Cattle provide meat, milk, and leather. We get meat and leather from pigs, also. The horse has played an important role in human history. First, they furnished transportation over long distances on land. Then they were used as work animals on farms.

The primates. Lemurs, monkeys, apes, and human beings are all members of the order of mammals called *primates.* The primates are mainly tree climbers, with grasping hands and a high level of intelligence. Physically, people are primates, but their mental abilities and their ways of living set them apart from all other species.

HISTORY OF THE MAMMALS

The Age of Mammals is thought to have started 65 million years ago. At that time there existed pouched mammals like the opossum. These survive today mainly in Australia. There were also some early placental mammals. Today we would call them shrews. They were small insect eaters that lived on the ground.

When most of the reptiles died out, these early shrews spread out into many different environments. Through natural selection, they are thought to have developed adaptations that allowed them to live in these different areas. They could have become the many different orders of placental mammals that we see in the world today. Fossils have been found showing many of the stages in the development of the mammal orders.

Early primates. Primates are thought to have developed from those same early shrews. We get some idea of how this might have happened when we look at the living groups of primates. Some species do not change very much, even over long periods of time. They are well adapted to where they live, so they remain unchanged. These "living fossils" give us some idea of how ancestors of other groups may have looked in the past.

Another line of evidence is found in actual fossils. Many fossils of early primates have been found. These fossils tell a story about the early primates.

primates

Assignment: Have students report on the field work of primate researcher Jane Goodall.

Fig. 32-9
The tapir. This tropical relative of the horse and the rhinoceros is a member of the odd-toed order of hoofed mammals. (W. Suschitzky)

Basically, the primates are tree climbers. It is thought that an early group of shrews began climbing trees to find food. In this new environment, mutations for better tree climbing ability were useful. The old shrew-type genes, for ground-living traits, were harmful ones. Gradually, through natural selection, the new genes remained and the old genes disappeared. These animals would no longer have been shrews. They would have become tree-living primates. A small animal called the tree shrew lives in Southeast Asia. It looks very much like a shrew, but it has enough of the primate traits to be called a primitive primate. The earliest primates must have looked very much like this tree shrew.

It is believed by some scientists that lemurs developed from tree shrews. In appearance, a lemur is about half-way between a tree shrew and a monkey. See Fig. 32-10. Some lemurs still live in Madagascar. Monkeys could have developed from lemur-like ancestors. They are very well adapted to life in the trees.

Fig. 32-10

From left to right; a tree shrew, a lemur, and a monkey. Do you notice any similarities among them? (Stouffer Enterprises/Animals Animals; The San Diego Zoo; George Roos/Animals Animals)

SUMMARY

Mammals are warm-blooded vertebrates whose bodies are covered with hair. All but one order bear their young alive. All of them care for their young and feed them milk.

Some scientists believe the present orders of mammals may have developed from earlier types after the reptiles died off about 65 million years ago. The present orders include egg layers, pouched mammals, shrews, bats, rodents, flesh eaters, whales, elephants, hoofed animals, and primates.

Activity

Studying Mammals

A. If there is a zoo in your area go there and observe the mammals. If you cannot go to the zoo, observe all the mammals you see in your neighborhood, or even on television, for a few days.

1. List the mammals that you see. Try to arrange the list so that the members of each order are listed together.

2. Opposite each name give your observations about the animals. These could include the food it eats, its adaptations for food-getting, its means of defense, and anything else that you find interesting.

Word Quiz

In your own words write a definition for each of the following:

mammary glands   rodents   primates

Check Your Facts

Science Reading Skill

3. Compare and Contrast

1. What are the main characteristics of the mammals?
2. Which of these characteristics give mammals an advantage in competing with the reptiles?
3. List the orders of mammals described in this chapter and name some animals that belong to each group.
4. Why are rodents a successful group of mammals?
5. List some specific characteristics of each of the 10 orders of mammals discussed in this chapter.

Thought Questions

Science Reading Skills

1. & 2. Problem Solving

1. Many species of African animals are able to live together because each kind eats slightly different kinds of food. This reduces competition between species. Is there any connection between this and the fact that no elk, buffalo, bear, or wolves live in the settled farmlands of North America? Explain.
2. If a land connection suddenly formed between Asia and Australia, what effect would it have upon Australian mammals?

The greater your qualifications, the better your chances in the market.

It is possible to work in the field of animal husbandry without a college degree, but any technical training you are able to obtain would help to give you a competitive edge in the market. There are, for example, approximately 197 zoos and 39 aquariums in the United States. To qualify as an animal keeper—someone who maintains the surroundings, feeds the animals, and observes behavior in order to be aware of any irregularities—extended experience with animals through summer work on a farm or with a veterinarian, and willingness to do minor chores may be enough. Each year, however, there are only a few openings. Obviously, the greater your qualifications in terms of formal education, the better your chances in the market. For further information, write to:

The American Association of
Zoological Parks and
Aquariums
Oglebay Park
Wheeling, W. VA 26003

Animal keepers are also used by circuses and carnivals. In these situations staff usually is hired for maintenance work

Norman Owen Tomalin/Bruce Coleman, Inc.

only. Research laboratories have staff for these purposes as well. While the work may not be exciting, it can offer an opportunity to gain experience and to learn about animal behavior through observation.

Retail businesses of all kinds exist to support and maintain our domestic animals. There are feed companies, nonfood pet

supply stores, pet grooming salons, talent agencies which promote animal acts, animal photographers, animal trainers, and animal therapists. Public agencies such as the Humane Society or ASPCA have shelters which are, in part, staffed by volunteers. Members in these organizations are also involved in publicity and fund raising for

their causes. Public relations and marketing skills as well as the ability to be resourceful are of primary importance to these jobs. For more ideas, read:

Your Career in Animal Services by Sam Kohl and Tom Riley New York : Arco Publishing Company, Inc.   1977

Some jobs involve working in cooperation with an animal. For instance, mounted police officers work with their horses in order to accomplish their jobs, particularly in maintaining crowd control. They learn to care for the horses, and the animals, in turn, become an important extension of their work. Joining the police department can offer opportunities not usually associated with the profession. For example, because harbor areas are as subject to crime as any other area, police departments near waterways train harbor patrols and have a scuba team that dives for evidence. While the patrols maintain constant surveillance, the scuba team is not always active.

Where there are horses, there are farriers. A farrier is a person who shoes horses. There are several schools in the United States that teach this skill. While it is not a growth industry, it is an important reminder that services of all kinds represent job opportunities and that not all services are directed toward humans.

Interest in marine animals and marine life lead people to a number of associated careers. There are, of course, animal curators and breeders in aquariums. These people take courses

in animal husbandry, genetics, and animal behavior and nutrition, among other subjects, to earn a degree in zoology or biology. Marine biologists also spend time in research and underwater exploration and may work for industry or the government as well as teach in a university or college.

Underwater explorations of varying types have generated professions concerned with diving. Scuba divers teach and guide people who enjoy this

M. Timothy O'Keefe/Bruce Coleman, Inc.

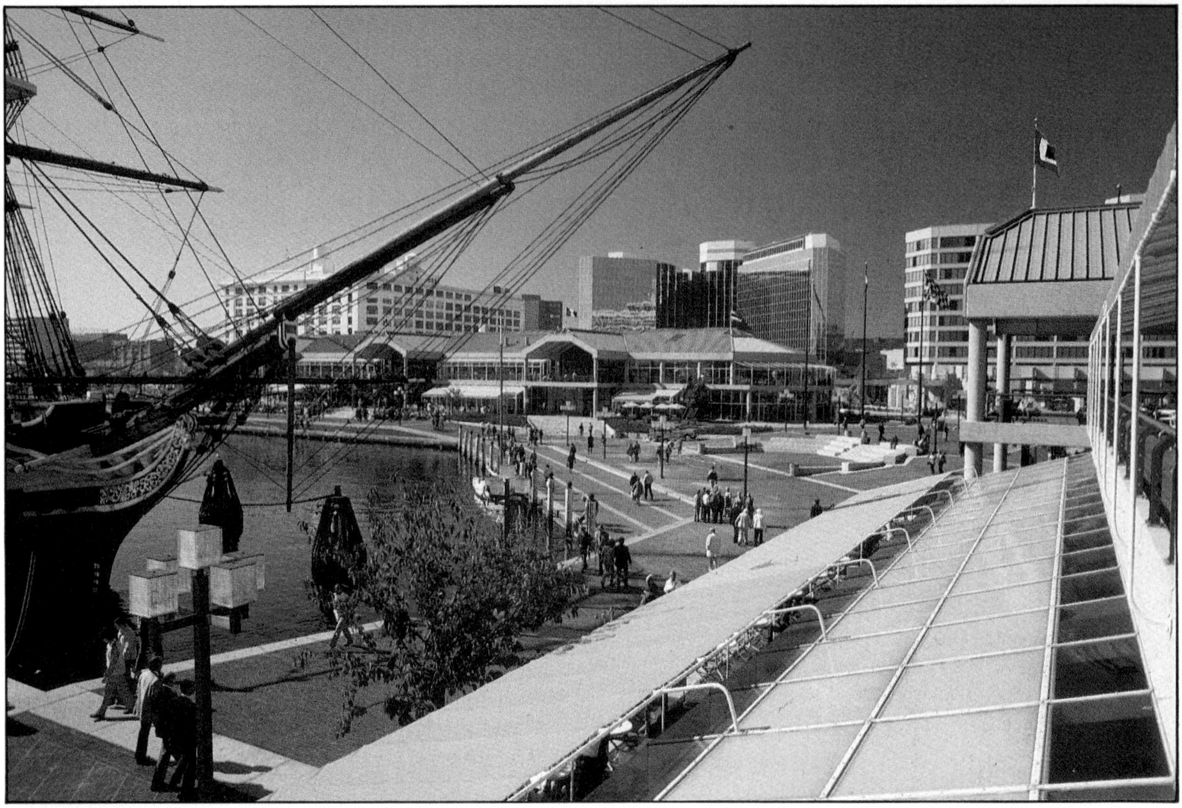
Stuart L. Craig, Bruce Coleman, Inc.

form of recreation. Other divers work for commercial companies where sponges and abalone, a kind of shellfish, are the target. Still other divers do deep-sea salvage and construction work. There are varying levels of expertise and experience needed for these jobs. Salvage work can be the most lucrative and also the most dangerous.

Many jobs are related to the fishing industry, including commercial fishing, fish "farming," and acting as a fishing guide for tourists. Seaport marketplaces, such as those in Baltimore, New York, and Boston, do a thriving business. These centers employ people in restaurants, in retail and wholesale fish markets, and in shops where crafts and nautical items are sold.

"Going to sea" is no longer among the most common apprenticeships for a young person, although, of course, many young adults join the Navy or Coast Guard. Harbors and rivers also need civilian pilots for the ferries, tugboats, barges, and other craft on these waterways. Positions are competitive, often requiring a civil service exam and extended seatime. The number of years of seatime required depends on the type of position you are aiming for. Apprenticeship is usually a period of low pay if it includes room and board on the vessel.

Lonny Kalfus: International Stock Photo

Boat repair is also done near most large bodies of water, and since increasing numbers of people own leisure crafts, a service industry has grown up for that purpose. Carpentry experience plus knowing how marine conditions react with shipbuilding materials are essential.

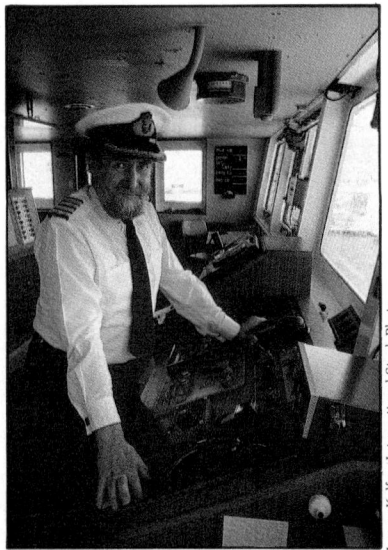

Sharing what you know with others is a way of organizing the information you have.

If you want to learn about animals and gather experience for a future career, volunteer work is a possibility. Many veterinarians offer opportunities for students to act as assistants. Some animal clinics do so as well. College and university animal laboratories may also be approached. If you're interested, don't be afraid to offer your services. Zoos, aquariums, and parks may also have volunteer or "friends" groups—that is, people who meet in order to help the institutions in any way they can. Sometimes these people become knowledgeable enough to give guided tours.

There are groups that lobby for the interests of animals as well as groups that monitor the existence of various species. The administrative, clerical, and public relations tasks necessary to the work can be extensive. Participating in this way can enable you to find out what kinds of work and positions exist in the field, and what the necessary requirements are. Valuable job contacts can be made.

Sometimes we don't identify recreation as a source of learning, but people have often turned their recreations into occupa-

New York Zoological Society

tions. Fishing, sailing, scuba diving, and horseback riding are all skills that may be shared or sold as services. Training your dog for obedience, teaching your bird to talk, raising a lamb to

show at a fair, or keeping spiders provides more than useful experiences. While all of these may lend themselves to future job possibilities, they are also skills and abilities which other people might enjoy knowing. Children's homes, institutions for the elderly, and church groups may all be interested in having a demonstration of your specialty. Moreover, sharing what you know with others is a way of organizing the information you do have. It is a way of learning to be an expert: in the business world that is called being a consultant.

Bruce Fritz, Monkmeyer Press Photo Service

The Oregonian

Fishing industry seeks to woo American tastes

By JIM KADERA

Changing the eating habits of Americans is one of the fishing industry's biggest challenges.

The U.S. annual per capita consumption of commercial fish products has stagnated at 10 to 12 pounds since the government began keeping records in 1909. Fish consumption peaked at 13.4 pounds in 1978 and gradually declined to 12.3 pounds in 1982, according to annual reports of the National Marine Fisheries Service.

However, the picture is even more bleak for U.S. fishermen and processors. About 60 percent of the seafood was imported from Canada, Japan, Iceland and scores of other nations. That leaves the average U.S. citizen eating only about 4½ pounds landed by the American fleet.

Higher prices of canned tuna and fear of contamination of Alaska canned salmon were blamed for declines in 1982.

The Portland-based West Coast Fisheries Development Foundation has embarked on an energetic program to both increase harvesting of under-utilized species and to put more fish into diets.

With about 90 processors, distributors, fishermen associations and others as members, the non-profit foundation is funded primarily by federal income from tariffs on fish imports. The foundation has a budget of about $860,000 for fisheries and market projects, and administrative costs in 1984.

The fishing industry is trying to develop ways to encourage greater fish consumption.

As we see from this article, bringing fish to the table involves researchers, salespeople, marketing specialists, lawyers who specialize in Fishing rights, and Wildlife managers helping to decide which fish will be commercially available and which will not. Can you think of other jobs along the path from the catch to the consumer? Could you design a campaign that would encourage the American public to eat more fish?

Los Angeles Times

Animals Help Draw Elderly Patients Out of Their Shells

By RUTH YOUNGBLOOD,
United Press International

MEDFIELD, Mass.—An 81-year-old woman confined to a nursing home bed became so withdrawn she refused to speak. Then, a kitten cuddled up on her lap, unleashing a torrent of buried affection.

Other patients who had to be coaxed into participating in any activities other than meals now comfort frightened baby rabbits, cradle puppies and delight in the antics of a de-scented skunk or a tame owl.

The animals "are more human than some people," said resident Laura Robinson with an armful of 6-week-old pups.

Even the most withdrawn patients have responded to the project at the Med-Vale nursing home. It was triggered by the exuberance over a mutt brought in by a member of the staff last fall.

Since then, Medfield animal officer Karen MacGregor has aroused the awareness of a community in her search for lovable creatures that awaken memories and suppressed emotions in the elderly residents.

"I never realized so little would accomplish so much," MacGregor said as the residents hugged the puppies, laughing as the animals licked their hands and faces.

The men and women, ranging in age from 58 to 97, are equally delighted with the guinea pigs, woodchucks, birds, opossums, lamb, and goat that have been brought into their world.

Being able to communicate and share with beings that can respond is very important to our mental health. However, sometimes human beings have needs for sharing and intimacy that are not met. Animals then become the "friends" that comfort and play with us and help us to exercise our needs for giving as well as receiving affection.

Have you ever had a pet? Were you responsible for feeding and grooming the animal? How did it respond to you? Did you learn anything about human beings by taking care of your pet?

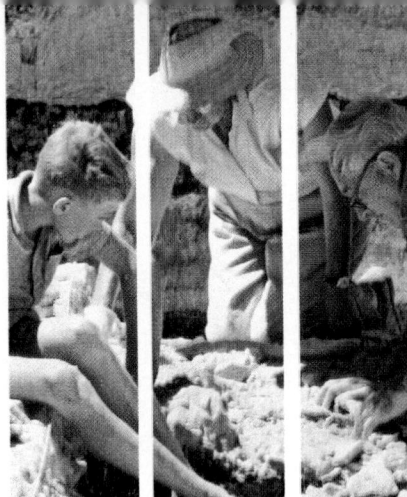

CHAPTER 33 HUMAN HISTORY

objectives

After you read this chapter, you should be able to:

__**Explain** how fossils give us information about the past

__**Describe** the probable appearance of the early ground-living primates

__**List** the traits that are unique to modern humans

__**Define** the term culture

__**Explain** why all races belong to *Homo sapiens*

Enrichment: Emphasize to students the scientific method that fossil hunters use to reconstruct an organism based on their discovery of a few bones.

Just imagine that you are climbing some rocky cliffs in Tanzania, East Africa. As you pick your way over the rough ground, you see an unusual object imbedded in the rocks. You bend down to examine the object and notice that it is an old bone hardened by mineral matter. You have just found a fossil in much the same way that Louis and Mary Leakey did. Dr. and Mrs. Leakey were a husband and wife team of professional fossil hunters. They spent more than 40 years on their hands and knees collecting fossils from the rocks in places such as Tanzania. The picture at the beginning of this chapter shows the Leakeys at work.

THE FOSSIL HUNTERS

The fossils found by the Leakeys help tell the story of human development. Over the past one hundred years, many other investigators have found the fossilized bones and teeth of humans and what are believed to be our primitive ancestors. A great number of these fossils have been found in Africa, Europe, and Asia. Piece by piece scientists study and classify these fossil remains. Scientists are trying to fit together pieces of the puzzle of human history on earth.

The work that professional fossil hunters do is painstaking and demanding. For instance, in July, 1959, Mary Leakey found a piece of a skull bone and two teeth that resembled human teeth. Working together at the place where these were found, the Leakeys discovered many other bits and pieces of this skull. It took them more than a year to assemble the 400 fragments of this fossil skull. Scientists now believe that it is the skull of a very early human.

You may wonder how scientists get information from fossil bones. First, they make a judgment about the part of the animal's body from which the bone came. This can be done by simply comparing the fossil *find* with bones of a closely related modern animal. (For example, there are strong similarities among hipbones of humans and other primate species. See Fig. 33-1.) Next the scientist tries to find out the approximate size of the prehistoric animal. The length, width, thickness, and weight of the bone provide some clues. A large animal will have big, heavy bones. A jawbone can provide information about the kind of teeth the animal had and the type of food it ate. The size of the brain, the shape of the forehead, and the position of the eyes can be estimated from even an incomplete skull. Bones of the feet, legs, and hips tell how the animal walked; whether on two legs or four. Fossil bones of the hands, arm, and shoulder tell if these limbs were used for carrying things or swinging from trees.

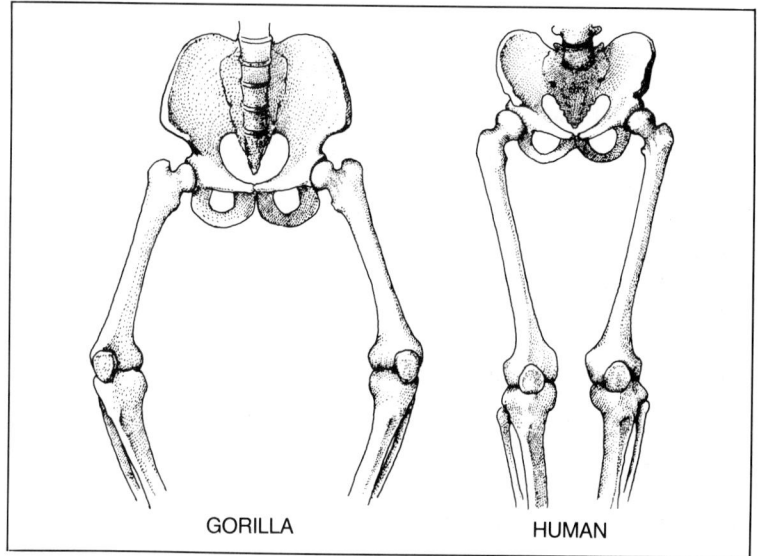

GORILLA   HUMAN

Science Reading Skill: *Compare and Contrast*—Describe the similarities and differences among humans and other primate species.

Fig. 33-1
What differences are there between the bones of the hips, legs, and feet of the gorilla and the human?

When bone fragments are put together to form a nearly complete skull, shoulder, or leg, we say that the part has been *reconstructed*. From these reconstructions, scientists can judge how the muscles were joined to the bones and how the flesh must have covered the skeleton. When we look at drawings of prehistoric humans, we are looking at an artist's conception of how this ancient being must have looked. The artist gets ideas from the scientist.

Professional fossil hunters direct a great deal of effort toward trying to find out the age of old bones and other fossils. Estimating the age of a fossil is known as ***dating***. Three methods are commonly used to date a fossil.

dating

333

One method of dating depends upon finding out the age of the rock bed in which the fossil was found. To find the age of rock, scientists compare the amount of a substance called uranium-238 with the amount of lead found in that rock. Uranium is a radioactive element. This means that it gives off radioactive rays for a certain period of time. When uranium no longer gives off these rays, it becomes lead. Scientists know how long it takes for a certain amount of uranium to become lead. By measuring how much uranium and how much lead are in a rock layer, scientists can estimate the age of the rock and the fossils found in it.

Another radioactive element found in preserved fossil remains of plants and animals is carbon-14. This can be used to estimate the age of fossils by a process similar to the one that involves uranium.

Very often many fossil bones are found in a single rock layer or within the same general region. A scientist may want to know if these bones are the same age. To find out if bones or teeth came from animals that lived about the same period, a method called *fluorine dating* is used. If the same amount of the element fluorine is found in fossils, it can be assumed that they are of a similar age. The longer a bone remains in the ground the more fluorine it absorbs.

All of these methods of measuring, comparing, and dating give scientists clues as to what might have been. These methods help scientists form theories. A theory is not an absolute fact.

A THEORY OF HUMAN DEVELOPMENT

Primate ancestors. Many scientists believe the theory that humans developed from earlier forms of life. You may choose not to accept this theory. However, by studying this theory you can better understand the traits that are unique to modern humans. Scientists do not believe that the human species descended from any of the present day apes. In 1871, a scientist named Charles Darwin proposed a theory that stated that humans and apes are all descended from some earlier primate ancestor. The paragraphs that follow describe how some scientists think this could have happened.

Changes in species may result from changes in the environment. When an environment changes, species living there must adapt or they will die. All changes in body form and behavior come about through long periods of mutation and natural selection. Some scientists think that two major changes took place during the development of humans. The first change produced the human type of body. Our body structure allows us to walk upright on two legs. Our hands and arms are free to do work.

Reference: You may wish to refer to the following for background reading: "The Early Relatives of Man," *Scientific American*, July, 1964; "The Antiquity of Human Walking," *Scientific American*, April, 1967; "*Homo erectus*," *Scientific American*, Nov., 1966; "Population Genetics and Human Origins," *Scientific American*, Jan., 1972; and *Origins* by R. Leakey.

Fig. 33-2
A photograph of a savanna, the type of environment where our ancestors are believed to have lived. (Timothy Ransom/Woodfin Camp)

The second change produced the intelligent human brain. Our highly developed brain makes us different from all other members of the Animal Kingdom.

Scientists are not sure where early human development took place. Some scientists think it was probably in Africa because many fossils are found there. In any case it was probably in the type of environment called a *savanna*. See Fig. 33-2. A savanna is a grassland with a scattering of trees. This theory of human development states that our primate ancestors probably lived in a forest where the climate was becoming drier. The trees were gradually dying out and being replaced by grass. These primates had to find food on the ground or die.

savanna

The change in environment from a forest to a grassland started changes in species. Under these new conditions, genes for ground-living may have allowed certain primates to adapt and survive. Such genes appeared as chance mutations. Through natural selection they replaced the old genes for tree-living traits. The feet could have become adapted for walking instead of climbing. These *ground-living primates* were not apes, but they were not human beings either.

ground-living primates

The ground-living primates were about 1½ meters tall. Their bodies were similar to ours, with long straight legs for walking upright. Their arms were thin. Their hands were much like ours and could be used for handling and carrying things. They probably used tools. A sharp stick could be used for digging roots or for defense against enemies.

The ground-living primates ate more meat than other primates. They hunted small animals such as turtles and mice. They probably ate the meat of large animals killed by lions and other flesh eaters. Scavengers still do this. This hunting and use of tools sounds quite human, but these primates did not have very large brains. Their brain size was slightly larger than that of the modern ape. But minute changes in the brain made these organisms behave differently. Apes seldom use tools. The ground-living primates must have depended upon tools to survive.

The ground-living primates survived because of special inherited traits. Group travel is thought to have been a trait inherited from primate ancestors. On the ground the danger from flesh eaters was very great. Therefore, group cooperation in meeting danger must have been important. Fossils suggest that these ground-living primates probably had ape-like heads and faces on human-type bodies. Some scientists believe that their behavior would be partly human and partly animal. There would be shouts and calls but no real language. The young would watch the old to learn tool-making and other useful things. Learning had to become an important part of the social structure. Learning ability is an important trait. It allows individuals to adapt quickly to new situations.

Fossils indicate that development slowed down after the ground-living primate level had been reached. There was very little change for the next 2 million years. One group of these ground-living primates developed larger bodies and heavier teeth. They seem to have eaten mainly plant foods, which require more chewing. This group died out completely a little over one million years ago.

About two million years ago, stone tools first began to be made. At first, they were simple "pebble tools." These were stones with one side broken off to give a sharp edge. Such a stone can scrape a stick to a sharp point. Or it can cut up a large animal. It can separate and scrape fibers to be twisted into rope or woven into baskets. As time went on, the tools got better. A common one was a type called a hand ax. It was chipped on all sides to form a tool with many uses.

The making of stone tools seems to mark the point where ground-living primates began to develop toward the human level of behavior. At first the brain was little larger than before, but fossils show a steady increase in size from this time on. It is probable that big game hunting became a way of life. At least, we find more bones of large animals along with the bones and tools of these primates. We can consider these early stone age hunters to be the first truly human species to have lived upon the earth. This species is **Homo erectus** (ee-REK-tus). See Fig. 33-3.

The fossils of *Homo erectus* show a body like ours and a primitive human face. The jaws were heavy. The chin sloped back, and there was a heavy ridge above the eyes. The forehead was

Enrichment: Emphasize the importance of natural selection in determining the direction of human change.

Homo erectus

336

Fig. 33-3
These footprints were discovered by Mary Leakey in Tanzania. They are believed to have been made by a ground-living primate about 3½ million years ago. (UPI photo)

low. The brain was about two-thirds as large as that of modern humans. The brain of *Homo erectus* developed more in the places where the centers of speech are located. *Homo erectus* is thought to have been able to talk. If this is true, members of this species could plan things together. They could have passed on information to their offspring. Language would have been especially important to their survival. At this point cultures may have come into existence. Today we say that the ***culture*** of a group of people is their total set of ideas, knowledge, and ways of doing things.

culture

Homo erectus were big game hunters. Some probably hunted as a team while others were gathering roots, nuts, and insects for food. Such a division of labor would allow all members of a group to contribute to its survival. Greater intelligence made human species more adaptable to changing conditions. Walking on two legs allowed a greater freedom of movement. They could live in different kinds of environments. *Homo erectus* spread out over Africa and the warmer parts of Europe and Asia.

Modern humans. The species to which modern humans belong is ***Homo sapiens*** (SAYP-ee-ins). Some scientists think that a gradual change took place from *Homo erectus* to *Homo sapiens*. The human brain seems to have reached its full modern size by about 100 thousand years ago. A famous early form with a full-sized brain was the group known as ***Neanderthals*** (nee-AN-der-thalls). Fossils show that the Neanderthal form lived over a wide area of Europe, south-western Asia, and northern Africa. The last of the Neanderthals died out about 30 thousand years ago. If some of them were still here, we would probably list them among the modern races of human beings. The Neanderthal face looked more like that of *Homo erectus* than ours does. The Neanderthal brain was slightly larger than ours. But this slight increase in size did not make the Neanderthal more intelligent than modern man. See Fig. 33-4. Some scientists suggest that the bulging forehead of the Neanderthal forms might

Homo sapiens

Neanderthals

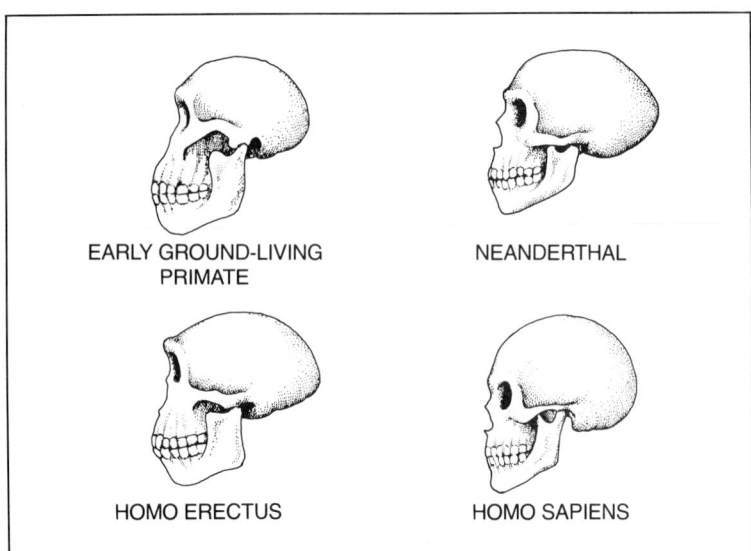

EARLY GROUND-LIVING PRIMATE

NEANDERTHAL

HOMO ERECTUS

HOMO SAPIENS

Fig. 33-4
The skull shapes found in four different levels of development in the primate order of mammals.

have been due to vitamin D deficiency. They lived in a region where there was hardly any winter sunshine. This absence of sunlight could cause vitamin D deficiency.

The earliest fossils of humans exactly like those of today are about 40 thousand years old. Fossil evidence shows that modern humans gradually replaced the Neanderthals. Today, there remains only one species of human being, *Homo sapiens*. We may not all agree that humans and apes descended from a common ancestor. However, we do know that modern man is quite different from other animal species.

HUMAN ADAPTATIONS

What makes a person different from other species in the Animal Kingdom? Think of all of the things that you can do that a bear cannot. Your body is built in such a way that you can perform many different kinds of activities. People can walk, run, ride a bicycle, roller skate, toe dance, climb, jump rope, and make various movements using the legs and feet. Your hands can be used to write, knit, carve, draw, hammer, grasp, play the piano, and accomplish many skills that require fine use of the fingers. The many different kinds of tasks that people are able to do are directly related to the structures that made the human body one of a kind.

Humans are really quite special among the animals. This is due to their body structure. All of the changes that made the human form possible were those that contributed to upright posture and two-legged walking. Fig. 33-5 shows the skeletons of a human and a gorilla. Refer to this illustration as you read

Science Reading Skill: *Reading Illustrations.* The text describes the similarities and differences between the human and gorilla as students refer to Fig. 33-5.

further. Imagine the weight distribution problem that had to be solved in the development of a two-legged body. As we examine the skeleton in Fig. 33-5, we will see how the problems of weight and balance were worked out. Of course, these changes in body form are thought to have taken place over millions of years.

The human head is quite different from the gorilla head. The human face does not stick out, but is moved back into the bony structures of the skull. The top of the skull is larger and the forehead broader. Both of these characteristics allow the development of a larger brain. The head is balanced on a flexible neck so you can turn your head from side to side. Imagine trying to play volleyball with a stiff neck!

As we continue to look at the skeleton, we see other differences. The human chest bones are flatter. The human spinal column is curved and stiffened. These structures help to distribute the weight evenly along the length of the body, back toward the spine. The hip bones of humans are shorter and broader. Humans have stronger hip bones, which allow the legs to grow straight and sturdy. In the early ground-living primates, the hip bones were much longer and narrower. This made them walk with their knees bent and their faces thrust forward. The legs of humans are longer and the weight of the lower body is supported by the knees. Two legs can hold the body weight.

Fig. 33-5

The structure of the skeleton of the gorilla (left) and the human (right). Notice the differences required for walking upright.

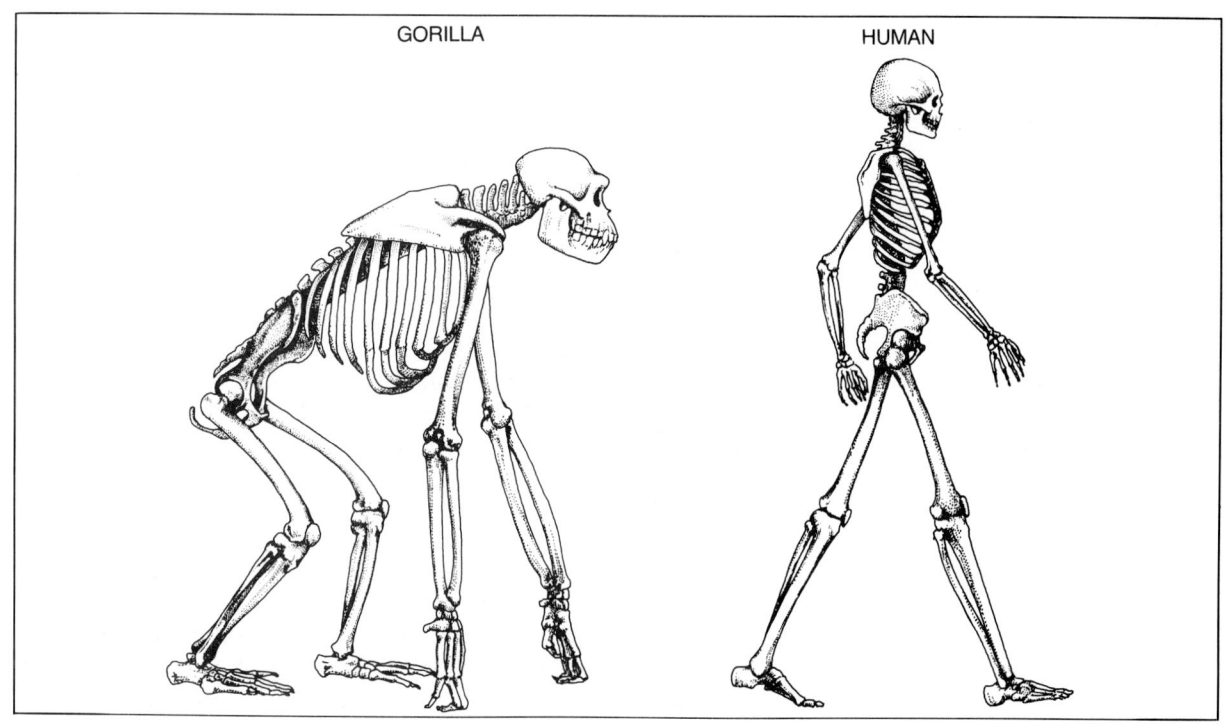

GORILLA   HUMAN

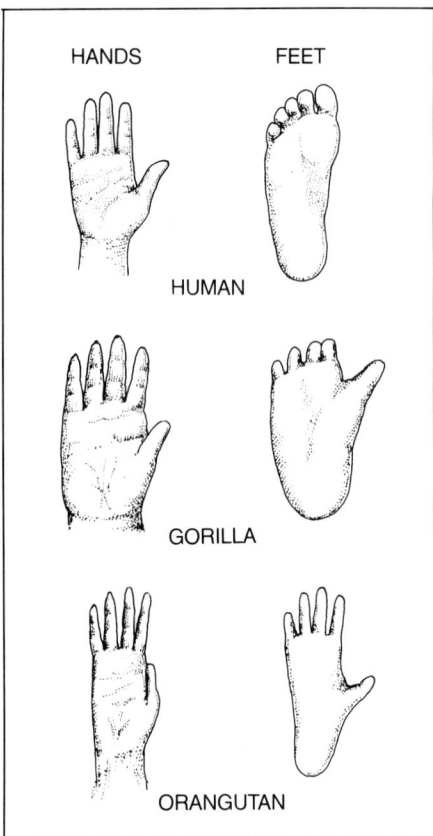

HANDS   FEET

HUMAN

GORILLA

ORANGUTAN

Fig. 33-6
How are the hands and feet of the human better adapted for walking and fine movements than those of the gorilla or the orangutan?

Enrichment: Review the definition of species to show why all races belong to *Homo sapiens*.

The hands and feet of *Homo sapiens* are truly unique. See Fig. 33-6. The thumb is opposite to the index finger. This is a primate characteristic. But in humans the thumb is lengthened and the finger joints are flexible. Such hand structure allows people to do a great deal of fine work. We can unscrew a bottle top, thread a needle, and assemble fine parts in watches. Human feet are longer and narrower than those of other primates. The toes are merely movable extensions on the feet to aid with balance.

Not all differences of the human body can be seen on the skeleton. For example, most of the body hair is not easily visible. The human body has developed thousands of sweat glands. Sweating is really a means of cooling for humans. The complex development of the brain made language, memory, and the ability to work out mathematics problems possible. So we can say that as brain size increased, so did intelligence. *Homo sapiens* is the only species that developed cultures based on language, memory, and learning. Other primate species follow a way of life that is controlled by the patterns of behavior they are born with. Human cultures have spread all over the world. When people move from one continent to another, they take with them part of their learned cultures.

RACES OF HUMANKIND

The origin of race. Modern people have spread out all over the world. They have developed cultures that enable them to live all the way from the tropics to the Arctic and from the rain forests to the deserts. As with any widespread species, the local populations have developed their own collections of genes. One group has more genes for curly hair. Another has more genes for straight hair, and so on. This makes the people of one population look somewhat different, on the average, from those of another population. We call these variations *racial differences*.

These racial differences are not great. People everywhere have about the same intelligence, the same range in size and strength, the same basic human nature, and get many of the same diseases. For any two races we can find some people with traits in between. It is hard to say where one race ends and another begins. There are many variations within each of the races. Scientists do not agree on how many "races" there are. The races developed when people were few and widely scattered. A mutation might take place in one group, but not in another. Natural selection would favor one gene in one environment, but not in another. So racial types developed as adaptations to different environments.

The main races living today are those groups that happened to develop successful cultures. Their cultural advantage over other groups enabled them to spread over wide areas and to build large populations. As we have noted, scientists do not always agree on the number of races. But six groups that are often described are presented here.

Race and history. The *Australoids* (AWE-struh-loids) are the people who first lived in Australia. Many still live like old Stone Age hunters, though some are now becoming cowhands or entering other modern occupations. They have heavy brow ridges. Their hair is wavy, and the men have heavy beards and hairy chests. Most Australoids have dark skin. See Fig. 33-7.

The *Capoids* (KAY-poidz) are the Bushmen of South Africa. They are called Capoids after the Cape of Good Hope at the southern tip of Africa. Originally they occupied most of Africa south of the Sahara Desert. They are rather small people with yellowish skin, tightly curled hair, and narrow eye openings. Most of them are hunters, living in the Kalahari (kal-uh-HAHR-ee) Desert. See Fig. 33-7.

The *Negroids* (NEE-groidz) are thought to have first appeared on the open plains of West Africa south of the Sahara Desert. Most of their racial traits are explained as adaptations to the intense sunlight and heat of this environment. The dark skin does two things. It reduces the possibility of sunburn. It also prevents too much vitamin D from being produced in the skin by sunlight. See Fig. 33-7.

Fig. 33-7
Left to right: An aborigine (Australoid), a South African (Capoid), and a Black American (Negroid). (Australian News and Information Service; South African Tourist Bureau; Richard Haynes Jr.)

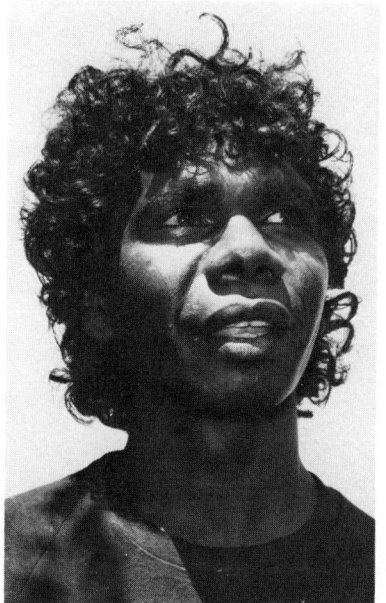
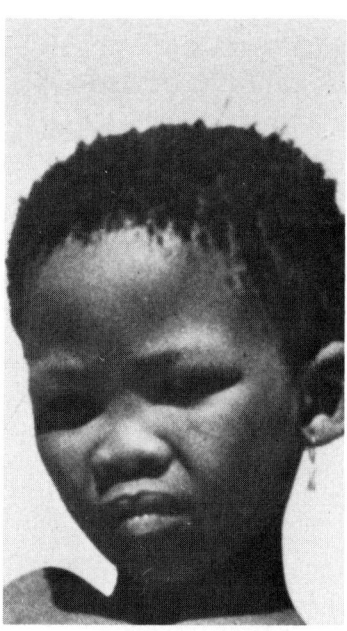

341

Negroid people may often have long arms and legs and a less than average amount of fat under the skin. This was probably an adaptation for exposing more skin surface to the air for cooling the body. The Negroids include the tallest and the shortest groups on earth, the Watusis and the Pygmies (PIG-meez). The average height of the Watusis is about 2.1 meters. The Pygmies are small people whose average height is about 1.2 meters.

In the last few thousand years the Negroid people have spread out from their West African homeland. Some even reached islands in the South Pacific. Others went eastward and southward in Africa. Finally, during the last century, they met Europeans spreading northward from the Cape of Good Hope. Descendants of the West African Negroids live in North America, South America, the West Indies, and Europe.

The *Caucasoids* (KAW-ku-soids) compose the race that is often called the white race. This is not a very good description because actually many of the Caucasoids are dark skinned. The Caucasoids are thought to have originated in southwestern Asia although the exact place is unknown. From this rather small area they spread out about 30 thousand years ago, in much the same way that the Negroids did later. The Caucasoids went westward replacing Neanderthal types who still occupied Europe and North Africa. The Caucasoids also went southeastward, occupying Persia and India. In southern India they seem to have mixed with earlier, Australoid types. See Fig. 33-8.

Fig. 33-8
Caucasoids from Northern Europe. (Jack Fields/Photo Researchers, Inc.)

Caucasoids vary a good deal. They tend to have rather wavy hair, an average build, a fair amount of fat development under the skin, and large noses. They and the Australoids are the two races today that have the most body hair.

The *Old Mongoloids* (MONG-ah-loids) are not very different from the Caucasoids. They may have somewhat darker skin color. Their hair is black, coarse, and straight. The inner corner of the eye tends to turn down slightly. These people originally spread over most of Asia and into Alaska. About 12 thousand years ago some of them got past the melting glaciers and occupied all of North and South America. These people became the American Indians. See Fig. 33-9.

The *New Mongoloids* are the present day Asiatics (ay-zhee-AT-iks). One group of Old Mongoloids became trapped north of the Himalaya Mountains during the last push of the Ice Age. They could not escape to a better climate because of mountains and other natural barriers. This is a region where even today the temperature falls below −40°C in the winter. Think how cold it would have been during the Ice Age! Many people must have died. Those that lived developed adaptations to combat the cold. We recognize these traits today as racial characteristics. This group of Mongoloids tends to have rather compact

Fig. 33-9

Left: An American Indian (Old Mongoloid); Right: An Eskimo (New Mongoloid). (John Running/Black Star; Bill Strode/Black Star)

bodies with an evenly developed fat layer under the skin, which holds in heat. This fat covers the cheek bones, giving the face a flatter than average appearance. The insulation over the face reduces sinus infection. A fold extends downward over the upper eyelid (the Mongolian fold). This fold gives the eyes a slanting appearance, although the eyes do not really slant at all. The fold protects the eyelids from freezing. See Fig. 33-9.

When a warmer climate returned, the New Mongoloids spread out and occupied all of eastern and southeastern Asia. Koreans are probably the best example of the New Mongoloid type today. The Chinese and Japanese are mainly New Mongoloids. Other groups show more or less mixing with Old Mongoloids or with Caucasoids. The New Mongoloids, like the Caucasoids, show a range of skin color. Northern Chinese are as light in skin color as many Europeans. Some tropical groups have brown skins.

Races today. The racial types get less and less clear-cut as time goes on. In Europe, for instance, there were repeated invasions by Mongoloids. Their genes are still there, mixed in with the local, Caucasoid genes. Europeans also have a little Negroid ancestry. A good many Negroids were brought across the Sahara by the Carthaginians (Car-tha-GIN-ee-ans), Egyptians, and Romans. Their genes are also still there, mixed with the Caucasoid ones.

The Negroids did not replace all the Capoids who lived in the Sahara. There was some mixing, and traces of Capoid traits can be seen in many Negroids. Arabs and Berbers of North Africa are Caucasoid types. There has been contact between them and the Negroids. Some Caucasoid traits are visible in some of the modern African tribes.

We have already shown how new and old Mongoloids are mixed in Asia. In America, Caucasoids, Negroids, and Old Mongoloids are all present. Already Black Americans carry about 25 percent of Caucasoid genes on the average.

So, you see, "pure" races are a myth. All races today are mixed. The old advantages of the racial characteristics in special environments are no longer very important. A pale-skinned Caucasoid can survive under the hot African sun by wearing a hat and light-colored clothing. Proper clothing allows Negroids to function in cold weather as well as warm. As a matter of fact, it was a Black American who first set foot on the North Pole. His name was Matt Henson. He was with Admiral Peary on his famous journey. When they reached the North Pole, men of three races stood on the floating sea ice in that terrible environment. They were Mr. Henson (Negroid), Admiral Peary (Caucasoid), and four Eskimos (New Mongoloid).

Enrichment: Discuss why skin color alone cannot determine racial origin.

SUMMARY

Some scientists support the idea that humans developed from earlier forms of life. Scientists do not believe that human beings descended from any of the species of modern apes. Fossil studies provide evidence that all primates may have had a common ancestor. However, the pathway of development for humans is considered to have differed from that of other primate species. Gene changes and natural selection resulted in the development of a primate species that walked on two legs.

Scientists have named the early humans, *Homo erectus*. *Homo erectus*, as the name tells us, had erect posture, a high degree of intelligence, made better tools, and had the ability to communicate through speech. They spread out through Africa, Asia, and Europe. *Homo erectus* gave rise to modern humans, *Homo sapiens*.

Today, all living humans belong to the species *Homo Sapiens*. Within *Homo sapiens* six populations or racial groups have been identified. All of these populations have basic human traits in common. These traits are human intelligence, skillful use of the hands, erect posture and two-legged walking, communication by speech, and development of the memory centers of the brain. The populations of *Homo sapiens* have developed cultures that are adapted to meet the needs of the environments in which they live.

Activity

Science Reading
Skill: *Compare and Contrast*

What Bones Tell Us

One of the methods used by scientists to get information is comparison. Fig. 33-5, page 339 compares the bone structure of a gorilla with that of a human. Study both drawings carefully.

A. By looking at these drawings compare each body region.

1. What differences do you see in the following: head, shoulder bones, chest, hip bones, limbs? Use single descriptive words such as thick, long, wide, short, etc. Record your observations.

B. Using a metric ruler, measure the length of each of the body regions mentioned above.

2. Add the measurement information to your recorded observations.

3. Based on this information, try to explain why humans are able to walk on two legs.

Word Quiz

From the new words you have learned in the chapter write the word that correctly completes each statement. Do not write in this book.

A grassland that replaces a tropical forest is called a(n) _____. It was in this kind of changing environment that the _____ are thought to have appeared. Organisms that some scientists believe to have been the first species of human beings have been given the name _____. The set of ideas by which humans live is known as their _____. All groups of modern humans belong to the species _____. Some scientists believe that modern humans descended from a type of human named _____. Finding the age of fossil remains is known as _____.

Check Your Facts

Science Reading Skill

3. Compare and Contrast

1. Describe the changes in the environment believed by some scientists to have led to erect posture and two-legged walking.
2. What environmental changes cause the formation of a savanna?
3. In what ways were the early ground-living primates different from other primates?
4. Why is the species *Homo erectus* considered by some scientists to have been human?
5. List the adaptations of *Homo sapiens* that set this species apart from other primates.
6. Why are people of all races classified as members of *Homo sapiens*?
7. How has the developed brain of humans helped them to satisfy their curiosity about the world?

Thought Questions

Science Reading Skills

1. & 2. Problem Solving

1. Do apes have a culture? Explain your answer.
2. Why are tool-making and use of tools a sign of developed intelligence?

CHAPTER 34 Food and Nutrition

All living things require food for life. Plants make their own food. Animals eat plants and other animals. In the last chapter you read how hunting and food gathering shaped the cultures of the early humans. Even today, the cultures of all people are involved very closely with food.

Food is important in our daily lives. It is the source of materials for building protoplasm in our cells. The energy needed to do the work of the body is supplied by many different foods. Some foods contain substances that regulate the chemical changes that go on in cells. You can understand that the study of food is important. The proper kinds and amounts of food help the body keep good health.

ENERGY AND CALORIES

You know that certain food molecules in protoplasm combine with oxygen and produce energy. Scientists have developed a method for measuring amounts of energy in foods. To do this they use a unit of measure called a *Calorie*. A Calorie is really a unit of heat. Using a special piece of equipment, scientists can measure the heat energy in foods. The energy in food is commonly spoken of in terms of Calories. A Calorie table tells you the number of Calories or heat energy in different kinds of foods.

The energy we need is obtained from the Calories in the food we eat. People differ from each other and therefore require different amounts of Calories. Many conditions affect the energy needs of the body. Among these are a person's sex, size, age, and activity. You can understand that a large person uses more energy than a small person and needs more food. Building new protoplasm is a process that uses energy. Young people who are growing need extra food, both to supply materials for growth and to supply energy to build tissues. As people become

objectives

After you read this chapter, you should be able to:

__**Explain** how Calories are used in the body

__**List** five classes of nutrients and tell what they do

__**Explain** the meaning of deficiency disease and give some examples

__**Describe** a balanced diet

Calorie

Science Reading Skill: *Cause and Effect*—Many concepts in this chapter reinforce this skill. Students should be able to explain how each nutrient contributes to good health. Where applicable, they should also be able to describe how an excess or lack of a nutrient affects the body such as in deficiency diseases.

Enrichment: 1,000 calories equals 1 Calorie. The instrument used to measure heat energy is called a calorimeter.

Fig. 34-1
Do you think a lot of energy is being used up here? Why? (Peter Miller/The Image Bank)

older, their bodies use less energy. Heavy physical activity requires much more energy than light work or exercise. A football player certainly must eat more than someone who exercises very little. It is also agreed that per kilogram of body weight men tend to use more energy than women.

You can see that the number of Calories needed by a person is an individual matter. Table 34-1 provides information about the average daily Calorie requirement. It is estimated that the average 15 year old boy needs about 3,000 Calories per day. The average 15 year old girl needs about 2,500 Calories each day. But these figures are modified by our individual needs. If you are neither fat nor very thin, if you are growing normally and feel well, you are probably receiving the right number of Calories.

If the food that we eat contains more energy than the body uses, the extra energy is stored as fat. If the food we eat has less energy than needed, the body must oxidize materials already in storage. This causes a person to lose weight. When people gain weight, they are probably eating too much.

Table 34-1   AVERAGE DAILY CALORIE NEEDS

Type of Person	Calories per Day
Child under 2 years	1,000
Child 2 to 5 years	1,300
Child 6 to 9 years	1,700
Child 10 to 12, or woman not working	2,000
Girl 12 to 14, or woman doing light work	2,200
Boy 12 to 14, girl 15 to 16, or man not working	2,600
Boy 15 to 20, or man doing light work	3,000
Moderately active person	3,200
Farmer in busy season	3,500 - 4,500
Person at hard labor	4,000 - 5,000

THE NUTRITIONAL NEEDS OF THE BODY

Nutrition is the use of food by cells. Nutrition is made possible by compounds that are present in food. These special com-

pounds are the *nutrients* (NEW-tree-ents). Nutrients are substances that nourish the body. In general, they function in three
ways. Some nutrients provide the body with fuel for energy.
Other nutrients supply materials for the building of body tissues. Still others furnish materials that help the work of cells.

Each nutrient has its own special job to do. In addition, a
nutrient may do other work also. As you read this chapter, you
will learn how the nutrients support each other by carrying out
special jobs and doing other work, as well. Just as people may
function in various ways, so it is with the nutrients. They contribute to body health in many ways. There are five groups of
nutrients; carbohydrates, fats, proteins, vitamins, and minerals.

Carbohydrates. Starches and sugars belong to the family of
carbohydrates. Members of this group are made from simple
sugars like glucose. The body is able to take apart the large
carbohydrate molecules that we eat and break them down into
glucose. Glucose is oxidized in the cells and energy is released.
This energy is used to do the chemical work necessary to keep
organisms alive.

The carbohydrates are our main sources of energy. About
half of the Calories that we get from our diet are supplied by
foods that contain a large amount of starch and sugar. The heart
and other muscles in our body depend heavily on carbohydrates for energy. The nervous system needs constant energy,
too. If muscle and nerve cells do not have a constant supply
of energy, they cannot function properly.

Besides providing the body with energy, the carbohydrates
have other functions. They help the proteins and fats to work.
When the body has too little carbohydrates, proteins are used
for energy. This means that the proteins cannot do their own
special jobs. This can cause tissues to break down. If not enough
carbohydrates are eaten, fats are used for energy instead. When
too many fats are used for energy, an unhealthy condition can
develop. Carbohydrates also help to maintain the proper
amounts of water and salts in the body tissues.

Certain foods provide most of the carbohydrates the body
needs. For convenience, we divide these foods into four groups.
The first of these is the cereal group. Foods made from wheat,
rice, corn, and oats give us a very good supply of starch. You
can test for starch very simply. Put a drop of iodine on the food
you wish to test. If starch is present, the iodine spot will turn
blue-black.

The second food group that contains starch is the vegetables.
White potatoes, sweet potatoes, acorn squash, beets, dried
beans, and peas are a few examples of starchy vegetables. In
the United States, white potatoes are a very popular food. Ac-

Enrichment: Water, although not a
nutrient, is essential to life. Water
dissolves food and wastes that are
transported to and from body tissues. In
addition, cells cannot function and will
die if too much water is lost from the
body.

Reference: See *Food: Facts, Foibles,
and Fables* by A.J.W. Simeons, *The
Meaning of Human Nutrition* by M.
Lamb and M. Harden, and *How Did We
Find Out About Vitamins* by I. Asimov.

cording to the U.S. Department of Agriculture, each person in this country consumes about 259.6 kilograms of potatoes per year! Just think of the quantities of French fries and potato chips that you eat.

The third group of carbohydrate foods are the fruits. Bananas have a high starch and sugar content. Dates, figs, grapes, and plums have a great deal of sugar also. The concentrated sweets make up the fourth group of carbohydrate foods. Sugars, syrups, molasses, jams, jellies, candies, and honey sweeten our food. See Fig. 34-2. Americans eat too much sugar. Scientists estimate that for each man, woman, and child living in the United States, 226.6 kilograms of sugar are eaten per year. Later in this chapter, you will learn why eating too much sugar is not good for your health. You can test food for simple sugar. Heat a small sample of the food in some Benedict's solution. A color change in the solution from blue to brick red tells us that sugar is present.

Fats. Fat is an important nutrient in our diet. It is really a very concentrated form of energy food. Fats have twice as many Calories as carbohydrates. It is generally agreed by food scientists that Americans get too much of their Calorie intake in the form of fats. They would be better off if they reduced their fat intake and let carbohydrates furnish most of their energy needs.

One function of fat is to supply energy to cells. During cell respiration energy is released from fatty acid and glycerol molecules. A second function of fat is to build tissues. Cell membranes and other cell parts contain fat molecules. The cell could

Fig. 34-2

Can you tell to which carbohydrate group each of these belongs? (HRW photo by Russell Dian)

not function without fat. A third use for fat is that it serves as a dissolving agent. Fat dissolves some vitamins that are essential to our health. These vitamins cannot dissolve in water and would remain useless to the cells if it were not for fat. Finally, fat that is not used is stored in tissues. The body needs some fat. It helps to keep us warm. It protects certain important organs such as the kidneys and the reproductive system. Our hands, feet, and limbs are cushioned with fat. Too much fat, of course, is not healthy for us. It clogs our arteries making circulation difficult.

Very few foods are made only of fat. Usually fat is found together with other nutrients such as proteins, carbohydrates, and vitamins. Foods that supply fat come from both animals and plants. Meat, fish, poultry, milk and milk products, and eggs are fatty foods that come from animals. Plant fat is obtained from seed oils such as corn oil, margarine, nuts, olives, and avocados. Fatty foods rubbed on a piece of brown paper leave an oily spot. This is a very simple test for fat. Table 34-2 shows some sources of fat and other nutrients.

Table 34-2   SOME ENERGY FOODS

STARCH	SUGAR	FAT
Bread	Candy bar	Sausage
Macaroni	Honey	Butter
Potato	Grapes	Cooking oil
Banana	Cake	Peanuts

Proteins. Of all of the nutrients, proteins are the most important. This does not mean that we do not need the carbohydrates and the fats. We certainly do need them for energy. But proteins are the tissue builders. They build the protoplasm in cells. When proteins are taken into the body, they are broken down into amino acids. These amino acids are then used to build new proteins. Our bodies use about twenty different amino acids. Eight of these cannot be manufactured in our cells so we must eat them in our foods. The proteins in our food and the protein made by cells are giant molecules. One protein molecule may contain several hundred amino acid units. Another protein molecule may be built from several thousand amino acid units. You can see that protein molecules are quite complex.

About 20 percent of our body tissues are made up of proteins. There are many different kinds of proteins. Each of these does

Enrichment: Amino acids that must be included in the diet because the body cannot make them are called essential amino acids.

a special job. Some proteins build muscles and bones. Other proteins form important parts of the blood. Amino acid molecules are used by cells to build cell structures and to manufacture other chemical compounds needed by the body.

The best food sources of proteins are animal products such as milk, eggs, cheese, meat, poultry, and fish. Vegetables are good sources of proteins, too. Gelatin is an animal product also, but it does not have many useful proteins. Nuts, beans, peas, and cereal grains provide many of the proteins that we need. See Fig. 34-3.

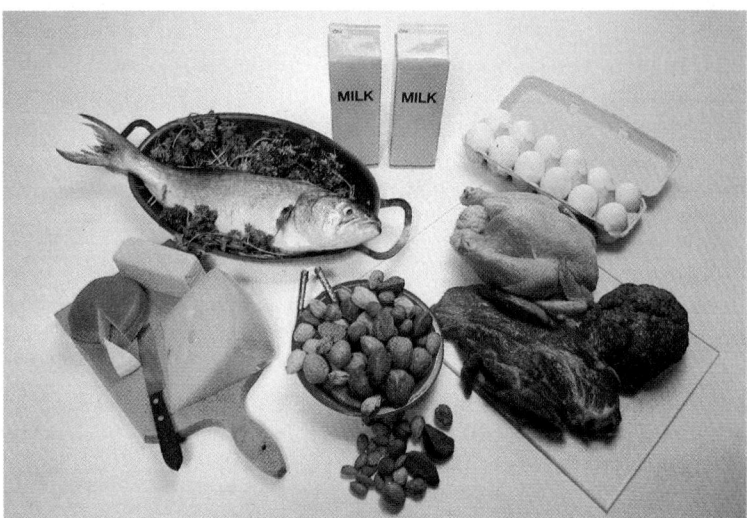

Fig. 34-3
These are some of the foods that supply us with proteins. (HRW photo by Russell Dian)

malnutrition

deficiency disease

MALNUTRITION

The condition of poor health that is caused by too few nutrients in the diet is known as *malnutrition*. Children in many of the African countries are fed diets high in carbohydrates and low in proteins. Their one meal a day may be a porridge made from starchy cereal grains and water. Their families cannot provide them with the milk, meat, and vegetables that contain proteins. In many countries there is not enough food.

When people get sick from lack of a particular nutrient, we say that they have a *deficiency* (dee-FISH-en-see) *disease*. Deficiency means lack of something. These children we have just described lack protein in their diets. The name given to this protein deficiency disease is *kwashiorkor* (kwahsh-ee-OR-core). It is a word borrowed from the country of Ghana and means "little red boy." It is easy to see the signs of kwashiorkor. The hair becomes a reddish-orange. Its texture changes, too. The child's liver swells and the stomach balloons out with water.

The limbs become thin and wasted. There is poor body growth and poor brain development. Kwashiorkor children have no energy and lie in one spot all day. If better nutrition is not given to them, they die. One cup of whole milk a day in their diets can help to prevent this wasting disease. See Fig. 34-4.

VITAMINS

Vitamins are chemical substances produced in plants and sometimes in animals. People need them only in very small amounts. A few milligrams per day will supply you with all you need of most vitamins. Yet without them you would develop a vitamin deficiency disease. Many vitamins are known to be needed by the human body. Although vitamins have very little in common, they act in the body in two general ways. First of all, they prevent deficiency diseases. Secondly, they regulate the work of other nutrients. We will discuss some of the more important vitamins.

Vitamin A. Vitamin A is needed for health of tissues lining the throat and eyelids. It is also needed for normal vision. A lack of vitamin A produces *night blindness*. This is a condition in which the person sees poorly in dim light. Some serious automobile accidents have taken place because the drivers could not see at dusk. A continued shortage of vitamin A causes very serious eye trouble that finally may result in blindness. Vitamin A deficiency may also reduce resistance to colds and other throat infections.

Vitamin A is found in foods such as liver and kidneys, whole milk, butter, eggs, tomatoes, and nearly all green and yellow vegetables. The body can store it, so extra amounts eaten at one time can be used later.

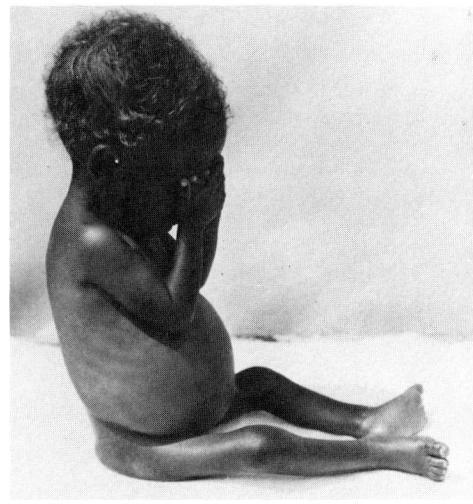

Fig. 34-4
When a child does not get enough protein, the stomach fills with water and the body becomes wasted away. One cup of milk a day can prevent this. (UNICEF)

vitamins

Fig. 34-5
The rat on the left has a typical eye condition caused by a lack of vitamin A in its diet. On the right, the same rat has normal eyes as a result of eating foods containing a rich supply of vitamin A. (Squibb Division/Olin)

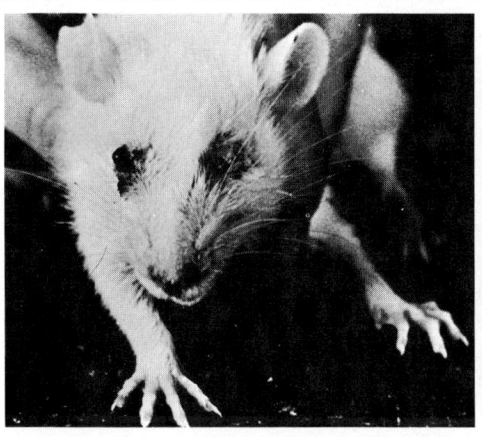

Enrichment: The subject of adding vitamins to one's diet is a controversial one. Recent estimates of the vitamin business in this country are in excess of $500 million a year. Most experts in this field feel that a balanced diet requires no additional vitamins or food supplements.

Assignment: Have students research the history of the discovery of vitamins. Students can also prepare reports on deficiency diseases.

Vitamin B₁. Vitamin B_1 is also known as *thiamin* (THIE-uh-min). This vitamin helps to control the oxidation of glucose in the cells. As we have said, oxidation is the process that releases energy from certain nutrients. A lack of vitamin B_1 leads to poor health. Persons who are lacking some thiamin in their diets feel tired. If they are lacking much thiamin, the result is a deficiency disease called *beriberi*. This disease leads to paralysis of the legs and in its final stages involve the heart. Beriberi occurs frequently in the Philippines and in parts of Asia where polished rice is the major part of the diet. Brown rice has just enough vitamin B_1 in its seed coats to prevent this disease. When the seed coats are polished off to make the rice white, the vitamin is lost.

We polish our wheat to make white bread, just as the Asians polish rice to make white rice. But many Asians eat almost nothing but rice. We eat many other things in addition to bread. These other foods provide us with the vitamins that are lost when the wheat is polished.

All whole grain foods contain some vitamin B_1. Vitamin B_1 is present also in meats, fish, milk, and most vegetables. Because thiamin is present in so many foods, you might think that we get it easily in our daily diet. This vitamin is not only damaged by heat, but dissolves in water also. So it is destroyed by long cooking and lost in the cooking water. We can save a certain amount of thiamin by shortening the time that we cook vegetables, using as little water as possible.

Vitamin B₂. Vitamin B_2 is also called *riboflavin* (RYE-buh-flay-vin). Vitamin B_2 is needed for respiration in cells. It combines with enzymes to help them do their jobs. A lack of this vitamin results in sores of the mouth and tongue. It may also cause reddening of the eyes. The food sources for riboflavin are about the same as those for vitamin B_1.

Niacin. Niacin (NIE-uh-sun) is a vitamin belonging to the vitamin B family, too. It is needed for the proper use of carbohydrates in the body. Niacin is present in animal protein foods and in many vegetables. A lack of this vitamin is the cause of the deficiency disease called *pellagra* (puh-LAY-gruh).

Pellagra used to be a problem in some parts of the United States. It is practically nonexistent today. It appeared among poor farmers and mill hands of the South, where corn, molasses, and fat pork were often the main diet. These foods are very poor in protein, and they lack niacin almost entirely. Pellagra produces a skin rash, upset stomach, paralysis, and mental disturbance. One method of combating pellagra has been to encourage people to grow their own garden vegetables. Adding niacin to the cornmeal sold in stores has helped also.

Vitamin C. Vitamin C is known as *ascorbic* (uh-SCORE-bik) *acid*. Vitamin C is needed to bind cells together. Without vitamin C there are defects in bones, teeth, connective tissue, skin, and blood vessels. The deficiency disease, *scurvy*, is caused by a lack of vitamin C. Scurvy produces painful swelling of the tongue, blackened lips, bleeding gums, and sometimes loss of teeth. It was once common among sailors.

Some vitamin C is found in most fruits. The best sources are oranges, limes, grapefruit, tomatoes, strawberries, green peppers, cabbage, lettuce, and other salad greens. Vitamin C is destroyed by exposure to air. The living cells of raw vegetables and fruits preserve the vitamin C content. When we cook such foods, the vitamin molecules escape through the cell membranes and come in contact with the air. Orange juice exposed to air also loses its vitamin C.

Vitamin D. Vitamin D is one of the few vitamins made in the body. With the aid of sunlight, it is produced in the skin. You may wonder how we can produce it in the winter, when the sun is low in the sky and our bodies are covered with heavy clothing. Actually, vitamin D, like A, can be stored in the body. We usually produce enough of it in the summer to fill our needs during the winter. Vitamin D controls the deposit of calcium in bones and teeth. Children must have more vitamin D than adults because they are forming new bones. When a child does not have enough vitamin D, a disease called *rickets* develops. In this disease the bones are soft and deformed. Fewer children have rickets today because most of the milk sold has vitamin D added to it. In addition, babies are usually given multiple vitamins.

Fig. 34-6
Certain vegetables are rich in vitamin C. When these are cooked, the vitamin C escapes into the air. (HRW photo by Russell Dian)

MINERALS

The nutrients already discussed are all manufactured by plant cells. Vitamin D is produced in the human body. Minerals are not produced by living cells. They are drawn from the ground by plants. We get most of the mineral compounds from the food we eat. A few may come from our drinking water. There are over a dozen different elements that the body must have in the form of minerals. We will study three of the most important ones.

Calcium and phosphorus. Two important minerals that your body must have are *calcium* and *phosphorus*. These two elements combine with oxygen to make the hard material in bones and teeth. Calcium compounds are also important in blood. Protoplasm contains phosphorus. ATP, the energy carrier of the cell, is a compound containing phosphorus. DNA, which makes

Fig. 34-7
These are some sources of calcium.
(HRW photo by Russell Dian)

up the genes, contains phosphorus also.

The best source of calcium and phosphorus is milk. Eggs, cabbage, oranges, and several other foods also contain calcium. Meat, whole grain cereals, and green leafy vegetables also contain calcium and phosphorus. See Fig. 34-7.

Iron. Another mineral that is necessary to life is *iron.* Iron is present in every cell of the body. It is part of an important compound that is necessary for the oxidation of energy nutrients. Iron is present also in the compound that makes blood red. This red substance carries oxygen around the body. If the iron should suddenly disappear from your body, you would die. Your body needs iron to carry oxygen to the cells. When people do not get enough iron in their daily food intake, a deficiency disease called *anemia* (ah-NEE-mee-uh) results. The blood of people with anemia cannot carry the proper amount of oxygen. Anemia that results from an iron deficiency can be cured if the person is given iron.

The best food sources for iron are liver, red meats, egg yolks, dried fruit, and green vegetables. There is no iron in milk. Babies should eat other foods besides milk. People on vegetarian diets may need special sources of iron.

FOOD FOR BETTER HEALTH

Better nutrition. We have mentioned the malnutrition in African children that is caused by too few proteins in the diet. Many Americans are malnourished also, but in a different way. In general, we have enough food. But we must improve our use of food.

One sign of poor nutrition is overweight. This is our most common health problem. A person gains too much weight by eating foods that contain too many Calories. Concentrated sweets, candies, and sugars give us "empty Calories." This means that sweets have a great many Calories, but they do not provide vitamins, minerals, or proteins. We get little useful nutrition from concentrated sweets.

We also add to our weight by eating too many fatty foods. Most people are aware of the fat content in butter, bacon, lard, cream, and salad dressing. But they forget that a great deal of fat is contained in foods such as French fried potatoes and onion rings, doughnuts, pastries, cake, olives, and cheeses. These foods contain "hidden fats." Overweight in many young people comes from eating snack foods that contain hidden fats. By being careful about what types of food you eat for meals and snacks, you can control your Calorie intake. Table 34-3 shows some snack foods that can be nutritious, if eaten wisely.

Table 34-3 NUTRITIOUS SNACK FOODS

Cereal Foods	Milk Products	Meats	Vegetables and Fruits
dry cereal	milk shakes	hamburger	raw carrots
cereals with milk	cheese	frankfurter	celery
crackers	ice cream	cold cuts	green pepper
sandwiches (bread)	buttermilk	fish sticks	fresh fruits
oatmeal cookies	cottage cheese	egg rolls	fruit juices
pizza		tacos	peanut butter
			nuts

Our diets require fiber, somtimes called the "forgotten nutrient." Fiber is not really a nutrient. It does not provide substances for energy or for tissue building. The body cannot digest fiber. Fiber is the cellulose in vegetables, fruits, and whole grain cereals. We use this fiber as *roughage*. Roughage helps the body to eliminate waste materials. Diets containing too much meat and too many canned foods provide little fiber.

Enrichment: Americans eat large quantities of processed foods instead of fresh fruits, vegetables, and whole grain foods. This results in a lack of roughage in the diet.

roughage

Careers: Discuss the work of dieticians and opportunities for employment in clinics and institutions and in the fields of research and education.

A balanced diet. A balanced diet is one that gives the body everything it needs in the right amounts. Scientists who study nutrition divide foods into four groups. They tell people to eat some of each group every day. This is an easy way to determine what foods you should eat.

Fig. 34-8

This is an example of a balanced meal. Can you tell what group each food comes from and what type of nutrient each contains? Can you suggest other balanced meals? (HRW photo by Russell Dian)

The milk group. Young people should have three or more glasses of milk a day. Your diet should include other dairy foods such as cheese, ice cream, and butter. A serving of ice cream or cheese may occasionally take the place of milk.

The meat group. Two or more servings of meat, fowl, eggs, or seafood should be eaten daily. Dried beans, peas, and nuts may be substituted occasionally.

The vegetable-fruit group. You should have four or more servings from this group every day. At least one serving should be foods containing vitamin C such as oranges, grapefruit, tomatoes, berries, or salad greens. One serving should be dark green or deep yellow vegetables such as spinach or carrots. These contain vitamin A.

The bread-cereal group. This group includes whole grain or enriched bread or cereals, macaroni, and spaghetti. You should have two or more servings from this group every day.

If you have understood everything in the chapter, you can explain which food values are represented in each of these groups. You will also see places where one food can be substituted for another. For instance, if you eat more potatoes, you will need less bread. You can eat cottage cheese instead of meat. Our modern knowledge of food has already brought good results. Deficiency diseases are less common. There is less illness overall. Young people are growing more rapidly and getting taller than their parents and grandparents. This is due mostly to improved diet.

Ideally, all three meals should be well balanced and about equal in size. This may surprise you. Perhaps you have been skipping breakfast. Breakfast is a very important meal. It gives the body energy to run on for the whole morning. Remember that you have not eaten for many hours when you wake up in the morning. There is no food in the intestine to be absorbed by the blood. Going without breakfast can weaken your resistance to disease, especially to the common cold and tuberculosis.

In planning your own diet, remember that you need more or less energy according to your age, sex, weight, and activity. In general, young people need more proteins, vitamins, and minerals than adults do. Remember that some food value will be lost if food is improperly cooked.

Assignment: Have students plan balanced diets.

SUMMARY

All living organisms require food. Food provides cells with the energy to do work and with the materials to build tissues. The energy in food is measured by a unit called a Calorie. Calories that are not changed to chemical energy in cells are stored in the body as fat. The amount of Calories needed by people differ.

There are special compounds in food called nutrients that enable the body to meet its needs for growth and repair of tissues. Carbohydrates and fats are the fuel foods. They supply the cells with energy. Proteins are the tissue builders. Vitamins and minerals help to regulate the chemical activities of cells.

Lack of certain nutrients can cause deficiency diseases. Such nutritional disorders can be cured by giving the person foods with the needed nutrients. To avoid malnutrition, we must eat a balanced diet. A balanced diet contains foods from each of the four food groups spread over the three meals of the day.

Activity

Calorie Use Record

You can calculate approximately how many Calories you use each day. This information will help you to judge whether or not your Calorie intake is adequate.

A. Make an activity table of all of the things that you do each day. Estimate the number of hours that you spend in each activity. Be sure to include all activities, such as the time going from class to class.

B. Locate the column on the Calorie-activity record, Table 34-4, that is provided for your age and sex. You will use this information to calculate the number of Calories that you use for each activity. To do this, multiply the number of hours you spent doing each activity times the number of Calories per kg. per hour for your age and sex. For example, if you are a 14 year old girl and sleep 8 hours, $8 \times 1.1 = 8.8$, the number of calories you used in sleeping for each kilogram of body weight.

C. Add up the Calories you use per kilogram of body weight in 24 hours. Multiply this total by your weight in kilograms. This will give you an estimate of the Calories you use per day. The answer that you get is a *guide only*, and not an absolute one.

Table 34-4. CALORIE-ACTIVITY RECORD

	Est. Cal. per Kg. per Hr.			
	Girls		Boys	
Ages	11-14	15-18	11-14	15-18
TYPES OF ACTIVITY				
SLEEPING	1.1	0.9	1.5	1.1
SITTING QUIETLY (reading, writing, eating, studying, watching TV)	2.0	1.8	2.4	2.0
LIGHT EXERCISE (dressing, typing, driving car, playing piano)	2.6	2.4	3.0	2.6
MODERATE EXERCISE (bicycling, walking, bending, stretching)	4.2	4.0	4.6	4.2
HEAVY EXERCISE (dancing, skating, playing table tennis, horseback riding)	6.0	5.7	6.4	6.0
VERY HEAVY EXERCISE (running, swimming, basketball, football, tennis)	10.3	10.1	10.8	10.3

Word Quiz

Use each of the following words in a sentence to show that you know the meaning.

Calorie, malnutrition, deficiency disease, vitamin, roughage

Check Your Facts

1. Name the five groups of nutrients. Name three foods from each of these groups.

2. What is the difference between Calories and energy?

3. What does the body use carbohydrates for? Fats? Proteins?

4. List five vitamins and tell what they do.

5. List three minerals and tell what they do.

6. What is meant by malnutrition? How can we prevent malnutrition?

7. What is meant by a balanced diet?

Science Reading Skills
3. & 6. *Cause and Effect*

Thought Questions

1. Why should we be concerned about our daily nutrition?

2. How can the wise choice of snack foods improve your diet?

3. Why are infants given additional vitamins?

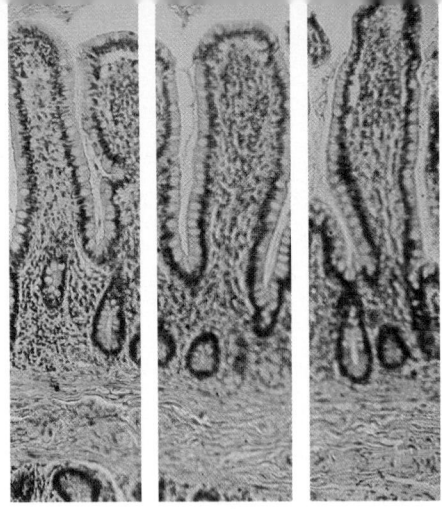
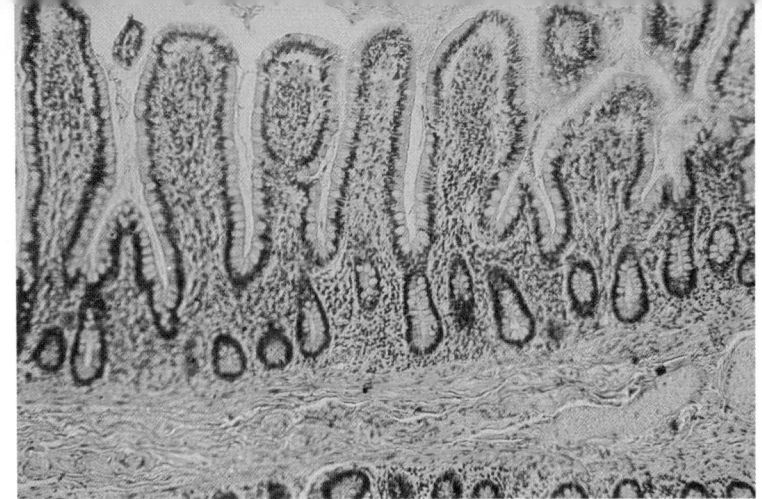

CHAPTER 35 Digestion

objectives

AftER you REAd tHis chAptER, you should be Able to:

__**Describe** the structure and function of teeth

__**Name** the parts of the digestive system

__**Explain** the work of each part of the digestive system

__**Explain** the purpose of digestion

villi

Science Reading Skill: *Sequencing*— Have students trace the passage of food through the digestive system and describe the digestion process.

You might think that the inside of your stomach looks like the inside of a balloon and your intestine is just a hollow tube that attaches to it. These structures are not really that simple. The photograph above shows a magnified view of structures called *villi* (VILL-eye); singular, villus. These are actually tiny, finger-like bulges covering the inside wall of the small intestine. As you read this chapter, you will find out the function of these villi and the other parts of the digestive system.

MECHANICAL DIGESTION

When you eat, food first enters your mouth. Inside the mouth your cheeks and gums are lined by a smooth tissue. This delicate tissue is the mucus membrane that lines all of your organs of digestion. The cells of this lining give off a thick liquid called mucus. This mucus helps to keep the lining soft.

While you are chewing, the tongue pushes the food in between the teeth and around the mouth. In this way the food is ground into smaller pieces. Liquids given off by the mucus lining and by glands under the tongue help to soften the food. This grinding and softening of the food is known as *mechanical digestion*.

The teeth. You have four kinds of teeth. Look at Fig. 35-1. Notice that this diagram has been divided into quarters. This is the way that a dentist maps teeth. Up front in each quarter you have two *incisors* (in-SIE-sors). These teeth have sharp edges for biting off pieces of food. Right behind them is a strong, pointed tooth that can tear food too tough for the incisors to bite off. This tooth is called the *canine* tooth. Next come the chewing teeth. Two *premolars* are used for light grinding. The three *molars* are used for heavier grinding. In each quarter there are eight teeth. In a full set there are 32 teeth: 8 incisors, 4 canines, 8 premolars, and 12 molars.

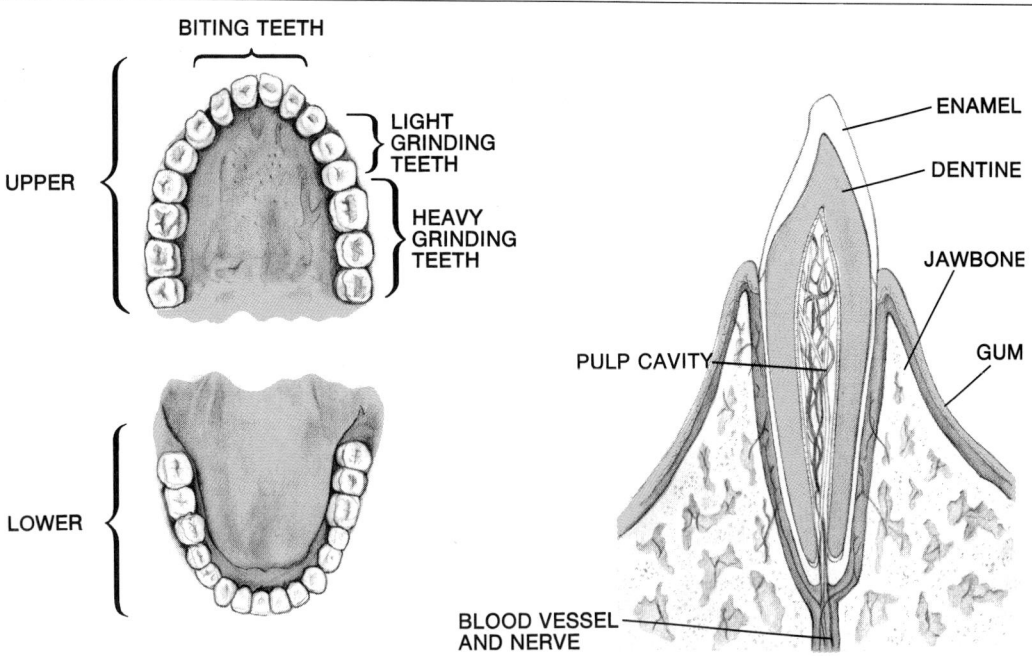

Small children need teeth, but their mouth is not big enough to hold 32 full-sized ones. First they get a set of 20 baby teeth, ten in the upper jaw and ten in the lower jaw. The baby teeth consist of 8 incisors, 4 canines, and 8 molars. Later, as the jaw grows, each tooth is replaced by a permanent tooth. The premolars grow in also. The last of the molars does not appear until about age 18. These are often called "wisdom teeth."

Teeth have structure. The main body of a tooth is a hard, bone-like substance called *dentine* (den-TEEN). The part of the tooth that shows is covered with an extra hard white material, the *enamel*. The root of the tooth fits into a socket in the jawbone. A hollow space in the center of the tooth, the *pulp cavity*, contains blood vessels and nerves. See Fig. 35-1.

Enamel for some of the permanent teeth is already forming before we are a year old. Good teeth cannot grow unless we eat tooth building foods that contain calcium, phosphorus, protein, vitamin C, and vitamin D. Even good teeth are in danger from tooth *decay*.

Tooth decay refers to the breakdown of the enamel and the dentine. These hard parts of the teeth are destroyed by certain bacteria that can reproduce rapidly in the mouth. The action of these bacteria can result in unattractive teeth, unpleasant breath, pain, and loss of teeth.

Brushing your teeth soon after eating helps to reduce the time that the bacteria have to damage your teeth. You cannot always

Fig. 35-1
Compare your teeth with those in this diagram. Can you identify the different types of teeth in your mouth?

Enrichment: Discuss the importance of good dental care.

brush your teeth after you eat. But you usually can rinse your mouth with water.

At the present time, fluorine treatment of teeth reduces the amount of decay. Fluorides added to toothpaste and to city drinking water help also. It is a wise practice to go to your dentist for regular check-ups.

CHEMICAL DIGESTION

Food is not useful to the body until it is changed. Solids cannot enter body cells. It is the job of the digestive system to change solid foods into dissolved molecules that the cytoplasm can use. Enzymes help to change carbohydrates into simple sugars. Fats are broken down into fatty acids and glycerol. Proteins are changed to amino acids. This breaking down of the food molecules is called *chemical digestion*. It is in these simple forms that nutrients can enter cells. In the cells, glucose, fatty acids, and glycerol are used for energy. Amino acid molecules are used to build tissues and other proteins. All of this is made possible through the process of digestion.

Chemical digestion in the mouth. *Saliva* is mixed with food during chewing. Saliva comes from three pairs of glands located in or near the mouth. These glands are masses of special cells. The saliva flows into the mouth through little tubes. A tube that leads from a gland is called a *duct*. One duct enters the mouth on the inside of each cheek. The other four ducts pour their saliva out under the tongue.

Saliva contains an enzyme that can digest starch. However, food is not in the mouth very long. Only a small amount of chemical digestion takes place there. Saliva also moistens the food so it will not stick to the teeth and will slip down the throat during swallowing. The action of saliva stops when food reaches the stomach.

The esophagus and stomach. When food is swallowed, it passes down the *esophagus* (eh-SAHF-uh-gus). The esophagus is the tube leading from the throat to the stomach. You can see it in Fig. 35-2. Food is pushed along as muscles in the wall of the esophagus contract.

The *stomach* is a muscular sac. Its main function is to store food. If there were no such storage organ, you could not eat a whole meal at one time. The stomach lining also contains a great many tiny glands that pour out a clear yellow liquid that mixes with the food. This is the *gastric juice*. It contains an acid that softens fibers in the food and kills many bacteria. Gastric juice also contains two enzymes that begin the digestion of proteins.

saliva

Enrichment: You may want to introduce the term salivary glands.

duct

Enrichment: You may want to teach the names of the enzymes involved in the digestive process.

esophagus

stomach

Assignment: Have students find out about different types of ulcers and the effect of gastric fluid on ulcers.

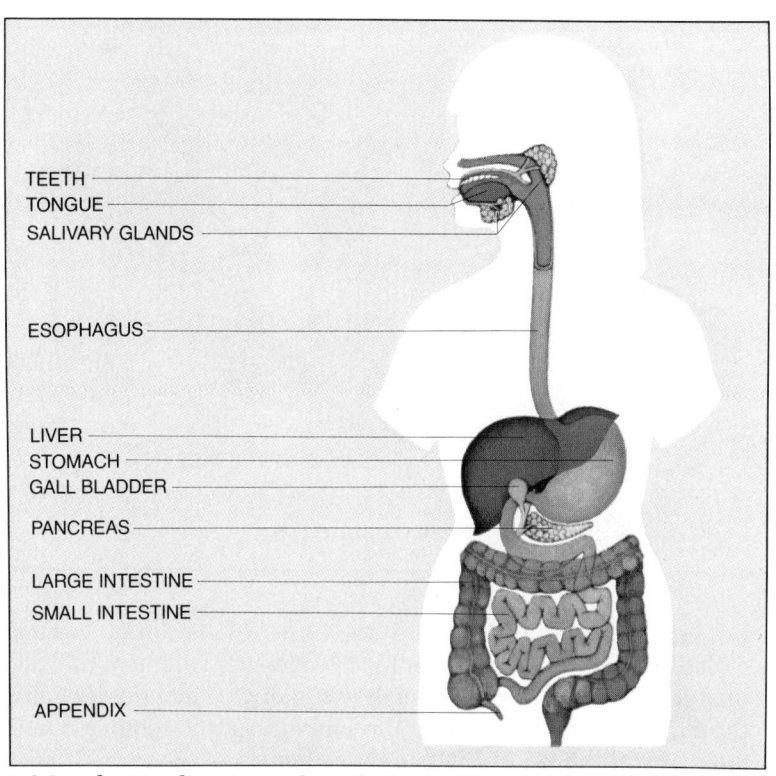

TEETH
TONGUE
SALIVARY GLANDS

ESOPHAGUS

LIVER
STOMACH
GALL BLADDER

PANCREAS

LARGE INTESTINE
SMALL INTESTINE

APPENDIX

Fig. 35-2
The digestive system has several parts.
What is the function of each organ?

Muscles in the stomach wall keep contracting and relaxing. This squeezes the food and mixes it with the gastric juice. The food particles break into smaller and smaller pieces. The proteins become partly digested. The food then passes into the intestine a little at a time. It takes about two hours for the stomach to empty itself completely.

The small intestine. The organ where most digestion takes place is the *small intestine*. It is a tube about 3 centimeters across and 7 meters long. As you can see, it is called small because of its width and not its length. Food takes several hours to pass through the small intestine.

small intestine

While in the small intestine, food is acted upon by three digestive juices. Two of these juices contain several enzymes and water. *Intestinal juice* is produced by glands in the lining of the intestine. *Pancreatic* (pan-kree-AT-ik) *juice* flows in from an important gland lying near the stomach. This gland is the *pancreas* (PAN-kree-us). Its duct enters the small intestine near its upper end. See Fig. 35-3. A juice called *bile* also enters the intestine at this same place. Bile is produced in the liver. Bile is not an enzyme. It is used to help break down the surface area of fats.

Enrichment: Cirrhosis of the liver causes destruction of the liver. This serious disease is associated with excessive alcoholic consumption and malnutrition.

pancreas

bile

The *liver* is the largest gland in the body. It has several functions that are necessary for life. One of these functions is to

liver

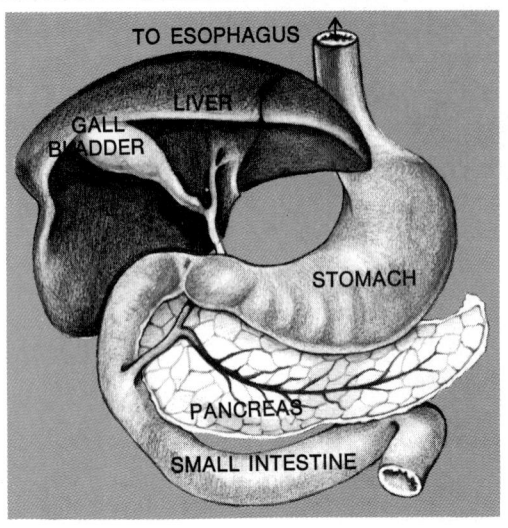

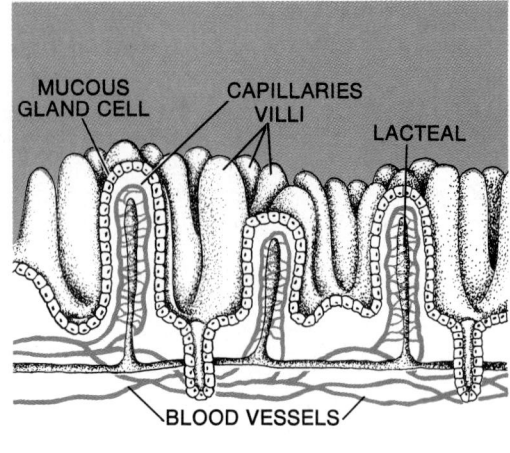

Fig. 35-3
(Left) The relationship of certain diges-tive organs. The liver has been pulled upward to uncover the gall bladder.

Fig. 35-4
(Right) The villi increase surface area for absorbing digested food. Villi are rich in capillaries so that the blood car-ries food to every body cell.

gall bladder

large intestine

produce bile. It does this all of the time. Bile is stored in a sac called the ***gall bladder***. When food leaves the stomach, the gall bladder empties its bile through a bile duct. The bile duct and the duct of the pancreas enter the small intestine together. Their juices enter the small intestine together.

The enzymes in pancreatic juice and in intestinal juice digest all of the food nutrients. This means that the carbohydrates and proteins are broken down into the molecules from which they are built. Before digestion can take place, fat is separated into tiny droplets by bile. This must be done before fat digesting enzymes can begin their work. Muscular movements in the wall of the intestine stir the food and move it along slowly. Digestion is completed in the small intestine.

The small intestine is not only the major organ of digestion, it is also the main organ of absorption. Dissolved molecules of digested food diffuse across the membranes lining the intestine and enter the bloodstream. There are ridges in the lining of the intestine that add extra surface area for absorption. Even more surface is provided by the *villi*. These are shown in the pho-tograph at the beginning of this chapter. Villi absorb dissolved food in somewhat the same way that root hairs in plants absorb water and minerals. Each villus contains tiny blood vessels that take in the dissolved food nutrients. Blood carries the food away to other parts of the body.

The large intestine. The organ known as the ***large intestine*** is about 1.5 meters long. It is called the large intestine because it is *wider* than the small intestine. The materials that go into the large intestine are waste materials. These consist of cellulose from plant foods and other substances that we are not able to digest. These wastes are mixed with water. The function of the large intestine is to absorb most of this water.

Wastes are held in the large intestine for several hours while water is absorbed back into the bloodstream. This leaves wastes in a more solid form. They become *feces* that must pass out of the body through the *anus*. The feces contain more than just undigested materials from food. A great many bacteria are present. The feces also contain mineral salts that come as wastes from the blood.

The appendix. The *appendix* is a small tube about as thick as your little finger and usually about 9 centimeters long. One end is closed. The other end opens into the large intestine, just below where the large intestine and small intestine connect. See Fig. 35-5. The appendix has no known function in humans. In some plant-eating animals, such as rabbits, the appendix is a useful organ. It helps the large intestine absorb water and functions in digesting the cell walls of plants. In these animals the appendix is large, but in humans it is small.

When the human appendix is infected by bacteria, the disease is called *appendicitis* (uh-pen-duh-SITE-us). This condition is very dangerous, because the infection can cause the wall of the appendix to break down. If this happens, pus forming bacteria spread all through the body cavity. Death can result. Doctors prevent the bursting of the appendix by removing it when it is infected.

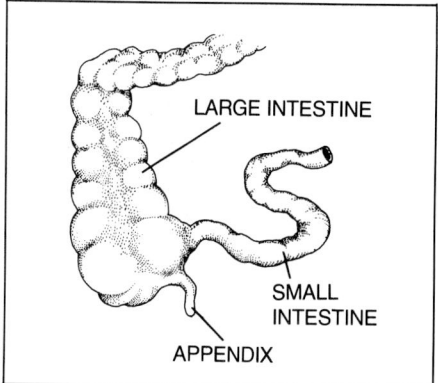

Fig. 35-5
The appendix is attached to the large intestine near the point where the small intestine enters.

appendix

Enrichment: Constipation is a problem that many people have. However, in many cases, laxatives become psychologically addictive and can result in harm to the functioning of the digestive tract.

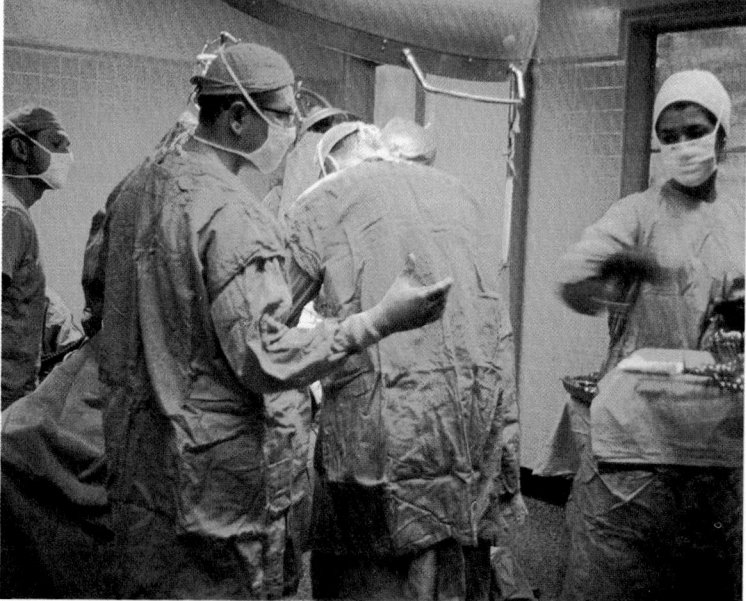

Fig. 35-6
If a person's appendix becomes infected it must be removed. (Dan McCoy/Rainbow)

SUMMARY

The purpose of the digestive system is to make solid food soluble. Only dissolved nutrients can enter cells.

Digestion begins in the mouth. Teeth grind the food, making it possible for enzymes to act on small particles. The glands under the tongue and in the jaw send saliva into the mouth. The saliva contains an enzyme that begins the digestion of starch.

From the mouth food travels through the esophagus into the stomach, where the digestion of protein begins. Food leaves the stomach and goes to the small intestine. In this organ the major digestive processes occur.

Digested food is absorbed by the villi. These are very tiny bulges in the lining of the small intestine. Waste materials are temporarily stored in the large intestine. Large amounts of water are absorbed by this organ. The waste material leaves the body in the form of feces.

ACTIVITY

Digestion of Starch

A. Place about one-half teaspoon of cornstarch in a test tube. Fill the test tube about three-fourths full with warm water from the tap. Roll the test tube between your hands to mix the starch and water. Remove 10 drops of the starch suspension and place this into the empty test tube. Add one drop of iodine. Rotate to mix.

1. What color change occurs?

B. If you have a water bath, it is a good idea to keep the starch suspension at 37°C. Now add as much saliva as you can to the test tube that contains only the starch suspension. (This is the test tube that does not contain any iodine.) Mix the contents of the tube by rolling the test tube between your hands. Every 5 minutes remove a 10 drop sample from the test tube. Test each sample with one drop of iodine.

2. What color changes do you see at each testing?

3. How do you explain these changes? (Hint: Remember what is contained in saliva.)

4. What is the final product of starch digestion?

Word Quiz

Describe the function of each of the words listed below. Use one sentence for each answer.

saliva, duct, esophagus, stomach, small intestine, pancreas, bile, liver, gall bladder, villi, large intestine, appendix

Check Your Facts

1. Using your own words, explain the process of digestion.

2. Name the parts of a tooth. Why are the teeth important to the process of digestion?

3. Explain the functions of saliva.

4. List the parts of the digestive system.

5. Why must food be digested?

6. Where do digested foods go?

7. What is the difference between mechanical digestion and chemical digestion?

Science Reading Skill
1. *Sequencing*

Thought Questions

1. Are teeth living structures? Explain your answer.

2. Milk is the first food given to infants. What kinds of enzymes must babies have? (Hint: Think of the kinds of nutrients in milk.)

3. Why does the small intestine need so much surface area for diffusion?

Science Reading Skills
1. & 3. *Problem Solving*

369

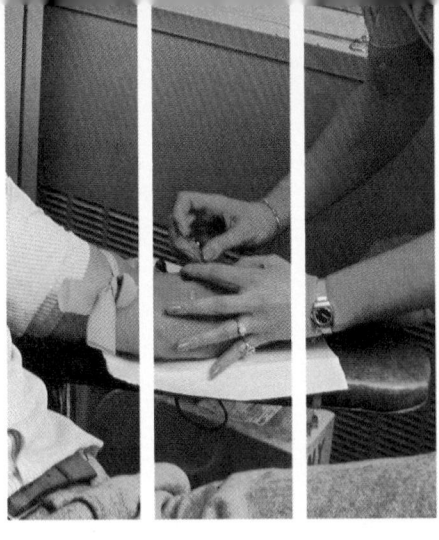
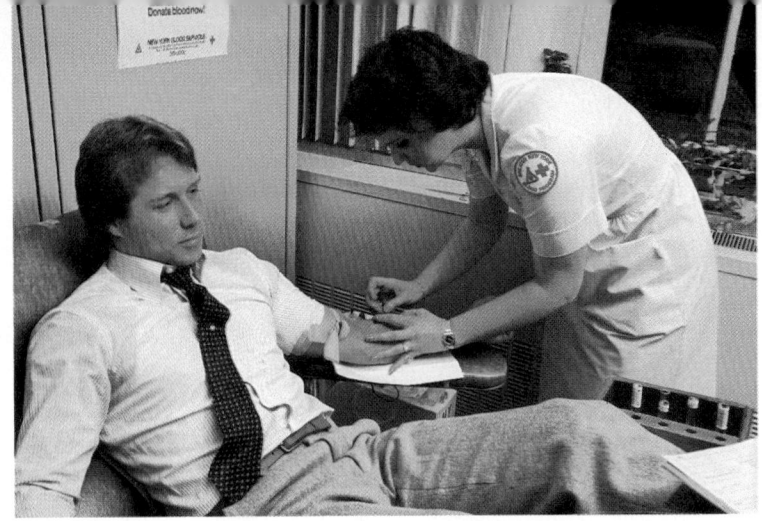

CHAPTER 36   Blood and Circulation

objectives

After you read this chapter, you should be able to:

___**Name** the parts of the blood and give the function of each part

___**Explain** the function of the circulatory system

___**Describe** the structure and function of the heart and blood vessels

___**Discuss** the importance of the lymph system

plasma

Blood is sometimes called the river of life. It flows through the body bringing cells food and oxygen and carrying away their wastes. Humans and other animals are dependent upon the work of the blood for survival. Because it is so necessary for life, scientists have found ways to preserve blood outside the human body. This blood is used to restore health to the sick. Without blood we could not survive.

THE COMPOSITION OF BLOOD

Plasma. Blood is a tissue. Unlike other tissues, blood cells are carried in a liquid. The liquid part of the blood is called *plasma*. About 92 percent of the plasma is water. In it are dissolved a number of salts, including common table salt, calcium salts, potassium salts, and others. Plasma contains food molecules on their way from the small intestine to the body tissue cells. It also carries carbon dioxide and other waste products. Plasma contains special molecules known as *blood proteins*.

Blood proteins have many functions. They help to hold water in the blood vessels, so that it does not pass out through the vessel walls. Some blood proteins destroy disease germs. Others carry food molecules around the body. One blood protein can turn into a sticky, jelly-like form under certain conditions. This jelly-like substance tangles with other material in the blood, forming a *clot*. Clots plug up cuts and stop bleeding. Without clots forming we would be in danger of bleeding to death from any small wound.

Blood cells. Floating in the plasma are three kinds of blood cells. These are the *red blood cells*, the *white blood cells*, and the *platelets*. There are more red blood cells than any of the other blood cells. A drop of blood the size of a pinhead contains 5 million red blood cells. Fig. 36-1 shows you that red blood cells are disc-shaped. They have thin centers and lack nuclei. Red blood cells contain the red iron compound, *hemoglobin* (HE-ma-glow-bun). This protein gives blood its red color. But the function of hemoglobin is more important than just giving blood its color. When red blood cells reach a place where there is plenty of oxygen, their hemoglobin unites chemically with this oxygen. This happens in our lungs. It also happens in the gills of fish. Later, when these same red blood cells pass through a tissue that contains little oxygen, the hemoglobin releases its oxygen atoms. This oxygen can then be used by the cells in that tissue. Then the red blood cells are carried back to the lungs by the flowing blood. There they take on more oxygen to be carried to the body cells.

red blood cells

white blood cells

platelets

hemoglobin

Word Study: "hemo-" means blood

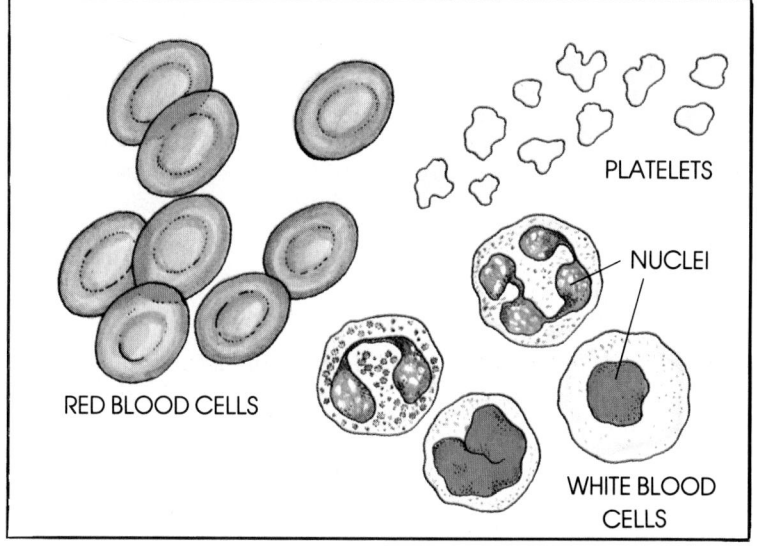

Fig. 36-1
A diagram of the three different types of human blood cells.

White blood cells look different from red blood cells. White blood cells are larger in size and have large nuclei. See Fig. 36-1. There are several kinds of these white blood cells. Some look like amebas. In a drop of blood there are about 5 thousand of these cells. They are clear, colorless, and some have no definite shape. They move about under their own power just as an ameba does. There are always some white blood cells in the blood, but many others are found in the body tissues. They can pass between the cells of small blood vessel walls. In this way they enter or leave the blood stream.

While the white cells roam about the body in this way, they do some very useful work. Some of them destroy bacteria or other disease germs. This prevents the bacteria from multiplying and causing infection. Wherever germs enter the body, white cells swarm the area and attack them. In these battles

Reference: For students who are interested in learning more about hemophilia, see *Journey* by Robert and Suzanne Massie.

many white cells are killed by the bacteria. If the bacteria succeed in growing and multiplying, the person becomes sick. The white cells, helped by blood proteins, destroy the bacteria and the person becomes well again.

The *platelets* are small pieces of cells. They do not have nuclei. See Fig. 36-1 on page 371. They are known to have two functions. Platelets stick to the walls of injured blood vessels. In this way they help to prevent blood from escaping. They also start the clotting process. When they contact the rough edges of a wound, the platelets break open and release a certain enzyme. This enzyme reacts with the calcium salts and blood proteins in the plasma that we mentioned before. The protein turns jelly-like, producing a clot.

Blood cells do not live long. Red blood cells wear out in about 90 to 120 days. Platelets live less than 10 days. Certain kinds of white blood cells last only one day. New blood cells must be made to replace the cells that die. Red blood cells, platelets, and many of the white blood cells are produced in the *marrow* found in the center of the bones. Marrow cells are able to divide. New cells produced by this division become specialized to form the different kinds of blood cells. Some kinds of white blood cells are produced in another kind of tissue that will be discussed later in this chapter.

CIRCULATION OF BLOOD

Blood vessels. Blood is pumped by the heart. Blood flows through the body in closed tubes known as the *blood vessels*. Blood vessels that carry blood away from the heart to the body tissues are called ***arteries***. Blood vessels that bring blood back to the heart from the body tissues are called ***veins***.

arteries

veins

As an artery leads away from the heart, it branches into smaller and smaller arteries. The smallest branches reach into all parts of the body. The job of the arteries, then, is to carry blood into the body tissues.

Many small veins gather blood from the body tissues. These small veins join to form larger ones. These join to make still larger ones, and so on. The largest veins empty into the heart.

capillaries

The smallest arteries do not join the smallest veins directly. They are connected by a network of very tiny blood vessels called ***capillaries*** (CAP-uh-ler-ees). They are the smallest of all the blood vessels. Many are so small that red blood cells must go through them in single file. Every cell in your body is close to some capillary. A single pin prick will break through several of them. The smallest branches of the artery system empty into the capillary networks. See Fig. 36-2. Blood from the capillaries flows into the smallest veins.

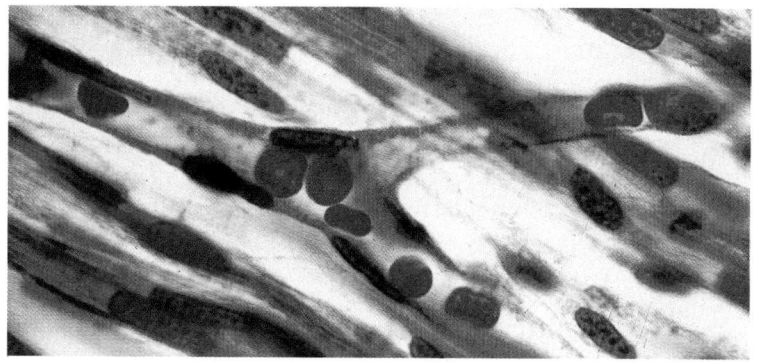

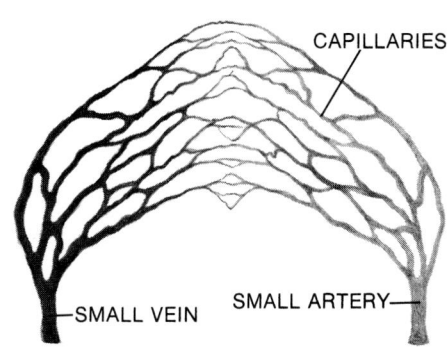

Capillaries are the vessels that allow an exchange of molecules between the blood and the body cells. The walls of the capillaries are very thin. In fact they are just one cell layer thick. This allows molecules in the blood to diffuse across the capillary walls. In a like manner, molecules from cells can diffuse into blood that is in the capillaries.

Exchanges between the blood and other tissues. While blood is in the capillaries it exchanges different materials with the cells of other tissues. Blood gives tissue cells food molecules and oxygen. It takes waste products away from the cells. These are carbon dioxide, dissolved nitrogen compounds, mineral salts, and other materials.

Exchanges between the blood and the cells vary in different tissues. The cells in muscle, skin, bone, and brain tissues receive oxygen and food molecules from the blood. These tissues release into the blood carbon dioxide and nitrogen-containing wastes. The cells in lung tissue receive food molecules and carbon dioxide from the blood. Lung tissue cells send oxygen and nitrogen-containing wastes into the blood. Blood in the capillaries in the small intestine releases oxygen. It takes on food molecules, carbon dioxide, dissolved mineral salts, and nitrogen wastes. Blood that goes to the kidneys releases oxygen, food molecules, and nitrogen-containing wastes. Blood removes carbon dioxide from the kidneys.

Circulating blood is really a pick-up and delivery service for the body cells. In general, the blood brings oxygen and food molecules to the cells. It removes from the cells carbon dioxide and nitrogen-containing wastes.

THE HEART

Structure. The heart is an organ that pumps blood. It is made of muscle and works by contracting and relaxing. You can think of this as a squeezing and releasing movement. Each contraction

Fig. 36-2
Right: Small arteries branch off to form capillaries, which empty into tiny veins; Left: A photograph showing red blood cells moving through a capillary. The capillaries are so small that the cells must move through one at a time. (Lennart Nilsson)

Assignment: Heart disease is one of the leading causes of death in this country. Have students write to the American Heart Association to find out more about the causes of heart disease and what can be done to prevent it.

of the heart is followed by relaxation. Under normal circumstances, your heart beats about 70 times per minute. Each time the heart beats, blood is pumped.

The heart is a hollow organ that fills with blood. Every time heart muscles contract, blood is forced out of the heart into the arteries. Whenever the heart muscle relaxes, more blood is brought back to the heart, filling it once again.

Actually your heart is a double pump. The right side and left side pump blood separately. The blood in the two sides does not mix while in the heart. Each side of the heart has two hollow spaces known as *chambers*. In all, the heart has four chambers. Each upper chamber is called an ***atrium*** (AY-tree-um); plural, atria (AY-tree-uh). The atria are the chambers of the heart that receive the blood. Each lower chamber is a ***ventricle*** (VEN-tri-kul). The ventricles are the pumping chambers. Large veins feed blood into the atria. When the atria are full, they force the blood down into the ventricles. The ventricles then pump the blood out into the arteries. Both atria contract at the same time, as do both ventricles. The pulse that you can feel in an artery is the rush of blood sent out each time the ventricles contract. See Fig. 36-3.

atrium

Word Study: "atrium" means room

Enrichment: Atrium is the preferred term, although auricle is often used.

ventricle

Activity: Have students use a stethoscope to compare their heartbeat rate at rest and during exercise.

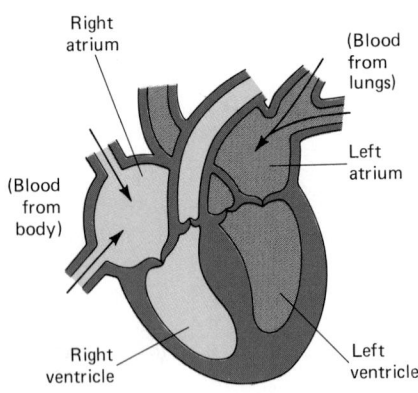

Atria fill with blood

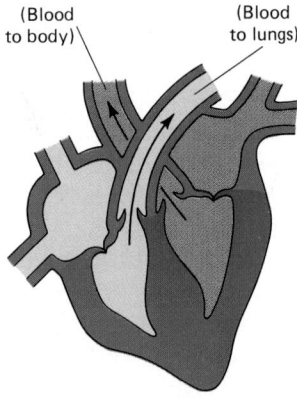

Atria contract

Ventricles contract

Fig. 36-3
The heart is a muscular pump. Blood is moved by the contractions of the atria and the ventricles.

Like any other pump, the heart must have *valves* to keep the blood from backing up. The edges of the openings between each atrium and ventricle are made of thin, tough sheets of connective tissue. When blood passes downward into the ventricle, these flaps of tissue are pushed aside. If blood starts upward through the opening, the flaps are pushed inward and meet. They close the opening like one-way swinging doors. There are tough fibers attached to edges of the valves that keep them from swinging too far.

Another kind of valve occurs where the large arteries join the heart at the two ventricles. These valves are formed by little pockets of thin tissue that are pushed aside when blood goes outward. When blood tries to back up into the heart, the pockets fill with blood and plug the artery.

When muscles contract, they shorten and get thicker. These thickened muscles squeeze against the veins in that region. This puts pressure on the veins, caving them in and forcing the blood along. The blood can go only one way because of the valves in the veins. These flaps of tissue are located every so often along the veins all through the body. See Fig. 36-4.

The path of the blood through the body. There are approximately 5.8 liters of blood in the human body. The beating heart keeps the blood in constant circulation. Blood enters the right side of the heart from all over the body. It is sent from there to the lungs. Blood from the lungs enters the left side of the heart. From there it is pumped back out to all parts of the body.

Any one drop of blood might go through capillaries in the leg on one trip around the body. The next time it might go to an ear, hand, foot, or eye, depending on which artery branch it happened to enter. But all of the blood goes through the lungs and the heart on every trip around the body. Study Fig. 36-4 as we explain how blood is pumped.

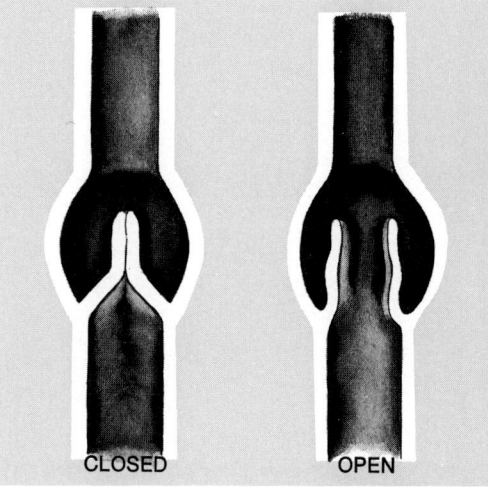

CLOSED   OPEN

Fig. 36-4
A valve in a vein. Which way is the blood moving?

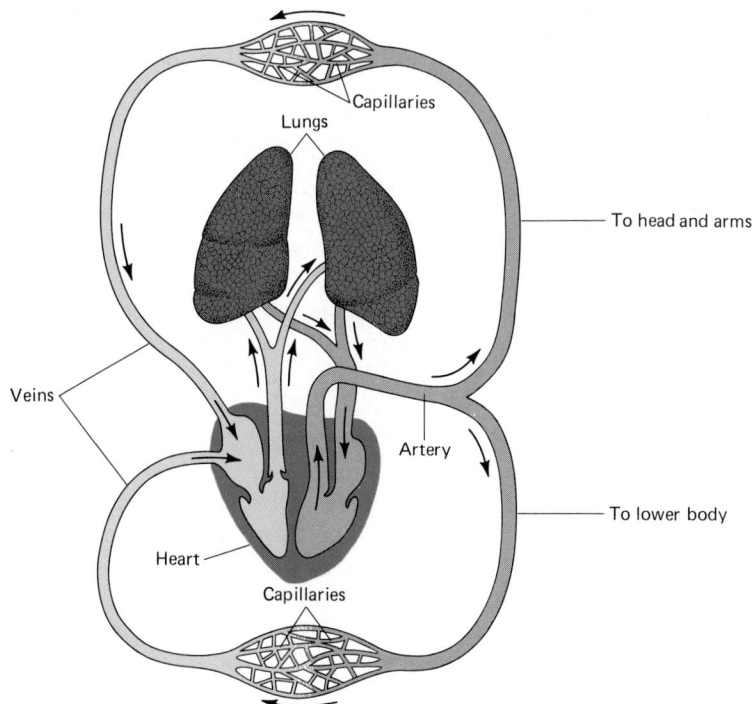

Capillaries
Lungs
To head and arms
Veins
Artery
To lower body
Heart
Capillaries

Fig. 36-5
The pathway of circulation. Blood containing carbon dioxide enters the right side of the heart and is sent to the lungs. After the exchange of gases takes place, blood containing oxygen is sent back to the heart and then pumped to the rest of the body.

375

Two large veins enter the right atrium. One vein brings in blood from the head and arms. The other vein brings in blood from the lower part of the body. When the right atrium fills, the blood is squeezed into the right ventricle. From there, the blood is pumped into an artery, which takes it to the lungs. There the blood receives oxygen. Blood from the lungs is returned to the heart through the left atrium. Then this chamber sends the blood on to the left ventricle. From here the blood enters a very large artery and is pumped into smaller branching blood vessels that carry the blood to all of the body cells. Nutrients and oxygen leave the blood and pick up carbon dioxide and other wastes. The blood is then returned to the heart. It enters the right atrium and the process continues. See Fig. 36-5.

FILTERING THE BLOOD

The lymph system. Blood is not the only liquid that moves in the body. There is a clear liquid present in all the tissues of the body. It fills any spaces that may be between the cells. This liquid is called *tissue fluid*. This is what fills blisters in the skin. It is the clear fluid that oozes out over the surface of a skinned knee or elbow.

Tissue fluid is much like plasma, but more watery. It is formed from molecules of water, salts, and other materials that leave the blood. They pass through the capillary walls into spaces between the body cells. The fluid bathes the cells, keeping them moist and giving them nourishment.

Some of the tissue fluid returns to the bloodstream by way of little tubes called **lymph vessels**. After the fluid enters these lymph vessels it is called **lymph**. The lymph vessels start out in the body tissues, and carry lymph toward the general region of the heart. At first the lymph vessels are no bigger than capillaries. They contain valves much as veins do. Exercise moves lymph through these vessels just as it moves blood through veins. The lymph vessels empty into the bloodstream at two places in the large veins just above the heart.

Every so often along the way, lymph vessels pass through little gland-like swellings. These are **lymph nodes**. They contain large numbers of white blood cells. White blood cells destroy germs or poisons that have entered the lymph from infected body tissues. The lymph nodes are filters that clean up the lymph before it returns to the blood. See Fig. 36-6.

The lymph node tissue has two other functions. One is that it produces some of the body's white blood cells. The other is that lymph node tissue produces some blood proteins, especially those that combat certain diseases.

Assignment: Have students make a diagram of the heart showing the flow of blood through the heart and labeling its parts.

lymph vessels

lymph

lymph nodes

Fig. 36-6
The lymph system acts as a filtering system. Lymph nodes are located throughout the body.

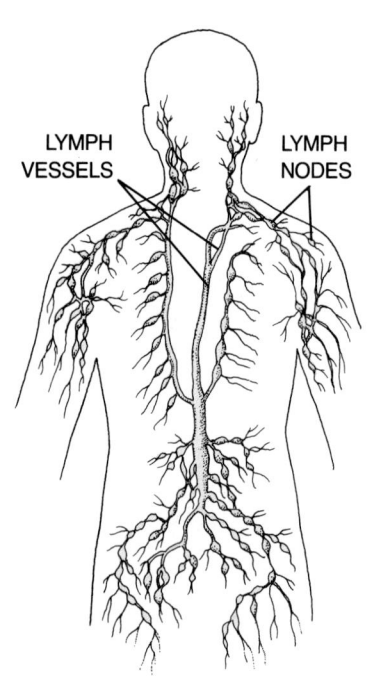

LYMPH VESSELS   LYMPH NODES

The liver and spleen. Like lymph, the blood also has filters to remove germs and other harmful materials. These are the liver and the *spleen*. The spleen is smaller than the liver. It is located in the upper left part of the abdomen, in back of the stomach. The liver and spleen destroy germs, worn out red blood cells, and some poisons that enter the bloodstream. The spleen also stores extra red blood cells. In emergency situations, the muscle walls of the spleen squeeze red blood cells into the circulation.

spleen

Blood transfusion. We said before that the human body needs about 5.8 liters of blood for proper circulation. Through accident or illness, a person can lose blood. Lost blood is restored by the process called *blood transfusion.* This means that blood is taken from one person and given to another. Not just any blood can be put into the veins of a person. The blood received must be the same type as what the person already has.

There are special proteins on the red blood cells and in the blood plasma. These proteins are not the same in all people. This is why there are different blood types. If a wrong type of blood is given during transfusion, the blood cells will stick together. Blood then cannot flow freely through the capillaries. Oxygen cannot be delivered to the cells.

HEALTH OF THE CIRCULATORY SYSTEM

Diseases of the circulatory system are quite common today. The heart and arteries are often the first organs to "break down" as people get older. Some general rules of good health will help to make your heart and arteries last longer. A balanced diet helps any system of the body. You need iron, proteins, and several vitamins to build red blood cells. It is believed that too

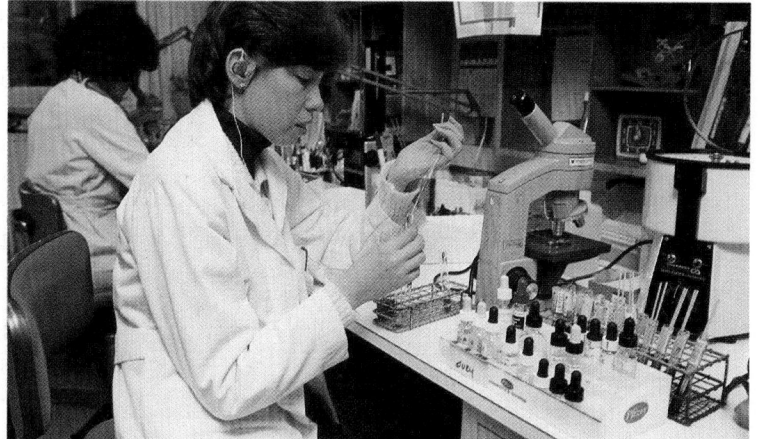

Fig. 36-7
Blood bank technicians work at typing blood and preserving it for use in transfusions.

much fat in the diet causes artery damage. Smoking damages the heart also. You need daily exercise and the proper amount of rest. Nervous tension leads to several kinds of diseases and heart trouble is one of them. A relaxed outlook on life is important for good health.

SUMMARY

The blood is composed of the liquid, plasma, and three types of blood cells. In general, blood carries oxygen, food, and other molecules to the cells. It carries away from cells the wastes of cellular respiration. The task of carrying molecules is the function of the red blood cells. The white cells are specialized for acting against disease-producing bacteria. The platelets function in blood clotting.

The heart is a four chambered muscular pump. Arteries are blood vessels that take blood away from the heart. Veins carry blood back to the heart. Exchanges of materials between the blood and the body cells takes place in the capillaries.

Humans have a system that carries lymph. This is a clear fluid that bathes the body cells. Lymph nodes are filters for germs and are places that white blood cells develop. The spleen and the liver serve as filters for the blood.

Each person belongs to one of the four blood groups. Blood typing to determine the blood group must be carried out before transfusion.

ACTIVITY

Circulation in the Tail of a Goldfish
Do not take too long with the exercise. You want to return the fish to the aquarium alive and healthy.

A. Wrap a goldfish in a thin layer of wet, absorbent cotton. Let the tail stick out. Lay the fish in an open Petri dish and spread the tail out flat. If the tail moves about, place a clean glass slide over it. Remove the clips from the stage of the microscope. Place the Petri dish on the stage and examine the tail under low power. You will be able to see blood flowing in the thin membrane of the tail.

1. Does the blood in all of the blood vessels travel in the same direction? Explain.

2. How can you tell the arteries from the veins?

3. Describe how blood travels through a capillary.

Word Quiz

Make a list of all the new words found in the margins in this chapter. Circle those words that relate to organs in the circulatory system. Underline those words that are fluids. In your own words, write a definition for those words that are neither circled nor underlined.

Check Your Facts

1. What is the function of each of the following parts of the blood: plasma, red blood cells, white blood cells, platelets?
2. How do the blood cells differ from each other in appearance?
3. Name three types of blood vessels. What is the function of each?
4. Describe the structure of the heart. In your description use the new words that you have learned.
5. Trace the path of the blood through the heart and then through the body.
6. What exchanges are made as blood in capillaries passes each of these organs: lungs, small intestine, kidneys?
7. What is the purpose of lymph?
8. Why is it important to type the blood?
9. In what ways can you guard the health of your circulatory system?

Science Reading Skills
2. Compare and Contrast
5. Sequencing

Thought Questions

1. Why is the circulation of blood important?
2. Are the lymph nodes necessary for life? Explain.
3. Let us suppose that a person loses 2 liters of blood. Why should this blood be replaced by transfusion?

Science Reading Skills
1. & 3. Problem Solving

CHAPTER 37 Breathing and Respiration

Science Reading Skill: *Sequencing—* Trace the pathway of air through the respiratory system.

nasal passages

voice box

windpipe

air sac

Learning how to breathe efficiently is a major part of an athlete's training. Active sports require a great use of energy. To produce this energy, muscle and brain cells must receive a constant supply of oxygen. In Chapter 2 we explained how oxygen is used by cells to release energy from glucose molecules. This process is called *respiration*. The energy released during respiration is used to move muscles and do other work of the body. In the last chapter you learned that red blood cells carry oxygen from the lungs to all body cells. In exchange the cells give up carbon dioxide and water vapor. Now you are ready to learn how air is taken into your lungs.

THE RESPIRATORY SYSTEM

The path of air. The work of the human respiratory system is threefold. First, air must be drawn into the lungs. Second, the lungs must filter oxygen from the air into the bloodstream. Third, waste gases of respiration must be collected by the lungs and forced out of the body. To do these things special organs are needed. Let us follow the pathway of a sample of air.

Air is drawn in through the nose. Then the air travels through the ***nasal passages***; spaces above the roof of the mouth. From there the air goes down past the back of the mouth through the *throat* and into the ***voice box***. From the voice box, air travels through the ***windpipe***. The windpipe divides into two branches, one going to each lung. In the lung there is further branching into smaller and smaller passageways. Each of these fine tubes ends in an ***air sac***. The air sacs are the places where oxygen and carbon dioxide are exchanged. You can use Fig. 37-1 to trace the path of the air into the lungs.

The structure of the respiratory system. The nasal passages are convenient by-passes around the mouth. They make it pos-

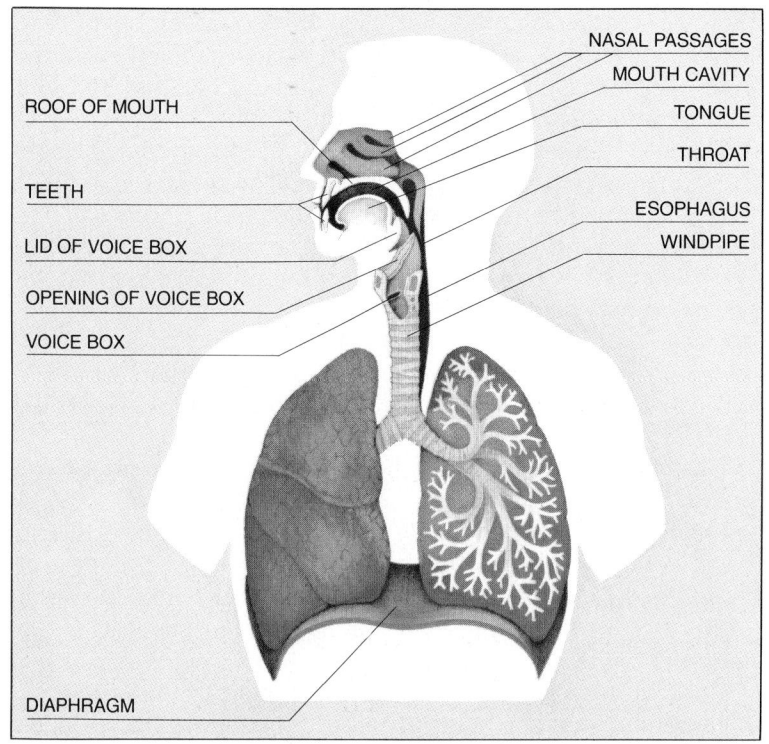

ROOF OF MOUTH

TEETH

LID OF VOICE BOX

OPENING OF VOICE BOX

VOICE BOX

NASAL PASSAGES
MOUTH CAVITY
TONGUE
THROAT
ESOPHAGUS
WINDPIPE

DIAPHRAGM

Fig. 37-1
The parts of the respiratory system. Can you explain the function of each?

Reference: See *The Story of Your Respiratory System* by E. Weart.

Enrichment: You may want to introduce the terms larynx, trachea, bronchi, bronchial tubes, and alveoli.

Fig. 37-2
Cells lining the air passage. Notice how they are covered with cilia and a film of mucus. When dust sticks to the film what will happen to it?

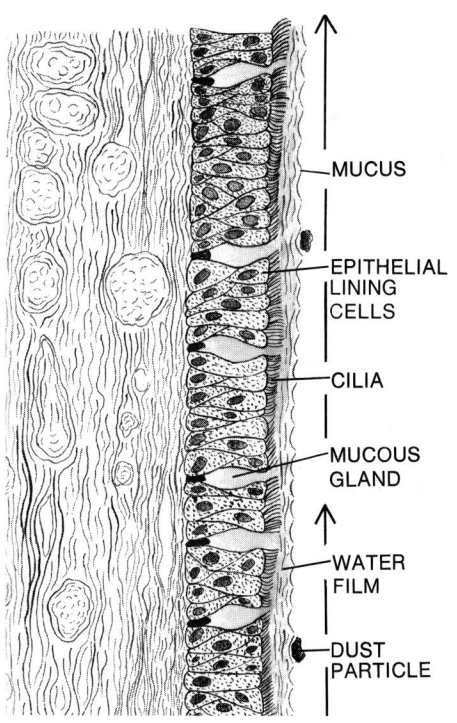

MUCUS

EPITHELIAL LINING CELLS

CILIA

MUCOUS GLAND

WATER FILM

DUST PARTICLE

sible to breathe and chew at the same time. They also produce changes in the air that passes through them. The tissue lining the nasal passages is always covered with a film of moisture. Air is warmed and moistened as it passes through. Cold, dry air could harm the delicate tissues of the lungs.

Many of the dust particles in the air stick to the moist lining of the nasal passages producing a filter. This filtering action also protects the lungs. The nasal passage lining is covered with cilia. These are the same hair-like structures that help the *Paramecium* to swim. These beating cilia cause the moisture film to move backward through the nasal passages in the direction of the throat. The dust is carried with the film. The lining of the windpipe also catches dust and moves it up to the throat. The dust particles in the throat are usually swallowed. Then they pass harmlessly through the digestive system. They would do damage if they piled up in the lungs. See Fig. 37-2. Short, soft hairs are located in the front of the nostrils. They help to prevent some larger dust particles from entering the back of the nostrils.

The voice box is hollow and its walls are stiffened by sheets of cartilage. It has a lid that is pushed down to close the opening to the voice box when we swallow. This keeps food from going down the windpipe. The esophagus lies just behind the wind-

vocal cords

Enrichment: Some people have had to have their voice boxes removed because of cancer. Many of these people have learned to speak by a special method of "burping" air.

pipe. Food entering the esophagus slides over the opening to the voice box. If this opening did not close when you swallow, your air passages would become clogged with food and you would suffocate.

There are flaps of strong tissue on each side of the voice box. These are the *vocal cords*. They can be moved against each other to close the passage, or they can be pulled back to leave a large opening. When their edges are stretched and air is forced between them, they move back and forth rapidly. We say that the vocal cords *vibrate*. This produces sound. In singing a high note, the vocal cords are stretched tightly. For low tones they are stretched less tightly. Men's voices are lower than women's because their voice boxes are larger. Their longer, heavier vocal cords vibrate more slowly, producing lower tones.

The vocal cords have another function besides producing sound. When food sticks in our throats, they clamp shut to prevent any solids or liquids from entering the air passage. Vocal cords aid also in coughing. When we cough, the vocal cords close the air passage and then open it suddenly, allowing air to rush out. Coughing serves to clear the air passages in the neck and chest.

The windpipe is a tube about 12 centimeters long and about 2.5 centimeters wide. It is in front of the esophagus and leads from the voice box down into the chest. It has C-shaped rings of cartilage in its walls that hold it open at all times. At its lower end the windpipe divides, sending one branch to each lung. See Fig. 37-1, page 381. The branches from the windpipe have the same basic structure of cartilage rings in their walls. As these branches become smaller, they have less cartilage. The finest of the branches have no cartilage at all. These air tubes do not permit any exchange of gases. Their function is to bring the air to the air sacs inside the lungs.

One lung fills each side of the chest cavity. There are about a billion air sacs in both lungs. These air sacs look like clusters of grapes. See Fig. 37-3.

The air sacs have thin, moist membranes that are one cell layer thick. The walls of the air sacs are surrounded by networks of capillaries. Remember that capillaries are the smallest of the blood vessels. These are also one cell layer thick. It is in the air sacs that the exchange of gases takes place.

Question: Ask students to explain oxygen exchange between the air sacs and capillaries in terms of diffusion.

The exchange of gases. Oxygen molecules from the air in the lungs travel along the branches from the windpipe into the air sacs. Then the oxygen molecules pass through the walls of the air sacs and into the capillaries. See Fig. 37-4. Red blood cells absorb this oxygen and carry it to the cells of the body tissue. At the same time, carbon dioxide from the blood passes into the air that is within the air sacs.

The air we breathe in contains about 20 percent oxygen and just a slight trace of carbon dioxide. The air we breathe out contains about 16 percent oxygen and 4 percent carbon dioxide. In other words, we trade carbon dioxide for oxygen in about equal amounts.

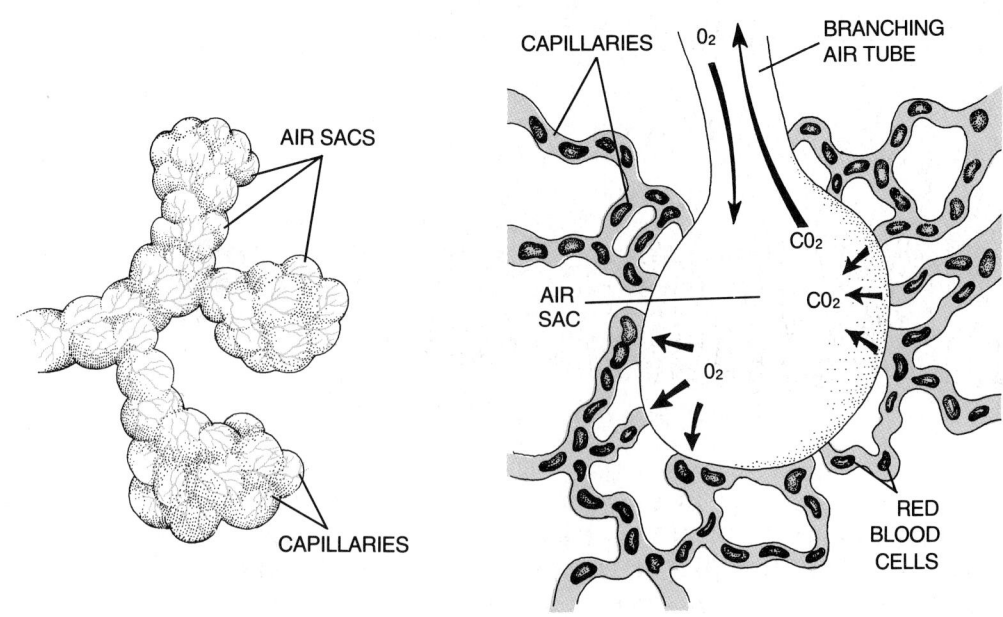

THE PROCESS OF BREATHING

The chest cavity. In humans and other mammals the body cavity is divided into two parts. The upper part of the body is called the *chest* cavity. Below this is the *abdominal* cavity. These two cavities are separated by a tough, muscular partition called the **diaphragm.** The lungs are in the chest cavity. The bones of the ribs, the body wall, and the diaphragm seal off the chest cavity so it is airtight. Tough membranes separate the lungs from each other.

Breathing. The act of moving air into and out of the lungs is called **breathing**. The diaphragm forms the floor of the chest cavity. When muscle fibers in the dome-shaped diaphragm contract, it is pulled downward. The stomach, liver, and other digestive organs are pushed down ahead of it, bulging the front wall of the abdomen outward. Muscles in the abdomen must relax to allow this to happen. At the same time that the diaphragm moves downward, the rib muscles contract and lift the

Fig. 37-3

(Left) A cluster of air sacs at the end of a tiny air tube in the lung. Notice the network of capillaries around each sac.

Fig. 37-4

(Right) The exchange of oxygen and carbon dioxide between the air sac and the capillaries. Where does each of these gases come from? Where does each of them go?

diaphragm

breathing

Activity: Obtain the lungs of an animal from a butcher or biological supply house and inflate them using a small pump.

Science Reading Skill: *Cause and Effect*—Explain what causes air to enter the lungs when the diaphragm is pulled down.

ribs outward and upward. Both the action of the diaphragm and that of the ribs increases the size of the chest cavity so that air rushes into the lungs. When people take air into the lungs, we say that they *inhale*. Fig. 37-5 shows a model of the chest cavity. When the rubber sheet "diaphragm" is pulled downward, the balloon "lungs" fill with air.

When the rib and diaphragm muscles relax, all of these parts move back into place, pushing air back out of the lungs. When air leaves the lungs, we say that a person *exhales*. Fig. 37-5 shows the position of the diaphragm when the lungs are not full of air. When we wish to talk, sing, or breathe deeply, we must actually force air out of the lungs. This is done mainly by contracting the muscles of the abdominal wall. This action moves the digestive organs back up against the diaphragm, forcing it upward against the lungs and pushing air out of the windpipe.

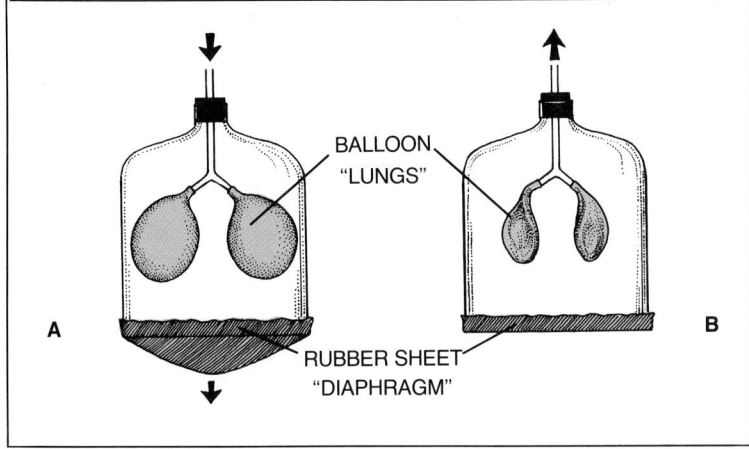

Fig. 37-5
How does this model demonstrate how the chest cavity functions during breathing?

Assignment: More than 20 thousand Americans die from emphysema each year. Heavy smoking and air pollution are contributing factors in this disease. Have students report on emphysema, bronchitis, lung cancer, and tuberculosis.

Care of the respiratory system. The lungs and air passage linings are moist and delicate. They are easily attacked by viruses and bacteria. These disease germs cause colds, flu, and pneumonia. Rest and proper diet are our best protection against infection of the respiratory system.

Foreign materials that are inhaled into the lungs cause tissue damage. Particles from cigarette smoking and air pollution can cause the air sacs to break down. This condition leads to emphysema. Asbestos fibers, the dust particles from sand blasting rock, and coal dust can cause cancer of the lungs. Dangerous fumes that are breathed in by workers in rodent and pest control industries cause lung damage also. Workers so exposed should wear filter masks to prevent injury by foreign particles and poisonous fumes.

SUMMARY

The respiratory system has three main functions; taking air into the lungs, giving off oxygen to the blood, and removing wastes such as carbon dioxide and water vapor from the body.

Air is warmed and moistened as it goes through the nasal passages. It then travels through the windpipe, which branches into the lungs. Carbon dioxide and oxygen are exchanged between the tiny air sacs in the lungs and the capillaries.

The movement of air into and out of the lungs is called breathing. This is accomplished with the help of the muscles in the rib cage and the diaphragm.

Sound is produced by the vibration of the vocal cords. These are located in the voice box just above the windpipe.

People should try to avoid smoking and inhaling foreign materials such as asbestos and coal dust. These and other materials have been found to cause lung disease.

Activity

Measuring Your Lung Capacity
You can measure the amount of air that your lungs can hold.

A. Use a graduate cylinder to fill up a 4 liter bottle.

  1. How many milliliters of water does the bottle hold?

B. Mark off 10 centimeters on the outside of a bucket. Put in water up to this level. Now hold your hand over the mouth of the 4 liter bottle and turn it upside down into the bucket. The mouth of the bottle must be underneath the water in the bucket. Tilt the bottle slightly. Insert one end of a rubber tube that is about 45 centimeters long, into the bottle. Take in as deep a breath as you can. Now blow into the free end of the tube.

  2. What happens to the water in the bottle? What is taking the place of the water?

  3. How much water did your breath force out of the bottle? How can you measure this amount?

Word Quiz

Use each of the words that follow in a sentence to show that you understand the meaning:

nasal passages, voice box, windpipe, air sac, vocal cords, diaphragm

Check Your Facts

Science Reading
Skills
3. *Sequencing*
7. *Cause and*
Effect

1. List the organs that make up the respiratory system.
2. What happens to air that enters the nose? How do the cells with cilia help the process of respiration?
3. Trace the path of oxygen from the air as it goes from the nose into the blood.
4. Describe the structure of the lungs.
5. What is the function of the air sacs?
6. Distinguish between breathing and gas exchange.
7. Look at Fig.37-5. Why does air fill the balloon "lungs" when the diaphragm is pulled down? Explain inhaling and exhaling on the basis of this diagram.
8. What kinds of diseases affect the lungs?

Thought Questions

Science Reading
Skills
1. & 2. *Problem*
Solving

1. Why should we not give young children small beads to play with?
2. If you were running in a race, what would begin to happen to your rate of breathing as you ran? Why does this happen?

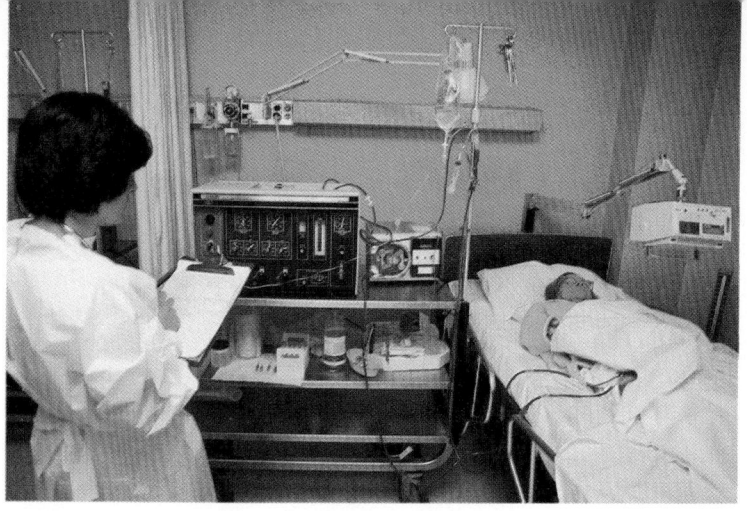
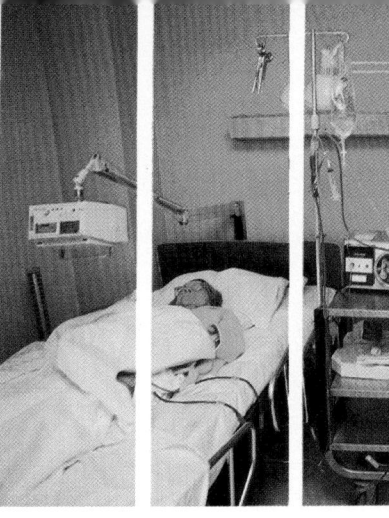

CHAPTER 38 Excretion

A kidney machine removes waste materials from the blood. In healthy people, this job is done by living kidneys. When the kidneys do not work, a person's blood must be filtered by a kidney machine or the patient will die. The body has special structures that remove the wastes produced by cells.

ORGANS OF EXCRETION

Excretion. As you know, the work of cells involves the use of food molecules. Each time cells use nutrients, waste products are formed. After cells break down molecules of carbohydrates and fats for energy, carbon dioxide and water are left over as wastes. The waste products of protein breakdown are nitrogen compounds called *nitrogen-containing wastes*. As protoplasm is built up and broken down, mineral salts are left as wastes.

All of these waste products are carried away from the body cells by the circulating blood. If they accumulate in the blood, waste materials become poisonous to body tissues. Now you can understand why it is important for wastes to be removed from the cells and the body regularly. The process by which carbon dioxide, water, nitrogen-containing wastes, and certain mineral salts are gotten rid of by the organism is called *excretion* (ex-CREE-shun). The organs of excretion are the lungs, the liver, the kidneys, and the skin.

The lungs. The lungs help in excretion. The blood picks up carbon dioxide from your cells and takes it to the lungs. Whenever you exhale, you are getting rid of carbon dioxide.

Blow your breath on a mirror or on a cool glass surface. What happens? The fog that appears on the mirror is water in the form of gas. We call this *water vapor*. Water vapor becomes a part of our breath in the same way as carbon dioxide. When

objectives

After you read this chapter, you should be able to:

__**List** the organs of excretion

__**Name** the parts of the urinary system

__**Explain** the function of the urinary system

__**Describe** the structure of skin

Careers: For more information about the work of a dialysis technician and other opportunities in the allied health professions, see *Medicine's New Technology: A Career Guide* by J. Nassif.

excretion

Activity: Have students test for the presence of carbon dioxide in their breath. This can be done by using a straw to bubble breath into a solution of bromothymol blue or limewater.

cells oxidize glucose, water is one of the waste products formed. Extra water in the cells diffuses into the blood. Some of this water passes into the air sacs and is breathed out through the nose and mouth. You may wonder why the exhaled water is in the form of a gas or vapor. Blow your breath on the palm of your hand. Does it feel cold or warm? The water is warmed as it passes through the air tubes. The warming process changes liquid water into water vapor. This is the same thing that happens when you boil water and it becomes steam.

The liver. The liver is a gland that has many functions. All of its functions are necessary for life in human beings. You learned that the liver produces bile, a substance needed to make fat ready for digestion. The liver serves in the process of excretion, too. As we said before, nitrogen-containing wastes, formed when proteins are broken down, are removed from the cells by the blood. The blood carries these nitrogen compounds to the liver. Here they are changed into *urea* (you-REE-uh), a nitrogen compound that is not harmful to cells. The liver also gets rid of bile salts formed during the breakdown of worn out red blood cells.

Bile salts and urea are not gases so they cannot leave the body through the lungs. They must leave the body through other means. Bile salts are sent from the liver through the bile duct to the small intestine. They are sent on to the large intestine along with undigested food, forming *feces*. The urea is excreted by another pathway.

urea

Fig. 38-1
Left: The parts of the urinary system; Right: A cut-away view through a kidney. Wastes diffuse into the many tubules and are later excreted.

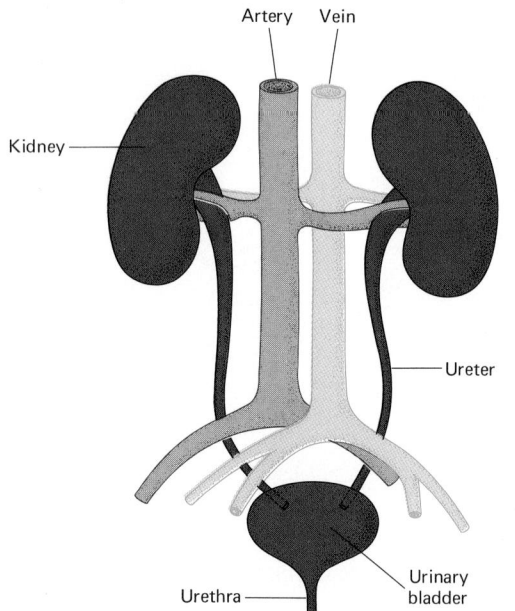

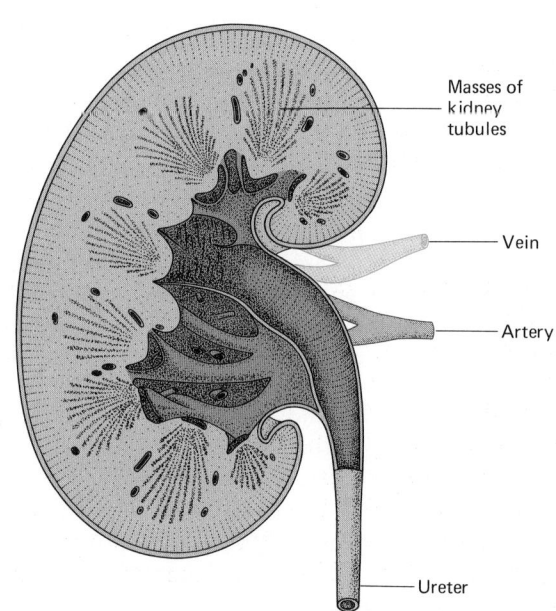

THE URINARY SYSTEM

The kidneys. Most of the work of excretion in humans is carried out by the *urinary system*. Fig. 38-1 shows you the organs that make up this system. There are two kidneys. From each of these extends a tube called the *ureter* (YOUR-eh-ter). The ureters lead into the *urinary bladder*. A single tube called the *urethra* (you-REE-thrah) leads out of the bladder.

The two kidneys are bean-shaped organs about 11 centimeters long. They lie just inside the back body wall under the last two pairs of ribs. Each kidney is protected by a surrounding layer of fat. Fig. 38-1 shows a view of the structures inside of a kidney. A large artery enters the indented side of the kidney. The artery branches into fine capillaries, which empty into a vein. The vein leads out of the kidney.

If you were to look at a very thin slice of kidney tissue through a microscope, you would see something amazing. The kidney contains a mass of closely packed tubes. These tubes are so small that they can be seen only with the aid of a microscope. A tube that is microscopic in size is called a *tubule* (TOOB-you-ul). Each kidney has about one million of these tubules. Each tubule lies in close contact with a capillary. See Fig. 38-2.

Blood enters the kidneys through the arteries. Blood flows from the arteries into the branching capillaries. Blood in the capillaries carries with it both useful compounds and waste materials. Some of the useful compounds are glucose, water, amino acids, and some salts. The waste materials are excess water, urea, and certain mineral salts.

As blood circulates in the capillaries of the kidneys, the useful compounds and the waste materials pass into the kidney tubules by diffusion. The useful compounds are sent back from the tubules into the blood. The wastes travel from the tubules into the ureters and then into the urinary bladder. This mixture of water and wastes is called *urine*. This urine leaves the body through the urethra.

Importance of the kidneys. The kidneys are the most important organs of excretion. They remove nitrogen-containing wastes, excess water, and some salts from the blood. They filter out wastes and reabsorb useful compounds. The kidneys regulate the environment inside the body and keep it stable. It is only in a steady environment that cells can work properly.

Some disorders cause glucose and proteins to be used too slowly by the cells. When these molecules build up inside the body, they will eventually be excreted in the urine. By testing a person's urine, a doctor can sometimes find out that the body is not working properly.

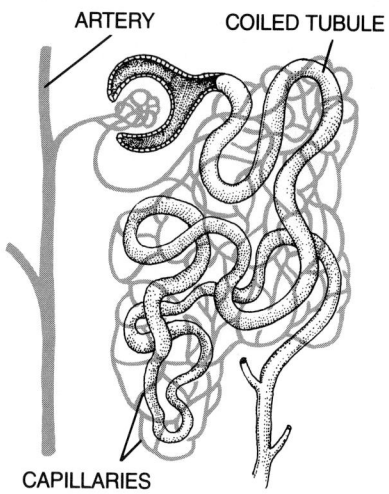

ARTERY   COILED TUBULE

CAPILLARIES

Fig. 38-2
A tubule from the inside of a kidney. Notice how the tubule lies in close contact with the capillaries. Each kidney has about one million of these tubules.

urinary system

ureter

urinary bladder

urethra

urine

Activity: Dissect a beef or pork kidney for your students.

Science Reading Skill: *Sequencing*—Trace the flow of wastes through the urinary system.

THE SKIN

Structure of the skin. The skin is another organ of excretion. It too, like the lungs and the liver, has other functions.

The skin has two layers. The outer layer is the *epidermis*. The inner layer of the skin is the **dermis**. The dermis is the thicker layer and the true skin. The cells of the dermis are living. They divide and produce new cells. Underneath the dermis is a layer of fatty tissue.

The surface layer of the epidermis is composed of flat dead cells. These cells are filled with a tough protein that makes them strong and scale-like. They protect the living cells underneath. The lower layer of the epidermis is made of living cells. These contain pigments or coloring materials. The pigment cells absorb the burning rays of the sun and protect the deeper cells from sunburn.

The dermis layer of the skin is made mostly of strong connective tissue. In this tissue are located blood vessels, nerves, hair-producing cells, oil glands, sweat glands, and muscle.

Functions of the skin. The skin has many functions. One of these is excretion. The skin reinforces the work of the kidneys. Deep in the dermis layer of the skin are the *sweat glands*. These glands have long, narrow ducts that extend to the epidermis of the skin. These ducts end in microscopic pores or openings in the skin. At their inner ends the glands are coiled into tight little balls. See Fig. 38-3. Capillaries wrap around these balls of tubing. The cells of the sweat glands extract salt water from the blood, together with traces of urea and other wastes. This mixture of water, salts, and urea released by the sweat glands is known as *perspiration*. This perspiration flows out on the skin, where it evaporates.

Perspiration helps to regulate our body temperature. Our bodies often need to lose heat. This heat is lost through the skin. The skin contains many more small arteries and veins than are needed just to service its cells. This large blood supply helps to regulate body temperature. Remember that we are warm-blooded animals. The human body temperature is about 37°C. If our body temperature drops or rises very many degrees, we could be in danger.

If too much heat is in the body, the blood is warmed up. When this blood flows through the skin, it loses heat to the air. The skin acts as a kind of radiator that gives off heat. The cooled blood then returns to the rest of the body and is heated again. When the body is warm, the walls of the blood vessels stretch and the vessels get wider. Extra blood flows to the skin. In this way heat is lost and the body gets cooler.

dermis

Activity: Have students observe the surface of the skin with a hand lens. Notice scars, fingerprints, pores, and patterns of hair growth.

Enrichment: The skin also tells of one's emotions at a particular time. Discuss flushing and blushing and what they indicate. Ask students to think about other things that the skin may indicate; moods, age, and general health.

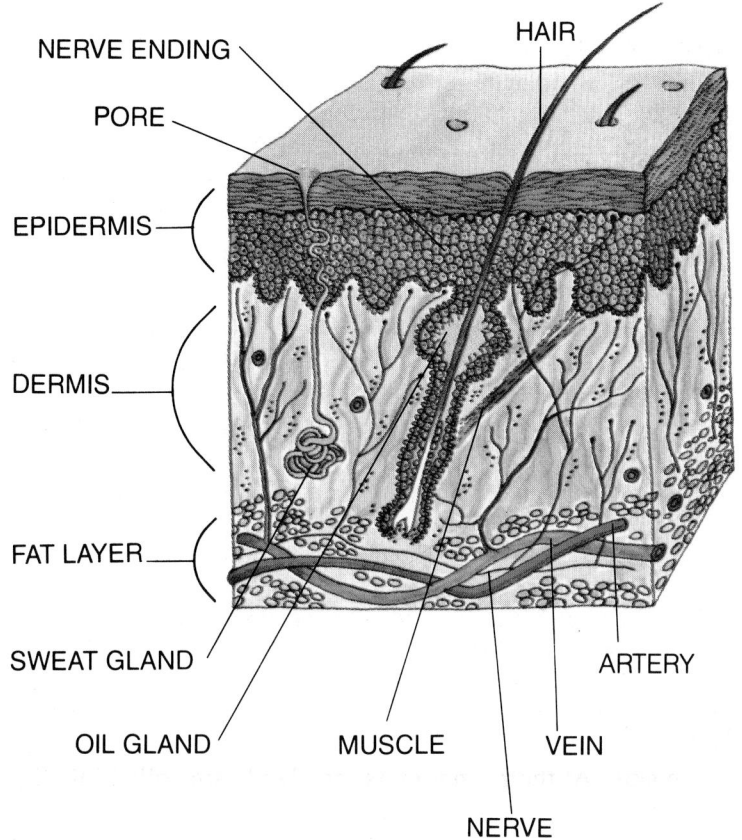

NERVE ENDING

PORE

EPIDERMIS—

DERMIS—

FAT LAYER—

SWEAT GLAND

OIL GLAND

MUSCLE

HAIR

ARTERY

VEIN

NERVE

Science Reading Skill: *Reading Illustrations*—Emphasize the layers and functions of skin for protection, sensation, and temperature regulation.

Fig. 38-3
The structure of the skin. How does the skin function as an organ of excretion?

Assignment: Students will probably have many questions concerning skin and hair care. Discuss these questions and have students do additional research on these topics.

When the body is chilled, blood vessels in the skin contract. Then less blood circulates to the skin. Most of the blood remains in deeper tissues where is cannot lose much heat. The fat layer under the dermis is good insulation and serves to hold in heat at these times. Races of people who live in northern regions have more evenly developed fat layers. This is part of their adaptation to the cold.

On a very hot day the air may be as warm as the blood is. Then the presence of blood at the skin does not cool our bodies. At such times our lives depend on our ability to perspire. The evaporation of this perspiration is what cools us.

When a liquid turns into a gas, it must absorb heat energy. When water evaporates, it turns into a gas. This means that water vapor contains more heat energy than water in liquid form. When we perspire, the water evaporates from our skins. Each water molecule carries off some of the heat. In this way, people's bodies stay at normal temperatures even when the air around them is warmer than they are.

Activity: Have students place a few drops of water on the back of one hand and a few drops of alcohol on the other. See which evaporates first.

SUMMARY

Excretion is the process through which organisms get rid of the wastes produced by cells. The organs of excretion are the lungs, liver, kidneys, and the skin. Carbon dioxide and water are released by the lungs. The liver changes nitrogen-containing wastes into the nonpoisonous urea. It also discharges bile salts from old hemoglobin into the intestine.

The kidneys are the major organs of excretion. They filter out urea, excess water, and mineral salts from the blood. A mixture of these compounds forms urine. Urine is temporarily stored in the bladder and then released from the body through the urethra.

The skin has many functions. Among these is the excretion of salt water and urea in perspiration.

ACTIVITY

The Effect of Evaporation on Temperature

A. Record the room temperature with a laboratory thermometer. Wrap the bulb of this thermometer with a small piece of absorbent cotton. Secure it with a rubber band. Dip the wrapped thermometer bulb in a container of water that has been warmed 10 degrees above room temperature.

1. What is the temperature of the water?

B. Hold the thermometer by the uncovered end and lift it from the water. Gently wave it back and forth in the air for 3 minutes.

2. What is the temperature now?

3. How did the evaporation of the warm water affect the temperature?

4. How does the evaporation of perspiration affect your skin?

C. Place the thermometer back in the container of warmed water for 2 minutes.

5. What is the temperature?

D. Now gently wave the thermometer back and forth in the container for 3 minutes.

6. Does the temperature change?

7. Does this tell you anything about why we feel "hotter" on a humid day?

Word Quiz

Use each of the words that follow in a sentence to show that you know the meaning:

excretion, urea, urinary system, ureter, urinary bladder,
urethra, tubule, urine, dermis

Check Your Facts

1. Name the wastes that are produced by cells.

2. In your own words, define the word excretion.

3. List the names of all of the organs that function in excretion.

4. How do the kidneys work?

5. What is the function of the bladder?

6. What is the difference between the ureters and the urethra?

7. Explain why the skin is considered to be an important organ of excretion.

Science Reading
Skills
4. *Sequencing*
6. *Compare and
Contrast*

Thought Questions

1. Why couldn't a person live without kidneys?

2. Can a person live without sweat glands? Explain.

Science Reading
Skills
1. & 2. *Problem
Solving*

393

CHAPTER 39 Bones and Muscles

A framework of steel and cement supports a skyscraper. A framework of bone, called a skeleton, supports your body. The bones that form the skeleton vary in size and shape. For instance, the bones of the middle ear are very small and delicate, measuring about 2 or 3 millimeters. Other bones, such as the thigh bones, are large and heavy. All of these bones contain living cells that aid in the growth and repair of the skeleton.

THE BODY FRAMEWORK

Functions of the skeleton. The bones of your skeleton give strength and form to your body. The bones are also part of the system that allows us to move around. The skeleton functions as protection for the delicate organs of the body. The skull forms a bony box around the brain. The heart and lungs are protected by the ribs, breastbone, and backbone. The spinal cord is also protected by the backbone.

In addition to protection and movement, the skeleton helps the body in other ways. It is a storehouse for calcium and phosphorus. These two elements are needed by cells for chemical activities. In the center of the long bones are places where red blood cells, some white blood cells, and platelets are formed. So you can see that the skeleton is a living framework for the body.

Bone and cartilage. The skeleton is made of *bone* and *cartilage*. These are both special kinds of connective tissue. In bone, a great deal of hard mineral matter is deposited between the living cells. This mineral is made of calcium and phosphorus compounds. These are bound together by proteins.

Bone has many passageways running through it. These contain blood vessels and nerves. There are also tiny passages

Enrichment: These passageways are called Haversian canals.

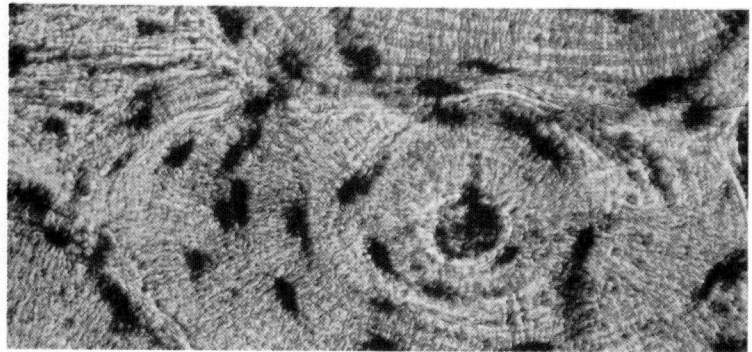

Fig. 39-1
A photograph of bone tissue taken through a microscope. Notice the many passageways through this tissue. (Courtesy Carolina Biological Supply Company)

between the cells. These allow tissue fluid to carry food and oxygen to the living bone cells. Each bone is covered by a tough sheath that is well supplied with blood vessels. See Fig. 39-1. If a bone is broken, cells from this sheath move into the break and repair it. The healing of bones is a slow process. Doctors usually place the broken part in a cast, so the ends of the bone will not move during healing.

Cartilage has a tough, flexible substance between its cells. We have cartilage in the walls of our voice box and windpipe, in our ears, in the lower part of our nose, and in other places. Cartilage is strong. It stiffens the structures that it supports and, at the same time, is quite flexible. Cartilage forms a smooth covering over the ends of the long bones. Discs of cartilage lie between the sections of the backbone. Strips of cartilage connect the ribs to the breastbone, making the chest wall flexible enough for breathing.

Structure of the skeleton. There are over 200 bones in the skeleton. The place where bones come together is called a **joint.** Some joints allow a great deal of movement. The shoulder, elbow, hip, and knee joints are examples. Some joints allow little movement. The joints between the ribs and the backbone are like this. Some joints allow no movement at all. This type of joint is found in the bones of the skull.

A movable joint is really quite complicated. The ends of the two bones connected by a movable joint are covered with smooth cartilage. This reduces friction. There is also a sheath surrounding the joint. The sheath contains a liquid that keeps the joint smooth and slippery. The two bones are bound together by fibers of strong, tough connective tissue. These are the *ligaments*. If we damage our ligaments, we say that we have a "sprain."

Look at Fig. 39-2. See how everything is centered around the backbone. This backbone is actually a whole row of bones

Activity: Fresh bones can be used to demonstrate the hard matter, the marrow, and the cartilage-covered ends.

joint

Enrichment: Using a skeleton is the most effective way to teach the major bones. Use a wall chart if a skeleton is not available.

ligaments

395

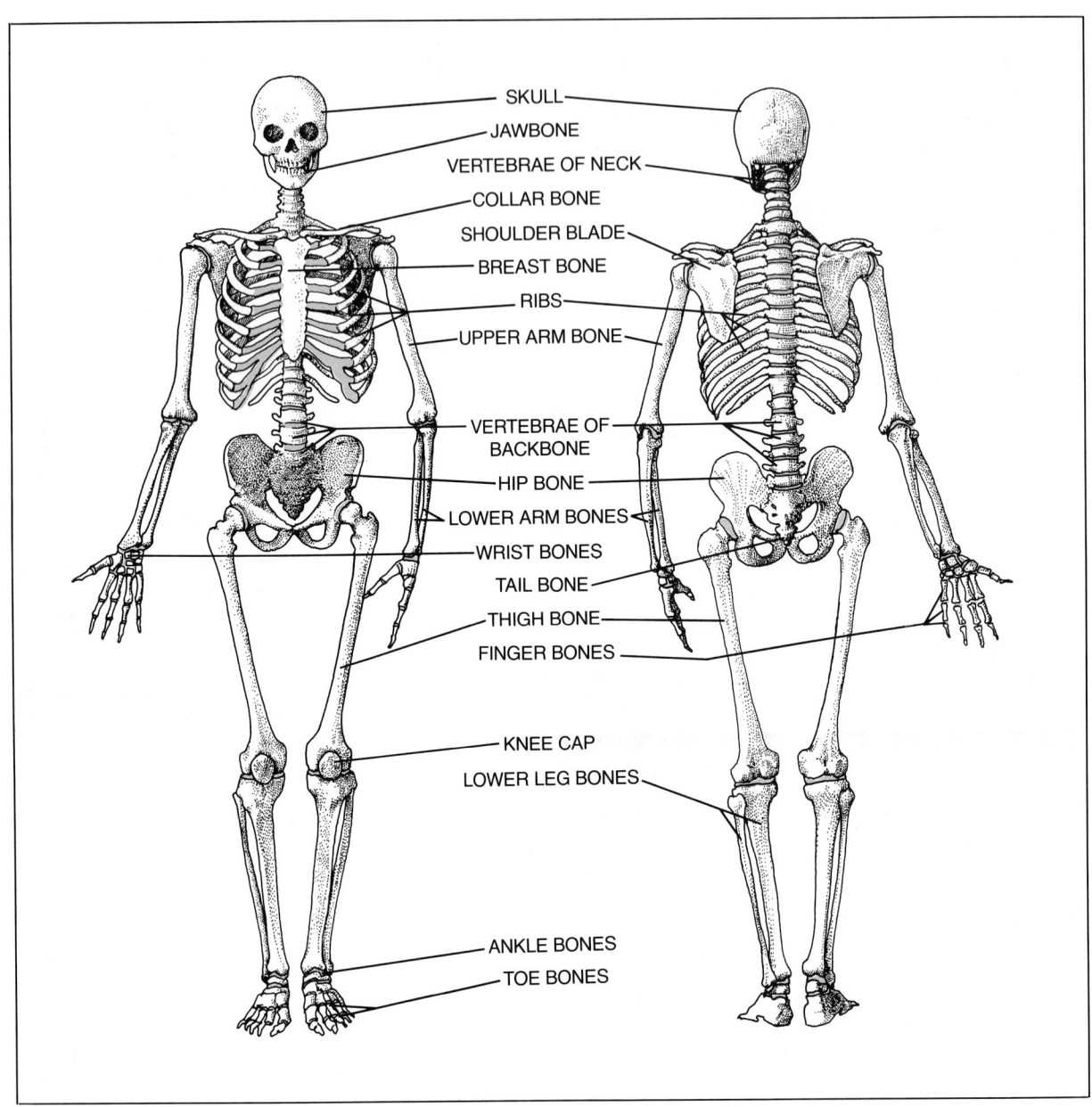

SKULL
JAWBONE
VERTEBRAE OF NECK
COLLAR BONE
SHOULDER BLADE
BREAST BONE
RIBS
UPPER ARM BONE
VERTEBRAE OF BACKBONE
HIP BONE
LOWER ARM BONES
WRIST BONES
TAIL BONE
THIGH BONE
FINGER BONES
KNEE CAP
LOWER LEG BONES
ANKLE BONES
TOE BONES

Fig. 39-2
*The human skeleton. The colored areas
indicate where cartilage is found.*

vertebrae

called *vertebrae* (VURT-uh-bray); singular, *vertebra* (VURT-uh-bruh). Each vertebra has a disc-like central part that supports the body weight. A bony arch is attached to the back of the central part. See Fig. 39-3. The entire row of these arches forms a long, protected cavity in which the spinal cord lies. The vertebrae have spines sticking out from them to which the back muscles are attached.

The skull includes many bones that have grown together. The lower jaw is the only movable part of the skull. Some of the skull bones form the brain case. The rest make up the bones of the face region. There is a movable joint between the base of the skull and the first vertebra. This allows you to move your head around.

The hip bones and the shoulder bones form supports for the legs and arms. The two bones of the lower arm can rotate around each other. This makes it possible to turn the hand. The human arm can move freely in all directions. The ability of a human hand to do fine finger movements is also unusual.

The skeletons of all land vertebrates are similar. The mammals, especially, have skeletons very much like ours. The same bones are present, but they have different shapes and sizes. See Fig. 39-4. These differences adapt them to different ways of life.

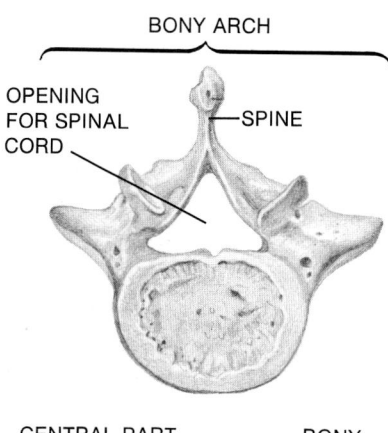

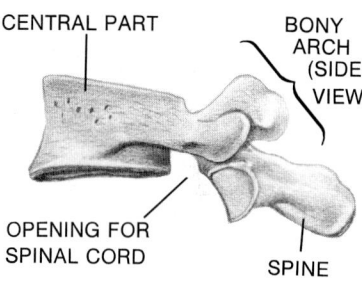

Fig. 39-3
Two views of one vertebra from the backbone. How are the vertebrae connected together?

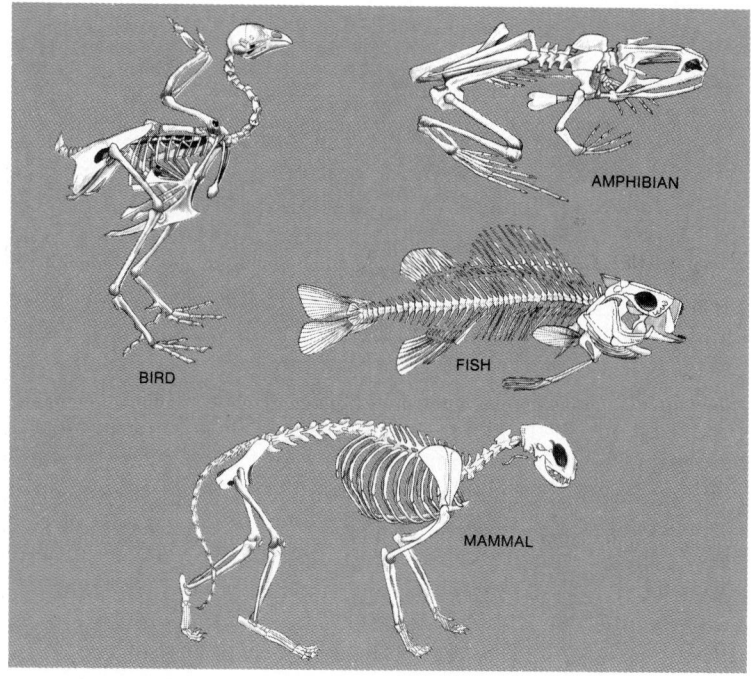

Fig. 39-4
The skeletons of various vertebrates. Notice the similarities.

THE MUSCLES

How the muscles work. Think what a good puppet you could make out of a skeleton. You could tie strings in the right places and make it walk, wave its arms, or wiggle its jaw. This is similar to what does happen when you move around. "Strings" pull your bones back and forth. These strings are tough cords

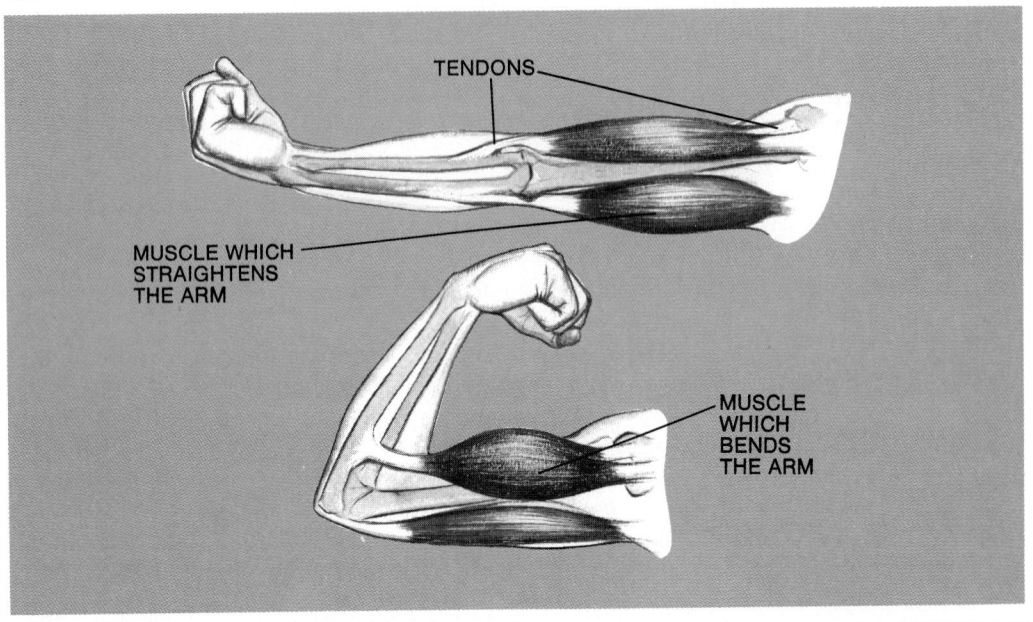

TENDONS

MUSCLE WHICH
STRAIGHTENS
THE ARM

MUSCLE
WHICH
BENDS
THE ARM

Fig. 39-5
Each muscle pulls in only one direction, therefore muscles work in pairs. Notice that one muscle contracts while the other relaxes.

tendons

Activity: A chicken leg can be used to demonstrate the action of tendons.

of connective tissue, called *tendons*. One end of a tendon is fastened to a bone. The other end connects with a muscle. When the muscle contracts, it pulls on the tendon. This moves the bone. In your muscle and bone system you are like a puppet with built-in motors that pull strings.

Look at Fig. 39-5. It shows you how the elbow is bent or straightened. One muscle on the front of the upper arm is fastened at its upper end to the shoulder. At the lower end is a tendon that extends across the inside of the elbow joint and fastens to one of the bones of the lower arm. When this muscle contracts, it becomes shorter. This pulls the arm into a bent position. Another muscle on the back of the upper arm pulls the arm out straight again. Of course, both muscles do not pull at once. Their movements are controlled by the nervous system.

A muscle only pulls, it cannot push. There are over 700 muscles attached to the skeleton. They pull the bones in many directions. Fig. 39-6 shows you how they are arranged. Some muscles have long tendons, some short ones. Some muscles attach directly to the bone when no real tendon is present. The hands are especially complex. They have many tendons coming from the muscles in the lower arm. You can feel these tendons in your wrist and in the back of your hand.

Besides the muscles attached to the skeleton, there are many others. You know about some of them already. One type of muscle forms the wall of the heart. Another is found in the veins, the arteries, the stomach, the intestines, and the bladder.

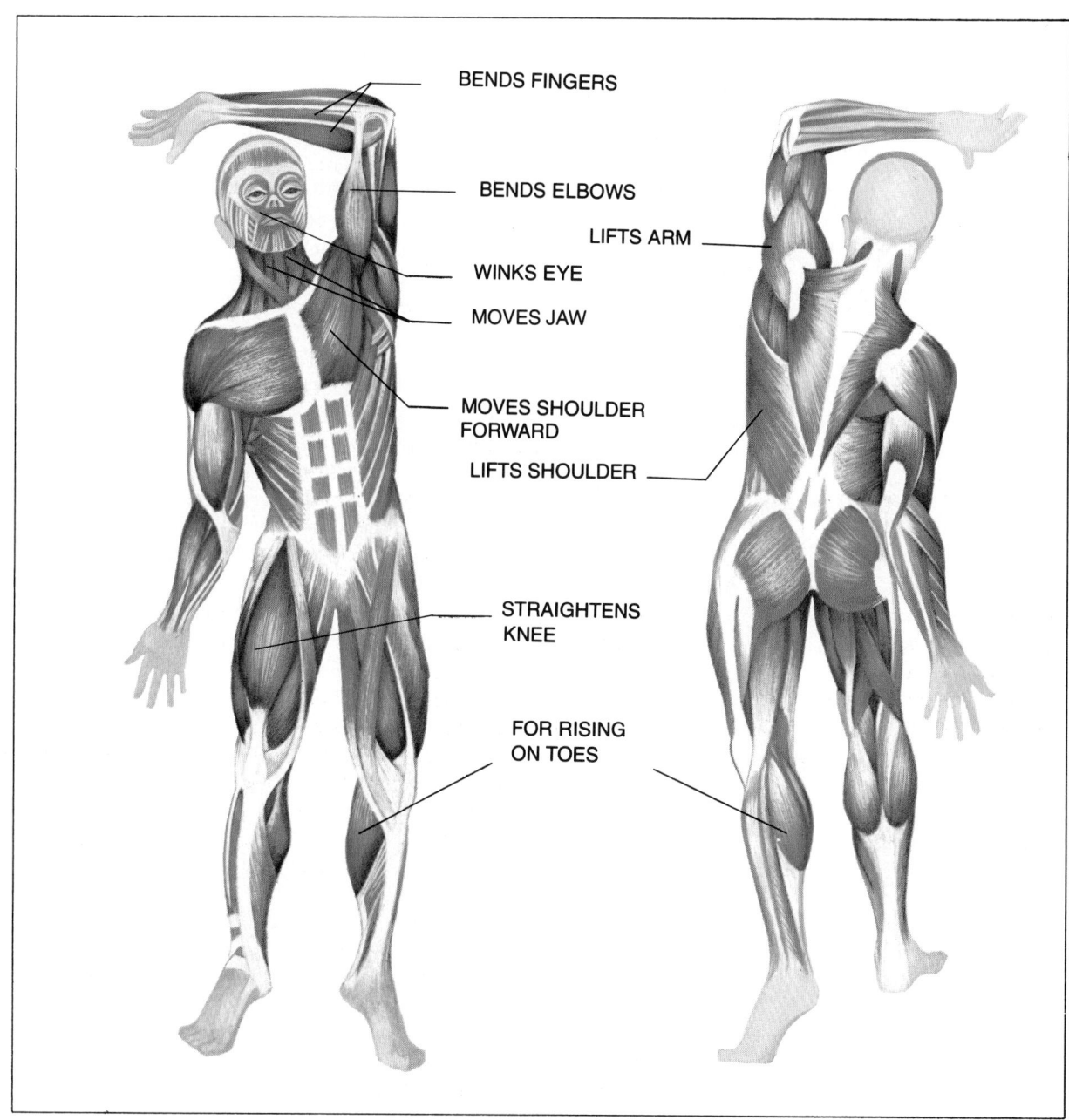

BENDS FINGERS

BENDS ELBOWS

LIFTS ARM

WINKS EYE

MOVES JAW

MOVES SHOULDER
FORWARD

LIFTS SHOULDER

STRAIGHTENS
KNEE

FOR RISING
ON TOES

How muscle cells work. There are three kinds of muscle cells, and they all work in the same way. They are able to contract because they contain two special kinds of proteins. These muscle proteins are in the form of long, slender molecules lying side by side. When a signal reaches a muscle from the nervous system, the muscle contracts.

Fig. 39-6
Human muscles. Some are labeled. Can you determine what the others do?

The process that makes muscles contract depends on some complicated chemical reactions. We will not explain these in detail here. You will only need to understand the general idea. In muscle cells, carbohydrates are oxidized by a special process that produces a great deal of energy. This energy is carried to the muscle proteins as ATP. The energy causes protein molecules of one type to slide in between the molecules of another type of protein. With millions of molecules all doing this at once, the muscle becomes shorter and thicker. It has now been learned that the cytoplasm of nearly all cells contains a network of fibers. These are made of the same proteins found in muscle, and they enable the cytoplasm to move.

During exercise, the rate of breathing and the rate of heartbeat speed up. The rushing blood carries the increased amounts of food and oxygen needed for muscle contraction. The same blood carries away the large quantities of heat and carbon dioxide that are produced by the working muscles.

Regular exercise causes muscles to grow thicker. The thicker a muscle becomes, the stronger it is. Exercise increases strength. This growth in thickness is not due to cell division. The number of cells stays about the same, but the amount of muscle proteins increases. Each cell becomes larger and stronger. A regular program of exercise keeps muscles in good condition and improves general health. Regular exercise also keeps the heart muscle in good, strong condition.

Of course, exercise should not be overdone. Sudden bursts of exercise after long periods of inactivity may cause damage. It is a good idea to consult your doctor before beginning a vigorous exercise program.

Science Reading Skill: *Cause and Effect*—Explain what causes muscles to contract.

Enrichment: You may want to discuss the differences between smooth, skeletal, and cardiac muscle.

SUMMARY

The skeleton gives the body strength and shape. The human skeleton is made of more than 200 bones. Each bone consists of living cells and mineral matter bound by proteins. Bone is hard and does not bend. The joints of bones are covered with cartilage, a kind of connective tissue that is tough and flexible. All movement is made possible by the working together of bones and muscles. Muscles contract because of special proteins in their cells that cause muscle fibers to slide over each other when powered with energy. Muscles are joined to bones by tendons. Bones are joined to other bones by ligaments.

Activity

Investigating the Structure of Bone
Bone is strong because it contains hard minerals bound together by flexible proteins. You can demonstrate the presence of minerals and protein in bone.

A. Place a chicken bone in undiluted vinegar. Make sure that the vinegar covers the entire bone. Let the bone stay in the vinegar for about 2 days. Remove it and then rinse.

1. How does the bone feel?

2. What happens when you bend it?

3. Describe the change that you see in the bone.

B. Heat another chicken bone in a large tin can. You will have to heat it for about an hour. **CAUTION:** Do not burn your fingers on the hot can. Use a heat-resistant pot holder if you have to handle the can.

4. How has the bone changed in color and in texture?

5. What materials were destroyed by heating?

Word Quiz

Use each of the following words in a sentence to show that you know the meaning:

joint, ligaments, vertebrae, tendons

Check Your Facts

1. Make a list of the kinds of activities carried out by bone cells. How do they get their nourishment? Fig. 39-1 can help with this answer.
2. What are three functions of the skeleton?
3. Explain the difference between the following pairs: bone and cartilage, ligament and tendon.
4. Are all muscles attached to bones? Explain.
5. Describe how a muscle works. You will find Fig. 39-5 helpful.
6. Look at Fig. 39-2. What are long bones? What are flat bones?
7. Explain how muscle cells contract.

Science Reading Skills
5. & 6. *Reading Illustrations*

Thought Questions

1. Why do weight lifters and body builders eat diets rich in protein?
2. Why does a runner have to do stretching exercises before running?

Science Reading Skills
1. & 2. *Problem Solving*

401

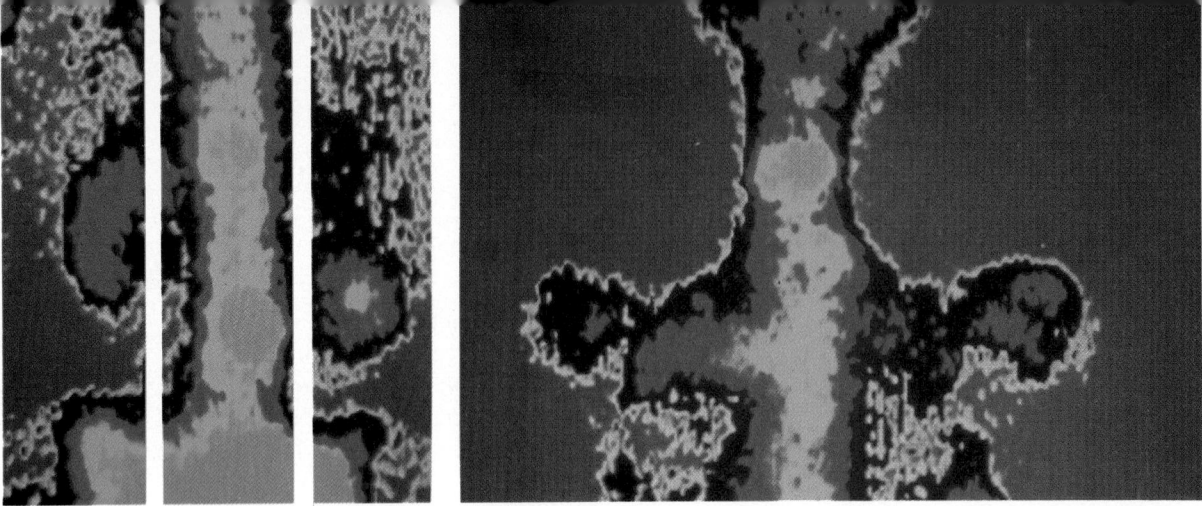

CHAPTER 40 The Ductless Glands

objectives

After you read this chapter, you should be able to:

__**Distinguish** between a duct gland and a ductless gland

__**Name** the ductless glands

__**Discuss** the function of hormones

__**Describe** diseases caused by ductless gland failure

Enrichment: Glands with ducts are called exocrine glands.

Enrichment: You may want to use the term endocrine glands.

References: See *Hormones: How They Work* by S. Riedman and *The Endocrine System: Hormones in the Living World* by A. and V. Silverstein.

Science Reading Skill: *Cause and Effect*—For each ductless gland, explain what happens to the body when the gland malfunctions.

ductless glands

hormone

secrete

The photograph above is called a nuclear medical scan. Doctors use unique pictures like this to detect tumors or diseases in various parts of the body. The scan above indicates a disease in a gland called the thyroid. In this chapter, you will learn about this and other ductless glands.

HOW DUCTLESS GLANDS WORK

Structure of ductless glands. A gland is an organ in which the cells manufacture a special chemical substance. This substance from glands is used to help the chemical activity of cells in another organ. We have already mentioned that saliva, gastric juice, and pancreatic juice are produced by glands. You have read about the intestinal glands and the sweat glands. All of these glands are duct glands. The chemical substances that they make flow out through a duct and empty into another organ.

There is a group of glands that have no ducts. These are called the *ductless glands*. The ductless glands manufacture chemical substances also. Capillaries surround the ductless glands. There is a reason for this. The substances made by the ductless glands cannot flow through ducts into other organs. Therefore they diffuse into the blood that is contained in the capillaries. The blood stream then carries the material from the ductless glands to all parts of the body.

Hormone action. The ductless glands are located throughout the body. Although these glands are not joined together, they are spoken of as a system. This is because, in general, the ductless glands work to regulate some body activity.

The chemical compound produced by a ductless gland is called a *hormone*. The ductless glands *secrete* these hormones. This means they are sent out to other parts of the body. Each

402

ductless gland makes its own hormone. A few of these glands manufacture more than one hormone. Each hormone has its own special job to do by regulating some activity of cells. Sometimes these are cells in one particular organ. Some hormones can regulate different types of cells. All hormones are carried by the bloodstream to the place where they will work.

Hormones are called *chemical messengers* because they stimulate an organ or a group of cells to work at a particular speed. The hormones are the regulators. They keep the body cells working together. The different organs of the body must perform their functions at the right times and at the right speed. A small amount of hormone is all that is needed to have an effect on a gland or another kind of organ. A hormone can either speed up or slow down a particular activity of cells.

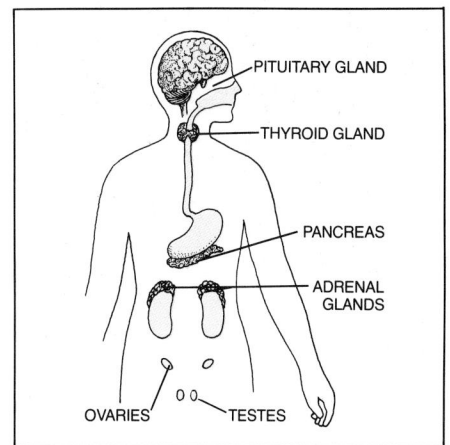

Fig. 40-1
The location of the ductless glands.

TYPES OF DUCTLESS GLANDS

Fig. 40-1 shows you where the ductless glands are located in your body. We shall discuss each of these.

The pituitary. At the base of your brain is a ductless gland that is the size of a large pea. The name of this gland is the **pituitary** (puh-TOO-uh-ter-ee) gland. Actually it is a double gland having a front half and a back half. Although small in size, the pituitary is one of the most important of the ductless glands. The front half of the pituitary secretes about seven hormones. The back half secretes two hormones. The hormones of the pituitary gland control the other ductless glands. Therefore, it is known as the "master gland" of the body. In addition to controlling other ductless glands, hormones of the pituitary affect the action of certain organs. These are the kidneys, the blood vessels, the uterus, and the milk-producing glands.

An important hormone made by the pituitary gland is called the **growth hormone**. This hormone stimulates body growth, especially the growth of the skeleton. The right amount of this hormone causes a person to grow to normal size. In some people, the pituitary gland fails to produce enough growth hormone. This results in an undersized person commonly called a dwarf.

Sometimes a tumor or an injury may cause the pituitary gland to manufacture too much growth hormone. Growth of the long bones in the affected person is overstimulated by the hormone. The person becomes abnormally tall. This type of person is called a giant. Giants may grow 2½ meters tall. Pituitary giants usually do not live very long. The heart and other vital organs cannot support a body that continues to grow very large.

pituitary

growth hormone

Fig. 40-2
When the pituitary gland does not produce the right amount of growth hormone, the person does not grow to normal size. (Mark Chester/Black Star)

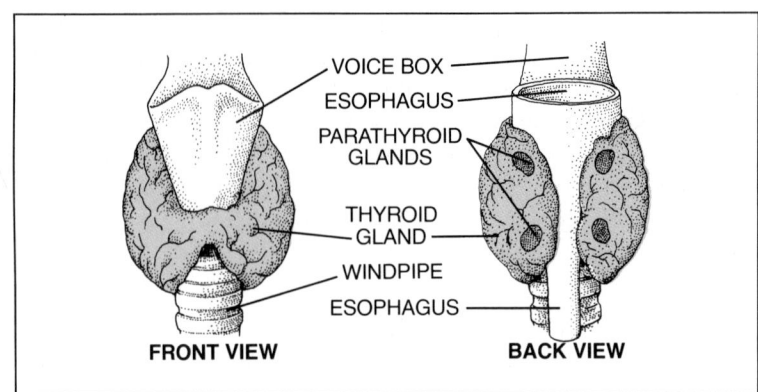

VOICE BOX

ESOPHAGUS

PARATHYROID GLANDS

THYROID GLAND

WINDPIPE

ESOPHAGUS

FRONT VIEW   BACK VIEW

Fig. 40-3
The thyroid and parathyroid glands.

thyroid

Enrichment: An overactive thyroid can be treated with radioactive iodine. This is absorbed by the gland that is bombarded by radioactivity. This destroys some of the gland the way surgery does.

The thyroid and parathyroid glands. The *thyroid* gland lies in front of the windpipe, below the voice box. You can see its shape and position in Fig. 40-3. The thyroid gland produces a hormone that controls the rate of chemical activities in the tissue cells. This means the respiration rate of cells is controlled by the thyroid gland. In normal amounts, thyroid hormone keeps all of the body cells working at just the right speed.

Sometimes the thyroid gland does not function properly. It may produce too much hormone or too little. In each case, the affected person becomes ill. When the thyroid gland produces too much of the hormone, it is said to be *overactive*. A person with an overactive thyroid may have a large appetite and eat a lot, but still lose weight. The heart beats faster and the person becomes restless and nervous. The person's cells are working too fast. To stop this condition, doctors may sometimes have to remove part of the gland.

An *underactive* thyroid gland produces too little hormone. Persons affected with this condition have little energy. Even though they may not eat a lot, these people may become overweight. The chemical activities of the body are slowed down. The respiration rate of the cells is too slow. Such conditions are treated with thyroid hormone.

Sometimes a baby is born with an inactive thyroid gland. The child gets no thyroid hormone at all. If left untreated, the child will become feebleminded, undersized, and will age prematurely. Early treatment with thyroid hormone can prevent this.

The thyroid manufactures its hormone from iodine. In certain areas, such as the Great Lakes region and the Pacific Northwest, there is little iodine in the soil. Food grown in these regions does not supply enough iodine to the people, so thyroid disease once was very common. This situation has been changed by the use of *iodized salt*. This is ordinary table salt to which a little iodine is added. See Fig. 40-4.

The *parathyroids* (par-uh-THIE-roids) are four small glands embedded in the thyroid gland. The hormone they produce is entirely different from that of the thyroid. The parathyroid hormone regulates the amount of calcium salts in the bloodstream. Normal amounts of these salts are needed for the muscles to function properly. A shortage of the hormone leads to painful muscle cramps. Too much parathyroid hormone robs the bones of calcium, thus making them too soft.

The pancreas as a ductless gland. The *pancreas* is a digestive gland with a duct. It also contains small masses of a different type of gland tissue. See Fig. 40-5. These are scattered throughout the gland and have no connection with the duct of the pancreas. These masses of tissue are ductless glands that produce a hormone called *insulin*. Insulin is needed for the cells to use glucose properly.

All of the carbohydrates that you eat are digested into the simple sugar, glucose. Only some of the glucose that is absorbed from the intestine stays in the blood. The rest is stored in the liver and muscles until needed. Insulin is needed to help control both the storage of glucose in the liver and its oxidation in the tissues.

Diabetes (die-uh-BEET-eez) is a common disease in which the amount of sugar in the blood remains too high. This is caused by the failure of the sugar to enter the cells. This disease affects about one percent of the population.

People suffer from diabetes because the pancreas does not produce enough insulin. Without this hormone, glucose cannot

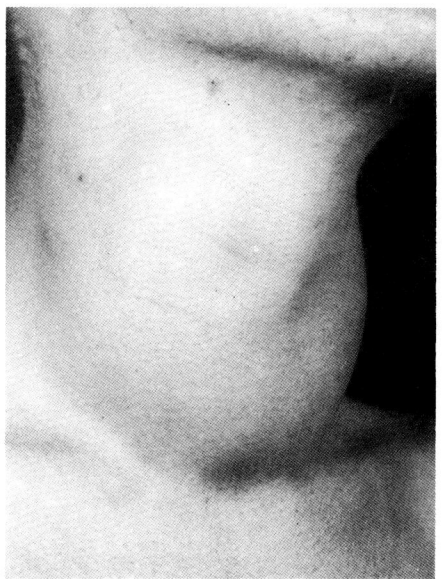

Fig. 40-4
An enlarged thyroid gland. A deficiency of iodine can cause this condition. What food substances are a good source of iodine? (Percy W. Brooks)

parathyroids

insulin

Assignment: Have students report on the discovery of insulin.

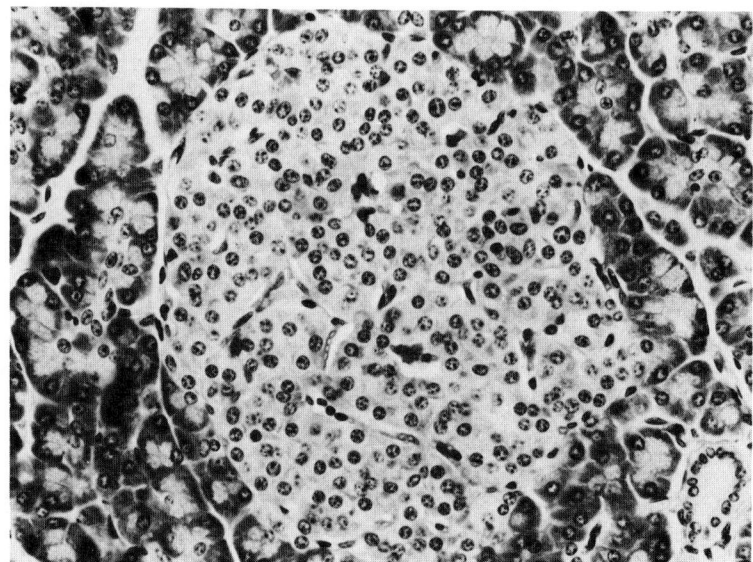

Fig. 40-5
A section through the pancreas showing ductless gland tissue. (D. W. Fawcett)

405

enter the cells of the liver and muscles for storage. Glucose also cannot enter the body cells where it is broken down for energy. Since glucose cannot enter these cells, it piles up in the blood. The excess sugar is excreted in the urine. Frequent urination and thirst are two of the symptoms of diabetes. Loss of weight is another. Since the cells cannot use sugar for energy, they break down fats and even use protein from the body tissues. The whole body chemistry is upset.

A person who has diabetes is called a *diabetic*. Diabetics are given a special diet and many take daily injections of insulin. As long as this treatment is kept up, diabetics get along very well and are able to lead normal lives.

A normal pancreas can secrete more or less insulin as it is needed by the body. This is controlled by the amount of sugar that is in the blood. A diabetic takes a certain amount of insulin each day. Sometimes the balance between the amount of sugar in the blood and the amount of insulin the diabetic has taken may be upset. This upset might be brought on by what the person has eaten or by the amount of physical exercise performed. There may be too much insulin for the amount of glucose in the body. This will cause a diabetic to seem to be drunk. The person may lose consciousness. This condition is known as *insulin shock*. Most diabetics wear a wrist band or carry a card explaining what should be done if this happens. You can help such a person by following these instructions. On the other hand, a diabetic may not have enough insulin in the body for the amount of sugar present. If this is allowed to continue, they may also become unconscious. This is known as *diabetic coma*.

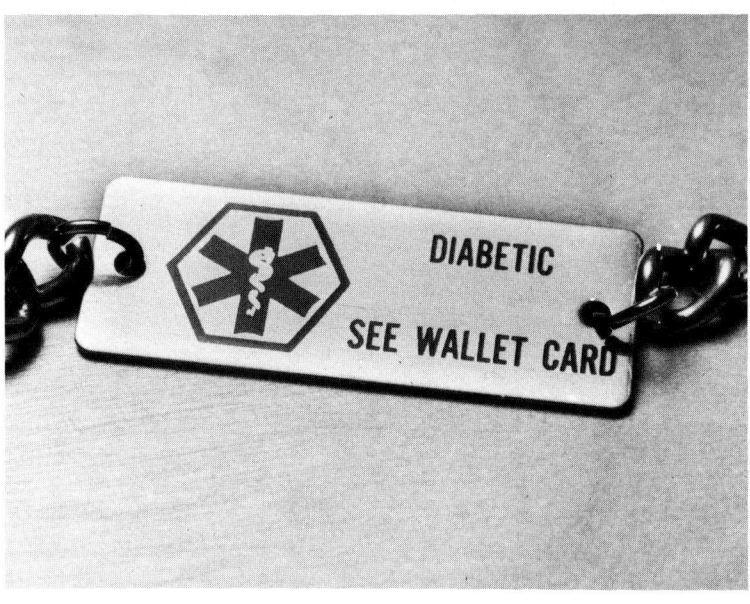

Fig. 40-6
This type of bracelet is worn by many diabetics. The instructions should be followed if the person is found unconscious. (HRW photo by Russell Dian)

The adrenal glands. The two *adrenal* (uh-DREEN-ul) *glands* fit like caps on the upper ends of each kidney. See Fig. 40-6. The adrenal gland is really two ductless glands in one. It consists of an outer layer and an inner core. Both of these sections manufacture and secrete their own hormones. The outer layer of the adrenal produces a group of hormones that control several activities in the body. One of these hormones affects the rate at which the kidneys do their work. Another influences the sex organs. Other hormones affect the balance of dissolved salts in the blood. Failure of this outer layer to produce these hormones will cause some salts to become scarce and others to become too abundant. The result is loss of weight, drying out of the tissues, weakness, and possible heart failure.

The core of each adrenal gland produces a hormone that you may have heard about. It is called **adrenaline**. The amount of adrenaline produced increases when a person becomes excited. When large amounts of adrenaline pour into the blood, many changes prepare the body for action. The heart beats faster. Blood pressure rises. Blood flows in large amounts to the muscles and nervous system, but not to the skin or digestive organs. Breathing is deeper and faster. Sugar enters the blood.

These changes give the muscles the extra glucose and oxygen needed to do hard work. People under stress often show strength they did not know they had. There have been cases where people of normal strength have lifted cars off accident victims, carried heavy loads out of burning buildings, and even killed leopards with their bare hands!

Doctors use adrenaline to stimulate a patient's heart when there is danger that it might stop beating. An injection of adrenaline may even start a heart that has already stopped beating. Adrenaline has been used also to relieve asthma attacks. It has the ability to open the small air tubules of the lungs and to stimulate breathing.

The sex glands. The ovaries and testes are sex glands. The ovaries produce egg cells and the testes produce sperm. The sex glands, like the pancreas, are both duct glands and ductless glands. Certain groups of cells in the sex glands produce *sex hormones*. The ovaries manufacture female sex hormone, and the testes produce male sex hormone.

The sex hormones control the development and functioning of the sex organs. They also control the development of *secondary sex characteristics*. These are characteristics that make males and females appear different. The soft facial features, the widening of the hips, and the development of breasts are female secondary sex characteristics. In males, the voice deepens, facial hair grows, the jawline hardens, and there is heavier muscle development.

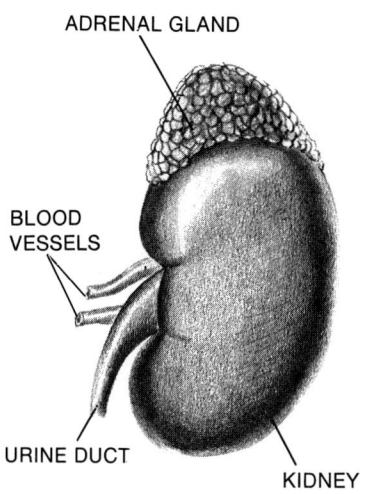

Fig. 40-7
An adrenal gland is located at the upper end of each kidney.

adrenal glands

adrenaline

407

Importance of the ductless glands. The secretions of the duct-less glands keep the blood and lymph in just the right condition so that the cells can live and grow normally. Whenever a duct-less gland fails to work as it should, the internal environment of the body is upset. If cells cannot function properly, disease conditions result. You have read about some of these diseases in this chapter. Many others are caused by the failure of the ductless glands to work properly. Today, it is possible for doctors to treat many of these ductless gland disorders.

SUMMARY

The ductless glands are located in the head and other regions of the body. The pituitary is the master gland of the body and controls the other ductless glands. These glands produce hormones that are secreted directly into the blood. The blood carries the hormones to all parts of the body. Some ductless glands produce only one hormone and others produce many. Each hormone is specialized to do a particular job on a particular group of cells. The general work of hormones is the regulation of cell activities. To do this work, only small amounts of hormone are needed. Disease conditions result from either too much or too little of these hormones.

ACTIVITY

The Effects of Feeding Thyroid Hormone to Tadpoles
You and your classmates can work together on this activity.

A. Label one fish tank "E" (for experimental) and another fish tank "C" (for control). Fill both tanks two-thirds full with pond water. Dissolve one-half of a thyroid hormone tablet in the E tank water. Do not add thyroid hormone to the C tank. Place five tadpoles in each tank.

1. Describe the appearance of the tadpoles. Estimate their length. Keep a written record of this.

B. Change the water in both tanks each day. Do not forget to add the one-half tablet of thyroid hormone to the E tank each time you change the water.

2. Record changes you observe between the two different groups of tadpoles.

3. How can these changes be explained?

Word Quiz

Choose the number from **Column B** that best matches each letter in **Column A.** Do not write in this book.

Column A	Column B
a. master gland	**1.** ductless gland
b. controls bone growth	**2.** hormone
c. secretes hormones	**3.** pituitary
d. regulates calcium salts in the blood	**4.** secretes
e. located on top of the kidneys	**5.** thyroid
f. chemical messenger	**6.** parathyroid
g. gives off useful chemicals	**7.** insulin
h. prepares body for emergencies	**8.** growth hormone
i. helps sugar cross cell membranes	**9.** adrenal glands
j. regulates cellular respiration	**10.** adrenaline

Check Your Facts

1. Explain the difference between a ductless gland and a duct gland.
2. What is the relationship between a ductless gland and a hormone?
3. Why are hormones called chemical messengers?
4. Why is the overproduction of hormones dangerous? Give specific examples.
5. Why is the underproduction of hormones harmful? Give some examples to illustrate.
6. Draw the outline of a human figure. Put the ductless glands in their proper position. Label each gland.

Science Reading Skills
1. Compare and Contrast
4. & 5. Cause and Effect

Thought Questions

1. Suppose that the pituitary gland of a person stopped working. List the number of things that would go wrong in that person's body.
2. Why is the bearded lady an example of ductless gland malfunction?

Science Reading Skills
1. & 2. Problem Solving

409

CHAPTER 41 The Nervous System

stimulus

response

References: See *The Nervous System: The Inner Works* and *Exploring the Brain* by A. and V. Silverstein.

What does it take to be a good tennis player? A person must have coordination, must be alert, and must be able to react quickly. These are characteristics a person can be born with. But a person is not born knowing how to play tennis. They must learn how to hold the racket, how to serve, and how to move to the ball.

What controls learning? What controls coordination and reaction? What enables a person to put all of these things together in the performance of a particular activity? In this chapter you will find out.

NERVE CELLS AT WORK

Stimulus and response. The nervous system adjusts the individual to the outside environment. It does so by controlling the actions of muscles and glands. It acts with glands to regulate the internal environment. The work of the nervous system helps us to receive and use information from both our internal and external environments.

To a biologist, everything an individual does is called *behavior*. This has nothing to do with right or wrong. It has only to do with the reactions of an individual. An organism reacts to changes in the environment. A change in the environment is called a *stimulus*; plural, stimuli. The reaction of the individual to a stimulus is called a *response*. Therefore, the behavior of organisms depends upon a stimulus and the response to the stimulus. The degree to which our nervous system is developed determines our behavior. In other words, the nervous system controls the kinds of responses that can be made to stimuli.

Let us see how three different kinds of organisms respond to the same stimulus. Suppose you are crossing a road and see a car coming. You step back to let it pass. The sight of the car is the stimulus. Stepping back is your response. If a rabbit sees a car coming, it may run straight down the road. It does not step back to let the car pass. The rabbit's eyes see the car just as yours do, and it can move even faster than you can. But the rabbit's nervous system is not as well-developed as yours. It is not able to predict that the car will go straight down the road. The rabbit responds as if the car were chasing it. If a caterpillar were crossing the road, it might not respond at all. It could not see the car coming, nor could it move fast enough to save itself. For the caterpillar, the car is not a stimulus.

Responses are not all the same. Responses depend upon the ability of the sense organs to detect the stimulus. They depend on the brain development of the animal. They also depend on how well the animal can use its muscles.

Structure of nerve cells. The cells of the nervous system are specialized. These nerve cells are of three basic types. Fig. 41-1 shows one type called a *motor nerve cell*. There are several parts to this nerve cell. The cell body contains the nucleus and cytoplasm. Extending from the cell body are short branching fibers. These are called *dendrites*. Notice that a long nerve fiber extends from the cell body also. This is the *axon*. The branching ends of the axon are the *end brush*.

Another type of nerve cell is shown in Fig. 41-2. It is a *sensory nerve cell*. It has the same parts as a motor nerve cell, but the arrangement is a little different. Notice that the dendrites of the sensory nerve cell are not attached to the cell body. They are connected directly to the axon and branch into the skin or into other *sense organs* such as the eye, ear, nose, or tongue. Each of these sense organs functions by picking up a particular stimulus from the outside environment.

We said before that nerve cells are specialized to carry messages through the body. What we have called a "message" is really a *nerve impulse*. The nerve impulse is a change in a tiny charge that flows along the nerve fiber. Salts inside of the nerve cell and in the fluid surrounding the cell cause this difference in charge. The rate at which a nerve impulse travels in us may be as slow as 3.6 kilometers per hour or as fast as 424 kilometers per hour! Its speed depends upon the size and structure of the particular nerve fiber.

Nerve cells are the longest cells known. Single nerve fibers can extend for the entire length of your arm. Fig. 41-3 shows a motor nerve cell and a sensory nerve cell with a *connecting nerve cell* between them. These three kinds of cells work together to control response.

A mass of nerve cell bodies is known as a *nerve center*. The nerve fibers leading out from a nerve center are grouped to-

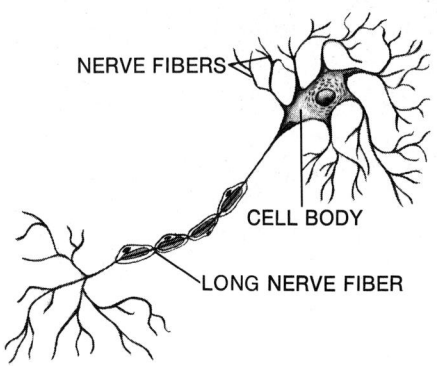

Fig. 41-1
A motor nerve cell.

Enrichment: Nerve impulses travel in a one way direction from dendrite to axon to end brush.

motor nerve cell

dendrites

axon

end brush

sensory nerve cell

connecting nerve cell

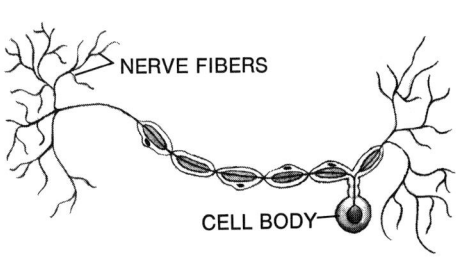

Fig. 41-2
A sensory nerve cell.

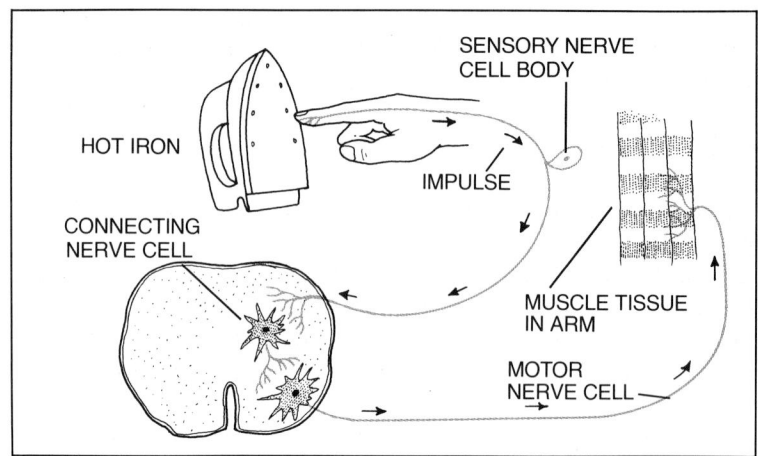

SENSORY NERVE
CELL BODY

HOT IRON

IMPULSE

CONNECTING
NERVE CELL

MUSCLE TISSUE
IN ARM

MOTOR
NERVE CELL

Fig. 41-3
A circuit made up of three kinds of nerve cells. What type of response does this represent?

nerve

Science Reading Skill: *Sequencing*— Trace the steps that take place in the nervous system during an automatic response.

reflex

gether in bundles with a protective covering around them. Such a bundle is called a *nerve*. **A nerve is not just one nerve cell. It contains many nerve fibers.**

An automatic response. Let us trace the path of a nerve impulse through the pathway shown in Fig. 41-3. Suppose your fingertip accidentally touches a hot iron. The heat of the iron is the stimulus. The dendrites of the sensory nerve cell in your skin translate the stimulus into an impulse. The nerve impulse travels through the sensory nerve to its end brushes. Then the impulse jumps onto the dendrites of the connecting cell. From this cell, the impulse jumps to the dendrites of the motor nerve cell. Next, the impulse speeds along the axon of the motor nerve cell to its end brushes. The end brushes of the motor nerve cell end in a muscle. The nerve impulse reaches the muscle and causes it to contract. The contracting muscle pulls your finger away from the iron. This all happens in a fraction of a second. Of course, there are many of each kind of cell working at the same time, but Fig. 41-3 shows only one of each kind. A quick, automatic, unlearned response like this is called a *reflex.*

Nerve cells are set up to cause a reflex in a particular way. You do not have to think about a reflex before you move. You become aware of it only after it happens. As a matter of fact when you burn your finger, you pull it back even before you feel pain. This is because a reflex takes place very quickly. It takes longer for the "pain message" to be sent to your brain and interpreted. We are not aware that anything has happened until after it has happened.

We have many reflexes, including those that blink the eyes, those that jerk hands and feet, and those that cause muscles to tighten when they are stretched. These stretch reflexes keep the muscles tense enough to hold us erect even when we are not thinking about it. The movements of muscles in the internal organs are also reflexes. Simple reflexes can be controlled by centers in the spinal cord, or in the brain.

THE CENTRAL NERVOUS SYSTEM

The brain. The central nervous system consists of the brain and the spinal cord. See Fig. 41-4. There are three main parts of the brain. See Fig. 41-5. In human beings, the *cerebrum* (suh-REE-brum) is the largest part of the brain. However, in lower vertebrates this is not the case. The cerebrum of a frog or a fish is very small. The more intelligent animals have larger cerebrums.

In humans, the outside surface of the cerebrum is very much folded. It is composed mostly of nerve cell bodies. The folding greatly increases the amount of surface area. This makes room for many more nerve cell bodies than there could be in a flat surface. Similar folding of the cerebrum surface is found in other mammals, but in humans the folding is much greater.

Cell bodies near the surface of the cerebrum have fibers that extend deeper into its structure. Some of these fibers connect with other centers in the cerebrum. A number of fibers connect with centers in other parts of the brain. Some fibers pass out of the brain to other parts of the body. Others connect centers in the cerebrum with centers in the spinal cord. The cerebrum

Activity: Demonstrate some human reflexes: knee jerk and touch sensation with a feather.

cerebrum

Fig. 41-4
(Left) The brain and spinal cord make up the central nervous system.

Fig. 41-5
(Right) The three major parts of the brain.

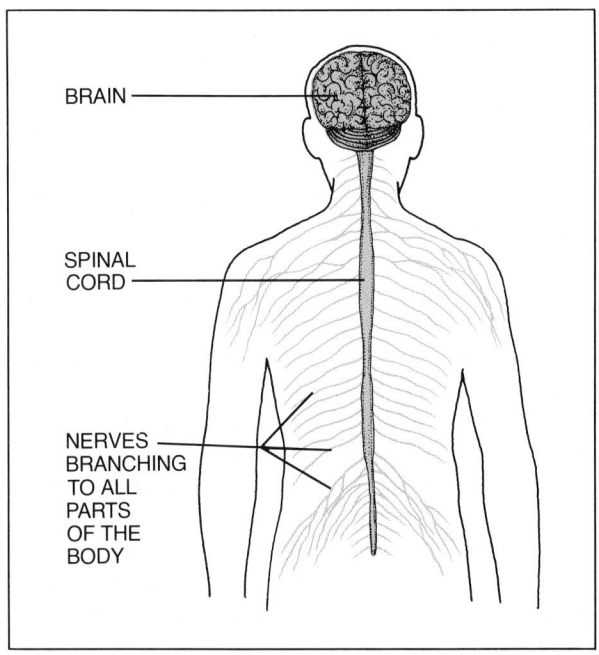

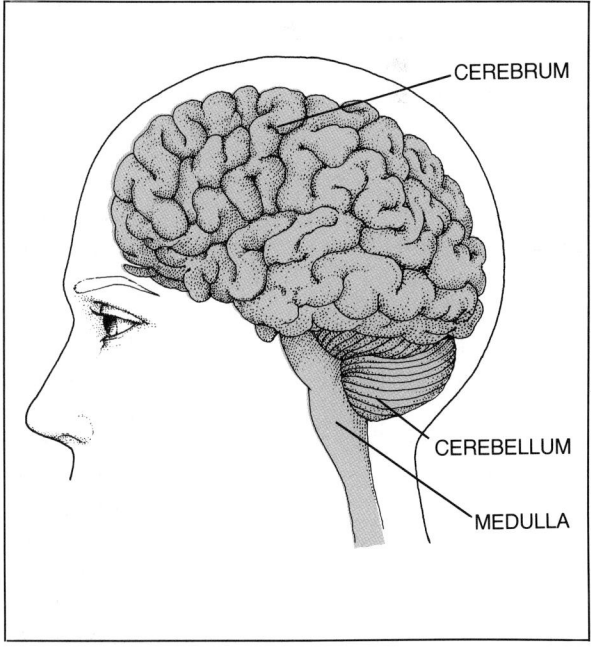

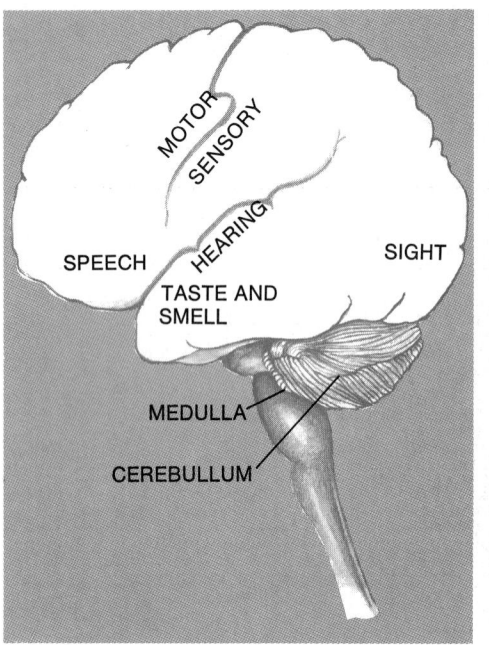

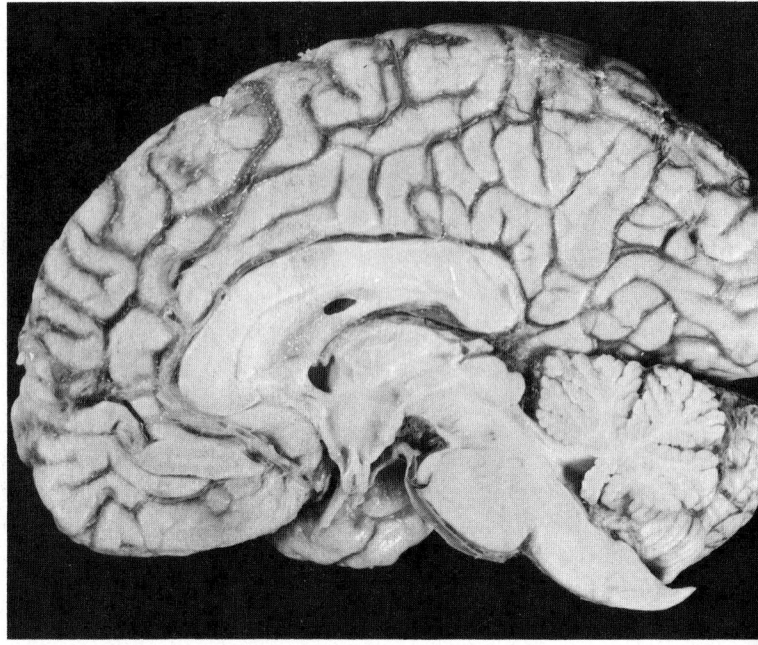

Fig. 41-6

Left: A diagram of the sensory centers of the cerebrum; Right: A photograph of a brain. Where are the sensory centers? (The American Museum of Natural History)

Enrichment: Areas of the brain have been mapped according to the function or functions they control. Students may be interested in studying more about this and what it means to modern medicine and the correction of certain disorders.

Enrichment: Recent studies have found direct biochemical evidence of damage to brain cells due to malnutrition during the few weeks before and immediately after birth.

cerebellum

is a unique structure. Its function is quite amazing. The entire cerebrum is specialized for receiving nerve impulses and putting them together in all sorts of combinations. In some way this all adds up to make what we call the human mind. Memory, learning, thinking, and conscious acts are all functions of the cerebrum.

The cerebrum is an engineering marvel that is far more efficient than any human-made computer. It is always receiving impulses from the sense organs. The cerebrum has the enormous task of sorting all of these impulses and sending them to the proper centers to be interpreted. The *sensory centers* in the cerebrum then send these impulses to the proper motor neurons for responses. Fig. 41-6 shows some of the sensory centers in the cerebrum, where there is a center that is responsible for sight. Another center controls touch. Similar areas interpret pressure and pain. There is a center for speech and another for taste and smell. The *motor centers* control muscular movement. You can understand that some centers in the cerebrum receive incoming impulses. Other centers interpret these impulses. Still others sort them out. The motor centers control the responses to be made.

The middle section of the brain is called the ***cerebellum*** (ser-uh-BELL-um). The cerebellum helps the cerebrum to control the muscles. It also helps to maintain our sense of balance. Cooperation among the muscles would be impossible without the cerebellum.

The section of the brain that is closest to the spinal cord is the *medulla* (muh-DULL-uh). The medulla controls the vital functions of the body. These are the automatic and involuntary activities, such as heartbeat, movement of the muscles of the digestive tract, breathing, and blood pressure. The medulla has reflex centers for coughing, sneezing, swallowing, and the production of saliva. It is also a pathway between the spinal cord and the rest of the brain. The medulla is necessary for life.

The spinal cord. The spinal cord lies along the back. It connects with the brain through a large hole in the base of the skull. Together, the spinal cord and the brain contain about 10 billion nerve cells organized into many nerve centers. These centers connect with one another by means of nerve fibers.

From the brain and spinal cord, nerves pass out to all parts of the body. Most nerves coming from the brain connect with parts of the head and face. Those from the spinal cord connect with all other parts of the body. Injury to the spinal cord is very serious. If the spinal cord is destroyed at the neck level, the internal organs lose their nerve connection with the brain. Death usually results. If the injury is lower, parts of the body

medulla

Fig. 41-7
Why is it so important to have qualified people carefully move a person who may have a back injury? (The American Red Cross)

may become paralyzed. This is why moving an injured person must be done so carefully. Cracked vertebrae can cut the spinal cord.

Besides the nerve centers in the brain and spinal cord, there are others in the chest cavity and abdomen. They lie on both sides of the backbone. The function of these centers is to help the centers in the brain and spinal cord control the internal organs. You may have had your "wind knocked out" at some time. This means the nerve centers that help to control breathing received a blow that caused the diaphragm to be paralyzed for a short time.

Question: Ask students how the structure and function of the nervous system is like a telephone network.

The entire organization of the nervous system is very complicated, but now you should have a general idea of what it is like. There are controlling centers in the brain, spinal cord, and elsewhere. These connect with one another and with all parts of the body. The whole system works together to receive sensation, interpret impulses, and control responses.

BEHAVIOR

Unlearned responses. Some responses are learned, and some are inborn. Some types of responses do not require a nervous system. The growth responses of green plants are controlled by hormones. Plant stems and leaves grow toward the light because of these hormones. *Paramecia* have no nervous systems yet they respond to light, chemicals, and touch. These responses are unlearned and take place because protoplasm has the ability to respond to stimuli.

Reflexes are the simplest responses controlled by nervous systems. Most movements of *Hydra* are reflexes. Their nervous system is just a loose network of nerve cells. There are no nerves or nerve centers.

Lower animals do many things by *instinct*. This is a complex pattern of behavior that is built into the nervous system. Nest building in birds is a good example of an instinct. Suppose you were to hatch a robin's egg in the house without a nest, and then raise the young bird as a pet. If you freed this robin the next spring, it would probably mate and build a regular robin's nest of mud and grass. Since the bird had never seen a nest before, we can be quite sure that its nest building was the result of instinct. When a spider spins a particular type of web, it does so by instinct. When a bee does a particular type of dance to show the location of pollen, it does so by instinct also.

Assignment: Have students report on instinctive behavior in animals.

All animals except humans have well-developed sets of instinctive responses to stimuli. Instincts are needed for survival in their natural environments. They have inherited nervous

416

systems that are all set to respond to stimuli in the environment. The instincts of animals are used to obtain food, escape from enemies, and reproduce.

Learned responses. The more highly developed an animal's cerebrum, the more it can learn. We call the ability to learn and use information *intelligence*. Of all the vertebrates, humans have the most well-developed cerebrum. Humans can learn more than any other animal.

In many vertebrates, learning is often a matter of *conditioning*. This means that an inborn reflex or instinct can be changed by learning. Wild dogs have instincts for hunting, defending the home range, and accepting leadership within the pack. Your tame dog has these same instincts, but through learning it applies them to local conditions. Your family is its "pack." Its master is the "lead dog." Barking at strangers is defending the "home range."

A *habit* is another type of learned, automatic response. You develop a habit by consciously doing the same thing over and over again until it becomes automatic. When you tie the laces of your sneakers, you are demonstrating a habit. All of the ordinary motions you make in sitting, walking, dressing, eating, and writing are done mostly by habit. A well-established habit is done without thinking. Habit formation makes our life much easier. We do not have to think and struggle to carry out some routine activity.

Of course you can also develop habits that are not helpful. Nail-biting, smoking, and gum-chewing are habits that do not improve the quality of life. Bad habits are learned through practice just as good ones are.

A great deal of the learning that we do is by *trial and error*. This means that we try to do a particular thing in several ways. We continue to use the way that gives us the best results and practice it until it becomes a habit. For example, suppose that the first time you use a certain tool you cut yourself. You know that you must hold the tool differently to avoid accidents. You try different ways of holding the tool and select the way that helps you to work most efficiently.

Memory is an amazing characteristic of the human brain. Although all higher vertebrates can remember certain things, the human brain has the best developed memory center. Think of the mass of information that we store in our brains. We remember names, faces, addresses, telephone numbers, dates, words, recipes, and so on. Interestingly enough, we can get this stored information whenever we need it.

Memory plays an important role in the ability to learn. We learn most things by consciously making an effort to remember.

Fig. 41-8
Someday this activity will become automatic. (HRW photo by Russell Dian)

You have probably noticed that some people seem to be able to learn faster than others. What makes people differ in this way? Our best guess is that learning ability depends partly on heredity and partly on the person's own will. We know that genes control the physical structure of the brain. So some people may inherit brains with better developed memory and reasoning centers. The ability to learn involves not only memory, but reasoning ability and insight as well.

Living in an environment where people are encouraged to be curious and to learn will develop the habit of learning in young children. Children who like to learn and can feel the benefit from it seem to continue learning throughout their lifetime. Not all people learn the same things at the same rate or with the same skill. Some people show special talent in music, while others learn mathematics more easily. Some people are good mechanics, while others learn languages readily. The special talents of many people are what help to build a balanced society.

SUMMARY

A change in the environment is a stimulus. The reaction to a stimulus is called a response. Our behavior is the sum total of the kinds of responses that we make to stimuli. The nervous system interprets the types of stimuli that are received by the brain and controls the responses made.

Sensory nerve cells carry the impulses to the brain or spinal cord. Motor nerve cells carry impulses to a gland or muscle. Impulses travel through nerve cells in a one-way direction.

The cerebrum is the largest part of the brain. It controls memory, learning, speech, sight, and other sensory centers. The motor centers are also located in the cerebrum. The cerebellum controls balance. The medulla controls the involuntary, vital activities such as heartbeat.

Intelligence means the ability to learn and to use information. Memory plays an important role in learning.

ACTIVITY

A Type of Human Behavior
Have one person serve as timekeeper for this activity.

A. In one minute, copy all of the sentences below. Write as quickly as you can. Do not cross any t's or dot any i's. Do not take time to make any corrections. "East Indians do not live in tents World Trade Center building. Tigers and lions are terrible animals. Yankee traders tempted the tourists with trinkets. Time and tide wait for no man. Travelers often tell thrilling tales of adventure and terrible trouble with lost travellers' checks.

Tea for Two was an old time tap dance song. The twenty-fifth knight thought it was time to retire."

1. How many t's are in all of the sentences you copied?

2. How many i's are in these sentences?

3. How many t's did you cross?

4. How many i's did you dot?

5. Why do you think you did this even though the instructions told you not to?

6. What kind of behavior is this?

Word Quiz

From the new words you learned in this chapter, write the word that correctly completes each statement. Do not write in this book.

The part of the brain that controls balance and coordination is the_____. A change in the environment is a(n)_____. A(n)_____ nerve cell branches into sense organs. A reaction to a stimulus is a(n)_____. Short branching fibers that extend from nerve cells are the_____. A group of nerve fibers is a(n)_____. The part of the brain that controls memory is the_____. Heartbeat and breathing are controlled by the part of the brain called the_____. The_____ is the long, thin fiber that extends from a nerve cell. The_____ nerve cell causes a muscle to contract. An automatic response to a stimulus is a(n)_____. The branching ends of the axon are the_____. A(n)_____ nerve cell carries the impulse from the sensory nerve cell to the motor nerve cell.

Check Your Facts

1. Explain the difference between a stimulus and a response.

2. How does a biologist define the word behavior?

3. How does the function of a sensory nerve cell differ from that of a motor nerve cell? How are the functions the same?

4. Draw a motor nerve cell. Use arrows to show the path that the impulse travels.

5. Name the three parts of the brain. What is the function of each part?

6. Where is the spinal cord? What is its function?

7. Explain what is meant by unlearned behavior. Give some examples.

Science Reading Skill
3. Compare and Contrast

8. Describe some types of learned behavior.

9. Define intelligence.

Thought Questions

1. Are most habits good or bad? Explain your answer.

2. Suppose that the cerebellum of a robin was injured. How would this affect the bird?

CHAPTER 42 THE SENSE ORGANS

In the photograph above, a deaf child is being taught to speak. Since he cannot hear sounds, he must learn to form words by seeing how his mouth should move. Hearing and sight are two of the *senses*. We could say that the child in the photograph is using his sense of sight to substitute for the sense of hearing that he does not have. In this chapter, you will learn more about how the senses work.

THE SENSES

A mother worried about her small child. He leaned his hand against a hot stove and laughed. His hand was badly burned, but he did not mind at all. Tests showed that this little boy was normal in every way but one—he had no sense of pain.

Maybe you think that this is a wonderful thing. But consider that if this child had felt pain, he would have learned not to touch a hot stove. He would not have injured himself so badly.

The ability to feel pain alerts us to dangers in the environment. Our nervous system helps us to make those responses that may allow us to avoid danger. The stimuli are received through the sense organs. Our senses are *sight, hearing, balance, smell, taste, and touch.* (This includes the ability to feel pressure, heat, cold, and pain.) The sense organs allow us to receive each kind of sensation. Sensory fibers carry impulses from the sense organs to the nerve centers. The cerebrum has sensory centers for receiving the impulses coming from each of the sense organs.

The sense of touch. The skin contains four kinds of tiny sense receptors. They are organs of touch, pressure, heat, and cold. Fig. 42-1 shows five kinds of sensory nerve endings in the skin. In each one, the nerve fibers are surrounded by a little capsule

objectives

AFTER YOU READ THIS CHAPTER, YOU SHOULD BE ABLE TO:

__**Explain** why we taste food flavors

__**Describe** the structure of the ear and tell how it functions

__**Describe** the structure of the eye and tell how it functions

Reference: See *The Sense Organs—Our Link with the World* by A. and V. Silverstein.

Enrichment: The human nervous system has approximately 3,000 taste buds, 100,000 hearing-, 130,000 light-, 250,000 cold-, and 500,000 touch-receptors.

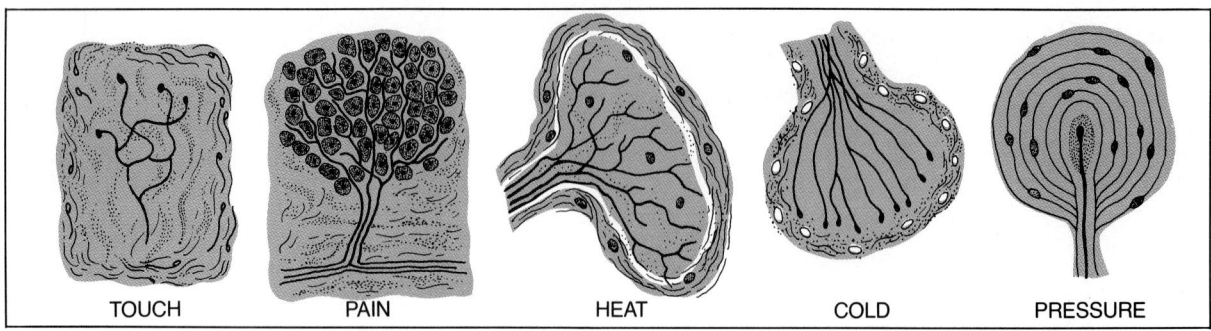

TOUCH   PAIN   HEAT   COLD   PRESSURE

Fig. 42-1
Five kinds of sensory nerve endings found in the skin.

taste bud

Fig. 42-2
A diagram of a taste bud found in the tongue.

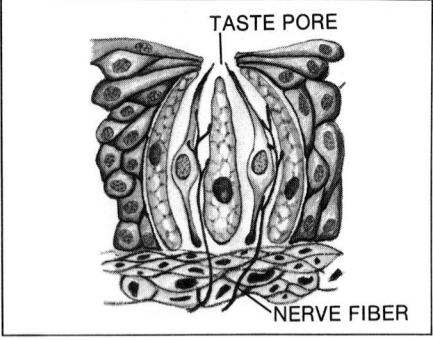

TASTE PORE

NERVE FIBER

of special cells. But there is no capsule around the nerve endings that detect pain. They are simply nerve endings in the tissue. Many of them are in the skin and other parts of the body.

The senses of smell and taste. Smell and taste detect the presence of chemicals. In the case of taste, the chemical must be dissolved in the mouth before it is sensed. In the case of smell, the chemical molecules must be in the air. These enter the nose during breathing. When liquids evaporate, their molecules enter the air. Many solids also lose a few molecules to the air.

When these molecules make contact with nerve endings in our nasal passages, we experience the sensation of smell. No one is sure just how many odors we are able to detect. It may be that there are just a few basic ones and that all of the other odors are just combinations of these.

Our sense of smell is poor when compared with other mammals. Most mammals have a very keen sense of smell. They use it to detect things at a distance. A deer can smell another animal about 100 meters away if the wind is right. Dogs can smell the ground and tell if a rabbit has passed that way. They can follow a rabbit's trail.

The sense of taste is due to the work of special cells in the mouth. These cells come in clusters, and each little cluster is contained in a tiny pore in the tongue and the mouth lining. These taste cells are not nerve cells, but the ends of nerve fibers connect with them. The whole structure is called a *taste bud*. See Fig. 42-2. Certain chemicals cause the cells in the taste buds to send nerve impulses to the brain. There these impulses register taste.

There are only four sensations that can be detected by taste buds. They are sweet, sour, bitter, and salty. Of course more than one may be present at once, giving a combination of taste sensations. Much of what we think is the taste of foods is actually odor. We smell the food while we chew it. Smell and taste blend together to give the complete taste.

The sense of hearing. Sound is caused by the rapid vibration of matter. Sound travels through air, water, steel, rock, and many other materials. Our ears are adapted to hear sound in the air. When you slap the top of a table, air molecules in the area are disturbed. They hit other molecules of air that in turn hit more molecules. This disturbance of air molecules moves outward in all directions in what is called a *sound wave*. People hear sound because a membrane in their ears is made to vibrate by the vibrating sound wave.

The human ear is divided into three regions. These are the *outer ear*, the *middle ear,* and the *inner ear.* See Fig. 42-3. The outer ear includes the part of the ear that can be seen on the outside and also the passageway into the head. A thin membrane stretches across the inner end of this passageway. It is called the **eardrum.** The hollow space behind the eardrum is the middle ear. The inner ear is made of curving passages of hollow, tube-like bone. It lies in a bone of the skull in back of the middle ear.

The first part of the ear to respond to sound waves is the eardrum. The sound waves bounce against it, making the eardrum itself vibrate. The eardrum is connected to the inner ear by way of three tiny bones that are hinged together. They reach across the middle ear and carry the vibration to the inner ear.

eardrum

Fig. 42-3
The parts of the ear.

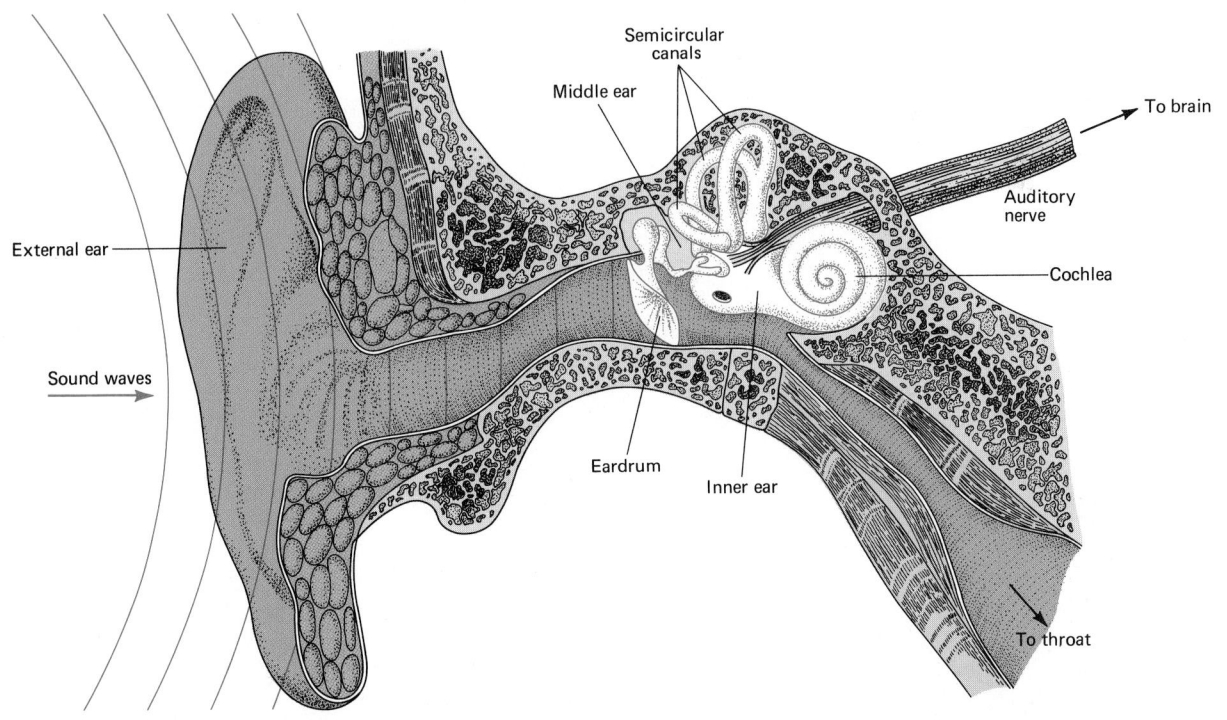

423

cochlea

auditory nerve

The hearing part of the inner ear is coiled and looks like a snail shell. This is called the *cochlea* (KOE-klee-uh). It is filled with fluid. The last of the middle ear bones vibrates against an opening in the wall of the cochlea. This causes the liquid inside the cochlea to vibrate. A row of special cells lies inside the cochlea. Sensory nerve cells connect with these special cells. Sensory nerve fibers connect with the brain by way of the *auditory* (AWD-uh-tore-ee) *nerve*.

The pitch of a sound depends upon the rate at which matter vibrates in the air. A slow rate of vibration gives a low pitch. A rapid rate gives a high pitch. Each special cell in the cochlea responds to a different pitch. Each reports to the brain by way of its connecting nerve fiber, which passes along the auditory nerve. Most sounds stimulate several of these nerve fiber endings at once. A musical note, however, has a fixed vibration rate, so we hear it as a definite pitch.

The pressure of the air around us is not the same at all times. This could cause trouble for the eardrum if the ear had no way of adjusting to the changes. High pressure would push the eardrum in. Low pressure would allow it to bulge out. Damage to the eardrum could result. This is prevented by a passageway that connects the middle ear with the top of the throat, just behind the nasal passages. Whenever you swallow, yawn, or shout, the end of this passageway opens allowing the outside and inside pressures to become equal.

Care of the ears. Hearing can be damaged by too much loud noise. Constant exposure to loud, highly-amplified music can make a person hard of hearing. People living near airports complain about the excessive noise of jet planes. Very loud, penetrating noises can puncture the eardrum. People exposed to constant loud noise, such as rock musicians, should protect their ears with ear plugs.

Ear infections can damage hearing. Bacteria or viruses present in the throat during a cold may pass up into the middle ear. When this happens, the germs grow in the middle ear and produce pressure on the eardrum, causing an earache. Ear infections can cause permanent scarring of the eardrums and make the person hard of hearing. It is important that a doctor treat ear infections at the first sign of earache.

The sense of balance. Besides the cochlea, the inner ear also contains a small, round, hollow space and three tubes that curve around in half circles. These are known as the *semicircular canals*. These structures are not involved in hearing. They are organs of balance. The round hollow space is lined with sensitive cells. In the middle are small pebble-like particles of

semicircular canals

Fig. 42-4
What structures inside the ear help us to keep our balance? What would happen if these structures were not functioning?   (Moose-Hake/Greenberg—Peter Arnold, Inc.)

mineral. The weight of these mineral particles causes them to press down on the sensory cells. Nerve impulses from these cells make it possible for the brain to tell if the body is upright.

The semicircular canals give us the sensation of movement. They are filled with liquid and they have nerve endings at each end. When the head is turned, liquid rushes to one end of the canal, pressing on the nerve endings. Turning the other way presses the liquid against the opposite nerve endings. The three canals are laid out in the three directions the head can move: sideways, nodding, or turning. No matter which way the head is turned, one or more nerve endings will be stimulated. The brain sends impulses to the muscles so that a person can walk upright without stumbling.

You also have the ability to know the position of your appendages without looking. If you hold your arm out sideways, you can feel its position even if you have your eyes closed. This ability to tell your own position in space is very useful. It is needed in order to move accurately when walking, jumping, or reaching out for things.

THE SENSE OF SIGHT

Sight is regulated by the eyes. The eyes are organs that detect light. Light reflects from all the objects around you. This reflected light enters your eyes and enables you to see those objects. Most animals can sense light. An earthworm senses light with its body surface. The eyespots of *Planaria* are light sensitive. Protists like *Ameba* and *Paramecium* also respond to light. None of these organisms can see objects. They are only able to distinguish light from dark.

Only three groups of animals have eyes that can make out the shapes of things. These animals are the squid class of mollusks, the arthropods, and the vertebrates. The arthropods have compound eyes. The eyes of the squid and those of the vertebrates are very much alike. In both types of eyes there is a *lens* to form an *image*, or picture, upon a **retina** (RET-un-uh).

The retina is a layer of nerve cells that are sensitive to light. Fig. 42-5 shows a diagram of the human eye. If you study it, you will see that the retina lines the inside of the eyeball at the back. The only way light can reach the retina is by coming in through a curved "window" in the front of the eye. This structure is the **cornea** (CORE-nee-uh).

The cornea is a clear membrane. Its surface is curved like a magnifying glass. The curved surface of the cornea is a transparent covering of the eye. Inside the eye is a transparent lens. This lens works with the cornea to form an image on the retina.

The nerve cells of the retina connect with the brain by way of the **optic nerve**. Each cell of the retina transmits information about the lightness or darkness of the particular part of the image that falls upon that cell. In the brain, these thousands of bits of information are combined and the people become conscious of what they are seeing. Blindness can result from damage to the eye, the optic nerve, or the sight centers in the brain. Each of them has its part in the process of seeing an object.

A lens cannot form a clear image of near and far objects at the same time. In a camera, the lens is moved closer to the film to photograph distant objects and farther from it for nearby objects. This is called *focusing* the camera. The eye focuses in a different way. It changes the shape of the lens. For seeing nearby objects the lens becomes thicker and more curved. For distant objects it flattens out into a less curved form. There is a ring of muscle fibers around the lens that does this focusing. When these muscles are at rest, the eye is focused for distance. When they contract, it focuses on nearby objects. This is the real function of the lens in the eye—to focus on nearby objects.

Just in front of the lens is a partition with a hole in the middle. This partition is the **iris** (eye-rus), the colored part of the eye.

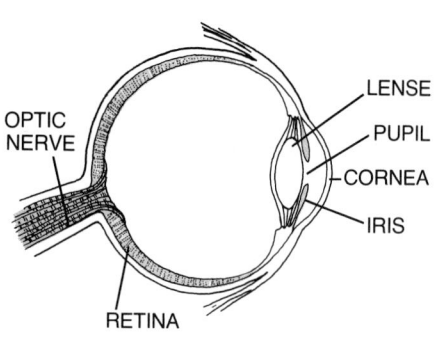

Fig. 42-5
The parts of the eye.

retina

cornea

optic nerve

iris

426

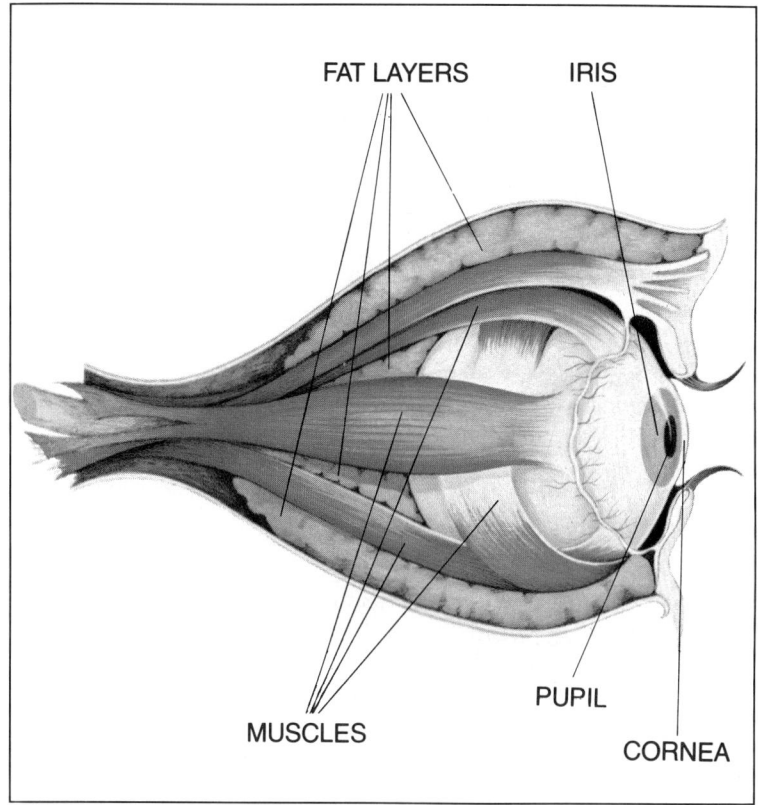

FAT LAYERS   IRIS

MUSCLES

PUPIL

CORNEA

Fig. 42-6
The eyeball is moved by several muscles.

The hole in the iris is the *pupil*. The iris is muscular and it controls the size of the pupil. In this way, the amount of light entering the eye is regulated. In dim light, muscles of the iris open the pupil wide, allowing more light to enter. In bright light, the pupil becomes small. This protects the sensitive retina from contact with too much light. The opening and closing of the pupil is a reflex act.

The inside of the eye is filled with a clear jelly. It is under enough pressure to stretch the tough outer covering and to hold the eyeball in shape. The bones of the skull protect the eye on all sides. The eyelids blink shut to protect the front. They also spread a fluid across the surface. This fluid comes from the tear gland under the upper edge of the eye socket. It washes out dirt and protects the eye against germs.

The eyes are moved in their sockets by six small muscles attached to the outside of each eyeball. See Fig. 42-6. These muscles keep both eyes looking at the same object. Although each of our eyes forms an image, our brain puts the two images together. Therefore we see only one picture. The slight difference between the two images is interpreted by the brain in a special way. It tells us whether an object is nearby or far away.

pupil

427

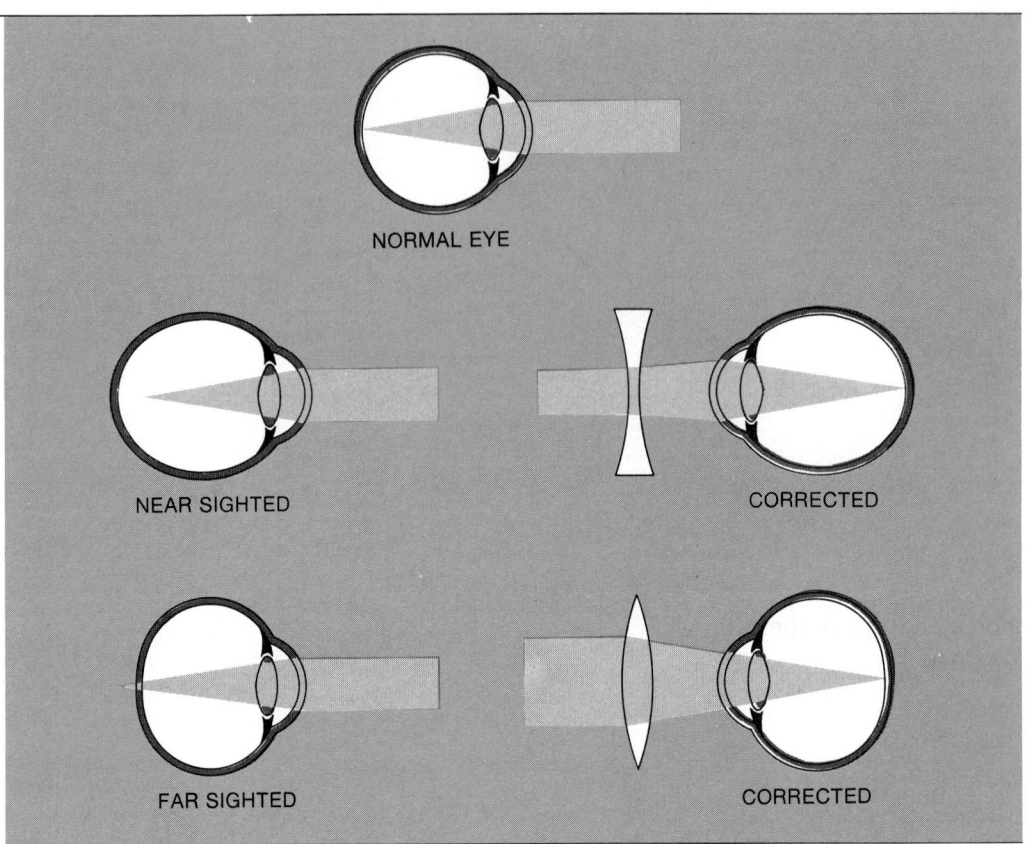

NORMAL EYE

NEAR SIGHTED

CORRECTED

FAR SIGHTED

CORRECTED

Fig. 42-7
Different types of eyeglass lenses can correct nearsightedness and farsightedness.

Careers: Discuss occupations in the field of eye care such as optometric technician, optometrist, optician, and ophthalmologist.

Assignment: Have students report on eye disorders such as cataracts, glaucoma, corneal scarring, and detached retina.

Defects of the eye. The shape of the eyeball causes nearsightedness and farsightedness. *Nearsightedness* is a condition in which there is too great a distance between the lens and the retina. When people who are nearsighted look at a distant object, the image comes to a focus in front of the retina instead of on it. The light reaching the retina is out of focus, and the person sees a fuzzy image. This condition is helped by wearing glasses with *concave* lenses in them. These lenses are thinner in the center than at the edges. They spread out the light before it reaches the eye. Then the light rays take longer to come into focus. They now produce a clear image. See Fig. 42-7.

Farsightedness is just the opposite. The distance between the lens and the retina is too short. The image is not yet fully formed when it reaches the retina. Such eyes cannot see nearby objects clearly. They are helped by *convex* lenses, which are thicker in the center than at the edges. These lenses bring the rays of light closer together before they reach the eye. See Fig. 42-6.

Astigmatism (uh-STIG-muh-tiz-um) is a condition in which the eyeball is a little out of shape in such a way that it forms an image that is out of shape, also. This makes small things hard to recognize at a distance, and the person may think he is nearsighted. If one eye has a different amount of astigmatism from the other, there is a feeling of eyestrain. Headaches may result. To correct astigmatism, we use lenses with a lopsided curve that is just right to correct the out-of-shape image produced by the eye.

SUMMARY

Our senses keep us in touch with the outside environment. The sense organs are specialized to receive certain stimuli.

Sensory nerve cells change these stimuli to nerve impulses that are taken to the sensory centers in the brain for interpretation. We feel pressure, pain, heat, and cold because of the special nerve endings in the skin.

Special senses in our muscles orient us in space. We experience taste and odor because of the smell receptors in the nose and the taste buds on the tongue.

The ear controls hearing and balance. The eye receives light and forms images. It also enables us to judge distances. These senses aid our survival.

ACTIVITY

Mapping Your Taste Buds
Buried in your tongue are taste buds that pick up the taste sensations of sweet, salty, sour, and bitter. The specific taste buds are not scattered through the tongue tissue but are located in specific areas. You are going to find the location of the centers for each of the taste sensations.

A. Draw a diagram that reasonably resembles the shape of your tongue. As you find the locations of the taste buds, mark them on the "tongue map."

B. Dip a wet cotton swab in some sugar. Touch various places on your tongue with it until you find the sweet-detecting taste buds. Mark your map. Use a fresh cotton swab and repeat for each taste sensation. Use salt to locate salt-detecting taste buds, lemon juice for sour-detecting taste buds, and quinine (tonic) water for bitter-detecting taste buds.

1. Where is each region for taste sensation located?

2. Why are we not aware of the separation of taste regions when we eat?

Word Quiz

From the new words you learned in this chapter, write the word that correctly completes each statement. Do not write in this book.

The nerve of sight is called the_____ nerve. The nerve of hearing is the_____ nerve. The colored part of the eye is the_____. Taste receptors in the tongue are the_____. A clear membrane in the eye is the_____. The coiled part of the inner ear is the_____. A thin membrane that separates the outer ear from the middle ear is the_____. Images form on the_____ of the eye. The structures in the inner ear that control balance are the_____. The hole in the colored part of the eye is the_____.

Check Your Facts

Science Reading
Skill
*5. Reading
Illustrations*

1. What kinds of sense receptors are embedded in the skin? Of what use are these?

2. What kinds of taste sensations can we detect?

3. Name the parts of the ear. What does each part do?

4. Name the parts of the eye. What is the function of each part?

5. Look at Fig. 42-5. Explain the causes of nearsightedness and farsightedness. How is each corrected?

Thought Questions

Science Reading
Skill
*2. Problem
Solving*

1. Why are sense organs connected with the cerebrum?

2. Certain "blind fish" have been found in some caves. Explain why these fish never develop normal eyesight.

CHAPTER 43 Disease

On a modern dairy farm, a great deal of attention is paid to keeping milk clean. Clean cows are milked by clean milking machines. The milk travels through pipes to a milk room where it is cooled, then heated, and cooled again. This process is used to destroy disease-producing bacteria. In this chapter, you will learn how disease is caused and what we can do to prevent it.

CAUSES OF DISEASE

Types of diseases. There are three general types of diseases. You have learned about *deficiency diseases*. You may remember that these are caused by the lack of a particular nutrient, usually a vitamin, in the diet. There are also the *functional diseases* that occur when some body part does not work. Heart trouble, kidney ailments, and diabetes are examples. *Cancer* is the too rapid growth of cells. It is sometimes considered a functional disease because it interferes with the function of body organs. But, until the actual cause of cancer is known, it cannot definitely be put into any one group.

Finally, there are the *infectious* (in-FEK-shus) *diseases*. An infectious disease is caused by certain germs entering the body. These germs can be any of several different types of organisms.

Kinds of germs. Most diseases are caused by parasites that are too small to be seen by the naked eye. We need a microscope to see them. Most of these small organisms belong to various groups of single-celled organisms. Parasites that cause disease are known as germs. They include forms from the following groups.

Viruses are a type of germ organism, although they are not really cells. In order to reproduce, a virus must be inside a living cell. Some viruses cause diseases in humans, such as chicken pox, rabies, influenza, mumps, and polio.

objectives

AfTER you READ THiS chAPTER, you should bE AblE To:

___**List** three types of diseases and their causes

___**Explain** how disease is spread

___**Describe** methods of disease prevention

___**Explain** how the body defends against diseases

functional diseases

infectious diseases

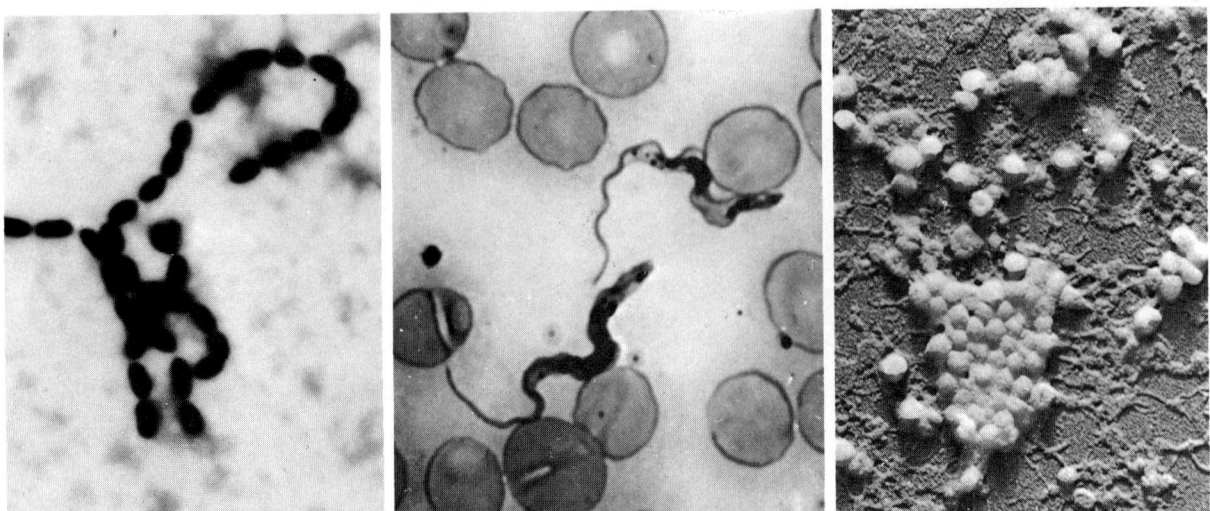

Fig. 43-1

Some agents that cause disease. From left to right: influenza virus, Type B, the protozoan that causes African sleeping sickness, the bacteria that cause pneumonia. (Lederle Laboratories; Eric V. Grave; Lederle Laboratories)

Some *bacteria* enter the body and cause disease. They may attack and destroy body tissues. This happens in tuberculosis. In other cases, bacteria give off poisons that make people sick, as in diphtheria. Scarlet fever, typhoid (TIE-foid) fever, most pneumonias, and many other diseases are caused by bacteria.

Rickettsias (rik-ET-see-uhs) are smaller than bacteria, but are like bacteria in some ways. They carry on more of the chemical activities of protoplasm than a virus can, but they do not totally function like regular cells. Rickettsias, like viruses, can multiply only in the cells of a host. Rocky Mountain spotted fever, typhoid fever, and Q fever are some diseases caused by rickettsias.

Protozoa do not cause as many diseases as the bacteria and the viruses. The diseases that they do cause are quite troublesome. These include malaria, African sleeping sickness, amebic dysentery (DIS-un-ter-ee), and several others.

Certain species of *fungi* can infect humans. There are yeasts that attack the lungs and other tissues. Athlete's foot is caused by a mold infection. Ringworm and junglerot are examples of other fungus diseases. There are also mold-like fungi that affect the ears and the lungs.

METHODS OF SPREADING DISEASE

Food and water. In general, when germs leave one host they must quickly enter another host because they cannot live very long outside the body. There are exceptions to this. For instance, the bacteria that cause typhoid fever can live outside the body for months. Typhoid fever is a disease of the intestines,

Science Reading Skill: Cause and Effect—Describe the ways that disease can be spread.

and the bacteria are present in the feces of a person who has the disease. If sewage is not properly disposed of, some of these germs may enter the water of wells, lakes, or streams. Anyone drinking such water may get typhoid. Swimming in polluted water or eating raw clams or oysters that grew in this water can also spread disease. Cholera and dysentery are two other diseases that are spread in water. We have almost eliminated these diseases here. If you travel to less sanitary parts of the world, you may need to get an injection to prevent typhoid and cholera. When in less sanitary places, do not eat raw food. You should also boil the water before drinking it.

Food as well as water can be a carrier of disease-producing organisms. The same forms of disease that are carried in water may also be present in foods. Vegetables grown on ground polluted by sewage often carry parasites such as *Ascaris*, and the germs of cholera, dysentery, and typhoid. This is most common in countries where human sewage is left on the surface of the ground. In the Far East, this sewage is used as fertilizer on the fields.

Bacteria may actually live and multiply in food. For example, many kinds of germs can grow in milk. Custards, which are made with milk, also serve as food for bacteria. Such foods must be produced under clean conditions to keep germs out. Refrigeration is used to keep foods cold so that bacteria do not grow rapidly in them.

Some diseases are spread by arthropods. Some germs are carried by bloodsucking arthropod parasites. The parasite takes in the germ when it sucks the blood of a sick person. Later, when it bites another person, the germ is passed on to that

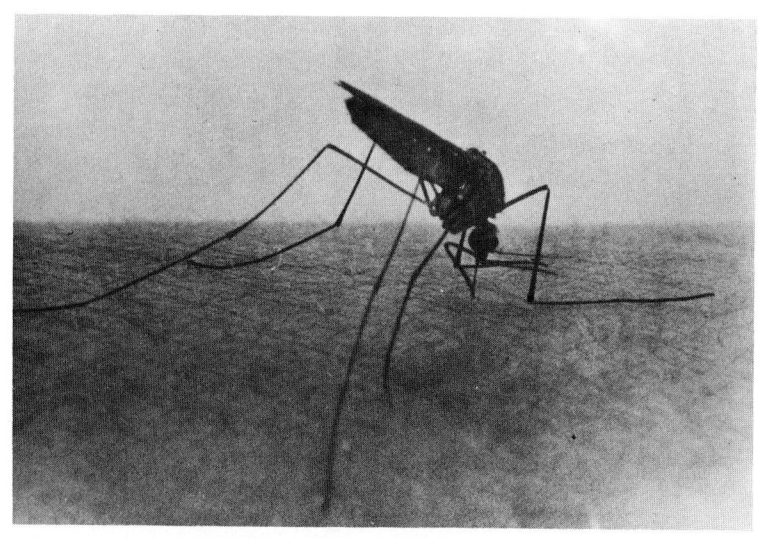

Fig. 43-2
How can this arthropod spread disease? (USDA)

person. Malaria and yellow fever are carried in this way by mosquitoes. African sleeping sickness is carried by the *tsetse* fly. Typhus fever is spread by lice, ticks, mites, and fleas.

Plague is spread by rat fleas. You may have read about the "black death" in your history books. It has killed as many as half the people in a country in a single year. The disease is carried by rodents, especially rats. Fleas then carry the germ from rats to people. Modern, sanitary ways of living have made this disease much less common. Also, the brown rat has replaced the old roof rat that more often carried the plague. The germ, however, is still present in some parts of our land. A few hunters get plague from wild rodents each year.

Some insects spread germs without biting anyone. The common housefly is an example. Flies breed in garbage and in damp manure. The larvae use these wastes as food. When the larvae become adults, they fly from their breeding places into people's houses and walk on the food. The harmful germs on their bodies are transferred to the food. Typhoid and dysentery germs are carried on the bodies of these flies. It is believed that cockroaches carry germs in a similar way. They hide in sewers, drains, and behind baseboards during the day. At night they come out and feed on human food. The bacteria that they tracked out of the sewers are left on the food for people to eat the next day.

Wound infection. Some types of germs enter the body through breaks in the skin and multiply in the tissue around the wound. These wound infections can be very dangerous. Deep, narrow wounds are especially bad. They are difficult to clean and air does not go down into them. Pus-forming bacteria, *gangrene*-producing bacteria, and the bacteria causing lockjaw (tetanus) live in deep wounds that are not exposed to air. All of these bacteria are so harmful that they can kill an individual.

Airborne infection. In many diseases, the germs are present in the saliva of the sick person at some time during an illness. These diseases spread through the air. Any time a person talks, coughs, or sneezes, tiny drops of saliva are sprayed into the air. The germs in these drops may live long enough to be breathed in by another person. All of the common respiratory diseases are spread in this way. Diseases such as colds, tuberculosis, and influenza travel from person to person by means of small drops of saliva. This is called *droplet infection.*

droplet infection

It is not possible to avoid these airborne germs. Any time people meet and talk they trade germs. A person's good nutrition and general good health help to ward off infection. Tuberculosis used to cause many deaths. It is an airborne disease that may attack any part of the body. It most often settles in the

Fig. 43-3
How is this person spreading infection?
(HRW photo by Russell Dian)

lungs. Tuberculosis is still a health problem. There are a good many cases, but not many deaths occur. A number of new drugs makes it possible to treat people at home.

In the past, many children died from diphtheria every year. Diphtheria germs grow in the throat and can cause it to swell so much that it almost shuts. These germs also cause poisons to spread throughout the body and cause death. Some types of pneumonia are caused by bacteria, others by viruses. These organisms are carried through the air and spread by droplet infection, also.

DISEASE PREVENTION

The body's defenses. The body is protected in many ways against infection by bacteria and viruses. The skin is a barrier against germs. As long as it remains unbroken, most infectious organisms are kept out. The linings of the mouth, throat, and intestines form protective barriers against germs. These linings are sticky mucous membranes that are usually able to trap bacteria and viruses. Some germs are destroyed by the acid secretions of your stomach. These are the germs that slip past the linings of the nose and throat. Germs that manage to enter the body tissues are destroyed by white blood cells. These cells surround invading bacteria and prevent them from spreading throughout the body.

The body produces several kinds of protein molecules to fight germs. Some destroy viruses. Some kill bacteria and others destroy the poisons produced by bacteria. Germ-fighting proteins are found in the blood, lymph, tears, saliva, and in the body cells themselves. *Antibodies* make up one of the group of

References: See *Wonder Drugs: A History of Antibiotics* by H. Boettcher; *Viruses and Colds: The Modern Plague* by J. Adams; *Famous Firsts in Medicine* by G. and C. Crook; and *Trial by Fury: The Polio Vaccine Controversy* by A. Klein.

antibodies

germ-fighting proteins. Antibodies are produced in the blood in response to harmful bacteria or viruses that enter the body. Each antibody acts against one particular type of infectious organism or the poison that it produces. For example, the antibody that destroys the smallpox virus has no effect upon the virus of mumps or that of chicken pox. It will only work against smallpox.

Our bodies cannot produce antibodies unless the disease-producing organism is present in the body. When germs enter the blood, antibodies begin to be produced in the lymph nodes. This often takes several days. During that time, the white blood cells may not be able to keep all of the germs from growing and reproducing. The person then becomes sick. If the germs grow and reproduce too rapidly, they can do much damage and the person may die. The antibodies are usually produced soon enough to save an individual. When the antibodies and white blood cells destroy all of the germs, the person becomes well.

immune

After an illness, the antibodies may continue to exist in the blood for some time. During this time the person will be *immune* to the disease. To be immune means that the person cannot become ill with the disease again, even if the germs are present. This immunity lasts only a short time for some diseases, such as the common cold. In other cases it may last for years or even for life, such as with chicken pox. Each time the germ enters the body more of the antibody is made. This strengthens the immunity.

Heredity has an important influence upon disease resistance. Some people seem to be naturally immune to certain viruses. The same virus will make some people mildly ill, and still other people very ill. These differences seem to be inherited. Whole populations may differ in disease resistance. Smallpox was a serious disease for Europeans who settled in America. It was an even worse disease for the Indians. Entire tribes were wiped out. Jewish people seem to have more than average resistance to tuberculosis. Eskimos are unusually sensitive to it. Can you explain how natural selection might account for such differences in populations?

Enrichment: The importance of immunization cannot be stressed too frequently. Point out to students that in 1969, for example, an outbreak of German measles was responsible for 30,000 miscarriages and the birth of 20,000 infants with congenital defects.

Immunization. In the past, many people became infected with the smallpox virus. Smallpox is a terrible disease. It makes a person break out with foul-smelling blisters all over the body. When the blisters heal, they leave pitted scars. Smallpox makes a person very sick and, in many cases, those infected do not live. Today nobody gets smallpox. A worldwide vaccination program has eliminated this disease from the whole earth.

A doctor can make you immune to smallpox and certain other diseases by using something that causes the body to produce

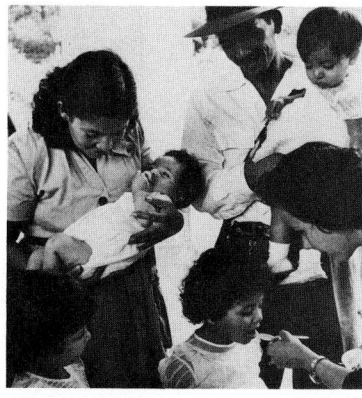 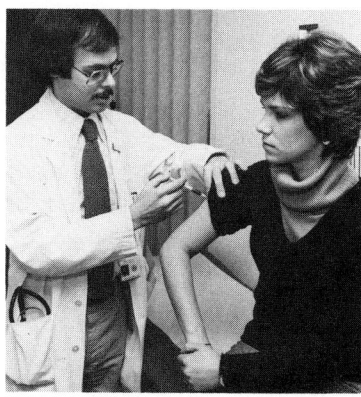

Fig. 43-4
Two different methods of receiving vaccine. Can you explain these? (WHO photo by H. Page; HRW photo by Ken Karp)

the right kind of antibodies. A process called ***vaccination*** (vak-sin-NAY-shun) is used against smallpox. The virus of a similar disease called *cowpox* is scratched into the skin. This cowpox virus is not able to make a person really sick. It just forms a sore in the skin at the place of the scratch. The body forms an antibody to kill the cowpox virus. This same antibody makes the person immune to smallpox. It does this because the two viruses are so much alike. The same antibody will kill both of them.

The use of cowpox to prevent smallpox was discovered by an Englishman name Edward Jenner. He noticed that people whose work was milking cows did not get smallpox. He reasoned that the mild illness called cowpox, which they got from cows, made them immune to smallpox. Just as he had thought, when he gave people cases of cowpox on purpose they became immune to smallpox.

A substance such as the one Jenner used is called a ***vaccine*** (VAK-seen). Other vaccines are made by weakening or killing the germ with chemicals. A mutated form of the germ is often used. In still other vaccines, a weakened form of the germ's poison is used. Some of these treatments immunize for only a short time. Some are good for several years or for life.

Doctors immunize babies against several diseases during their first year of life. These include tetanus, whooping cough, measles, diphtheria, and polio. At one time babies were also immunized against smallpox. Since the disease has been wiped out in all countries of the world, smallpox vaccinations are no longer given.

Besides the immunizations just mentioned, people should have others if they are going to travel in countries where sanitary conditions are poor. What they need depends on where they go. They are likely to need the typhoid vaccine. They may also need protection against cholera, typhoid fever, and plague. Before traveling in the tropics, yellow fever shots should be

vaccination

Assignment: Have students research the discoveries made by Reed, Pasteur, and Koch.

vaccine

Enrichment: A schedule of immunizations should be presented and students should be urged to check on their own immunization history. Some states now require a complete immunization history to be updated yearly for entrance to school.

taken. You must remember that being immune to one disease does not give you immunity to another.

Clean water. Sanitation is a very practical way of preventing disease. Typhoid fever has become a rare disease in this part of the world. We have brought this about by preventing the spread of the disease germ.

The best way to prevent the spread of disease germs that live in water is to provide a pure water supply. Most often a city depends upon a nearby river for its water supply. The river water may be polluted. This dirty water must be purified before it can be used. This is how it is done:

River water is allowed to lie quietly in tanks that are much larger than swimming pools. This is called *settling* because mud settles to the bottom. The water is then sent to other tanks where chemicals are added. These form sticky flakes that settle to the bottom and take fine mud particles with them. The water that now looks clear, is allowed to soak down through layers of clean sand and gravel. These layers are several feet thick. They act like a huge sieve, which filters out any solid particles that may be left in the water. Most of the germs are filtered out. A small amount of liquid chlorine may be mixed with the water. The chlorine kills any germs that have not been filtered out. Sometimes the water is then sprayed into the air by many nozzles that look like lawn sprinklers. This is called *aeration* (air-AY-shun) and it is used to improve the taste of the water by dissolving air in it. Oxygen from the air may also kill some types of germs.

Fig. 43-5
Why would it not be a good idea to drink directly from this body of water? (HRW photo by Ken Karp)

438

Many cities still dump their sewage into lakes or rivers. Since many of the most dangerous waterborne germs come from sewage, it is important to treat sewage before disposing of it. Many cities now have modern sewage disposal plants.

Protection of food. You know that food can carry disease germs. One of the important jobs of public health departments is to see that this does not happen. In most communities, milk cannot be sold to the public unless it has been produced in clean barns and handled in a clean manner in milk houses and dairies. Before it is put into bottles, milk must be *pasteurized* (PASS-ture-ized).

In one pasteurization process the milk is heated, but not boiled, and then it is suddenly chilled. This kills any dangerous germs the milk might contain. It also kills many of the harmless bacteria that turn milk sour. Pasteurized milk is not only safer to drink, but it also keeps longer than unpasteurized milk. The process has little effect upon flavor or food value.

Public eating places are also inspected by health departments. They see that restaurant owners obey the laws that require cleanliness in preparing food and washing dishes. People who work as food handlers, cooks, waiters, and waitresses, must be checked by public health doctors. It is important that food handlers are not carriers of typhoid or tuberculosis.

Insect control. As you know, several diseases are carried by insects. Different kinds of insects have different habits, so we must fight them in various ways.

Mosquitoes have been one of the biggest problems. Mosquito larvae, called "wrigglers," live in water. Even the water in old tin cans and leaf-clogged gutters on houses can produce certain kinds of dangerous mosquitoes.

Diseases such as malaria, *encephalitis* (in-sef-uh-lite-US), and yellow fever are carried by mosquitoes. Malaria has been the most common human disease. This is hard to realize if you live north of the malaria regions. However, in warmer parts of the world, large numbers of people still have this disease. Some of these people even die from it. The germs of malaria are protozoans and they are carried by certain kinds of mosquitoes. Encephalitis is a disease caused by a virus that can result in coma and death. There is hardly a region in the United States that is entirely free of encephalitis-carrying mosquitoes. Yellow fever is a tropical disease, but it has been known to occur as far north as Philadelphia. People of some tropical races have a considerable amount of natural resistance to yellow fever. Among northern races, two out of five people who get yellow fever die from this disease.

Activity: Have students heat some raw milk to pasteurization temperature (71.5°C for 15 sec). Place in a covered container in the refrigerator. Compare daily with a sample of raw, unheated milk.

439

To control the disease-carrying mosquitoes, attention must be paid to their breeding places. We can, of course, drain the ponds, lakes, and marshes where wrigglers live. But usually this is not practical or desirable. These wet lands often have value in flood control, recreation, or wildlife conservation. Draining wet lands near cities is sometimes done in special cases. Oil may be spread on the surface of water. This keeps the breathing tubes of the wrigglers from breaking through the surface. They get no oxygen and soon die.

The most successful attack on the malaria mosquito has been to spray poisons around buildings where people live. This kills the mosquitoes most likely to bite people. Poisons that break down quickly are most often used. Homes should be protected with screens. People who work or picnic out-of-doors should use insect repellants on their skin to keep from being bitten.

The breeding places of houseflies, which also carry germs, should be cleaned up. In cities, garbage should be wrapped and placed in containers that flies cannot enter. Then the garbage can be disposed of by burning. On farms, manure should be hauled out and spread on the fields frequently. Housefly maggots grow more in manure piles than in manure that has been spread out on fields.

Fig. 43-6
Why shouldn't the objects in this photograph be left around where water can collect in them?

Controlling airborne infections. As you have learned, it is not possible to avoid all airborne germs. Any person you meet may carry the virus of colds or influenza. But you can avoid having too many germs at any one time. Your body may be able to defend itself against a small number of germs. In fact, getting a few germs now and then may cause the body to produce more antibodies.

When too many germs enter the system at once, the body's natural defenses may be overwhelmed. There may not be enough white cells and antibodies to destroy the germs. When this happens, a person becomes sick.

People should wash their hands before eating or handling food. Persons with colds should cover their noses and mouths before sneezing. The germs that are sent out into the air infect others.

Care of wound infections. Whenever you get a small cut or scrape your skin, you should clean the wound carefully with soap and water. This will remove the dirt which might contain dangerous germs. The cut will heal best if it is left open to the air. The wound may be covered with a loose, porous bandage. This lets in air while keeping out dirt.

Any serious injuries should be treated by a doctor. A doctor is the only one who can clean out a deep wound. At times antibiotic drugs are prescribed to fight the germs of wound infection. You should keep your tetanus inoculations up to date. In this way tetanus infection can be avoided if you are injured.

WHEN YOU ARE SICK

So far, we have explained what disease is and how we prevent it. Sometimes prevention fails and we get sick. What can we do then?

If you do get sick, it is a good idea to go to bed for a while. Resting in bed gives your body the best chance to fight back against germs. Most of the serious infectious diseases cause a person's body temperature to rise. This is called a *fever*. Each family should have a thermometer. This permits you to measure body temperature. Conditions of fever should be reported to the doctor.

Contacting the doctor early in an illness is important. Many diseases are much easier to treat early than they are after they have become worse. Taking medicines at home without a doctor's advice is not a good idea. You know very little about medicines and illness. You are likely to take the wrong drug.

Enrichment: Many individuals have allergic reactions to certain antibiotics. These people should carry identity cards with them at all times to warn doctors in case emergency treatment is needed.

441

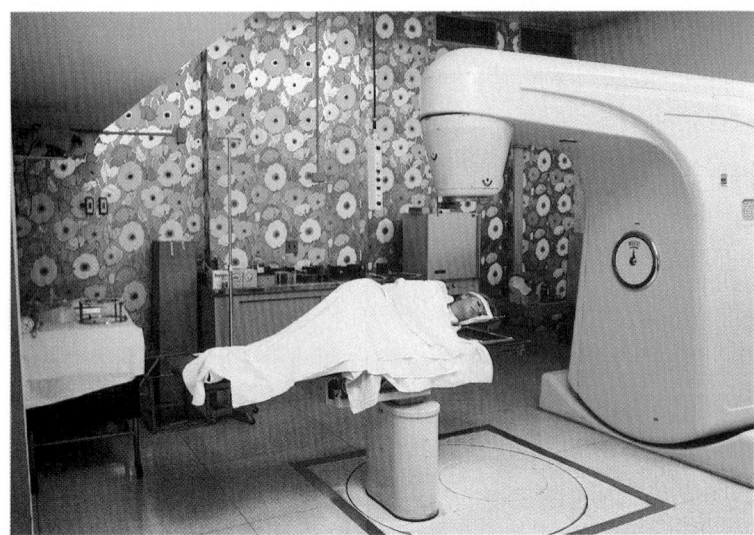

Fig. 43-7
One method of treating cancer is by radiation therapy. (American Cancer Society)

Treating functional diseases is a little different. There are no disease germs to be gotten rid of. Medicines help some functional diseases. Insulin is used in cases of diabetes. Other hormones are used in treating problems related to other glands, including the thyroid, pituitary, adrenal, and reproductive glands. Many functional problems can be helped by surgery.

Cancer may be thought of as a functional disease. It is believed that certain types of cancers are caused by infectious viruses. Certain chemicals have been found to cause other types of cancer. Exposure to nuclear radiation and excessive exposure to the sun's radiation will cause cancer also.

The three most useful treatments for cancer are chemotherapy, surgery, and radiation. In *chemotherapy* (KEY-moe-THER-ap-ee), a person is given some chemicals which will either kill the cancer cells or at least slow down their growth. These chemicals must not be too harmful to normal body cells. In surgery, the surgeon attempts to remove all of the cancer cells before they spread to other parts of the body. In radiation therapy, the affected site may be bombarded with radioactive rays. In any case, cancer must be treated early. For this reason, any unnatural swelling, bleeding, or pain should be reported to the doctor promptly.

General health rules. You can do a great deal to maintain your good health. Go to the doctor and dentist for regular checkups. Have the doctor give you all the vaccines needed. Many diseases have become rare today because people do this.

Keep yourself, your clothing, your house, and your neighborhood clean. Eat the right foods. You cannot develop a healthy body on just a few snacks of soda pop and potato chips.

Careers: Discuss some of the over 200 opportunities in health, such as medical technician, practical nursing, industrial hygienist, histologic technician, and emergency medical technician.

Get enough sleep so that you can feel rested and ready to get up in the morning. The average person needs about eight hours of sleep each night. Drink plenty of water. Get some exercise every day.

SUMMARY

For the most part, infectious diseases are caused by bacteria, viruses, protozoans, and rickettsias. These organisms cause disease by either attacking body tissue or sending out poisons that destroy body cells. Disease-producing organisms are spread by unclean water, contaminated food, or in the air by droplet infection.

There are certain body defenses against disease. The unbroken skin is one barrier to disease. The white blood cells are able to kill off certain disease organisms. Antibodies are protein molecules that are manufactured in the lymph nodes and carried by the blood. These molecules can fight off certain diseases and give us immunity.

The best protection that we have against disease is a clean environment. Clean drinking water, proper disposal of sewage, and control of disease-carrying insects help to prevent the spread of disease. Immunization against diseases and good health practices are important means of preventing disease.

ACTIVITY

The Effect of Antiseptics on the Growth of Bacteria

In this exercise you will find out if antiseptics decrease the growth of certain bacteria.

A. Put ten dried lima or kidney beans in each of four test tubes. Label the tubes A, B, C, D. Half fill each test tube with water. Plug test tube A. Put one tablespoon of alcohol in test tube B and then plug. Put one tablespoon of iodine in test tube C and stopper with the cotton plug. Add one tablespoon of peroxide to test tube D. Then plug this tube also. Allow these tubes to stand in a warm place.

B. After 4 days, remove the cotton plugs and note by the odor the tubes in which bacteria have been active. The stronger the odor, the greater the activity of the bacteria.

1. How do the odors of these tubes differ?

2. Which of the antiseptics is most active against dried bean bacteria?

3. Why is cotton used as a plug? Hint: It has something to do with air.

Word Quiz

In your own words, write a definition for each of the following:

1. functional diseases
2. infectious diseases
3. droplet infection
4. antibodies
5. immune
6. vaccination
7. vaccine

Check Your Facts

Science Reading
Skill
1. *Cause and
Effect*

1. List the three main types of diseases and explain the causes of each type.
2. Define the word parasite.
3. Why are infectious organisms called parasites?
4. Name some diseases caused by each of the following organisms: bacteria, virus, protozoan, fungus, rickettsia.
5. In what ways are diseases spread?
6. What steps have been taken in the United States to stop the spread of certain diseases?
7. What can your body itself do to prevent disease?
8. What contribution to disease prevention did Edward Jenner make?
9. Why is the disposal of sewage in large cities an important process?
10. In what ways can harmful, disease-producing insects be controlled?
11. What are some health practices that you and your classmates should observe?

Thought Questions

Science Reading
Skill
2. *Problem
Solving*

1. Your friend is taking a trip to Mexico. What advice would you give concerning drinking water, milk, meat, and the eating of raw vegetables?
2. Should a person who has emphysema avoid contact with other healthy people? Why or why not?

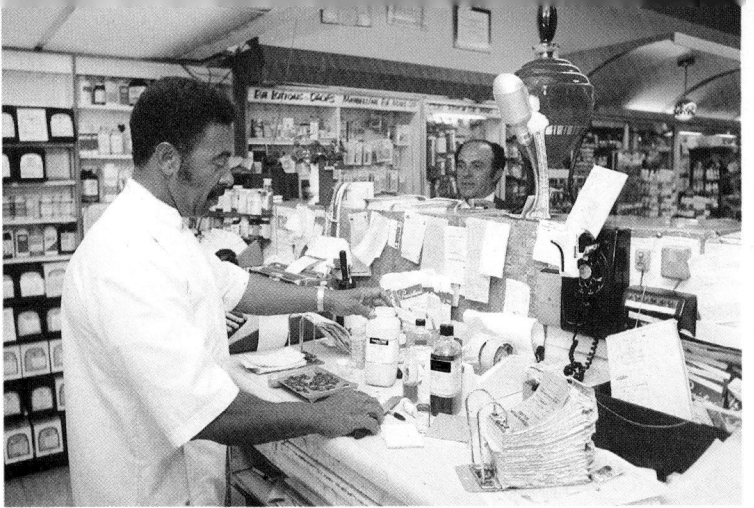

CHAPTER 44 Drugs, Alcohol, and Tobacco

In the photograph above, you see a person preparing a prescription. People who are trained to do this kind of work are *pharmacists*. These pharmacists can give good advice about taking medicines.

Each medicine is made to treat a specific illness. Some medicines are prescribed by doctors. Others, such as cold tablets, are sold over the counter. The quality of medicines must always meet an approved standard. People use these medicines to correct health conditions. We commonly refer to medicines as drugs.

You are probably aware that not all drugs are always used as medicines. There are some chemical substances that people use for reasons other than treating illness. These chemicals, or drugs, are able to create certain feelings in those who take them. Some drugs calm a person down and others make them feel excited. There are drugs that make a person feel light-headed. A few drugs give the user a false sense of courage and strength. All of these drugs are *mood-changers*. They are able to change a person's emotions and alter the way in which they sense what is happening around them. Those under the influence of mood-changing drugs are chemically controlled and cannot act or respond according to their own will. You can see that such drugs can have a serious effect on the body and the mind. In this chapter, we shall learn how dangerous the misuse of drugs can be. This misuse of drugs is called *drug abuse*.

HALLUCINOGENS

The **hallucinogens** (huh-LOOSE-un-uh-juns) are drugs that affect the sensory centers of the brain. These drugs cause people to have *hallucinations*. An hallucination is a false idea of what is going on around you. It might also be thought of as

After you read this chapter, you should be able to:

___**Describe** the general effects of the hallucinogens

___**List** the dangers of narcotic drugs

___**Describe** the effects of depressants and stimulants on the nervous system

___**Discuss** ways alcohol can harm the body

___**Discuss** the health hazards of tobacco

mood-changers

hallucinogens

Science Reading Skill: *Cause and Effect*—Describe the effects of various drugs on the body.

445

Fig. 44-1
Marijuana is made from the dried leaves of the hemp plant. (Courtesy Carolina Biological Supply Company)

a "twisted vision." People who are hallucinating are given false signals by their sense organs. For example, they may have hallucinations about spiders that are not really there. Big spiders may seem to be all over the walls and ceiling. Or they may feel spiders crawling all over them. They may *hear* the spiders, *smell* them, or simply *believe* these spiders are present. Some drugs produce mild hallucinations, other produce more serious ones.

Marijuana. *Marijuana* (mar-uh-WAHN-uh) is made from the shredded leaves and stems of the hemp plant. It can be eaten, but is usually smoked. A stronger form of the drug, called *hashish*, is made from the flowers of the same plant. The plant must grow in a hot, dry climate to produce large amounts of the drug.

The active chemical in marijuana is called *THC*. In large doses it is a powerful hallucinogen. The marijuana smoker generally gets a small dose. In these small amounts, it acts more like a *sedative*

sedative

. A sedative is a drug that has a quieting effect on a person. The effects are felt quickly and last for a few hours after smoking. But THC may remain in the body for a week.

The effects of marijuana smoking include the appearance of silliness. Some people say they are "feeling good." Many people experience a changed sense of time. (Time seems longer than it really is.) People under the influence of marijuana do not hear well, and their ability to judge distances is faulty. They lose the ability to concentrate and their short-term memory is poor. Marijuana causes body changes also. There is an increased pulse rate, lowered blood pressure, and the eyes become red. All of these effects become stronger as the use of marijuana is increased. Very large amounts of the drug produce hallucinations.

Recent studies show that marijuana is not a harmless drug. There is mounting evidence that marijuana smoking can have very serious effects on people. It produces steady changes in the personality. Some people have developed a strong pattern of usage, smoking five or more "joints" a day. These individuals appear in a trance-like state and are known as "burn-outs." They have lost all desire to do anything but smoke marijuana. Studies further indicate that the continued use of marijuana may lower the body's resistance to disease. It seems to interfere with the formation of antibodies.

Scientists are now investigating the effects of marijuana on the pituitary gland. It is very possible that marijuana may affect the hormones produced by this gland. If this is so, marijuana can have serious effects on development and reproduction. We do know that marijuana smoking hinders a person's judgment, and therefore, their driving ability. A large percentage of traffic accidents, many of them serious, have been caused by drivers who had been smoking marijuana. People under the influence of marijuana should not drive.

Angel dust. Many of the abused drugs are manufactured illegally. Dishonest people make them in so-called underground laboratories. This means that there are no standards for the chemicals used and no controls for quality and safety. A particular drug made by the same person at different times can vary in content and strength. This means its effects on a user can vary also. This is one reason why it is dangerous to use these chemicals.

PCP or "angel dust" is one of the drugs that is made illegally. It is an hallucinogen that appears in several forms. In pill or capsule form, it is swallowed. As a powder, it may be sprinkled on marijuana and smoked. Some people inhale it through the nostrils in a process called "snorting." In liquid form, PCP is either injected into the veins or soaked into marijuana leaves. No matter what form it is taken in, the effects of this drug can be extremely bad. In small doses, it makes a person feel "like they are floating on air." Strong doses have made users violent and destructive of property and life. Some people seem to develop a split personality.

PCP is a *mind-altering* drug. It has its greatest effect on brain cells. Scientific evidence shows that PCP is stored in the fat tissues of the brain. It can be broken down by the liver only at a very slow rate. Because it is stored in brain tissue for a long time, it affects the user in a very serious way. It causes *flashbacks*. A flashback is the return of symptoms although a person has not used the drug for several weeks or months. A flashback experience keeps a person in a state of fear. One never knows when these symptoms will return. In addition to flashbacks

References: See *The Student Biologist Explores Drug Abuse* by G. Edwards.

flashbacks

Fig. 44-2
What does this message mean to you? (The National Institute of Mental Health)

and hallucinations, the abuser of PCP may feel very tired and lose the ability to pronounce words or to express an idea. They also may have difficulty in thinking and concentrating and become overly concerned with death. Some even commit suicide or try to kill others.

Mescaline and LSD. At one time *LSD* was the most powerful hallucinogen sold on the illegal drug market. It was called a *psychedelic* (sie-kuh-DELL-ik) drug because its users saw bright colors when in an LSD trance. Sometimes the hallucinations produced by LSD were pleasant. At other times they were terrifying. LSD is no longer as available as it once was.

Mescaline is an hallucinogen found in the *peyote* (pay-OTE-ee) cactus. When eaten, it produces hallucinations. Its effects are very similar to those produced by LSD. These same effects are produced by several species of wild mushrooms that grow in various parts of the world.

NARCOTICS

narcotics

Drugs that produce sleep and at the same time relieve pain are called **narcotics**. Because narcotics are able to decrease pain in suffering people, doctors use certain of these drugs as pain killers. However, narcotic drugs are habit forming and their usefulness in medicine is limited. The narcotic drugs include *opium, morphine, heroin,* and *methadone.*

Opium. Opium is made from the juice of poppy pods. These poppies are grown in Asia, Turkey, Mexico, and several other warm countries. After the poppy flowers bloom, field workers slit the unripe seed pods. Through this cut oozes the gummy

Fig. 44-3
Poppies in bloom can be nice to look at, but the juice from their pods is addicting and dangerous. (Courtesy Carolina Biological Supply Company)

sap that is collected and pressed into bricks. In this form opium is processed in laboratories for use. At one time opium was legally used in medicines. Today there is no longer a need for it. Newer, less harmful drugs have been developed.

Opium is either eaten or smoked. It dulls the senses and causes a dreamy feeling. It causes people to lose consciousness, a condition that may last for several hours. Opium is addicting. It is a hard narcotic that depresses the central nervous system.

Morphine. Raw opium is a complex substance that consists of twenty different compounds. One of these compounds is morphine. Morphine is injected into the bloodstream with a hypodermic needle. Because morphine is a very effective pain-killer, doctors use this drug in special cases. The drug is used to reduce suffering in people who are in extreme pain. Doctors are very careful in their decisions to use morphine because, like the parent drug opium, morphine is very habit forming. Scientists are now trying to develop effective pain killers that are not habit forming.

Heroin. By changing the positions of some atoms in morphine, chemists produce heroin. Heroin is the most dangerous of the narcotic drugs. It is *not* an efficient pain killer. It *is* highly addicting. Heroin has no medical use. It is sold illegally.

Heroin may be taken into the body in three ways. It is most usual for heroin abusers to inject the dissolved drug directly into their veins. This is known as "mainlining." Injecting heroin under the skin is known as "skin popping." Heroin may also be "snorted." Once heroin reaches the bloodstream, it travels to every cell in the body. In some way it affects the chemical processes of cells.

Methadone. Methadone is not obtained from opium. It is a manufactured narcotic that resembles the heroin molecule in structure. Methadone is produced to treat heroin addicts. When a heroin addict takes methadone, the addict no longer craves heroin. Methadone does not cure heroin addiction. It substitutes methadone addiction for heroin addiction. Methadone is taken by mouth. It is usually given with orange juice.

Addiction. An *addict* is a person who cannot stop taking a drug without help. All of the narcotics are drugs that cause addiction. The people who use these drugs most often become addicted to them. This means that their bodies become adjusted chemically to the drug so that they feel a very strong desire to take the drug at regular times each day. Addicts become dependent on the drug both mentally and physically. If addicts fail to get the drug, they suffer *withdrawal symptoms*. They become sick with a severe headache and painful muscle cramps.

addict

withdrawal symptoms

449

Addicts are always afraid of withdrawal pains. They are caught in a drug trap that permits them to think only of their next "fix."

An injection or a "snort" of narcotic makes addicts feel drowsy. They do not speak clearly or pay much attention to what is happening around them. They feel that everything is all right. It was this "all right" feeling that made them like the drug in the first place. Soon this feeling passes and the sick feeling returns. Now the addicts begin to worry about getting the next dose. They really cannot think about much else. They

Fig. 44-4
Various types of barbiturates. These are addictive and dangerous. Withdrawal can be difficult. (U.S. Bureau of Narcotics and Dangerous Drugs)

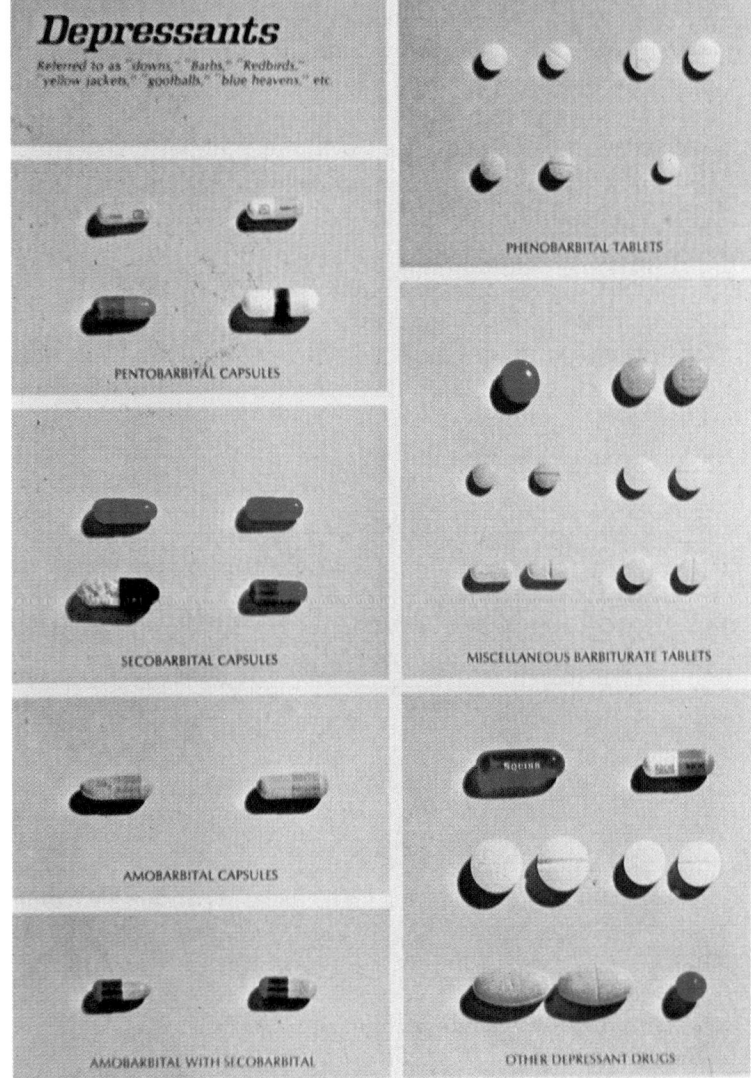

do not pay attention to their work so they are likely to lose their jobs. If they are students, their school work becomes very poor. Heroin addicts can be dangerous people. They will steal and even kill to support this terrible habit.

DEPRESSANTS

We have said that a sedative is a drug that has a quieting effect on people. Sedatives, which are drugs that slow down the nervous system, are also known as *depressants*. They include *tranquilizers, sleeping pills, bromides,* and *barbiturates*. For the most part, sedatives are made by legitimate drug firms. Somehow sedatives get into the hands of dishonest distributors who sell them illegally. There are many different kinds of sedatives. Each one may have several trade names. Sedatives temporarily relieve a person's anxiety. Besides having a calming effect, sedatives make a person feel drowsy and sluggish.

The abuse of sedatives does not solve problems. It permits a false and temporary escape and can lead to even more serious problems. Sedatives prevent people from facing their problems and doing something to solve them. In terms of health, some people become dependent upon these drugs and find it very hard to do without them.

Barbiturates. Barbiturates are especially dangerous because they are addicting in much the same way that narcotics are. Barbiturates affect the chemical machinery of muscle cells. Once a person becomes addicted to barbiturates, the withdrawal is quite difficult. If the barbiturate addict is not withdrawn from the drug under a doctor's care, the person can experience severe cramping of the heart muscle and die.

STIMULANTS

A drug that activates the nervous system is called a *stimulant*. The caffeine found in coffee, tea, and cola drinks is a very mild form of stimulant. *Cocaine* and *amphetamines* (am-FET-uh-meens) are much more powerful stimulants.

Cocaine. A fluffy white powder called cocaine is a product of the leaves of the coca trees. These trees grow high in the Andes Mountains. Cocaine is no longer used in medicines but is sold in great volume on the illegal market. Today it ranks high among the abused drugs. Cocaine is taken into the body by injection or snorting. It is a very powerful stimulant that works on the nervous system.

Cocaine causes a feeling of extra excitement in the user. The person feels happy and ambitious at first. Cocaine also raises the pulse rate and causes the user to breathe rapidly. It tightens

depressants

Enrichment: The combination of depressants (sedatives) and alcohol can kill. When taken together, each one doubles the effect of the other.

stimulant

the blood vessels and opens the pupils of the eyes. After the light-headed feeling wears off, the user often feels depressed and may become irritable and nervous. Sometimes cocaine uncovers a troubled personality. Users may blame others for trying to persecute and abuse them. This change in personality can make an abuser very aggressive and possibly dangerous.

Amphetamines. Amphetamines are a group of drugs with many chemical names, and even more trade names. They are made in legal drug houses, but may be sold on the illegal market. They are usually referred to as "pep" pills.

Fig. 44-5
Various types of amphetamines. When their effects wear off, the user can become very depressed. (U.S. Bureau of Narcotics and Dangerous Drugs)

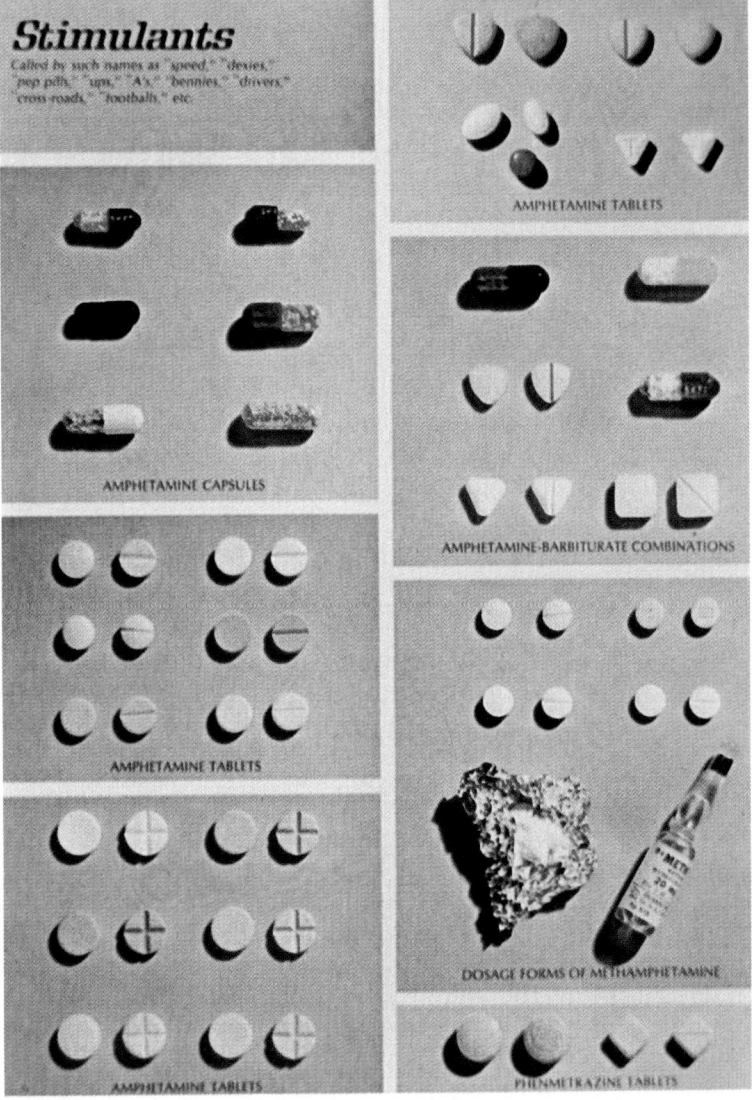

When taken in small quantities, amphetamines appear to have pleasant effects. They seem to increase alertness and decrease the appetite. They make a person feel happy and give them a feeling of contentment. However, these nice feelings do not last for long. From a happy mood, a person is plunged into depression. This causes a person to take more amphetamines to relieve depression. This becomes a cycle. Recent studies have shown that amphetamine abuse can lead to addiction. The withdrawal process is painful and often makes users so depressed that they want to commit suicide. The Federal government has restricted the amount of amphetamines that drug companies may manufacture.

VOLATILE CHEMICALS

A *volatile chemical* is one that gives off fumes. Many of the volatile chemicals are used as solvents to dissolve things that do not dissolve in water. You can usually recognize them because they have a strong odor and they dry up quickly. They are liquids in products such as model glue.

volatile chemical

Some people have discovered that they can sniff the fumes from volatile substances and become "high." The "high" is eventually replaced by slurred speech, double vision, and staggering. The sniffing of volatile chemicals damages the lungs, liver, blood cells, and brain tissue. In several cases it has even caused death.

Another substance that is sometimes sniffed is *butyl* (BYOOT-ul) *nitrate*. On the street, capsules of this chemical are called "poppers." Sniffing butyl nitrate produces rapid heartbeat, makes the person extremely active, widens the blood vessels, and decreases the blood pressure. The "high" from this chemical is often followed by headaches, nausea, and vomiting. Sometimes the person becomes dizzy and faints. This can cause serious injury.

ALCOHOL

For many centuries in the history of humanity, alcohol has been made and consumed. It is part of many religious rituals and is widely accepted as a social beverage. Alcohol is produced legally in the United States and sold in many communities. However, alcohol presents a problem. It is believed that more than 5 million persons living in the United States have a problem with alcohol. They are problem drinkers.

Effects of alcohol. Many people believe that alcohol is a stimulant. It is not. It is a depressant and decreases the normal activity of the nervous system. The target organ is the brain.

Fig. 44-6
Alcohol is often used during the celebration of special occasions. However, for some people alcohol can be a serious problem. (HRW photo by Russell Dian)

But remember that the effect that alcohol has on the brain of any one person depends on many things. One consideration is the *strength* of the alcohol that the person drinks. In other words, the *alcohol content* is an important factor. Another factor for consideration is *how much* the person drinks. One serving of white wine is certainly not equivalent to five glasses of white wine. Thirdly, the circumstances under which the alcohol is consumed and the nature of the drinker are also important. You probably know that many adults can have a "social" drink with very little effect. However, there are those people who cannot accept one drink. Their body chemistry is such that one drink causes them to drink without stopping. This condition will be discussed in more detail.

When alcohol enters the stomach, it is absorbed directly into the bloodstream. Usually, alcohol does not travel into the small intestine. Some of the alcohol in the circulating blood is taken into the liver. In the liver, alcohol is changed to harmless compounds that are later excreted by the kidneys. But there is one drawback to this process. The liver can handle only from 8 to 30 milliliters of alcohol per hour. Thirty milliliters is equal to the volume of one shot glass! So you see that the body can process alcohol only in very small amounts. Alcohol that is not broken down in the liver remains in the blood and circulates around the body until it reaches the brain.

Alcohol affects all of the centers of the brain. The medulla is one of them. The medulla controls heartbeat, breathing, size of blood vessels, and other vital activities. In response to alcohol, the capillaries expand and hold more blood. This causes the red noses and flushed faces of people who drink regularly. It also causes rapid loss of body heat. The kidney tubules and the bladder are controlled by the medulla, also. Alcohol weakens this control.

In general, the effect of alcohol is to interfere with the passage of impulses from one nerve cell to the next. This leads to a slowing down of all brain functions. The higher centers of the brain, such as those for complex thought, feel this effect first. Automatic reflexes take longer to be affected. An early effect is the slowing down of *reaction time*. This is the time that it takes to respond to a stimulus. For example, the average reaction time for stepping on the brake of a car when a green light changes to red is about one-half second. During that time the car continues to move at full speed. When people have been drinking, they will not respond as quickly. The car will travel farther before stopping. One glass of wine is enough to slow down reaction time.

Alcohol affects the judgment centers in the cerebrum so it often affects the behavior. Some drinkers have difficulty in telling the difference between right and wrong. You often read in the newspapers about incidents in which people blame their poor judgment and bad behavior on "too many drinks." In varying degrees, alcohol affects the mental processes. It can cause mental disorganization, slurred speech, staggering, and dropping off to sleep at the wrong time. Too much drinking of alcohol can even lead to coma and death.

Another effect of heavy drinking is a liver disease known as *cirrhosis* (suh-ROW-sus). Cirrhosis is a hardening of the liver tissues. When alcohol passes through liver cells too frequently, these cells and the tissues they form become damaged and unable to function. The liver cannot do its job of absorbing nutrients or breaking down the nitrogen-containing wastes. Blood does not flow easily through the liver. Cirrhosis is a fatal disease because a person cannot live without a functioning liver. Other organs such as the heart and kidneys are weakened and destroyed by too much alcohol in the blood.

Alcoholism. Alcohol is not addicting for most people. They can drink or not drink as they choose. About 5 million people in our country do become addicted. These people drink excessively because their body chemistry demands alcohol. They may or may not become drunk. People who are addicted to alcohol are called *alcoholics*. An alcoholic suffers from ***alcoholism,*** a disease that demands regular drinking. There is no cure for alcoholism. The only treatment is to stop drinking alcohol completely.

Heavy drinking continued for many years leads to the last stages of alcohol poisoning, a condition known as *delirium tremens* (di-LIR-ee-um TREE-munz). The alcoholic hallucinates and can imagine that insects are crawling on the walls and on

alcoholism

Enrichment: Literature and posters can be obtained from the National Council on Alcoholism, 733 Third Ave., New York, NY, 10017

455

Fig. 44-7
Warning signs of alcoholism. (The National Council on Alcoholism)

The illustrations within the figure are labeled:
- Difficult to get along with when he's drinking.
- Lies about his drinking.
- Drinks until he is "dead drunk" at times.
- Drinks to "calm his nerves."
- Can't recall some drinking episodes.
- Neglects to eat when he is drinking.
- Neglects his family when he is drinking.
- Drinks "because he is depressed."
- Hides liquor.

Enrichment: Alcoholism is a serious disease that affects the entire family and how they relate to each other and to society. Groups such as Alcoholics Anonymous, Al-Anon, and Alateen can be helpful to the alcoholic and family members.

his or her body. The person talks wildly and usually raves about animals. The muscles in the hands and arms contract causing them to shake. No doubt you have heard this referred to as "the shakes" or the "DT's."

The DT's are serious. Out of every hundred people who suffer from them, approximately fifteen die. This is due to the rapid change in the blood chemistry. The amount of mineral salts in the blood drops suddenly and dangerously. The person runs a high fever and loses large amounts of water through excessive sweating. The mineral loss leads to muscle cramping, which can affect the heart. The water loss can dry out the tissues. In this condition, the person requires immediate medical care.

Alcohol and the teenager. Most communities have laws to prevent the sale of alcohol and alcoholic beverages to young persons. Recently, the legal age for purchasing alcohol has been raised in some places. Too many young people are drinking and driving. When the percentage of alcohol in the blood is .05, intoxication begins. This means that alcohol begins to affect the brain centers. Under the law, persons with a blood alcohol concentration of .10 are judged intoxicated. The signs of intoxication are slurred speech and poor motor control. In addition, the reflexes are slowed and judgment is poor. A person who is under the influence of alcohol cannot judge distances accurately and cannot determine how fast they are driving. This is why a large number of people are killed in highway accidents. The misuse of alcohol is responsible for other types of accidental deaths such as those from drownings and falls. It is important for you to become aware that alcohol can become addicting. A person addicted to alcohol builds up a tolerance for it. The body will then require increasing amounts of alcohol.

I WAS IN LOVE WITH A GIRL NAMED CATHY. I KILLED HER.

"It was last summer, and I was 18. Cathy was 18 too. It was the happiest summer of my life. I had never been that happy before. I haven't been that happy since. And I know I'll never be that happy again. It was warm and beautiful and so we bought a few bottles of wine and drove to the country to celebrate the night. We drank the wine and looked at the stars and held each other and laughed. It must have been the stars and the wine and the warm wind. Nobody else was on the road. The top was down, and we were singing and I didn't even see the tree until I hit it."

Every year 8,000 American people between the ages of 15 and 25 are killed in alcohol related crashes. That's more than combat. More than drugs. More than suicide. More than cancer.

The people on this page are not real. But what happened to them is very real.

The automobile crash is the number one cause of death of people your age. And the ironic thing is that the drunk drivers responsible for killing young people are most often other young people.

DRUNK DRIVER, DEPT. Y*
BOX 1969
WASHINGTON, D.C. 20013
I don't want to get killed and I don't want to kill anyone. Tell me how I can help.* Youths Highway Safety Advisory Committee.

My name is____
Address____
City____ State____ Zip____

STOP DRIVING DRUNK. STOP KILLING EACH OTHER.

U.S. DEPARTMENT OF TRANSPORTATION • NATIONAL HIGHWAY TRAFFIC SAFETY ADMINISTRATION

Fig. 44-8
Even if you are not an alcoholic, drinking too much alcohol can cause serious problems.

SMOKING POLLUTES YOU AND EVERYTHING ELSE

American Cancer Society

Fig. 44-9
Cigarette smoking is a habit that affects everyone.

TOBACCO

Tobacco is made from the dried, shredded leaves of the tobacco plant. Before World War I, tobacco was used mainly in the form of pipe tobacco, cigars, snuff, and chewing tobacco. Cigarette manufacturing began early in the 1900s. By 1914, 4 billion cigarettes were manufactured in the United States per year. Today, over 600 billion cigarettes are manufactured and sold each year.

Several years ago, public health doctors began to notice the increase in the number of deaths from lung cancer. At one time, lung cancer was a rare disease. Then, during the 1930s it began to become much more common. By 1964, there were over seventeen times as many deaths per year from lung cancer as there had been in 1930. Studies were begun to find out why.

The best guess was that something being breathed into the lungs was causing these cancers. It had to be something that had not been there in the past. Many thousands of people were studied to find out what it was.

Medical scientists checked all kinds of air pollution. They found that factory smoke and car exhaust do irritate the lungs and do increase the lung cancer rate a little. But the main cause turned out to be the tars and other chemicals found in cigarette smoke.

Effects of cigarette smoke. When a person lights a cigarette, several things happen. The tobacco and the cigarette burn due to the oxygen in the air. The burning forms smoke and cigarette ash. You know that the cigarette ash is knocked off and does not enter the body, but the smoke does. It passes from a person's mouth through the throat and into the windpipe. The smoke travels through the branches of the windpipe into the lungs. Cigarette smoke contains nicotine, tars, and about fifteen different gases. Many of the gases irritate the tissue of the mouth, throat, and lungs. One of them also prevents red blood cells from carrying as much oxygen as they should.

Nicotine is a drug that occurs naturally in the tobacco plant. In large amounts it is a poison. In the small amounts that are taken in with each puff of a cigarette, nicotine affects the tissues slowly but steadily. Cigarette smoke diffuses from the lungs into the bloodstream. The blood takes some of the nicotine directly to the kidneys where it is filtered out and stored in the bladder. Then it leaves the body through the urine. The rest of the nicotine is taken to the liver. Here it is changed to a harmless substance. The compounds into which nicotine is changed are also excreted by the kidneys.

While nicotine is circulating in the blood, it touches many blood cells and tissues. It has an effect on many of these tissues.

For example, nicotine stimulates then depresses nerve activity. This causes the shaky hands of many smokers. It makes the pupils of the eyes larger and interferes with the normal reflexes of the eyes. It also speeds up heartbeat, increases blood pressure, and shrinks the size of the blood capillaries.

Remember that the air passages are lined with cilia, which remove dust particles from the lungs. Cigarette smoke kills these cilia, and the lungs are left without this protection. Nearly all smokers are affected in this way.

The air sacs of the lungs are irritated. This may cause them to enlarge and their walls to thicken. Such lungs cannot absorb enough oxygen. This condition is called *emphysema*. It can weaken people so that they can no longer work.

When cigarette smoke starts a cancer in the lungs, the person will die unless the cancer is removed. Often the surgeon must remove the entire lung.

Not every smoker gets lung cancer. But there is enough evidence to show that for people who smoke two or more packs of cigarettes a day there is a good chance that they will get lung cancer or emphysema or another cigarette-related disease. Many smokers develop cancer of the bladder from the nicotine that ends up there. Other types of cancer that affect smokers are cancer of the throat, the voice box, lips, tongue, mouth, esophagus, and stomach. Smokers also have more heart attacks than nonsmokers.

On the average, smokers die earlier than nonsmokers. How much earlier depends upon how early they start smoking and how much smoking they do. Smokers are subject to poor health, coughing, and a weakened sense of taste and smell. They always carry with them the unpleasant odor of stale tobacco smoke.

In all, the use of tobacco causes the deaths of about 380,000 persons in the United States each year. This is more than all of the deaths from accidents, infections, suicides, murders, alcohol, and stomach ulcers put together. Tobacco usage is listed by many health experts as the greatest health hazard today.

Assignment: Have students write to the local chapter of the American Cancer Society.

Fig. 44-10
Do you think cigarette smoking is worth the risks it involves? (HRW photo by Russell Dian)

AN OVERALL LOOK AT DRUG ABUSE

Drugs and the law. The sale of most of the drugs discussed in this chapter is regulated by national or local laws. Alcohol, tobacco, and most volatile substances are manufactured and sold legally. Barbiturates, amphetamines, and morphine have medical uses. They can be bought with a doctor's prescription, otherwise they are illegally sold and used. Heroin has never been legal. Heroin is manufactured in illegal laboratories. It is smuggled into the United States, Canada, and other areas. It is sold illegally on the streets. Marijuana laws vary a great deal from place to place. It too enters the United States and is sold illegally. Selling illegal drugs is a criminal offense. A person can be fined or sentenced up to forty years in prison. It depends upon the cirumstances and the local laws.

The drug supply. When people become drug abusers, a part of their lives is spent in activities that are outside of the law. How much of their day is spent performing illegal activity depends upon the extent of their drug habits. A heroin addict is totally dependent upon a drug "pusher" for the daily supply of drugs. People who have developed amphetamine or barbiturate habits are also dependent upon an illegal source of supply. Drug pushers are criminals who intend to make as much money as they can from their customers. They make big profits from selling all types of illegal drugs.

tissue tolerance

Addicts using heroin or barbiturates develop a *tissue tolerance* to these drugs. This means that they need larger and larger amounts of the drug to satisfy their daily craving. Soon the habit can cost over 100 dollars a day. This is a great deal of money for people who may have difficulty holding a job. They may turn to crime to get the money. Many robberies are committed by drug abusers who must get money to buy heroin or pills. So you see that a drug abuser is not only in personal danger, but is a threat to the well being of others also.

Dangers from drugs. People living outside of the law cannot expect protection from the law. This is very true regarding what a pusher sells to a drug abuser. There is no guarantee that the illegal drugs sold are what the pusher says they are. Very often marijuana is contaminated with insecticide. A pusher sells it anyway. Sometimes the marijuana is sprinkled with angel dust, giving the unsuspecting user a terrible experience. Samples of pills sold by pushers have been analyzed by government chemists. These pills contained such poisons as benzene, ether, chloroform, mercury, human urine, and floor cleaner. These materials get into the drugs because they are prepared in illegal laboratories by dishonest chemists.

Fig. 44-11
These are some of the substances that have been found mixed into illegal drugs. (HRW photo by Russell Dian)

Drug peddlers usually "cut" heroin and cocaine before selling them. That is, they mix these drugs with something cheap, like milk sugar, and then sell them for full price. Every once in a while addicts are sold a supply of heroin that is more concentrated than what they are used to taking. They have no way of knowing this. They take the usual amount and die of an *overdose*. This is a very common cause of death among drug users. At one time the death rate from "OD" had reached alarming proportions. As the use of heroin has decreased, so has death from overdose.

Another danger results from using dirty needles. Doctors using hypodermic needles are very careful to sterilize the skin and the needles first. They make sure that no germs will be injected along with the medicine. Drug users are not careful at all. They use and share dirty needles. As a result many addicts who inject themselves with drugs get an infection of the liver known as *hepatitis*. Many die from an infection of the heart muscle called *bacterial endocarditis* (en-doe-kahr-DITE-us). Others die of the very painful disease called tetanus.

Why drug abuse? Why do people start taking drugs? They usually know before they start that drugs are harmful but they do it anyway. There are probably several reasons. Young people may start because they are looking for thrills or because their friends are taking drugs. They want to be one of the crowd and do what everyone else is doing. During one's youth, the personality is still developing. Some young people may not be sure of themselves. As a result they may not think very much of

461

themselves. The more troubled they become, the more likely they may seek an easy and quick solution to their problems. They go ahead and try anything that might relieve their problem. Many times this solution comes in the form of drugs.

Even if they only use marijuana it is not good, as you learned earlier in the chapter. The use of drugs gives only a temporary and sometimes painful escape from problems. A person must cope with life instead of running away through drugs. The habit of running away from problems does not help a person develop a useful life.

Many people are misusing drugs. Many of these people are adults. They may be homemakers or business people. They may not take heroin, but they take alcohol every afternoon "to relax." They take sleeping pills each night so they can get to sleep. They take an amphetamine in the morning "to get going." In time, they become very dependent upon these drugs. They, of course, are harming themselves.

If people ever ask you to try a drug, do not be fooled by the words they use. They may call heroin by any number of different names—names such as horse, smack, white stuff, Harry, joy powder, junk, and many others. But it is still heroin, and it will still ruin your life. Amphetamines may be called bennies, speed, pep pills, wake-ups, uppers, bombido, or many others. They are still amphetamines and they are still dangerous to your health. Let us hope that you will decide that you are too important a person to throw your life away on drugs.

Treatment. Suppose people are already using drugs and wish to stop. What can they do about it? The first step is deciding that they really do want to quit.

That may be all there is to it in some cases. Drugs such as marijuana and amphetamines are not physically addicting. Neither is alcohol, unless the person is an alcoholic. Stopping the use of these drugs is a matter of deciding to stop and then doing it. Of course, often it is easy to develop a habit and difficult to break. If people have the habit of smoking marijuana or using amphetamines or alcohol, they will need to develop other, more positive, ways of getting satisfaction in life. These interests will become new habits that replace the old, bad habits.

Cigarette smoking is habit-forming for some and addicting for others. Some people stop smoking with relative ease. Others have a very hard time. If you are a smoker and wish to stop, pick a time when you are feeling happy and then quit—all at once. A good time to do this is when you will be doing something you like. This will keep your mind off your desire for a cigarette. Do not have any cigarettes in reach. If they are some distance away, or in the store, you will be less tempted. You

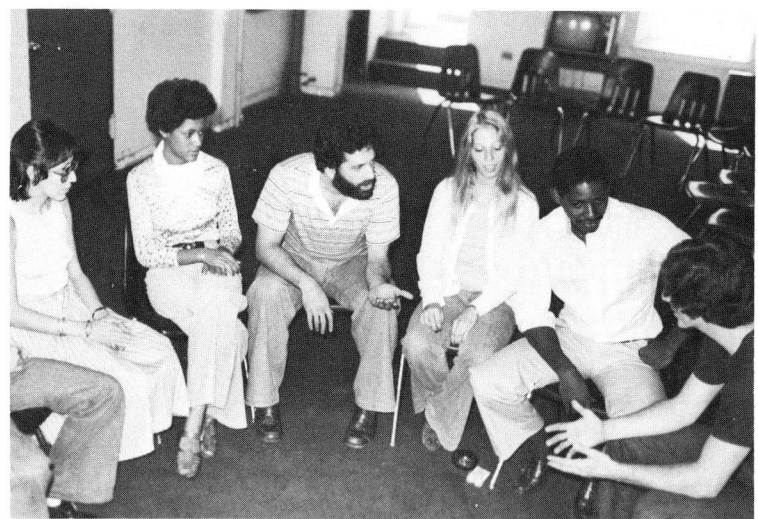

Fig. 44-12
There are many rehabilitation centers available for people who want to stop their drug habit. (Courtesy Phoenix House)

may have the desire to start smoking again, but your willpower can overcome the desire. This may be difficult at first, but after some time your desire to smoke should lessen.

The people with the biggest problem are those addicted to narcotics. If they quit all at once, they go into withdrawal sickness. They experience terrible pains, muscle and leg cramps. They vomit and very often injure themselves thrashing about. However, the sudden withdrawal from heroin or morphine is painful but does not kill. In three days the withdrawal symptoms are over. At this point a person can stay away from heroin, but most return to it because they also develop a psychological, or mind, dependency. They find it hard to give up the degrading way of life as an addict because they usually have lost the trust of their family and nondrug-using friends. Most heroin addicts need professional guidance and counselling.

Attempts have been made to treat heroin addicts with methadone. This just substitutes one drug addiction for another. In cities where many methadone clinics have opened, a new type of addict has appeared. This is the methadone addict who was never addicted to heroin. The heroin addicts sell their methadone to buy more heroin.

Barbiturate addicts may die if they withdraw suddenly. A person addicted to barbiturates should withdraw only under the supervision of a doctor. If a person continues to take barbiturates, the medulla of the brain will become hopelessly inactive. All vital activity of the body stops. The person dies.

The American Cancer Society has developed the slogan: "If you don't smoke, don't start. If you do, stop." To carry this idea further, the best time to stop taking any drug is before you start. Then you have no habit to break and no broken life to repair.

SUMMARY

Many drugs are mood-changers that control a person's emotions. This means that a person laughs, cries, or feels fright in response to a chemical, not in response to a real situation. Grouped among the mood-changers are such drugs as marijuana, PCP, mescaline, heroin, the amphetamines, barbiturates, and alcohol. Depending upon the amounts used and the person's own body chemistry, these drugs are capable of destroying the mind and body tissues.

Heroin and barbiturate drugs are addicting. A person cannot stop using them at will. They cause great harm to the body.

Tars and nicotine from tobacco lead to destruction of the lungs and other body organs. Lung cancer, emphysema, heart trouble, and many other diseases are related to tobacco use.

ACTIVITY

The Effects of Dizziness on the Ability to Function

Alcohol and certain drugs can cause a feeling of dizziness. In this activity a demonstration will be done to find out how well a person functions when dizzy. The feeling of dizziness will be caused by spinning around. Therefore, your teacher will choose a suitable volunteer for this demonstration.

CAUTION: You should not try this on your own. Dizziness can cause you to fall down and become injured.

A. Your teacher will arrange twelve paper clips side by side in a horizontal row. The student volunteer will then form a chain by attaching as many paper clips to each other as possible in 15 seconds. (Another student can act as a time-keeper.)

1. How many paper clips did the student attach?

B. Then the student will become dizzy by spinning around several times. If a revolving chair is available, the student should sit in this chair and be spun around by the teacher.

CAUTION: If the student is being spun around while standing up, other students should stand nearby to assist the dizzy student and prevent this person from falling. While dizzy, the student will again form a chain by attaching as many paper clips together as possible in 15 seconds.

2. How many paper clips did the student attach this time?

3. Explain the results of this demonstration.

4. Do certain drugs increase or decrease a person's ability to do things?

5. Why do the effects of drugs and alcohol last longer than the dizziness caused by spinning around?

Word Quiz

Choose the number from **Column B** that best matches each letter in **Column A.**
Do not write in this book.

Column A	Column B
a. a disease that requires regular drinking	**1.** mood-changer
b. a substance that gives off fumes	**2.** hallucinogens
c. need for a drug increases	**3.** sedative
d. a drug that causes fantasies	**4.** flashback
e. a drug that activates the nervous system	**5.** narcotic
f. a drug that has a quieting effect on the nervous system	**6.** addict
	7. withdrawal symptoms
g. a drug that slows down the nervous system	**8.** depressant
h. a person who cannot stop taking a particular drug	**9.** stimulant
	10. volatile chemical
i. a return of symptoms even though a drug has not been used for a while	**11.** alcoholism
	12. tissue tolerance
j. a drug that brings on sleep	
k. pains and distress caused by not taking a drug	
l. a drug that changes emotions	

Check Your Facts

1. What is the difference between medicine and an illegal drug?
2. Explain the meaning of drug abuse.
3. What is the difference between an hallucinogen and an hallucination?
4. List the names of those drugs that are classified as mood-changing. What is meant by mood-changing?
5. Give as many reasons as you can why one should not smoke marijuana.
6. What are the effects of sniffing volatile compounds?
7. Name some stimulants. How do stimulants affect the nervous system?
8. Name some depressants. How do depressants affect the nervous system?
9. In what ways does alcohol affect the body?
10. Why is smoking tobacco so dangerous? List as many harmful effects as you can.

Science Reading Skills
6.–10. Cause and Effect

Thought Questions

1. The newspapers often refer to cocaine as the "fun" drug. Why is this a misleading description for cocaine?
2. Drug abusers often say, "I can stop anytime I want." Why can't they usually keep this promise?

CHAPTER 45 Investigating the Frog

Suggestion: This frog investigation should take about three days. Use one day for the external characteristics, one day for the behavior portion, and another day for the dissection.

To do this investigation it would be helpful to use a live frog for the first part and a preserved or freshly killed frog for the dissection. However, if you do not have these materials available you can still gain much information by reading through this exercise and referring to the diagrams.

If you are working with a live frog you will also need the following materials: a living frog in a battery jar, a glass rod, vinegar, a wire, and a small piece of raw meat. The glass rod and the wire should be about 10 to 12 centimeters long.

Suggestion: Show students how to hold the frog so that it will not jump away.

Suggestion: If frogs are unavailable, charts and diagrams can be used.

THE LIVING FROG

Looking on the outside. If you are using a living frog, put it into a battery jar so that you can look closely without having the frog hop away. From time to time you will have to remove the cover so that you can touch the frog. You will be shown how to do this. You will learn also how to handle the frog so that it cannot get away from you.

A. Let us investigate some of the *external* characteristics of the frog. Look at the flattened *head*. It is shaped almost like a triangle. See Fig. 45-1. Notice how the head is attached to the body.

1. Does the frog have a neck?

The *eyes* seem to stick out from the eye sockets. Gently touch one of the eyes with the side of a glass rod. You should see two different movements of the eye.

2. Describe what happens.

B. Moving back from the eye you will see a brown, rounded area that is covered by a tightly drawn membrane. Can you guess what this is? It is the covering of the *eardrum*. It vibrates much like your own eardrum.

Fig. 45-1

Some external structures of the frog.

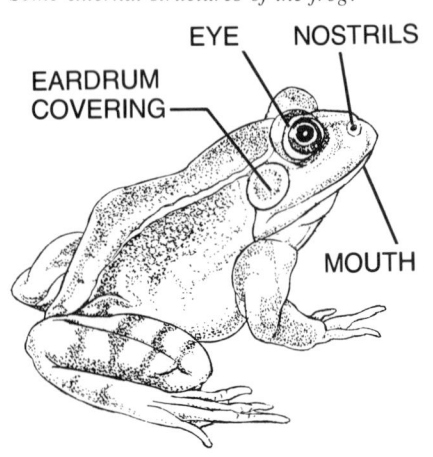

EYE   NOSTRILS

EARDRUM
COVERING

MOUTH

   3. What function does it serve?

C. In front of the eyes and behind the rounded end of the mouth
    are two holes. These are the external *nostrils*. These nostrils
    open into the mouth and are used for breathing.
D. Watch how the *throat* moves up and down. Air moves in
    and out of the mouth cavity. The frog takes in air through
    the nostrils. Throat movement forces air into the lungs.
E. Feel the *skin* of the frog. Notice that it is moist. This moist
    skin serves a useful function. Oxygen from the air dissolves
    in the skin's moisture and diffuses into the capillaries in the
    skin. The frog breathes through its skin as well as its lungs.
   4. Why do you think this is so?

F. Look at the two pairs of limbs.
   5. How do the *forelimbs* differ from the *hindlimbs*?
   The hindlimbs permit the frog to jump from place to place.
Now examine the feet.
   6. How many toes are there?
   7. Are the toes free or are they connected in some way?
We say that a frog has *webbed* feet. The thin skin between the
toes of the hindlimbs helps in swimming. Notice that the forelimbs and the hindlimbs have the same regions that we have
in our arms and legs.
   Fig. 45-2 illustrates the "hand" portion of a forelimb of a
male and female frog. Look at this diagram. Now look at the
forelimbs of your frog.
   8. Do you have a male or a female? How can you tell?
   9. Are there any claws or nails on the "fingers" of the frog?

Behavior. Let us investigate how a living frog behaves.

A. Turn the battery jar to the left.
   1. Which way does the frog turn in the jar?
       Turn the battery jar to the right.
   2. Which way does the frog turn now?
   If this frog were not inside of the battery jar, it would jump
in the direction opposite to which it is turned. This type of
unlearned behavior is a *reflex*.

B. Most of the frog's activities are reflexes. You know that a
    reflex is behavior that is inborn. Reflex behavior does not
    involve thought. Gently pinch the toes of the frog. Notice
    how the leg moves upward. Dip a glass rod into some
    vinegar. Apply it to the frog's thigh.
   3. What happens?

C. Attach a small bit of meat to a wire. Swing this in front of
    the frog as if it were a flying insect.
   4. How does the frog respond?

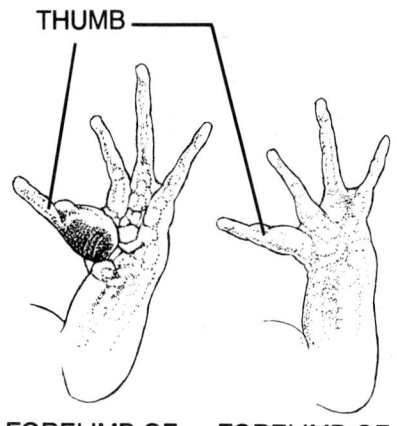

THUMB

FORELIMB OF   FORELIMB OF
MALE FROG   FEMALE FROG

Fig. 45-2
*How does the "thumb" of the male
frog differ from that of the female frog?*

If the frog jumps to grab the meat, notice how the tongue is used. The tongue flips out of the mouth and then is pulled back into the mouth again. The tongue seems to sweep the ''insect'' into the mouth. See Fig. 45-3.

FROG DISSECTION

Looking on the inside. If a preserved frog is available, you will *dissect* this frog. This means that you are going to cut it open to study the organs that are inside its body. If you do not have a frog, refer to the diagrams.

If you are using a preserved frog you will need a wax bottom pan. You will also need a scissors, a forceps (tweezer), some straight pins, and a probe.

A. Place the frog in the pan belly side up.

Study Fig. 45-4 as you prepare the frog.

Spread out the forelimbs and the hindlimbs and pin them to the pan, as shown. Note where you are to make the first cut. Use the forceps to pick up the skin between the hind legs at the position marked 1.

Cut through the skin in the direction shown on the diagram. This means that you will cut the skin in a straight line up to the frog's ''chin.'' The lines at position 2 show you where to cut the skin next. After that, cut where you are shown at line 3. Finally make the cut in the skin as indicated at line 4. Now you can pin back these flaps of skin.

Look at the inside of the skin. Observe the many small blood vessels.
  1. Why are there so many?

B. Now that you have pinned back the skin, you can see the muscles of the body. You will now cut through these muscles to expose the body organs. Using the same pattern of cuts as for the skin, cut through the muscle layer. **Use the forceps to lift the muscles so that the organs beneath will not be damaged.**

As your scissors move to the chin region of the frog, you will feel a bone. **Very carefully cut through the bone.** The heart lies underneath it. As you continue refer to Fig. 45-5. This will help you locate the internal organs. Pin back all of the muscle flaps. Now you can see the *heart*. A frog's heart has two atria and one ventricle. Carefully move the heart aside. Underneath it and off to each side you will see the *lungs*.

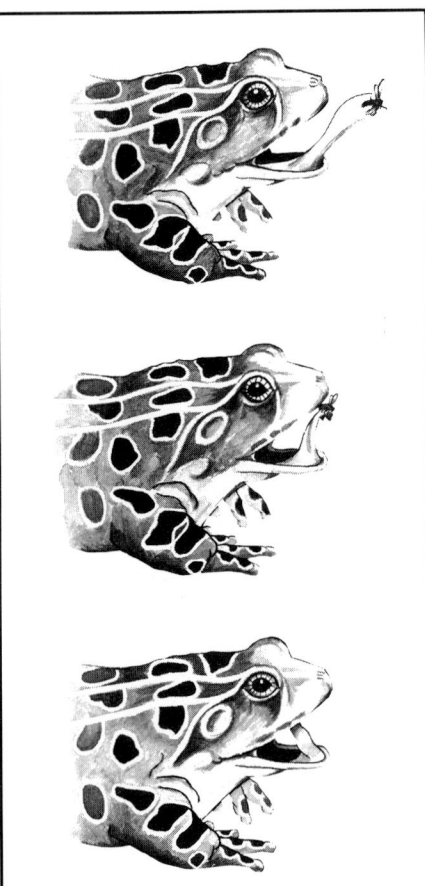

Fig. 45-3
A frog catching an insect.

Suggestion: Use chloretone or urethane to kill frogs. Both of these substances are poisons and should not be handled by students.

Suggestion: You may want to demonstrate a frog dissection before your students do it.

Suggestion: Students can expand the lungs by filling with water from a medicine dropper.

C. Partially covering the heart is the *liver*. It is the largest gland in the body. Lift the right lobe and look at the *gall bladder*.

   2. What color is the gall bladder?

Science Reading Skill: *Compare and Contrast*—Describe the similarities and differences between the frog's internal anatomy and that of a human being.

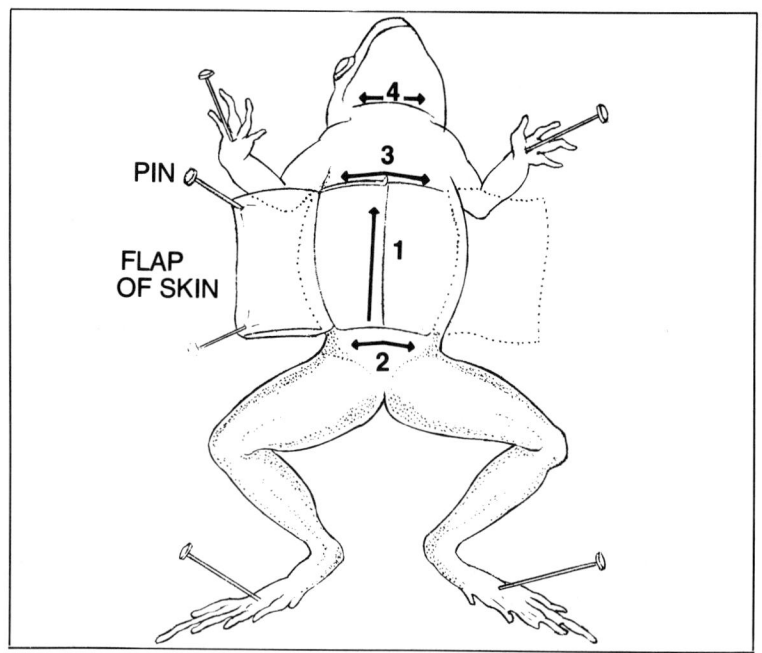

Fig. 45-4
Dissecting the frog.

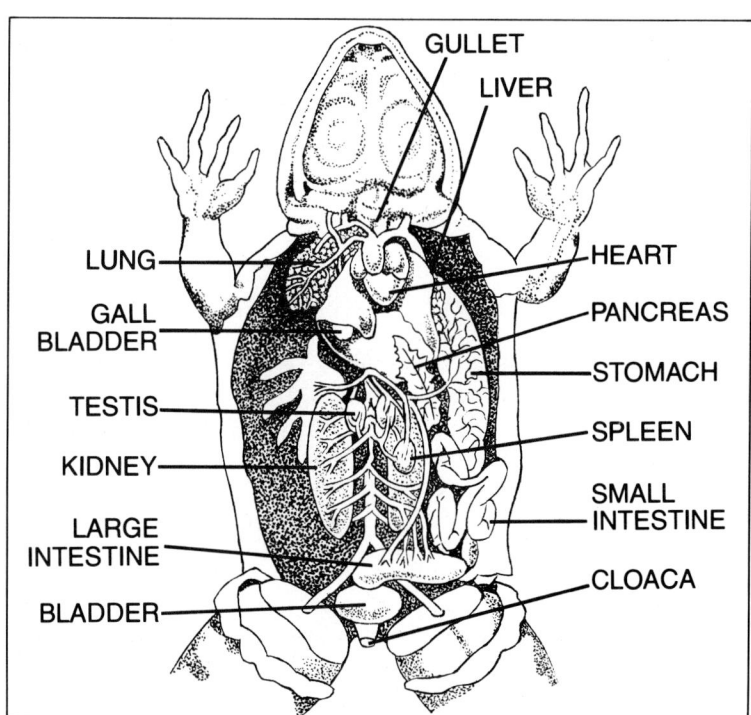

Fig. 45-5
The internal structures of the frog.

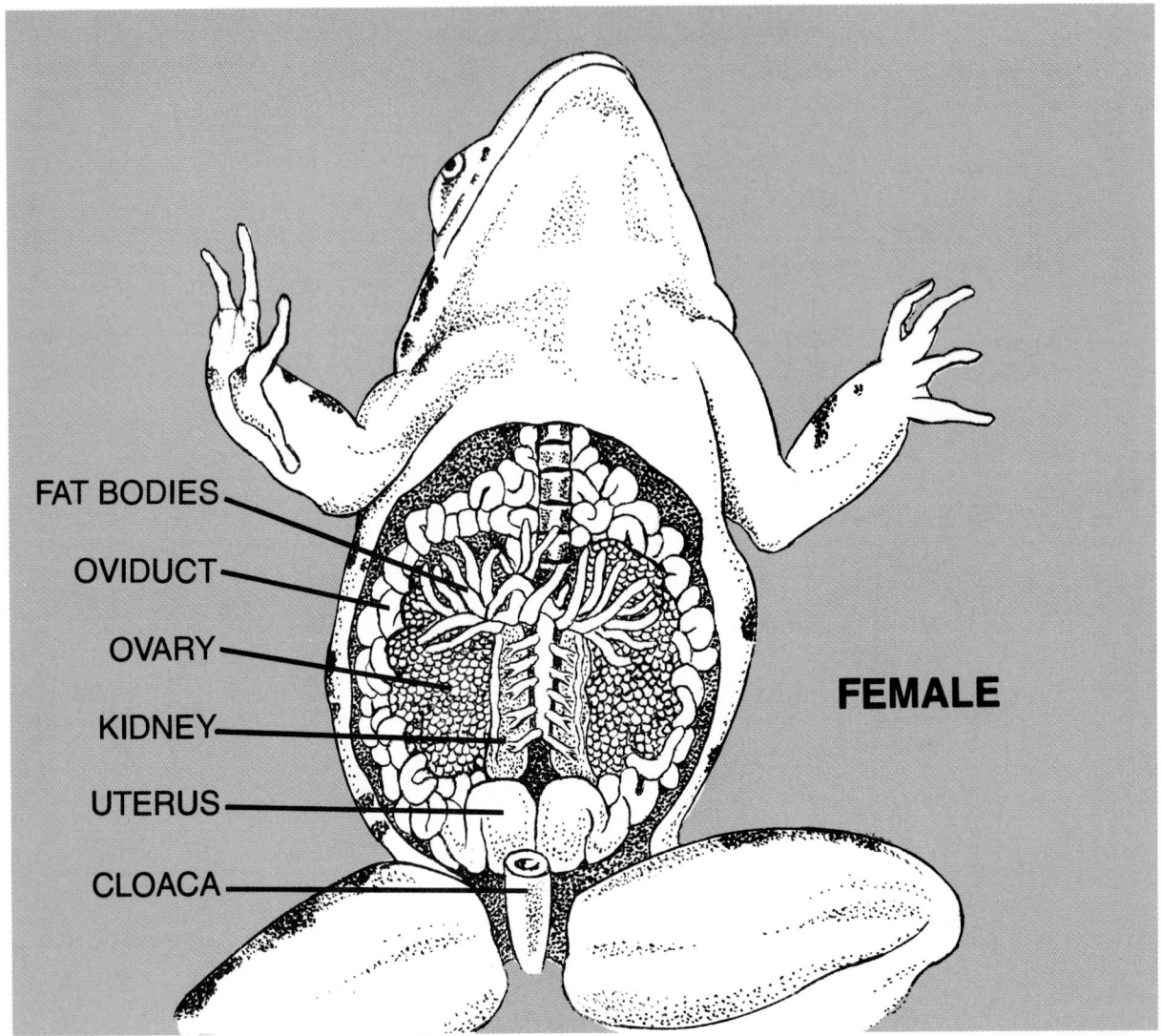

FAT BODIES

OVIDUCT

OVARY

KIDNEY

UTERUS

CLOACA

FEMALE

Fig. 45-6
Female frog: reproductive and excretory systems.

If you have a female frog, the body cavity may be filled with eggs. These look like small black and white beads. The egg mass grows directly from the *ovary*. See Fig. 45-6. You will have to remove the eggs and ovaries from the frog so that you can see the structures that lie beneath them.

Carefully remove the liver and place it in your dissecting pan. Now you have a better view of the digestive organs. Look for the whitish *stomach*. Leading downward from the stomach is a long, thin tube. This is the intestine. Look between the U-shaped loop made by the stomach and intestine. You will see a light-colored organ that has an irregular shape. This is the *pancreas*.

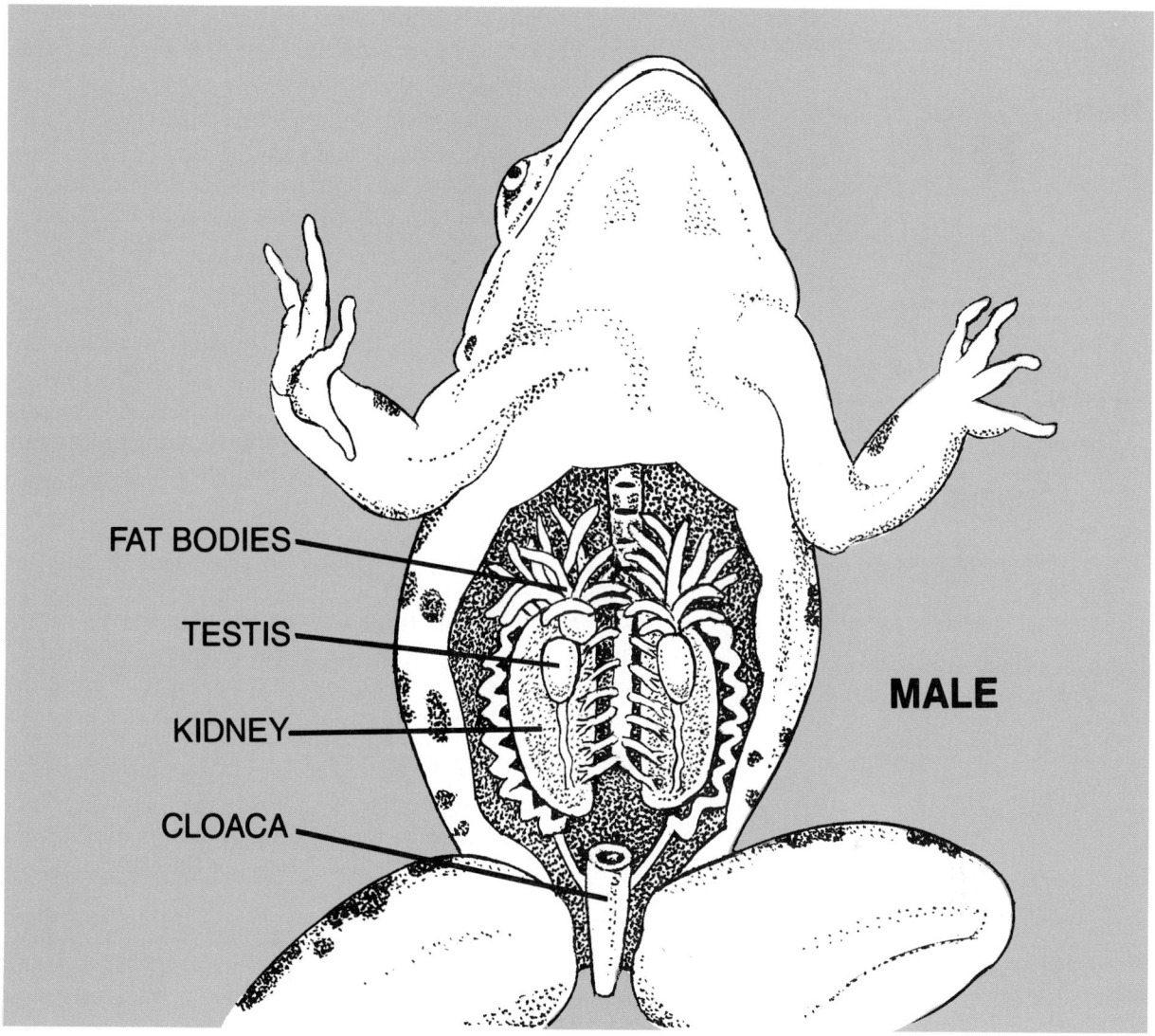

FAT BODIES

TESTIS

KIDNEY

CLOACA

MALE

Fig. 45-7
Male frog: reproductive and excretory systems.

D. Find the stomach again. Trace a path upward from the stomach. Find the esophagus. It is a very short extension from the mouth leading downward to the stomach.

Remove the entire digestive tract beginning with the esophagus and ending with the large intestine. Cut away the thin membranes that hold the organs of the digestive system in place. Lay the digestive tract in the pan next to the liver. Measure the length of the small intestine.

3. How many centimeters long is the small intestine?

The two reddish organs you now see on the body wall are the kidneys. Their function is the same as ours. See Fig. 45-7.

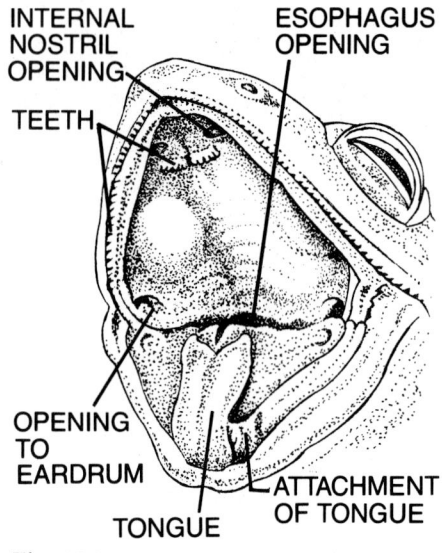

INTERNAL
NOSTRIL
OPENING

ESOPHAGUS
OPENING

TEETH

OPENING
TO
EARDRUM

TONGUE

ATTACHMENT
OF TONGUE

Fig. 45-8
*The structures found inside the mouth
of the frog.*

4. What do you see at the upper end of each kidney?

These yellow finger-like structures are the *fat bodies*. They contain stored energy in the form of fat.

E. Can you find a whitish, rounded body that lies on each kidney? If so, you have a male frog. These are the *testes*. These glands produce the sperm cells for reproduction. If your frog is a female, you removed the ovary with the mass of eggs.

F. Remove the kidneys. Behind these are the *backbone*. Inside of the backbone is the *spinal cord*. See if you can find some of the thread-like nerves that go to the muscles of the legs.

G. Now dissect the frog's mouth. Open the mouth. Clip through the jawbone if you have to. See Fig. 45-8. Notice how large the mouth is. The nostrils open directly into the mouth. Look at the opening at the back of the mouth.

5. To where do you think it leads?

The throat opens into the very short esophagus. In the frog this is really the uppermost part of the stomach. Notice that the tongue is attached at the front and loose at the back end. This is just opposite to the way in which our tongue is attached.

Do frogs have teeth? Run your finger around the rim of the lower jaw. Now do the same to the upper jaw.

6. What do you feel?

The frog has a large number of very tiny teeth growing from the upper jaw only. Notice the two rough bumps on the roof of the mouth. These are teeth also. Frogs use these teeth to grip their prey.

H. Now remove the skin from the legs. Make a lengthwise cut in the skin from the ankle upward. Carefully peel off the skin that covers the leg. Look at the muscles. They look very much like the ones in your legs.

Next cut the muscles off the leg and expose the bones. Look for the tough ligaments that bind the bones together. The ligaments are smooth bands of connective tissue that cover the joints and the bones.

7. Why do these ligaments have to be so strong?

This investigation has allowed you to study the behavior of the frog as well as its internal and external structures. You should have noticed that the frog is able to respond to stimuli, just as humans are. The internal organs of the frog are organized much the same way as in the human body also. Now you should have a much better idea of how your own body functions.

THE WORKING WORLD

**Extending the potential of the senses, training
the body, and assisting the grief-stricken are
all areas of career development in human biology.**

Jobs and careers related to the needs of human biology are so varied and numerous that it would be almost impossible to mention all of them. We will take note of only a few.

The powers of the senses can be enhanced through the skills of technicians, many of whom learn their trade on the job as assistants or apprentices. The developments in electronic music and advanced sound systems have created jobs in what could be called sound management. Electronic, acoustical, and auditory expertise is important. A sound technician must be able to understand and adjust the components of the system itself, must understand how sound moves within the design of a particular space, and must know how loud or soft a sound should be in order to keep sound distortion to a minimum and to keep from damaging the ears of the listeners. Training in electronics and knowledge of basic physics and the biology of the ear are helpful.

The services of the lighting technician in a theater, television, or film studio are geared to a similar purpose: to help the eye perceive as clearly as possible the effect intended by the artists

Yoram Kahana/Peter Arnold, Inc.

F. Siteman/Taurus Photos

and directors. Electrical knowledge and an understanding of color and lighting design are essential. On-the-job learning can begin by helping out in school and community theater productions.

Helping people who are missing senses is the focus of other professions. There are teachers who teach signing and the mechanics of speech to those who cannot hear. There are teachers of Braille, who help persons who are blind to read. And there are people involved in replacing body parts, lost through accident, injury, or illness. The making of prostheses—artificial devices to replace missing parts of the body—is a large industry. Different skills and backgrounds are necessary depending on what is being designed. An ocularist, for example, is a person who makes artificial glass or

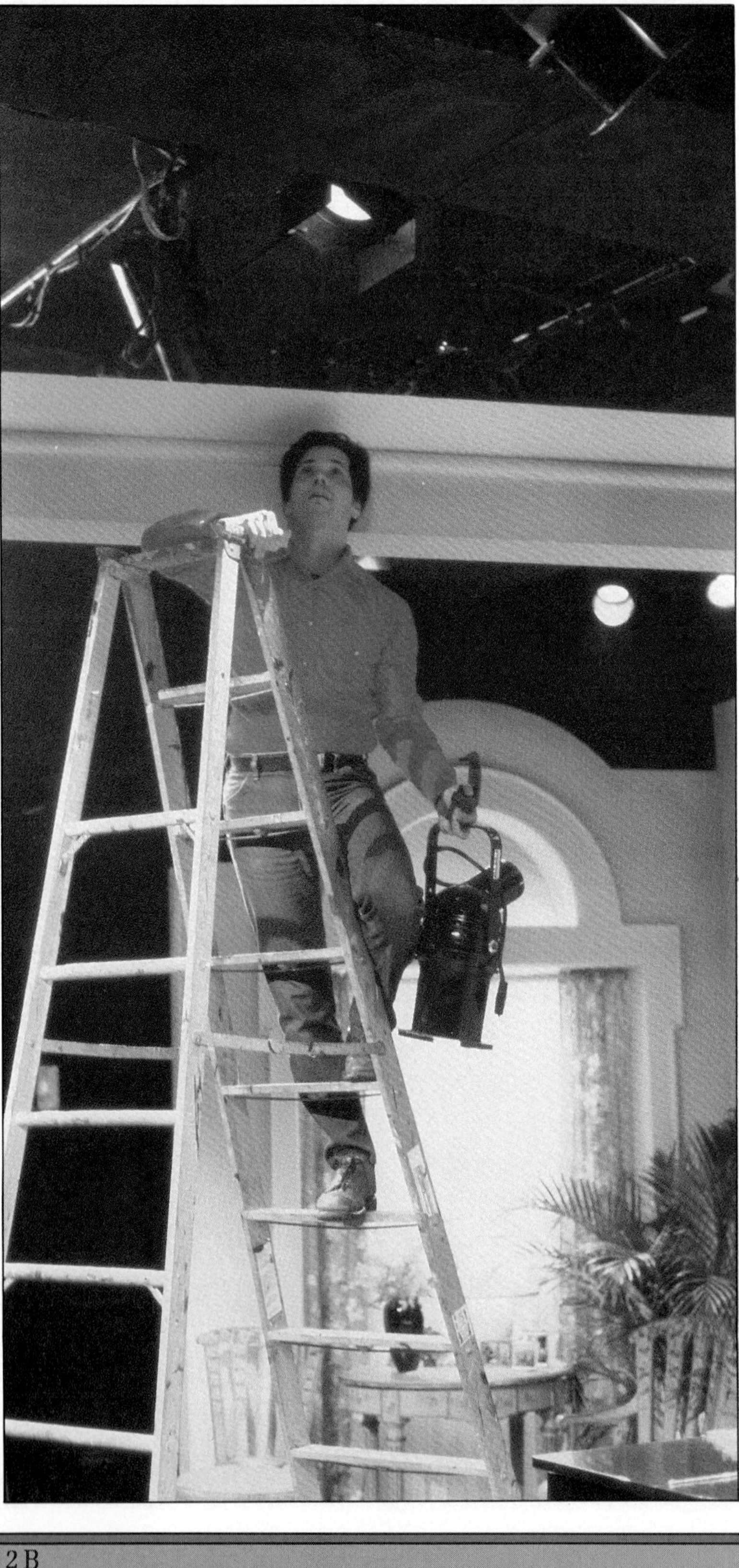

HRW: Ken Karp/Interior of Manhattan Theatre Club Scenic Designer, Ray Recht

HRW: Ken Karp

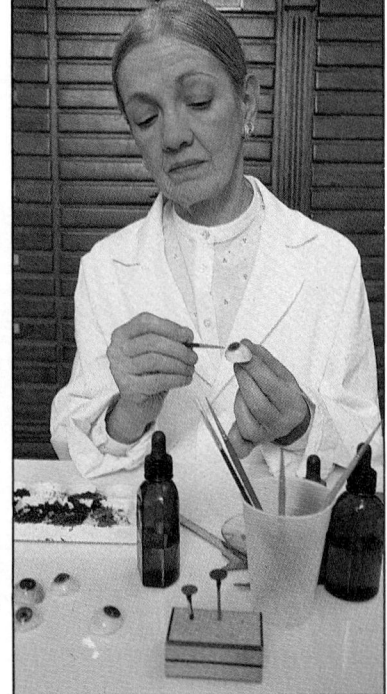

plastic eyes. The skills needed are artistic and mechanical abilities plus the ability to relate to persons in distress. Artificial eyes serve both a therapeutic and cosmetic purpose. They help a person appear "normal," and help to relieve feelings of depression or negative self-esteem created by the loss of an eye. The eye must be made to fit the measurements of the orb and it must realistically match the client's other eye or desired specifications. One learns this trade through apprenticeship, although some courses are given by the American Society of Ocularists. Ocularists may be certified through the National Examining Board of Ocularists. It is not compulsory that they take the exam to practice.

Developing the talents and abilities of the healthy body occupies the energies of a great many people. Teachers of all kinds are prime examples. Intellectual abilities are sharpened through discussion, talents are stimulated and trained in art and music programs, and bodies are developed in gymnasiums, health clubs, and community centers.

Qualifications for teaching vary, depending on what you teach and where. Public school teachers must have a college degree and, usually, must pass a state exam to be certified. Other qualifications may be necessary. A music teacher, for example, is usually an accomplished musician as well as an educator. This requires the self-discipline necessary for long hours of practice as well as skill in performance.

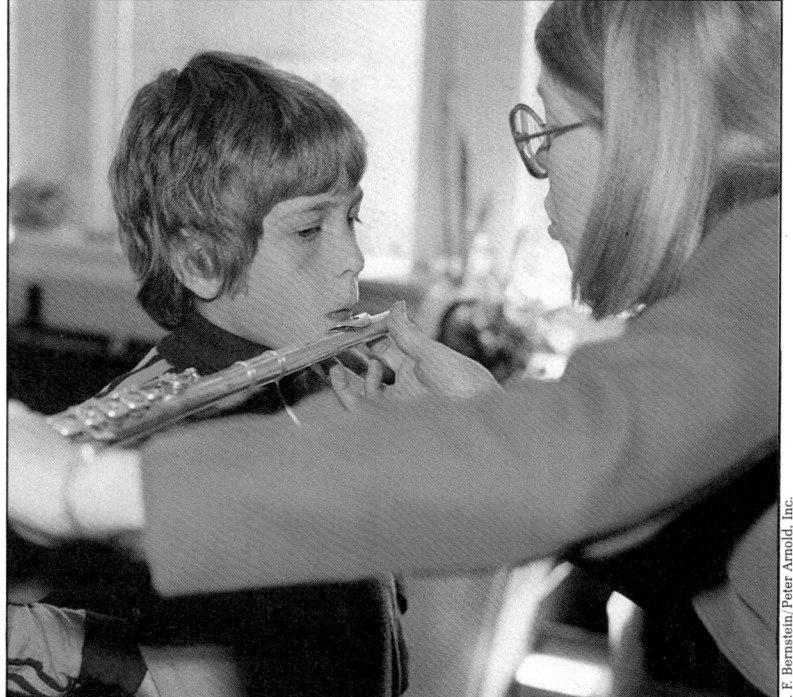

E. F. Bernstein/Peter Arnold, Inc.

Teaching in a community organization or private institution may not require an extensive formal education, but verification of your abilities will be important. Audi-tions, demonstrations, references, and resumes citing your accomplishments will help you obtain a nonschool teaching position. The growth of adult education alter-

E. R. Degginger/Bruce Coleman, Inc.

natives has created opportunities for anyone with a skill or expertise to teach. If you can demonstrate competence, formal degree requirements are not always as rigidly applied. Exercise teachers, for example, differ in their qualifications. Some are trained in physical education and have four-year degrees. Your gym teacher will most likely have a similar background. Many people working at fitness centers have a degree in exercise physiology. Openings at the supervisory and managerial level often require a formal degree in one of these areas. However, dancers and people with extensive training in yoga or the martial arts are considered well qualified to teach at fitness centers and recreation organizations if their skills can be demonstrated. In the more commercial teaching enterprises, marketing skills are very helpful since it often becomes the teacher's responsibility to insure enrollment.

The human life cycle is not completed until death. From beginning to end, there are related occupations. Mortuary science is not a field in which everyone would feel comfortable, but the care and preparation of the dead is a large business with various employment opportunities.

Embalmers and funeral directors obtain a degree in mortuary science, fulfilling requirements in anatomy and chemistry and learning the interpersonal skills necessary in dealing with people who have experienced recent death. This curriculum may be combined with other course requirements for a full college degree, but mortuary science is a course of studies varying from 9 months to 3 years depending on the state. An apprenticeship follows the schooling and state licensing is required for embalmers. There are many sales-related positions in this field. There are cemetery and crematorium caretakers, coffin and urn manufacturers, and sales-people for burial plots, crypts, and funeral clothing. While this is not a growth industry, different developments in our society could increase the complexity of tasks for this profession. More and more people, for example, are donating organs for transplant. This could create new administrative and mechanical problems to resolve. For further information, write to:

The National Funeral Directors
    Association of the U.S., Inc.
135 Wells Street
Milwaukee, WI 53203

An excellent source of information on almost all occupations mentioned here is

*The American Almanac of Jobs and
    Salaries*
by John Wright
New York: Avon Publishing
    Company, Inc.
1984, revised edition

**It's easy to find recreation and hobbies that
can increase your job skills and make you
a valuable member of the community in addition.**

Technical skills can be learned in an interesting setting. We have already mentioned working in theater productions. In fact, the easiest way to find out what an occupation is like is to volunteer to help someone who does it. Teachers' aides are a good example, as are hospital volunteers. There is usually a minimum age requirement, but where the position exists, it can provide you with a view of the environment in which you think you would like to work. The medical field, at least when viewed through the television screen, has a certain kind of glamour. Spending a summer as a hospital volunteer may not change your mind about wanting to be a medical technician, but it may give you a more realistic view of the work setting and its daily problems.

Volunteer activities may, of course, serve no ultimate career goal. Without volunteers, however, many human service organizations would be more understaffed than they already are. Volunteers read to the blind. Some people who have

Will McIntyre/Photo Researchers, Inc.

lost their sight have not learned Braille and, therefore, have no way to read themselves. Besides that, not all books have been translated into Braille. Sight-impaired students especially need readers. Some people have even completed law and medical degrees with the help of people who volunteered to read to them directly or to read their textbooks onto cassettes.

Paramedics are often volunteers. Though they undergo rigorous training, are reexamined for certification every few years, and must pass a state exam, a paramedic's services may be affiliated with volunteer ambulance and emergency service units. Others

work for municipal agencies and the fire or police department. In general, the paramedic is an emergency medical technician who works on a mobile intensive care unit under a doctor's supervision. Training periods of 800 to 15,000 hours plus six months of field experience are necessary.

Preparing for paid work is, of course, a basic life requirement. Preparing for volunteer work ensures that the social fabric is woven closer than it would be if basic necessities were our only human concern.

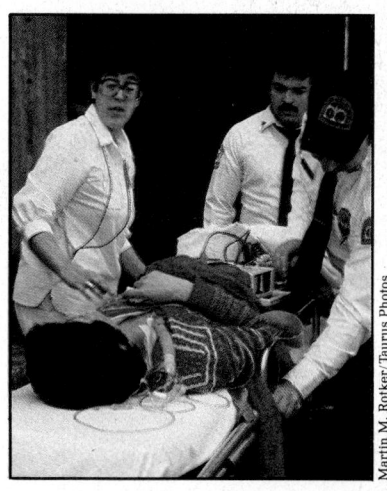

Martin M. Rotker/Taurus Photos

CONSUMER SCIENCE IN...

Richmond Times-Dispatch

New tests may hold key to evolution

From wire dispatches

LOS ANGELES—A group of young archaeologists has found new ways to analyze fossils that may ultimately resolve competing theories of how diet has affected human evolution.

The new techniques encourage archaeologists to analyze fossils chemically, rather than make inferences from the refuse associated with fossils.

Adherents to one school of thought on evolution say early humans had abandoned the ways of vegetarian, apelike ancestors to evolve into hunters with the instinct to kill one another as well as animals. Aggression, it is argued, is in our genes.

Another view is that man's ancestors were never more than moderately carnivorous and that eating meat was only one of many social changes that led, not to raw aggression, but to greater cooperation and organization. Sharing of food and jobs was considered the key to evolution.

The new research indicates the advent of agriculture 10,000 years ago was an event in economic, rather than dietary, history; that the Neanderthal diet differed little from that of modern humans; and that maize was eaten thousands of years earlier than previously thought.

Archaeologists study diet because food intake influences an animal's position in the world. Body size, population density and social behavior, for example, are linked to diet.

Traditional archaeological methods

deduce diet by examining such refuse as dung, seeds, animal bones, tools and the grooves on fossil teeth.

Many experts, however, say this method has a carnivorous bias: Stone tools used in hunting survive better than the digging sticks used to extract plant material.

The new techniques allow archaeologists to analyze fossils to obtain dietary information. One approach looks at the ratios of certain stable chemical isotopes in fossil bones.

Lincoln Journal

Drug triples heart attack survival rate

Journal Writers and News Wires

BOSTON—Chances for survival improve dramatically if doctors quickly inject a clot-dissolving enzyme into the hearts of victims of heart attack, the nation's leading killer which claims 550,000 lives a year, according to a study published today.

The physician who directed the research—Dr. J. Ward Kennedy of University Hospital in Seattle—said he thinks the therapy should be used routinely by all hospitals that are equipped to administer it.

The enzyme drug is called streptokinase. It has been used experimentally for several years, but the new report is the first major study to show that it actually saves lives.

Kennedy's research found that heart attack victims are almost three times more likely than those who get ordinary care to be alive one month after their seizures. And their survival is nearly four times better after six months.

In their study, conducted at 14 hospitals in western Washington and British Columbia, doctors treated 134 heart attack victims with streptokinase, while 116 others got standard therapy. Within the first month, 4 percent of the streptokinase patients died, compared with 11 percent of the comparison group. In a followup six months after treatment, no more streptokinase patients had died, but the death toll in the other group had risen to 15 percent.

Appendix A:
How to Use This Text

Table of Contents

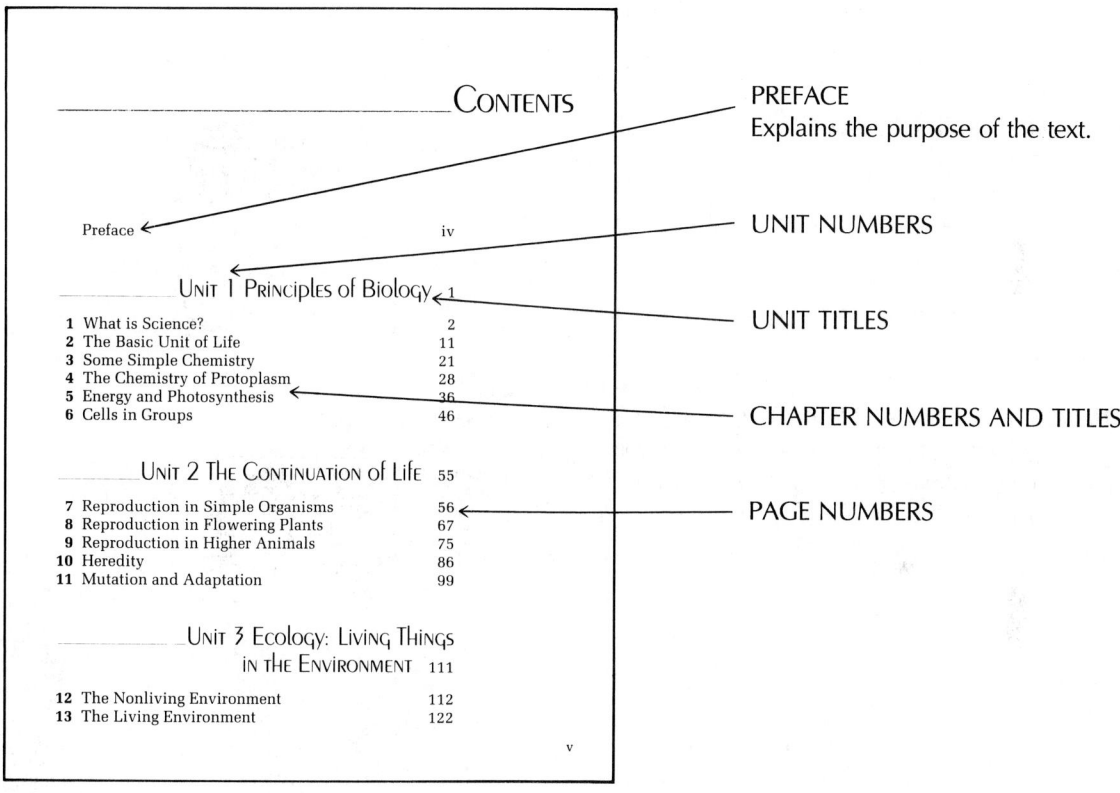

PREFACE
Explains the purpose of the text.

UNIT NUMBERS

UNIT TITLES

CHAPTER NUMBERS AND TITLES

PAGE NUMBERS

Exercise

Using the text, answer the following questions. Do not write in this book.
(Spaces for answers can be found on page 1 of _Exercises & Investigations for Living Things_.)

1. Can the Preface be found before or after Chapter 1?
2. What are the seven major topics discussed in this text?
3. How many chapters are in this text?
4. On what page does Chapter 34 begin?
5. How many pages are there in Unit 4?

Chapter Opener

PHOTOGRAPH ————————————————→
Each chapter opens with a color photograph that is
tied into the opening paragraph and related to the
contents of the chapter.

OBJECTIVES ——————————————————→
The objectives state the important concepts to be
learned in each chapter.

KEY WORDS ——————————————————→
New vocabulary is found in **boldface** type within
the text and is also listed in the margin so that
important definitions can be easily located.

CHAPTER 12 The Nonliving Environment

objectives

After you read this chapter, you
should be able to:

___**List** some conditions neces-
sary for life on earth

___**Diagram** the water cycle

___**Describe** how rainfall affects
the production of different landscapes

___**Compare** the effects of alti-
tude with the effects of latitude

___**Explain** the importance of light

ecology

The earth is not just made up of people, dogs, insects, trees,
and other living organisms. It is also composed of nonliving
things such as air, water, soil, and rocks. Each of the living
organisms on earth is affected by the nonliving things around
it and each particular living organism must be adapted to live
in its own particular environment.

THE EARTH'S ENVIRONMENT

No living thing can exist, even for a few seconds, without
being affected by its environment. You would realize this if
you tried holding your breath while you read the rest of this
chapter! Air is part of your environment. It affects you by supplying the oxygen you need. You affect it by removing some
oxygen from it and adding some carbon dioxide to it. In what
other ways do you and your environment affect one another?

The environment is everything that surrounds or affects the
individual, including such nonliving things as light, heat,
water, soil, and air. It also includes all the other living things
in a region. In this chapter, we shall study the nonliving environment, and in the next, the living environment. This whole
study of living things in relation to their environment is the
science of *ecology*.

Let us think about the earth as a whole. The earth is the home
of plants, animals, and people. What kind of environment must
these living things have if they are to succeed on the earth?

First, the environment for living things must contain *water*.
Protoplasm is made up largely of water. The other molecules
in this living material can only meet and react with each other
if they are dissolved in a liquid. There is probably no other
liquid in the universe that can take the place of water in living
things.

112

Exercise

Using the text, answer the following questions. Do not write in this book.
(Spaces for answers can be found on page 2 of *Exercises & Investigations for
Living Things*.)

1. When you finish studying Chapter 10, what will you be expected to be able
   to do?
2. What new terms are introduced on the opening page of Chapter 8? Why are
   they listed in the margin?
3. How can you tell what words will be defined within a chapter?
4. What is pictured in the photograph at the beginning of Chapter 18?
5. How is the photograph at the beginning of Chapter 18 related to the contents
   of the chapter?

TEXT PAGES

UNIT I   THE PRINCIPLES OF BIOLOGY

carried in the ATP molecules is used to manufacture proteins, vitamins, and other useful materials in the cytoplasm. The cell goes on growing. Both night and day the cell membrane uses energy to carry on its work of absorbing needed molecules from the pond water.

Molecules of waste produced by the cell pass through the cell membrane by diffusion. In the daytime, photosynthesis is going on very rapidly. Thus oxygen leaves the cell as a waste. At night, photosynthesis does not take place. It is at this time that oxygen enters the cell to be used in respiration. Carbon dioxide now leaves the cell as a waste. **Remember that the process of respiration goes on at all times in both plant and animal cells.**

DIFFERENT WAYS OF GETTING FOOD

BOLDFACE TYPE
Used to stress very important terms and phrases.

MAJOR HEADINGS
Introduce a general theme.

Other groups that reproduce by means of spores include the molds, mosses, and ferns. Perhaps you have seen small spots on the undersides of fern leaves. See Fig. 7-3. These are clusters of *spore cases*. Spaces form in these. When they are ripe, the spore cases split open, and the spores fall out.

The big advantage in the spore method of reproduction is that millions of spores can be carried long distances by the wind. Spores spread the organisms far and wide. The disadvantage of spores is that they are so tiny. They cannot carry enough stored food in their cytoplasm to give the new organism a good start. If a fern spore lands on moist soil, it may grow into a tiny fern plant. But if the surface of the soil dries out too soon, this plant will die. The underground parts will not be long enough to reach down to water. How do seed plants have an advantage in this respect?

You can see that conditions must be just about perfect if a spore is to grow. Because most spores do not land where there are such perfect conditions and they die, the organism must produce spores by the millions. Suppose all the spores coming from one fair-sized puffball, a fungus something like a mushroom, were to grow into new puffballs. These new puffballs, if put together, would make a pile bigger than the entire earth! Obviously, the spores cannot all grow. With so many spores being produced by simple living things, the air is always loaded with them. They are a part of the dust that is always present in the air. This very minute, as you are reading this page, you are probably breathing in many different kinds of spores.

Reproduction by vegetative means. *Vegetative reproduction* is another form of asexual reproduction. In higher plants, the flower is the reproductive structure. It produces seeds by a sexual process. The stems, roots, and leaves are called the *vegetative* parts of the plant. They have to do with carrying on the ordinary activities of the plant. If one of these parts grows into

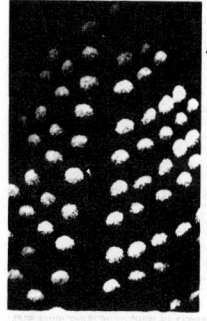

Fig. 7-3
The "spots" on the underside of this fern leaf are really groups of spore cases. (Hugh Spencer)

spore cases

vegetative reproduction

ILLUSTRATIONS
Help to make new ideas easier to understand. Each illustration has a figure number that begins with the number of the chapter it is found in.

SUBHEADINGS
Point out the main idea of each section.

ITALIC TYPE
Words or phrases related to the understanding of new material are printed in *italic* type.

Exercise

Using the text, answer the following questions. Do not write in this book. (Spaces for answers can be found on page 2 of *Exercises & Investigations for Living Things*.)

1. What is illustrated in Fig. 7–1?
2. List all of the major headings in Chapter 12. What is the purpose of these headings?
3. List all of the subheadings in Chapter 15. What is the purpose of the headings?
4. Why are some words printed in **boldface** type and some in *italic* type?
5. What is the most important concept stated on page 42?

Summaries and Activities

SUMMARY ─────────────
Reviews the main ideas covered in the chapter.

ACTIVITY
Further develops concepts covered in the chapter.
The title of the activity tells what the activity is
about. Each procedure is lettered and each
question is numbered.

NOTES OF CAUTION ───────
Warn of possible dangers that may result from
carelessness.

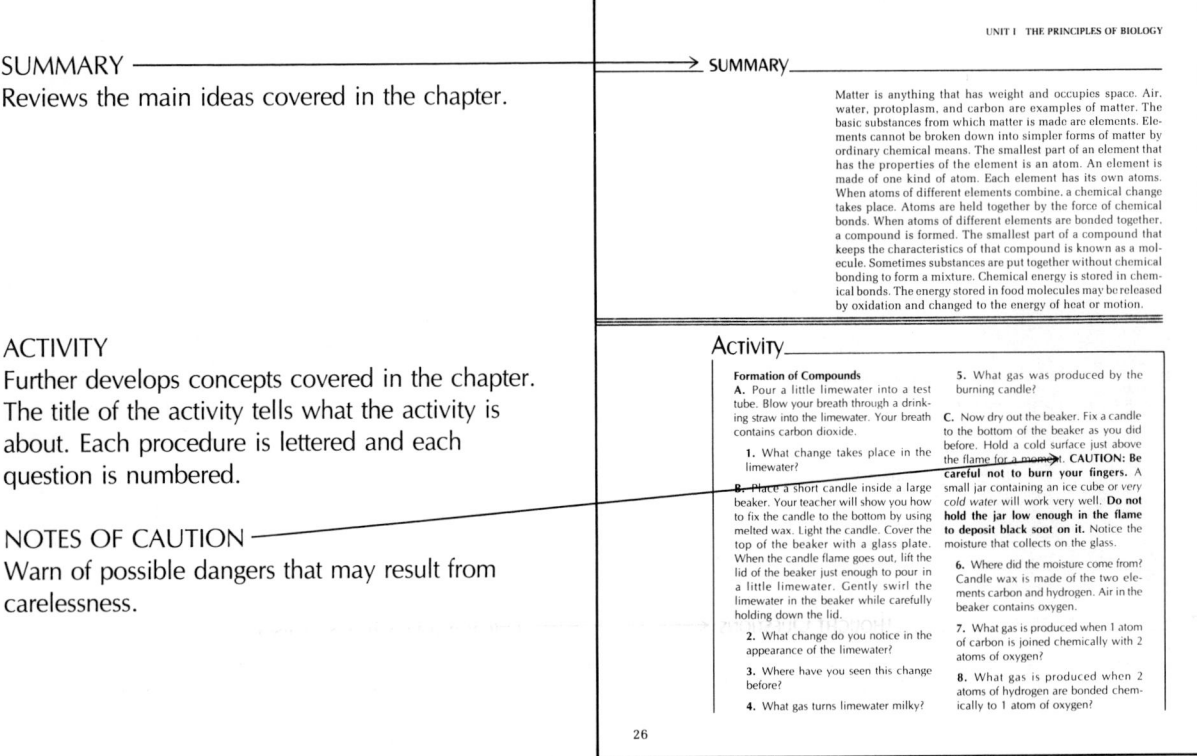

UNIT I THE PRINCIPLES OF BIOLOGY

SUMMARY

Matter is anything that has weight and occupies space. Air, water, protoplasm, and carbon are examples of matter. The basic substances from which matter is made are elements. Elements cannot be broken down into simpler forms of matter by ordinary chemical means. The smallest part of an element that has the properties of the element is an atom. An element is made of one kind of atom. Each element has its own atoms. When atoms of different elements combine, a chemical change takes place. Atoms are held together by the force of chemical bonds. When atoms of different elements are bonded together, a compound is formed. The smallest part of a compound that keeps the characteristics of that compound is known as a molecule. Sometimes substances are put together without chemical bonding to form a mixture. Chemical energy is stored in chemical bonds. The energy stored in food molecules may be released by oxidation and changed to the energy of heat or motion.

Activity

Formation of Compounds
A. Pour a little limewater into a test tube. Blow your breath through a drinking straw into the limewater. Your breath contains carbon dioxide.

1. What change takes place in the limewater?

B. Place a short candle inside a large beaker. Your teacher will show you how to fix the candle to the bottom by using melted wax. Light the candle. Cover the top of the beaker with a glass plate. When the candle flame goes out, lift the lid of the beaker just enough to pour in a little limewater. Gently swirl the limewater in the beaker while carefully holding down the lid.

2. What change do you notice in the appearance of the limewater?

3. Where have you seen this change before?

4. What gas turns limewater milky?

5. What gas was produced by the burning candle?

C. Now dry out the beaker. Fix a candle to the bottom of the beaker as you did before. Hold a cold surface just above the flame for a moment. **CAUTION: Be careful not to burn your fingers.** A small jar containing an ice cube or *very cold water* will work very well. **Do not hold the jar low enough in the flame to deposit black soot on it.** Notice the moisture that collects on the glass.

6. Where did the moisture come from? Candle wax is made of the two elements carbon and hydrogen. Air in the beaker contains oxygen.

7. What gas is produced when 1 atom of carbon is joined chemically with 2 atoms of oxygen?

8. What gas is produced when 2 atoms of hydrogen are bonded chemically to 1 atom of oxygen?

26

Exercise
Using the text, answer the following questions. Do not write in this book.
(Spaces for answers can be found on page 3 of *Exercises & Investigations for
Living Things*.)
1. What are the main ideas discussed in Chapter 1?
2. What skill would you be learning if you did Activity I at the end of Chapter 2?
3. How is the activity at the end of Chapter 1 related to the contents of the chapter?
4. How many procedures are there in the activity at the end of Chapter 22?
5. What should you be very careful of when doing the activity at the end of Chapter 7?

End of Chapter Questions

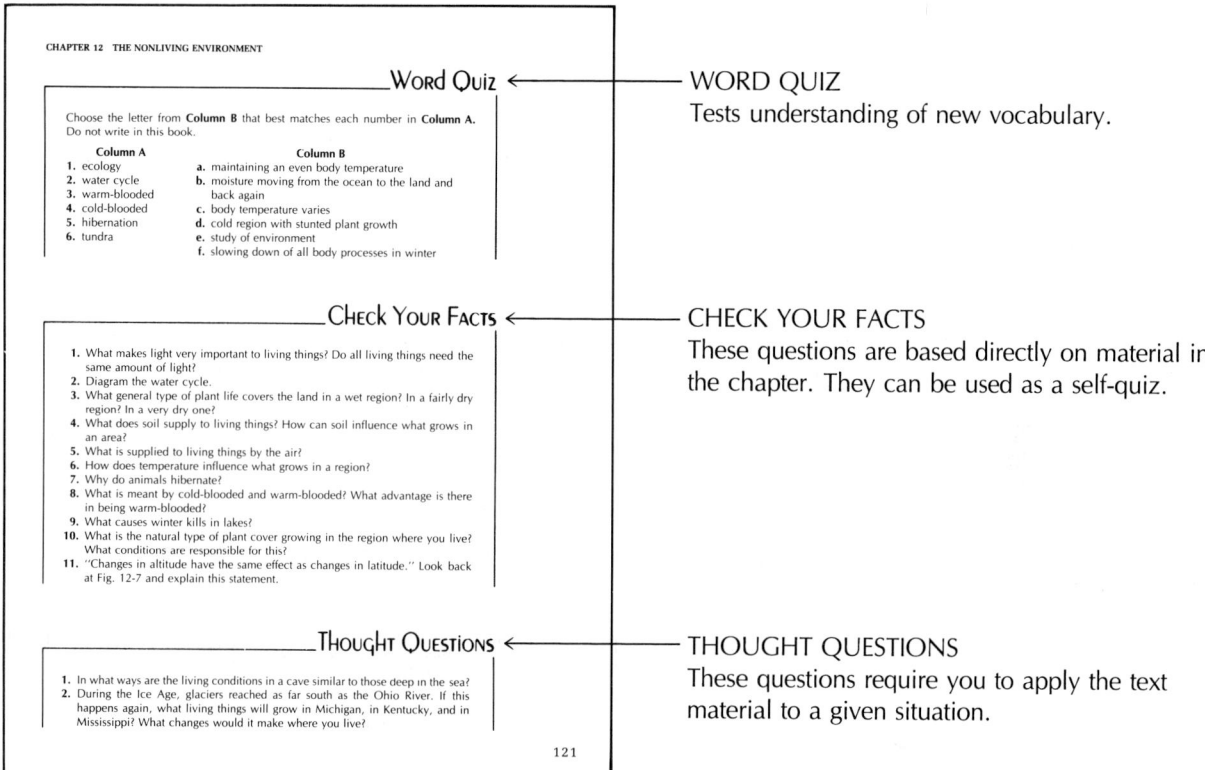

CHAPTER 12   THE NONLIVING ENVIRONMENT

Word Quiz

Choose the letter from **Column B** that best matches each number in **Column A**.
Do not write in this book.

Column A
1. ecology
2. water cycle
3. warm-blooded
4. cold-blooded
5. hibernation
6. tundra

Column B
a. maintaining an even body temperature
b. moisture moving from the ocean to the land and back again
c. body temperature varies
d. cold region with stunted plant growth
e. study of environment
f. slowing down of all body processes in winter

Check Your Facts

1. What makes light very important to living things? Do all living things need the same amount of light?
2. Diagram the water cycle.
3. What general type of plant life covers the land in a wet region? In a fairly dry region? In a very dry one?
4. What does soil supply to living things? How can soil influence what grows in an area?
5. What is supplied to living things by the air?
6. How does temperature influence what grows in a region?
7. Why do animals hibernate?
8. What is meant by cold-blooded and warm-blooded? What advantage is there in being warm-blooded?
9. What causes winter kills in lakes?
10. What is the natural type of plant cover growing in the region where you live? What conditions are responsible for this?
11. "Changes in altitude have the same effect as changes in latitude." Look back at Fig. 12-7 and explain this statement.

Thought Questions

1. In what ways are the living conditions in a cave similar to those deep in the sea?
2. During the Ice Age, glaciers reached as far south as the Ohio River. If this happens again, what living things will grow in Michigan, in Kentucky, and in Mississippi? What changes would it make where you live?

121

WORD QUIZ
Tests understanding of new vocabulary.

CHECK YOUR FACTS
These questions are based directly on material in the chapter. They can be used as a self-quiz.

THOUGHT QUESTIONS
These questions require you to apply the text material to a given situation.

Exercise

Using the text, answer the following questions. Do not write in this book. (Spaces for answers can be found on page 3 of *Exercises & Investigations for Living Things*.)

1. On what page could you check your answer to Check Your Facts question 2 in Chapter 22?
2. Are all of the new vocabulary words introduced in Chapter 16 tested in the word quiz for the chapter?
3. How can the information in Chapter 1 help you to answer thought question 1 on page 10?
4. Which questions in the Word Quiz on page 35 refer to the statement "A teaspoon of sugar is added to a glass of water"?
5. On what page in Chapter 7 would you find a definition of the word "grafting"?

Glossary and Index

NEW VOCABULARY
All new vocabulary words are listed alphabetically.

DEFINITIONS ——
Given for each new word.

PRONUNCIATION GUIDE
A pronunciation guide is given for difficult to pronounce words. Accented syllable is in caps.

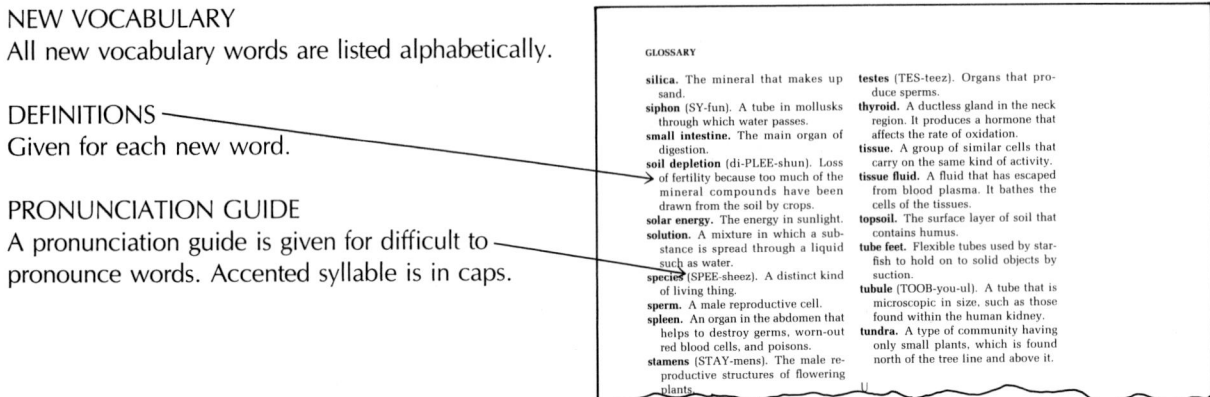

GLOSSARY

silica. The mineral that makes up sand.
siphon (SY-fun). A tube in mollusks through which water passes.
small intestine. The main organ of digestion.
soil depletion (di-PLEE-shun). Loss of fertility because too much of the mineral compounds have been drawn from the soil by crops.
solar energy. The energy in sunlight.
solution. A mixture in which a substance is spread through a liquid such as water.
species (SPEE-sheez). A distinct kind of living thing.
sperm. A male reproductive cell.
spleen. An organ in the abdomen that helps to destroy germs, worn-out red blood cells, and poisons.
stamens (STAY-mens). The male reproductive structures of flowering plants.

testes (TES-teez). Organs that produce sperms.
thyroid. A ductless gland in the neck region. It produces a hormone that affects the rate of oxidation.
tissue. A group of similar cells that carry on the same kind of activity.
tissue fluid. A fluid that has escaped from blood plasma. It bathes the cells of the tissues.
topsoil. The surface layer of soil that contains humus.
tube feet. Flexible tubes used by starfish to hold on to solid objects by suction.
tubule (TOOB-you-ul). A tube that is microscopic in size, such as those found within the human kidney.
tundra. A type of community having only small plants, which is found north of the tree line and above it.

TOPICS LISTED ALPHABETICALLY

PAGE NUMBERS ——
Indicate where discussions of the topics can be found. Pages in **boldface** type indicate where illustrations can be found. Pages in *italic* type indicate where definitions can be found.

INDEX

Exercise

Using the text, answer the following questions. Do not write in this book. (Spaces for answers can be found on page 4 of *Exercises & Investigations for Living Things*.)

1. What is the definition of the term "producer"?
2. On what page of the Glossary can you find the definition of the word "dentine"?
3. On what page of the text could you find a diagram of the internal organs of a frog?
4. On what pages of the text could you find a discussion on ciliates?
5. What syllable is accented in the word "mitochondria"?

Metric System Prefixes

Greater than 1
kilo (k) = 1,000
hecto (h) = 100
deka (da) = 10

Less than 1
deci (d) = 0.1
centi (c) = 0.01
milli (m) = 0.001
micro (μ) = 0.000 001

Commonly Used Metric Units

Length
Meter (m)
(about the distance from the floor
to a door knob)

1 kilometer (km) = 1,000 m
1 meter (m) = 100 cm
1 meter (m) = 1,000 mm
1 centimeter (cm) = 0.01 m
1 millimeter (mm) = 0.001 m

Mass
Gram (g)
(a paper clip has a mass of about
1 gram)

1 kilogram (kg) = 1,000 g
1 gram (g) = 1,000 mg
1 milligram (mg) = 0.001 g

Volume
Liter (L)
(slightly more than a quart)

1 liter (L) = 1,000 ml
1 milliliter (ml) = 0.001 L

Metric-English Equivalents

1 m = 39.37 inches
2.54 cm = 1 inch
1 km = .621 mile
1.61 km = 1 mile

1 kg = 2.2046 pounds
453.6 g = 1 pound
28.35 g = 1 ounce
1 L = 1.06 quarts
1 ml = 0.00106 quart

Appendix C:

Temperature Scales

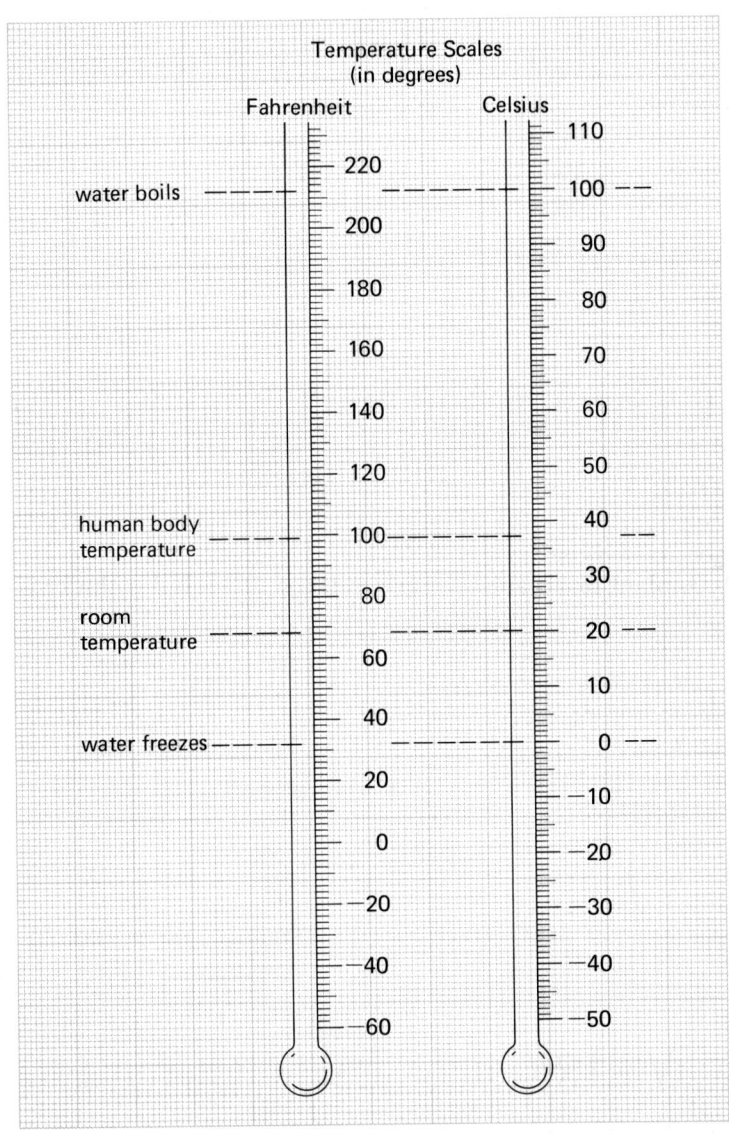

Temperature Scales
(in degrees)

water boils

human body
temperature

room
temperature

water freezes

Temperature Equivalents/Conversion Formulas
$1°F \cong 0.56°C$ $°F = 9/5°C + 32$
$1°C \cong 1.8°F$ $°C = 5/9(°F - 32)$

Glossary

A

abdomen (ab-DOH-men). In humans, the part of the body cavity below the diaphragm. In insects, the segments of the body behind the chest region.

adaptation. A structure or behavior that adjusts a plant or an animal to its environment.

adrenal (uh-DREEN-ul) **glands.** A pair of ductless glands. One lies against the upper end of each kidney.

adrenaline (uh-DREN-uh-lin). A hormone produced by the adrenal glands

air sacs. Tiny thin-walled sacs in the lungs containing networks of capillaries. Oxygen and carbon dioxide are exchanged in these sacs.

albino (al-BY-no). A plant or an animal lacking pigment, or coloring matter.

algae (AL-jee). Several groups of simple green plants, including kelps, rockweeds, *Chlorella*, and *Spirogyra*.

ameba (ah-MEE-bah). A single-celled animal-like organism belonging to the protist kingdom.

amino (a-ME-no) **acids.** Compounds found as essential components of protein molecules.

amphibians (am-FIB-ee-uns). Cold-blooded vertebrates such as frogs, toads, and salamanders. They are adapted partly to land and partly to water.

antibiotics (an-ti-by-OT-iks). Modern drugs that are taken from certain molds and bacteria. Some of them can also be made artificially.

antibody. A chemical substance in the body that destroys germs.

anus (AE-nus). An opening at the end of the intestine through which undigested wastes pass.

appendage. An arm, leg, antenna, or similar structure attached to the body.

appendicitis (uh-pen-duh-SITE-us). A disease in which the human appendix is infected by bacteria.

appendix. A structure attached to the large intestine in humans. It has no function.

artery (AR-ter-ee). A blood vessel that carries blood from the heart to body tissues.

arthropods (AR-throh-podz). A phylum of animals without backbones that are segmented and have jointed appendages. It includes the crustaceans, centipedes, millipedes, spiders, ticks, and insects.

asexual reproduction. Reproduction in which no sex cells are involved.

astigmatism (uh-STIG-muh-tiz-um). A condition in which the eye surface is out of shape and produces a poor image.

atoms. Tiny units of matter that make up the elements.

ATP. A phosphorous compound found in living cells. It provides the energy for chemical changes. Its full name is adenosine triphosphate.

atrium (AY-tree-um). An upper chamber of the heart.

auditory (AWD-uh-tore-ee) **nerve.** A nerve that carries sensory impulses from the ear to the brain.

axon. A long nerve fiber extending from a nerve cell.

481

B

bacteria (bak-TIR-ee-uh). One-celled living things that do not have nuclei. Many are parasites or feed on dead plants and animals.

bile. A fluid produced by the liver and stored in the gall bladder. It separates fats into small particles that can be digested.

bladder. An organ of the human excretory system that holds the urine until it can be discharged.

bone. Hard minerals deposited between living cells.

breathing. The act of moving air in and out of the lungs.

budding. A type of reproduction in which a bulge is developed on the parent plant or animal. The bulge grows into a new individual.

C

calorie. A unit of heat energy often used to describe the amount of energy in different foods.

cambium (KAM-bee-um) **layer.** A layer of growth cells in a plant stem.

capillaries (CAP-uh-ler-ees). Tiny blood vessels in the body tissues that form the connections between arteries and veins.

carbohydrates (kar-boh-HY-drayts). Compounds of carbon, hydrogen, and oxygen that are the basic food substances of the cell. They include various types of sugar and starch.

cartilage (KART-lij). A flexible tissue found in the voice box, ears, and nose.

cell membrane. A thin membrane that surrounds a cell and controls what enters and leaves.

cell wall. The outer, nonliving covering of many plant cells.

cells. Small units of living matter that make up the tissues of plants and animals. Some of them live independent lives as one-celled living things.

cerebellum (ser-uh-BELL-um). A part of the brain that helps to control the muscles and maintain balance.

cerebrum (suh-REE-brum). A part of the brain having to do with processes such as memory, thinking, and consciousness.

chlorella (klo-REL-ah). A single-celled plant-like organism belonging to the protist kingdom.

chlorophyll (KLOR-oh-fil). The green material in plants used in making food.

chloroplast (KLOR-oh-plast). The food-making unit of a cell.

chordate (CORE-date). The phylum that includes vertebrates and three related subphyla of animals.

chromosomes (KROH-muh-sohms). Structures in the cell nucleus that contain genes.

cilia (SILL-ee-uh). Short threads of protoplasm that extend out from some cells.

class. A classification group made up of related orders of plants or animals.

climax community. The type of community that becomes dominant at the end of a succession.

clot. A mass of blood that has become like thick jelly. It plugs wounds and stops bleeding.

cochlea (KOE-klee-uh). A coiled hearing structure of the inner ear.

coelenterates (suh-LEN-ter-ates). The phylum that includes *Hydra*, jellyfishes, corals, and sea anemones.

cold-blooded animal. An animal whose inside temperature changes with the outside temperature.

colony. A group of plants or animals living together.

community. All the plants and animals that live together in a local environment.

compound. Two or more elements united in definite amounts to form a substance that has its own special properties.

compound eye. A type of eye, made up of many small eye units packed together, found in some arthropods.

conservation. The wise management of natural resources.

consumer. A plant or an animal that cannot make its own food.

cornea (CORE-nee-uh). A transparent curved front part of the eye that acts as the outer lens.

crustaceans (kruh-STAY-shuns). A group of arthropods including crabs, lobsters, shrimps, prawns, and crayfish.

cutting. Part of a root, stem, or leaf from which a new plant may be grown.

cyst (SIST). A protective covering that some protozoans and parasitic worms form around themselves.

cytoplasm (sy-TOH-plaz-um). The mass of protoplasm between the nucleus and the cell membrane.

D

dating. Estimating the age of a fossil.

decomposers. Living things, such as bacteria and fungi, that cause dead plants and animals to decay.

deficiency (dee-FISH-en-see) **disease.** A disease caused by the lack of some vitamin.

dendrites. Short branching fibers extending from nerve cells.

dentine (DEN-teen). A hard, bone-like substance under the enamel and forming the main body of teeth.

dermis. The outer layer of the skin.

diabetes (die-uh-BEET-eez). A disease caused by lack of insulin, in which the body is unable to use sugar.

diatomaceous (die-ah-toe-MAY-shus) **earth.** A collection of the walls of dead diatoms on the floor of the ocean.

diffusion (di-FEW-shun). The movement of molecules from a place in which there are many of them to a place where there are few.

DNA. A nucleic acid found in the genes of cells. Its full name is deoxyribose nucleic acid.

dominant trait. A trait that develops even when only a single dominant gene is present.

drug addiction. Physical and mental dependence on narcotics.

duct. A tube through which liquids can flow from a gland.

ductless gland. A gland that discharges its secretion directly into the bloodstream.

E

eardrum. A membrane stretched across the opening of the middle ear. Sound waves cause it to vibrate.

ecology. The study of living things in relation to their environment.

egg cell. A female reproductive cell.

element. Any one of 107 basic substances found in nature. Each element is made up of atoms that are more or less alike and are different from the atoms in any other element.

embryo (EM-bree-oh). An early stage in the development of a complex plant or animal.

enamel. Extra-hard, white material on the outside of teeth.

end brush. The branching ends of an axon.

energy. The ability to do work.

environment. All the things and forces that surround an individual.

enzyme. A protein that controls some particular chemical change in a cell.

epidermis. The inner layer of the skin.

erosion (i-ROE-zhun). The carrying away of soil by wind or water.

esophagus (ih-SAHF-uh-gus). A tube leading from the throat to the stomach.

excretion (ex-CREE-shun). The process in which living things get rid of wastes.

external fertilization. A type of fertilization in which the sperm fertilizes the egg outside the female organism.

F

family. A classification group made up of related genera of plants and animals.

fats. Compounds found in protoplasm and often used as sources of energy.

feces (FEE-seez). Undigested wastes that leave the intestine.

fermentation. A type of respiration in which molecules are changed to release energy without the use of oxygen.

fertilization. The union of an egg cell with a sperm.

flagellates (FLAJ-uh-lates). Members of a phylum of protozoa that use a long whip-like tail (flagellum) to swim.

flagellum (fluh-JELL-um). A thin strand of protoplasm that extends out from a cell. Its lashing serves to move the cell in some cases.

flatworms. A phylum of animals that have flattened bodies and an intestine with a single opening; includes *Planaria*, flukes, and tapeworms.

food chain. The passing along of food as animal A eats plants, animal B eats animal A, and so on.

fossil. A record of past life such as an imprint, a bone, or a shell that has been preserved in the rocks.

fossil fuels. The remains of organisms that died millions of years ago. They include coal, natural gas, and oil and today supply us with most of our energy.

fruiting body. The spore-producing organ of a fungus.

functional disease. A disease in which some part of the body fails to function as it should but not because of an infection.

fungi (FUN-ji). A large group of non-green plants, including yeasts, molds, mushrooms, puffballs, and shelf fungi.

G

gall bladder. A sac in which bile is stored until it is discharged into the intestine.

genes. Units in chromosomes that affect heredity and control cell activities.

genus (JEE-nus). A classification group made up of related species of plants or animals.

geothermal (GEE-oh-ther-mal) **energy.** Energy from underground heat.

glucose (GLOO-kohs). The most important of the simple sugars found in living things.

grafting. Making the stem or bud of one plant grow on the stem of another plant.

growth ring. A ring of water-carrying tissue in a tree trunk. One ring is formed in each growing season.

guard cells. The cells surrounding a leaf pore. They contract and expand to close and open the pores, thus regulating the water loss.

H

hallucinogens (huh-LOOSE-un-uh-juns). Drugs that affect the functions of the sense organs.

hemoglobin (HE-ma-glow-bun). The iron compound found in human blood. This gives blood its red color.

herbicide (HUR-buh-side). A chemical used to destroy weeds.

heredity (huh-RED-ut-ee). The passing along of characteristics from one generation to another.

hibernation (hy-bur-NAY-shun). A period of inactivity in animals when the weather gets cold.

Homo erectus (ee-RECK-tus). The species of earliest human beings.

Homo sapiens (SAYP-ee-ins). The species to which modern humans belong.

hormone (HOR-mohn). A chemical compound produced by a ductless gland. It influences cells in other parts of the body.

host. A plant or an animal upon which a parasite feeds.

humus (HEW-mus). Decaying plant and animal material in soil.

hybrid. A plant or an animal having opposed genes for a given characteristic.

I

immunity. The state of being unable to get a disease.

incubate (IN-cue-bate). To keep warm until hatched or to encourage reproduction as in bacterial cultures.

infectious (in-FEK-shus) **disease.** A disease caused by certain germs entering the body.

insects. A class of arthropods. They have six walking legs, and their bodies are divided into a head, chest, and abdomen.

instinct. A complex series of reflex actions. It depends on the inherited nerve structure.

insulin (IN-suh-lin). A hormone produced by cells in the pancreas. It is needed for the body to use its sugar normally.

internal fertilization. A type of fertilization in which the egg is fertilized within the body of the female.

intestine. The main digestive organ of an animal.

iris (EYE-rus). A colored part of the eye in front of the lens.

K

kidney. An organ that removes water and waste materials from the blood.

L

large intestine. The section of the digestive system that absorbs water.

larva (LAR-va). Any young animal that is very different from the adult, such as a caterpillar or a tadpole.

lens. A structure in the eye that helps to form an image on the retina.

lichen (LIE-ken). A plant combination in which an alga lives with a fungus and both benefit.

life cycle. The stages in the development of an organism.

ligament. A strong tough connective tissue that holds two bones together.

lymph (limf). A clear liquid similar to plasma, found in the lymph vessels.

lymph nodes. Gland-like swellings in the lymph vessels. They contain white blood cells, which destroy germs.

M

malnutrition. The condition of poor health that is caused by too few nutrients in the diet.

mammals. Warm-blooded, air-breathing animals with backbones and bodies more or less covered with hair. They are the only animals that produce milk to feed their young.

mammary (MAM-ah-ree) **glands.** Glands in female mammals that produce milk to feed the young.

mantle. A thin layer of flesh that produces the shell in mollusks.

matter. Anything that takes up space and has weight.

medulla (muh-DULL-uh). A part of the brain that helps to control breathing, heartbeat, and blood pressure.

mitochondria (mite-oh-KON-dree-ah). A small structure in the cell that releases energy from the cell.

mitosis (MY-toe-sis). Nuclear division in which two equal groups of chromosomes are formed.

molecule. The smallest unit of a compound.

mollusks (MOL-usks). A phylum that includes clams, oysters, snails, slugs, squids, and octopuses.

motor nerve cells. Nerve cells that control gland and muscle action.

mutation (mew-TAY-shun). A new characteristic that can be inherited. It is caused by some change in a gene.

N

narcotics. Habit-forming drugs of the cocaine and morphine type. They produce sleep and at the same time relieve pain.

natural selection. The process by which animals adapt to their environment and produce young.

nematodes (NEM-uh-toads). Another name given to the roundworms.

nerve. A bundle of nerve fibers surrounded by a sheath.

nerve impulse. An electrical "message" carried along a nerve fiber.

niacin (NIE-uh-sun). A vitamin needed for the proper use of carbohydrates.

nitrates (NY-trates). Certain compounds containing nitrogen. Green plants use nitrates in making proteins.

nitrogen cycle. A cycle that takes place in nature. Bacteria of decay break down dead materials and produce nitrates. Other bacteria break down nitrates. Also, nitrogen of the air is used by nitrogen-fixing bacteria to make nitrates.

nitrogen-containing wastes. Nitrogen compounds that are the waste products of protein breakdown.

nitrogen-fixing bacteria. Certain bacteria that are able to combine nitrogen with other elements to form nitrates.

nucleic (NEW-clay-ik) **acid.** A substance that is either DNA or RNA. Both of these nucleic acids appear in many different forms.

nucleus (NEW-klee-us). A dense inner structure found in cells. It controls most of the cell's activities.

nutrients (NEW-tree-ents). Substances that nourish the body.

nutrition. The taking in and using of food.

nymph (nimf). A growth stage in the life of a grasshopper and some other insects.

O

optic nerve. A nerve connecting the nerve cells of the retina with the brain.

order. A classification group made

up of related families of plants or animals.

organ. A group of tissues that forms a working unit such as a stomach or an eye.

organism. A complete living thing.

ovary (OH-va-ree). An organ that produces egg cells.

oviducts (OH-ve-dukts). Tubes that lead from the ovaries to the outside.

oxidation (ox-suh-DAY-shun). A process in which a substance unites with oxygen.

P

pancreas (PAN-kree-us). A gland that produces a digestive juice and also the hormone insulin.

parasite. A plant or an animal that feeds on a host. It may live on the host or in it.

parathyroids (par-uh-THIE-roids). Four small ductless glands in the neck region that regulate the amount of calcium salts in the bloodstream.

particulates. Pollutants that enter the air.

pasteurization (pass-ture-ruh-ZAY-shun). The process of heating a substance enough to kill dangerous germs but not enough to change its flavor.

pellagra (puh-LAY-gruh). A deficiency disease caused by lack of niacin.

pesticide. A chemical used to kill troublesome insects or other pests.

phosphates (FOSS-fates). A type of salts used in making ATP and other important compounds in the body.

photosynthesis (foe-toe-SIN-the-sis). The process in which green plants make food.

phylum (FY-lum). A classification group made up of related classes of living things.

pituitary (puh-TOO-uh-TER-ee). A ductless gland at the base of the brain that produces hormones that affect the action of other glands.

placenta (pla-SEN-tuh). A structure that surrounds and nourishes some developing mammals before birth.

plasma (PLAZ-muh). The liquid part of the blood.

platelets. Tiny structures found in the blood. They break down when they touch a cut or rough surface. This breakdown causes the clotting of blood to begin.

pollen grains. Tiny grains produced by the stamens of a flower and containing the sperm cells.

pollination. The transfer of pollen to the pistil.

pollution. Impurity of water usually due to dumping sewage and factory wastes into lakes, streams, or the sea.

primates. An order of mammals, including monkeys, lemurs, apes, and human beings.

producer. A living thing that can manufacture its own food. Green plants are the most important producers.

proteins (PROH-teens). The most important compounds of protoplasm. They contain carbon, hydrogen, oxygen, and nitrogen and sometimes phosphorus and sulfur.

protists. Living things with simple body organization, many of which are single-celled.

protoplasm (PROH-toh-plasm). The living substance.

protozoa (prote-uh-ZOH-uh). A group of small, simple living things, many of which exist as single cells.

pupa (PYOO-puh). A resting stage in the development of some insects.

pupil. A circular opening in the iris of the eye through which light rays enter.

R

recessive trait. A trait that does not develop if the corresponding dominant gene is present.

red blood cells. Cells containing the red iron compound that carries oxygen.

reduction division. The separating of each pair of similar chromosomes in the formation of a sperm or egg cell.

reef. A shallow rocky ridge under water.

reflex. One type of automatic response to a stimulus.

reproduction. The process in which plants and animals produce more of their own kind.

reptiles. Cold-blooded, air-breathing animals with backbones and scales. Lizards, snakes, and turtles belong to this group.

respiration. The process in which living things release energy from food.

response. A reaction to a stimulus.

retina (RET-un-uh). A layer of nerve cells sensitive to the light in the back of the eye. The image forms on the retina, from which impulses are sent to the brain.

ribosomes (RIBE-oh-sohms). Structures in the cells that make protein.

rickets. A deficiency disease caused by lack of vitamin D.

rickettsias (rik-ET-see-uhs). A group of very small parasites that cause a number of diseases.

RNA. One of the nucleic acids. Molecules of DNA make corresponding molecules of RNA. The full name is ribose nucleic acid.

root cap. A mass of cells that covers and protects the growing tip of a root.

root hair. A hair-like part of a cell on the surface of a root that is well suited for absorbing water.

roundworms. A phylum of unsegmented worms with an opening at either end of the digestive canal. It includes *Ascaris*, hookworms, the trichina worm, and vinegar eels.

S

saliva. A juice produced by glands near the mouth. It moistens food and partially digests starch.

savanna. A grassland with a scattering of trees.

scavenger. An animal that eats dead animals.

scurvy. A deficiency disease caused by lack of vitamin C.

sedative. A drug that has a quieting effect on a person.

segmented worms. A phylum of worms that are divided into sections, which includes the earthworms, leeches, and sandworms.

selective cutting. The practice of cutting only the mature trees of a forest.

semicircular canals. Organs of balance in the inner ear.

sensory nerve cell. A type of nerve cell that receives a stimulus from the sense organs and relays it inward.

semen (SEE-men). In higher animals, the liquid that contains the male sperm.

sepals (SEE-puls). Leaf-like structures that cover a flower bud before it opens.

sex-linked trait. A trait carried only in the X chromosome.

sexual reproduction. Reproduction following the union of two sex cells.

sheet erosion. The wearing away of the entire soil surface.

silica. The mineral that makes up sand.

siphon (SY-fun). A tube in mollusks through which water passes.

small intestine. The main organ of digestion.

soil depletion (di-PLEE-shun). Loss of fertility because too much of the mineral compounds have been drawn from the soil by crops.

solar energy. The energy in sunlight.

solution. A mixture in which a substance is spread through a liquid such as water.

species (SPEE-sheez). A distinct kind of living thing.

sperm. A male reproductive cell.

spleen. An organ in the abdomen that helps to destroy germs, worn-out red blood cells, and poisons.

stamens (STAY-mens). The male reproductive structures of flowering plants.

stimulus. Anything that causes a response by a plant or an animal or by some part of the body.

stomach. A muscular sac whose main function is to store food.

strip cropping. The practice of alternating strips of different crops across the hillside.

succession. A series of different types of communities replacing one another in order. Finally, a more or less stable climax growth becomes dominant.

system. A group of organs working together to perform one general function such as digestion.

taste buds. Groups of cells on the top and sides of the tongue that are sensitive to taste.

tendons. Tough cords of connective tissue that attach muscles to bones.

testes (TES-teez). Organs that produce sperms.

thyroid. A ductless gland in the neck region. It produces a hormone that affects the rate of oxidation.

tissue. A group of similar cells that carry on the same kind of activity.

tissue fluid. A fluid that has escaped from blood plasma. It bathes the cells of the tissues.

topsoil. The surface layer of soil that contains humus.

tube feet. Flexible tubes used by starfish to hold on to solid objects by suction.

tubule (TOOB-you-ul). A tube that is microscopic in size, such as those found within the human kidney.

tundra. A type of community having only small plants, which is found north of the tree line and above it.

U

umbilical (um-BIL-i-kul) **cord.** A structure that connects unborn mammals with the placenta.

urea (yoo-REE-uh). A nitrogenous waste formed by the liver from ammonia wastes.

ureter (YOUR-eh-ter). A tube extending from a kidney.

urethra (you-REE-thruh). A tube leading out from the urinary bladder.

urine. A mixture of water and dissolved wastes that is drawn off by the kidneys and excreted.

uterus (YOU-ter-us). A structure in female mammals that holds the embryo during development.

V

vaccination (vak-sin-AY-shun). The process of introducing a small amount of virus under the skin in order to cause immunity.

vaccine (VAK-seen). A substance used to produce immunity to a given disease.

vacuoles (VAK-yoo-ohls). Spaces in the cytoplasm of cells that may contain stored food, water, or waste substances.

vascular (VAS-cue-lar) **bundle.** A duct made up of vascular tissue in a plant.

vascular plant. A plant having vascular tissue. Included in this group are club mosses, horsetails, ferns, and seed plants.

vascular tissue. In plants, tube-shaped cells that carry liquids.

vegetative reproduction. A type of asexual reproduction in which some part other than a seed or spore grows into a new plant.

vein. In animals, a blood vessel that carries blood from the tissues to the heart. In plants, a vascular bundle in a leaf.

ventricle (VEN-tri-kul). A chamber of the vertebrate heart that pumps blood to the lungs or other parts of the body.

vertebrae (VURT-uh-bray). Bones that make up the backbone.

vertebrates. Animals having backbones.

villi (VILL-eye). Small, finger-shaped bulges of the small intestine lining that absorb dissolved food.

virus. A tiny particle that contains DNA or RNA and is active as a parasite within a host cell. Viruses cause many diseases of plants, animals, or human beings.

vitamins. Compounds needed in our diet for normal body activities.

vocal cords. Structures in the voice box that vibrate to make sound.

W

warm-blooded animal. An animal that oxidizes its food so rapidly that its body is kept warm all the time.

water cycle. The constant movement of water from ocean to land and back to ocean again.

water table. The water level at the top of a water-soaked region in soil.

white blood cells. Several types of blood cells similar to amebas. Some of them destroy germs.

withdrawal symptoms. A reaction resulting from the lack of a drug to which the victim is addicted.

womb. Another name for the uterus.

woody stem. A type of stem that has water-carrying tissue in large masses of thick-walled cells. These cells become what we call wood.

Index

(Note: Page numbers in boldface type include illustrations and those in italic type include definitions.)

INDEX